Lecture Notes in Computer Science 11640

More information about this series at http://www.springer.com/series/7407

Ding-Zhu Du · Lian Li ·
Xiaoming Sun · Jialin Zhang (Eds.)

Algorithmic Aspects
in Information
and Management

13th International Conference, AAIM 2019
Beijing, China, August 6–8, 2019
Proceedings

 Springer

Editors
Ding-Zhu Du
The University of Texas at Dallas
Richardson, TX, USA

Lian Li
Hefei University of Technology
Hefei, China

Xiaoming Sun
Institute of Computing Technology,
Chinese Academy of Sciences
Beijing, China

Jialin Zhang
Institute of Computing Technology,
Chinese Academy of Sciences
Beijing, China

ISSN 0302-9743   ISSN 1611-3349 (electronic)
Lecture Notes in Computer Science
ISBN 978-3-030-27194-7   ISBN 978-3-030-27195-4   (eBook)
https://doi.org/10.1007/978-3-030-27195-4

LNCS Sublibrary: SL1 – Theoretical Computer Science and General Issues

This Springer imprint is published by the registered company Springer Nature Switzerland AG
The registered company address is: Gewerbestrasse 11, 6330 Cham, Switzerland

Preface

The 13th International Conference on Algorithmic Aspects in Information and Management (AAIM 2019), was held in Beijing, China, during August 6–8, 2019. The AAIM conference series, which started in 2005 in Xi'an, China, aims to stimulate the various fields for which algorithmics has become a crucial enabler, and to strengthen the ties between the Eastern and Western research communities of algorithmics and applications.

The topics cover most aspects of theoretical computer science and their applications. Both theoretical and experimental/applied works of general algorithmic interest were sought. Special considerations were given to algorithmic research that was motivated by real-world applications. Experimental and applied papers were expected to show convincingly the usefulness and efficiency of the target algorithms in practical settings.

We would like to thank the two eminent keynote speakers, Xiaotie Deng from Peking University, and David Woodruff from Carnegie Mellon University, for their contribution to the conference.

We would like to express our appreciation to all members of the Program Committee and the external referees whose efforts enabled us to achieve a high scientific standard for the proceedings. We would like to thank all members of the Organizing Committee for their assistance and contribution which attributed to the success of the conference. We would like to thank Alfred Hofmann, Anna Kramer, and their colleagues at Springer for meticulously supporting us in the timely production of this volume. Last but not least, our special thanks go to all the authors who submitted papers and all the participants for their contributions to the success of this event.

June 2019

Ding-Zhu Du
Lian Li
Xiaoming Sun
Jialin Zhang

Organization

Program Committee

Zhipeng Cai	Georgia State University, USA
Lichao Chen	Google, USA
Yongxi Cheng	Xi'an Jiaotong University, China
Ding-Zhu Du	The University of Texas at Dallas, USA
Andras Farago	The University of Texas at Dallas, USA
Xiaofeng Gao	Shanghai Jiao Tong University, China
Donghyun Kim	Kennesaw State University, USA
Lian Li	Hefei University of Technology, China
Minming Li	City University of Hong Kong, SAR China
Xiaoming Sun	Institute of Computing Technology, Chinese Academy of Sciences, China
Cong Tian	Xidian University, China
Guochuan Zhang	Zhejiang University, China
Jialin Zhang	Institute of Computing Technology, Chinese Academy of Sciences, China
Peng Zhang	School of Computer Science and Technology, Shandong University, China
Zhao Zhang	Zhejiang Normal University, China

Additional Reviewers

Chen, Hua	Wang, Changjun
Cheng, Yukun	Wu, Chenchen
Gan, Jinxiang	Xu, Chenyang
Hao, Jie	Xu, Yicheng
Khan, Fanid	Yao, Yuhao
Li, Minming	Zhang, Jia
Liu, Bei	Zhang, Ruilong
Mei, Lili	Zhao, Yingchao
Shan, Xiaohan	Zhu, Shenglong
Tao, Liangde	

Contents

One-Dimensional r-Gathering Under Uncertainty

Shareef Ahmed[1]([✉]), Shin-ichi Nakano[2], and Md. Saidur Rahman[1]

[1] Graph Drawing and Information Visualization Laboratory,
Department of Computer Science and Engineering,
Bangladesh University of Engineering and Technology, Dhaka, Bangladesh
{shareefahmed,saidurrahman}@cse.buet.ac.bd
[2] Gunma University, Kiryu 376-8515, Japan
nakano@cs.gunma-u.ac.jp

Abstract. Let C be a set of n customers and F be a set of m facilities. An r-gathering of C is an assignment of each customer $c \in C$ to a facility $f \in F$ such that each facility has zero or at least r customers. The r-gathering problem asks to find an r-gathering that minimizes the maximum distance between a customer and its facility. In this paper we study the r-gathering problem when the customers and the facilities are on a line, and each customer location is uncertain. We show that, the r-gathering problem can be solved in $O(nk + mn \log n + (m + n \log k + n \log n + nr^{\frac{n}{r}}) \log mn)$ and $O(mn \log n + (n \log n + m) \log mn)$ time when the customers and the facilities are on a line, and the customer locations are given by piecewise uniform functions of at most $k + 1$ pieces and "well-separated" uniform distribution functions, respectively.

Keywords: r-Gathering · Facility location problem

1 Introduction

The facility location problem and many of its variants are well studied [7]. In this paper we study a relatively new variant of the facility location problem, called the r-gathering problem [6].

Let C be a set of n customers and F be a set of m facilities, $d(c, f)$ be the distance between $c \in C$ and $f \in F$. An r-gathering of C to F is an assignment A of C to F such that each facility has at least r or zero customers assigned to it. The cost of an r-gathering is $\max_{c \in C}\{d(c, A(c))\}$ which is the maximum distance between a customer and its facility. The r-gathering problem asks to find an assignment of C to F having the minimum cost [6]. This problem is also known as the min-max r-gathering problem. The other version of the problem is known as the min-sum r-gathering problem which asks to find an assignment which minimizes $\sum_{c \in C} d(c, A(c))$ [8,11]. In this paper we consider the min-max r-gathering problem and we use the term r-gathering problem to refer the min-max version.

© Springer Nature Switzerland AG 2019
D.-Z. Du et al. (Eds.): AAIM 2019, LNCS 11640, pp. 1–15, 2019.
https://doi.org/10.1007/978-3-030-27195-4_1

Assume we wish to set up emergency shelters for residents C living on a locality so that each shelter can accommodate at least r residents. We also wish to locate the shelters so that evacuation time span can be minimized. A set F of possible locations for shelters is also given. This scenario can be modeled by the r-gathering problem. In this case, an r-gathering corresponds to an assignment of residents to shelters so that each "open" shelter serves at least r residents and the r-gathering problem finds the r-gathering minimizing the evacuation time.

For the r-gathering problem a 3-approximation algorithm is known and it is proved that the problem cannot be approximated within a factor less than 3 for $r > 3$ unless $P = NP$ [6]. Recently, the problem is considered in a setting where all the customers and facilities are lying on a line. An $O((n+m)\log(n+m))$ time algorithm [5], an $O(n + m\log^2 r + m\log m)$ time algorithm [9], an $O(n + r^2 m)$ time algorithm [12], and an $O(n+m)$ time algorithm [13] are known when all the customers and facilities are on a line. Ahmed *et al.* gave an $O(n + m + d^2 r^2 (d + \log m) + (r + 1)^d 2^d (r + d)d)$ time algorithm for the r-gathering problem when the customers and facilities are on a star [4].

In this paper, we consider the r-gathering problem when the customer and the facilities are on a line, and the customer locations are uncertain. Study of different problems under uncertain settings become much popular recently. Uncertainty in data usually occurs because of noise in measured data, sampling inaccuracy, limitation of resources, etc. Hence uncertainty is ubiquitous in practice and managing the uncertain data has gained much attention [1–3,15]. Different variants of the facility location problem have also been investigated under uncertain settings. Setting up a facility is costly and each facility is supposed to serve for a long period of time. On the other hand existence, location and demand of a client can change over time. Thus it is important to set up facilities by keeping the uncertainty in mind. For the detailed state of the art of uncertain facility location problem, we refer the survey of Snyder [14]. There are two models for uncertainty: one is existential model [10,18] and the other is locational model [1,2,16]. In the existential model, the existence of each point is uncertain. Thus each point has a specific location and there is a probability for the existence of each point. In the locational model each point is certain to exist, but its position is uncertain and defined by a probability density function. In this paper we consider the locational model of uncertainty. For customer locations, we consider two probability density functions: piecewise uniform function (histogram) and "well-separated" uniform distribution function.

When the customer and facility locations are deterministic and on a line, there is an optimal r-gathering where the customers assigned to each facility are consecutive on the line [12]. However, when the customer locations are uncertain, finding a suitable ordering of the customers is difficult. In this paper we give an $O(nk + mn\log n + (m + n\log k + n\log n + nr^{\frac{n}{r}})\log mn)$ time algorithm for the one-dimensional r-gathering problem when the customer locations are given by piecewise uniform functions of at most $k+1$ pieces, and an $O(mn\log n + (n\log n + m)\log mn)$ time algorithm for the one-dimensional r-gathering problem when the customer locations are given by well-separated uniform distributions.

The rest of the paper is organized as follows. In Sect. 2, we define the uncertain r-gathering problem and provide definitions of basic terminologies. In Sect. 3, we give algorithms for uncertain r-gathering problem when customer locations are specified by piecewise uniform functions and "well-separated" uniform distribution functions. Finally we conclude in Sect. 4.

2 Preliminaries

In this section we define the uncertain r-gathering problem and relevant terminologies.

Let $F = \{f_1, f_2, \cdots, f_m\}$ be a set of m facilities, and $\mathcal{C} = \{C_1, C_2, \cdots, C_n\}$ be a set of n customers where each C_i is a random variable. The probability density function (PDF) associated with customer C_i is denoted by $g_i(x)$. The expected distance between a facility f_j and an uncertain customer C_i, denoted by $E[d(C_i, f_j)]$, is $\int_{-\infty}^{\infty} d(x, f_j) g_i(x) dx$. An r-gathering A of $\mathcal{C}$ to F is an assignment $A : \mathcal{C} \rightarrow F$ such that each facility serves zero or at least r customers. A facility having one or more customers is called an *open facility*. $A(C)$ denotes the facility to which a customer C is assigned in an assignment A. The cost of a facility is the maximum expected distance between the facility and its customers if the facility is open, and zero otherwise. The cost of an r-gathering is the maximum cost among all the facilities. The *uncertain r-gathering problem* asks to find an r-gathering with minimum cost. Note that, the uncertain r-gathering problem is NP-Hard, since it contains the deterministic version as a special case.

3 One-Dimensional Uncertain r-Gathering Problem

In this section we give two algorithms for the uncertain r-gathering problem on a line.

Let $\mathcal{C} = \{C_1, C_2, \cdots, C_n\}$ be a set of n uncertain customer on a horizontal line where each customer C_i is specified by its PDF $g_i : \mathbb{R} \rightarrow \mathbb{R}^+ \cup \{0\}$, and $F = \{f_1, f_2, \cdots, f_m\}$ be a set of m facilities on the horizontal line. We consider the facilities are ordered from left to right. An r-gathering of $\mathcal{C}$ to F is an assignment $A : \mathcal{C} \rightarrow F$ such that each facility serves zero or at least r customers. The uncertain r-gathering problem asks to find an r-gathering such that the maximum among the expected distances between a customer to the assigned facility is minimum.

3.1 Histogram

In this section we give an algorithm for the uncertain r-gathering problem when each customer location is specified by a piecewise uniform function, i.e., a histogram.

We consider the PDF of each customer C_i is defined as a piecewise uniform function g_i, i.e., a histogram. The PDF of each uncertain customer is independent. We consider histogram model since it can be used to approximate

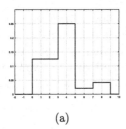 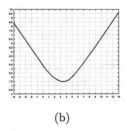

(a)   (b)

Fig. 1. (a) Illustration of a histogram and (b) corresponding function of expected distance

any PDF [1]. The histogram model is considered by Wang and Zhang [17] for the uncertain k-center problem on a line. Each g_i consists of at most $k + 1$ pieces where each piece is a uniform function. Each customer C_i has $k + 2$ points $x_{i0}, x_{i1}, \cdots, x_{i(k+1)}$, where $x_{i0} < x_{i1} < \cdots < x_{i(k+1)}$, and $k + 1$ values $y_{i0}, y_{i1}, \cdots, y_{ik}$ such that $g_i(x) = y_{ij}$ if $x_{ij} \le x < x_{i(j+1)}$. We consider $x_{i0} = -\infty$, $x_{i(k+1)} = \infty$, $y_0 = 0$, and $y_k = 0$. Figure 1(a) illustrates a histogram of 6 pieces. The expected distance $E[d(p, C_i)]$ from a point p to C_i is defined as follows.

$$E[d(p, C_i)] = \int_{-\infty}^{\infty} g_i(x)|x - p|dx$$

A function $h : \mathbb{R} \to \mathbb{R}$ is called a *unimodal function* if there is a point p such that $h(x)$ is monotonically decreasing in $(-\infty, p]$ and monotonically increasing in $[p, \infty)$. Wang and Zhang gave the following lemma [17].

Lemma 1 ([17]). *Let C_i be an uncertain point on a line which is specified by a histogram of $k+1$ pieces. Then the function $E[d(p, C_i)]$ for $p \in \mathbb{R}$ is a unimodal function consisting of a parabola in each interval $[x_{ij}, x_{i(j+1)})$. Furthermore the function $E[d(p, C_i)]$ can be explicitly computed in $O(k)$ time.*

Outline of the Proof. Without loss of generality, assume that $x_{it} \le p \le x_{i(t+1)}$. Then the function $E[d(p, C_i)]$ can be written as follows [17].

$$E[d(p, C_i)] = y_{it}p^2 + \left[\sum_{j=0}^{t-1} y_{ij} \left(x_{i(j+1)} - x_{ij} \right) - \sum_{j=t+1}^{k} y_{ij} \left(x_{i(j+1)} - x_{ij} \right) - y_{it}(x_{it} + x_{i(t+1)}) \right] p$$

$$+ \frac{1}{2} \left[\sum_{j=t+1}^{k} y_{ij} \left(x_{i(j+1)}^2 - x_{ij}^2 \right) - \sum_{j=0}^{t-1} y_{ij} \left(x_{i(j+1)}^2 - x_{ij}^2 \right) + y_{it}(x_{it}^2 + x_{i(t+1)}^2) \right] \quad (1)$$

Thus we can write $E[d(p, C_i)]$ as $a_{i1}(t)p^2 + a_{i2}(t)p + a_{i3}$ where each of $a_{i1}(t), a_{i2}(t), a_{i3}(t)$ depends on t satisfying $x_{it} \le p \le x_{i(t+1)}$. Note that if $y_{it} = 0$ then the function $E[d(p, C_i)]$ is a straight line in the interval $[x_{it}, x_{i(t+1)})$ which we consider as a special parabola. Figure 1(b) illustrates the $E[d(p, C_i)]$ function for the histogram in Fig. 1(a). We can compute the co-efficients $a_{i1}(j)$ for all j in $O(k)$ time. Moreover, the summation terms in $a_{i2}(j)$ and $a_{i3}(j)$ for all j can be computed in $O(k)$ time in total. Thus for all j, we can compute the $a_{i2}(j)$ and $a_{i3}(j)$ in $O(k)$ time. Hence the function $E[d(p, C_i)]$ can be computed explicitly in $O(k)$ time. □

We now give the following lemma.

Lemma 2. *Let C_i be an uncertain point on a line which is specified by a histogram of $k + 1$ pieces, and $F = \{f_1, f_2, \cdots, f_m\}$ be a set of m facilities on the line. We can compute the expected distances between all facilities and the uncertain point in $O(m + k)$ time. Furthermore the expected distances between the facilities and the uncertain point can be sorted in $O(m)$ time.*

Proof. We first precompute the co-efficients $a_{i1}(j), a_{i2}(j), a_{i3}(j)$ of function $E[d(p, C_i)]$ for all j in $O(k)$ time by Lemma 1. With the precomputed function $E[d(p, C_i)]$, the expected distance between the uncertain point and a facility f_u can be computed in $O(\log(k))$ time using binary search to find the $[x_{it}, x_{i(t+1)})$ where f_u is located. Thus the expected distance between all facilities and the uncertain point can be computed in $O(m \log k)$ time. However, we can improve the running time to $O(m + k)$ performing a plane sweep from left to right. We take the facilities from left to right, determine the corresponding interval $[x_{ij}, x_{i(j+1)})$, and compute the expected distance. Since both the facilities and the $x_{i1}, x_{i2}, \cdots, x_{ik}$ are ordered from left to right, the search for the interval in which f_u is located can start from the interval in which f_{u-1} is located. Hence each x_{ij} will be considered once. Thus the total running time is $O(m + k)$. We now show that the sorted list of the expected distances between the facilities and the uncertain point can be constructed in $O(m + k)$ time. Since $E[d(p, C_i)]$ is a unimodal function, there is a facility f_u such that $E[d(f_{v-1}, C_i)] \geq E[d(f_v, C_i)]$ for any $1 < v \leq u$, and $E[d(f_v, C_i)] \leq E[d(f_{v+1}, C_i)]$ for any $u \leq v < m$. Thus we have a descending list of expected distances for $f_1, f_2, \cdots, f_u$ and ascending list of expected distances for $f_{u+1}, f_{u+2}, \cdots, f_m$. We can merge these two lists into an ascending list of expected distances in $O(m)$ time. □

Corollary 1. *Let $\mathcal{C} = \{C_1, C_2, \cdots, C_n\}$ be set of n uncertain customers on a line each of which is specified by a histogram of $k + 1$ pieces, and $F = \{f_1, f_2, \cdots, f_m\}$ be a set of m facilities on the line. The expected distances between all pair of uncertain customers and facilities can be computed and sorted in $O(nk + mn \log n)$ time.*

Proof. By Lemma 2, we can compute n sorted list of expected distances between customers and facilities in $O(nk + mn)$ time. The n sorted lists can be merged into a single list using min-heap in $O(mn \log n)$ time. □

We first consider the decision version of the uncertain r-gathering problem on a line. Given a set of uncertain customers $\mathcal{C}$, a set of facilities F on a line, and a number b, the decision uncertain r-gathering problem asks to determine whether there is an r-gathering A of $\mathcal{C}$ to F such that $E[d(C, A(C))] \leq b$ for each $C \in \mathcal{C}$. The following lemma is known [17].

Lemma 3 ([17]). *Let C be an uncertain point on a line which is specified by a histogram of $k + 1$ pieces, and b is a number. Then the points p for which $E[d(C, p)] \leq b$ holds form an interval on the line.*

We call the interval which admits $E[d(C, p)] \leq b$ for customer C a (C, b)-interval and denote the interval by $[s_b(C), t_b(C)]$. Furthermore in any r-gathering A with cost at most b, $A(C)$ is in $[s_b(C), t_b(C)]$. Thus to find whether there is an r-gathering satisfying $E[d(C, p)] \leq b$ for each customer C, it is sufficient to solve the following problem. Given a set of facilities F on a line and a set of customers $\mathcal{C}$ where each customer $C \in \mathcal{C}$ has an interval $[s(C), t(C)]$ on the line, the *interval r-gathering problem* asks to determine whether there is an r-gathering A such that each facility $f \in F$ serves zero or at least r customers and for each customer $C \in \mathcal{C}$, $s(C) \leq A(C) \leq t(C)$ holds.

We now give an algorithm for the interval r-gathering problem. Let $F = \{f_1, f_2, \cdots, f_m\}$ be a set of facilities and $\mathcal{C} = \{C_1, C_2, \cdots, C_n\}$ be a set of customers on a line where each customer C_i has an interval $I_i = [s(C_i), t(C_i)]$. An interval I_i is called the *leftmost interval* if for each $C_j \neq C_i$, $t(C_i) \leq t(C_j)$ holds, and the customer C_i is called the *leftmost customer*. A facility f_u is called the *preceding facility* of C_i if $s(C_i) \leq f_u \leq t(C_i)$ and there is no facility f_v such that $f_u < f_v \leq t(C_i)$. Similarly a facility f_u is called the *following facility* of C_i if $s(C_i) \leq f_u \leq t(C_i)$ and there is no facility f_v such that $s(C_i) \leq f_v < f_u$. We call a customer C_j a *right neighbor* of C_i if $t(C_j) \geq t(C_i)$ and $s(C_j) \leq t(C_i)$.

Let $F = \{f_1, f_2, \cdots, f_m\}$ be a set of facilities and $\mathcal{C} = \{C_1, C_2, \cdots, C_n\}$ be a set of customers on a line where each customer C_i has an interval I_i. Let C_i be the leftmost customer, f_u be the preceding facility of C_i, and $\mathcal{C}_u$ be the set of customers containing f_u in their intervals. We now have the following two lemmas.

Lemma 4. *If there is an interval r-gathering of $\mathcal{C}$ to F, then there is an interval r-gathering with the leftmost open facility f_u. Furthermore, the customers assigned to f_u have consecutive right end-points in $\mathcal{C}_u$ including C_i.*

Proof. We first prove that there is an interval r-gathering with the leftmost open facility f_u. Assume for a contradiction that there is no interval r-gathering with the leftmost open facility f_u. Let A be an interval r-gathering with the leftmost open facility $f_v \neq f_u$. We can observe that $f_v \leq f_u$, since in each interval r-gathering C_i is assigned to a facility within the interval I_i and f_u is the preceding facility of C_i. Let $\mathcal{C}'_v$ be the set of customers assigned to f_v in A. For any customer C_j in $\mathcal{C}'_v$, we have $s(C_j) \leq f_v \leq f_u \leq t(C_i) \leq t(C_j)$, since I_i is the leftmost interval. We now derive a new interval r-gathering by reassigning the customers $\mathcal{C}'_v$ to f_u. A contradiction.

We now prove that the customers assigned to f_u have consecutive right end-points in $\mathcal{C}_u$. We call a pair $C_j, C_k \in \mathcal{C}_u$ a reverse pair if $t(C_j) < t(C_k)$, C_k assigned to f_u, and C_j assigned to $f_v > f_u$. Assume for a contradiction that there is no interval r-gathering where the customers assigned to f_u have consecutive right end-points in $\mathcal{C}_u$. Let A' be an interval r-gathering with minimum number of reverse pairs but the number is not zero. Let C_j, C_k be a reverse pair in A'

where $t(C_j) < t(C_k)$, and C_j is assigned to facility f_w, and C_k is assigned to f_u. Since $t(C_k) > t(C_j)$ and $f_w \geq f_u$, we get $s(C_k) \leq f_w \leq t(C_k)$. We now derive a new interval r-gathering with less reverse pairs by reassigning C_j to f_u and C_k to f_w, a contradiction.   □

Lemma 5. *Let C_j be the leftmost customer in $\mathcal{C} \setminus \mathcal{C}_u$, and $\mathcal{C}'_u \subseteq \mathcal{C}_u$ be the customers such that for each $C \in \mathcal{C}'_u$, $t(C) < t(C_j)$. If there is an interval r-gathering, then there is an interval r-gathering satisfying one of the following.*

(a) *If $|\mathcal{C}'_u| < r$, then the customers assigned to f_u are the r leftmost customers in $\mathcal{C}_u$.*

(b) *If $|\mathcal{C}'_u| \geq r$, then $\max\{|\mathcal{C}'_u| - r + 1, r\}$ leftmost customers of $\mathcal{C}'_u$ are assigned to f_u (possibly with more customers).*

Proof. (a) By Lemma 4, the customers assigned to f_u are consecutive in $\mathcal{C}_u$. Thus the leftmost r customers $\mathcal{C}^l_u$ in $\mathcal{C}_u$ are assigned to f_u. We now prove that there is an interval r-gathering where no customer in $\mathcal{C}_u \setminus \mathcal{C}^l_u$ is assigned to f_u. Assume for a contradiction that in every interval r-gathering there are some customers in $\mathcal{C}_u \setminus \mathcal{C}^l_u$ which are assigned to f_u. Let A be an interval r-gathering where the number of customers in $\mathcal{C}_u \setminus \mathcal{C}^l_u$ assigned to f_u is minimum, and C_k be a customer in $\mathcal{C}_u \setminus \mathcal{C}^l_u$ which is assigned to f_u. Since $|\mathcal{C}'_u| < r$, we get $t(C_k) > t(C_j)$. Let C_j is assigned to f_v in A. We now derive a new r-gathering by reassigning C_k to f_v, a contradiction.

(b) We first consider $r \leq |\mathcal{C}'_u| < 2r$. In this case $\max\{|\mathcal{C}'_u| - r + 1, r\} = r$. Hence by Lemma 4 the leftmost r customers in $\mathcal{C}_u$ are assigned to f_u.

We now consider $|\mathcal{C}'_u| \geq 2r$. In this case, $\max\{|\mathcal{C}'_u| - r + 1, r\} = |\mathcal{C}'_u| - r + 1$. Let $\mathcal{C}''_u$ be the leftmost $|\mathcal{C}'_u| - r + 1$ customers in $\mathcal{C}'_u$. Assume for a contradiction that there is no interval r-gathering where $\mathcal{C}''_u$ are assigned to f_u. Let A' be an interval r-gathering with maximum number of customers $\mathcal{D}_u \subset \mathcal{C}''_u$ assigned to f_u. Let $C_s \in \mathcal{C}''_u$ be the customer with smallest $t(C_s)$ which is not assigned to f_u. Let C_s is assigned to $f_v \geq f_u$. By Lemma 4, any customer $C_t \in \mathcal{C}''_u$ with $t(C_t) \geq t(C_s)$ is not assigned to f_u. We first claim that the number of customers assigned to f_v is exactly r. Otherwise we can reassign C_s to f_u and thus contradicting our assumption. Let $\mathcal{C}'_v$ be the customers assigned to f_v. We now claim that there is an interval r-gathering where $\mathcal{C}'_v$ consists of r customers having consecutive right end-points in $\mathcal{C}_u$. Assume otherwise for a contradiction. Let A'' be an interval r-gathering with minimum number of reverse pairs where a reverse pair is a pair of customer C_x, C_y with $t(C_x) \leq t(C_y)$, C_y assigned to f_v, C_x assigned to $f_w > f_v$. Since $t(C_x) \leq t(C_y)$ and $f_v \leq f_w$, we get $s(C_y) \leq f_w \leq t(C_y)$. We now derive a new interval r-gathering by reassigning C_x to f_v and C_y to f_w, a contradiction. Now since $|\mathcal{D}_u| < |\mathcal{C}'_u| - r + 1$, we get $|\mathcal{C}'_u \setminus \mathcal{D}_u| \geq r$. Thus $\mathcal{C}'_v \subset \mathcal{C}'_u$. We now derive a new interval r-gathering by assigning $\mathcal{C}'_v$ to f_u. A contradiction.   □

We now give an algorithm **Interval-r-gather** for the interval r-gathering problem.

We now have the following theorem.

Algorithm 1. Interval-r-gather($\mathcal{C}, F$)

Input : A set $\mathcal{C}$ of customers each having an interval and a set F of facilities on a line

Output: An interval r-gathering if exists

if $|\mathcal{C}| < r$ *or* $F = \emptyset$ **then**
$\quad\mid$ **return** $\emptyset$;
endif

$C_i \leftarrow$ leftmost customer in $\mathcal{C}$;

$f_u \leftarrow$ preceding facility of C;

$\mathcal{C}_u \leftarrow$ the set of customers containing f_u in their intervals;

$C_j \leftarrow$ leftmost customer in $\mathcal{C} \setminus \mathcal{C}_u$;

$\mathcal{C}'_u \leftarrow$ the set of customers in $\mathcal{C}_u$ having smaller right end-point than $t(C_j)$;

$F' \leftarrow$ the set of facilities right to f;

if $|\mathcal{C}_u| < r$ **then**
$\quad\mid$ **return** $\emptyset$;
endif

if $|\mathcal{C}'_u| < r$ **then**
$\quad\mid$ $\mathcal{D}_u \leftarrow$ the set of r leftmost customers in $\mathcal{C}_u$;/* Lemma 5(a) */
$\quad\mid$ $A \leftarrow$ Assignment of $\mathcal{D}_u$ to f_u;
$\quad\mid$ $Ans \leftarrow$ Interval-r-gather($\mathcal{C} \setminus \mathcal{D}_u, F'$);
$\quad\mid$ **if** $Ans \neq \emptyset$ **then**
$\quad\mid\quad\mid$ **return** $Ans \cup A$;
$\quad\mid$ **endif**
$\quad\mid$ **return** $\emptyset$;
endif

$\mathcal{D}_u \leftarrow$ the set of $\max\{r, |\mathcal{C}'_u| - r + 1\}$ leftmost customers in $\mathcal{C}_u$; /* Lemma 5(b) */

$A \leftarrow$ Assignment of $\mathcal{D}_u$ to f_u;

$\mathcal{C}''_u \leftarrow \mathcal{C}'_u \setminus \mathcal{D}_u$;

while $\mathcal{C}''_u$ *is not empty* **do**
$\quad\mid$ $Ans \leftarrow$ Interval-r-gather($\mathcal{C} \setminus \mathcal{D}_u, F'$);
$\quad\mid$ **if** $Ans \neq \emptyset$ **then**
$\quad\mid\quad\mid$ **return** $Ans \cup A$;
$\quad\mid$ **endif**
$\quad\mid$ $C_k \leftarrow$ leftmost customer in $\mathcal{C}''_u$; /* (possibly with more customers) */
$\quad\mid$ $A' \leftarrow$ Assignment of C_k to f_u;
$\quad\mid$ $A \leftarrow A \cup A'$;
$\quad\mid$ $\mathcal{D}_u \leftarrow \mathcal{D}_u \cup \{C_k\}$;
$\quad\mid$ $\mathcal{C}''_u \leftarrow \mathcal{C}''_u \setminus \{C_k\}$;
end

return $\emptyset$;

Theorem 1. *The algorithm* **Interval-r-gather** *decides whether there is an interval r-gathering of $\mathcal{C}$ to F, and constructs one if exists in $O(m+n\log n+nr^{\frac{n}{r}})$ time.*

Proof. The correctness of Algorithm Interval-r-gather is immediate from Lemmas 4 and 5.

We now estimate the running time of the algorithm. We can sort the customers based on their right end-points in $O(n\log n)$ time. For each customer we can precompute the preceding facility f_u in $O(n+m)$ time. For each facility f_u we can precompute the sets of customers C_u containing each facility and the leftmost customer C_j having left end-point on right of f_u in $O(n+m)$ time. In each call to Interval-r-gather, we need $O(|C_u|)$ time and at most r recursive calls to Interval-r-gather. Let $T(n)$ be the running time of the algorithm for n customers. We have $T(n) \leq O(|C_u|) + \sum_{i=1}^{r} T(n-r+1) \leq O(nr^{\frac{n}{r}})$. Thus the running time of the algorithm is $O(m+n\log n+nr^{\frac{n}{r}})$. $\square$

We now have the following theorem.

Theorem 2. *Let $\mathcal{C} = \{C_1, C_2, \cdots, C_n\}$ be a set of uncertain customers on a line each of which is specified by a piece-wise uniform function consisting of $k+1$ pieces, and $F = \{f_1, f_2, \cdots, f_m\}$ be a set of m facilities on the line. Then the optimal r-gathering can be constructed in $O(nk + mn\log n + (m + n\log k + n\log n + nr^{\frac{n}{r}})\log mn)$ time.*

Proof. We give outline of an algorithm to compute optimal r-gathering. We first compute the $E[d(p, C_i)]$ function for each $C_i \in \mathcal{C}$. This takes $O(nk)$ time in total. By Corollary 1, we compute the sorted list of all expected distances between customers and facilities in $O(nk + mn\log n)$ time. We find the optimal r-gathering by binary search, using the $O(m + n\log n + nr^{\frac{n}{r}})$ time algorithm for interval r-gathering $\log mn$ times. For each r-interval gathering problem, we compute the (C_i, b)-intervals in $O(n\log k)$ time. Thus finding optimal r-gathering by binary search requires $O(nk + mn\log n + (m + n\log k + n\log n + nr^{\frac{n}{r}})\log mn)$ time. $\square$

3.2   Uniform Distribution

In this section we give an algorithm for the uncertain r-gathering problem when each customer location is specified by a well-separated uniform distribution.

In the uniform distribution model, location of each customer C_i is specified by a function $g_i : \mathbb{R} \rightarrow \mathbb{R}^+ \cup \{0\}$ where $g_i(p) = 1/(t_i - s_i)$ if $s_i \leq p \leq t_i$ and $g_i(p) = 0$ otherwise. We denote the uniform distribution between $[s_i, t_i]$ by $U(s_i, t_i)$. The customer C_i having a uniform distribution $U(s_i, t_i)$ is denoted by $C_i \sim U(s_i, t_i)$. Figure 2(a) illustrates a uniform distribution where $s_i = 0$ and $t_i = 3$. The range of $U(s_i, t_i)$, denoted by l_i, is the value of $t_i - s_i$, and the mean of $U(s_i, t_i)$, denoted by μ_i, is the value of $\frac{s_i + t_i}{2}$. The uniform distribution model is a special case of the histogram model described in Sect. 3.1. We now have the following lemma.

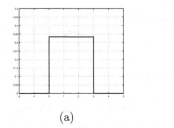 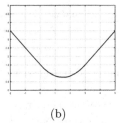

(a)   (b)

Fig. 2. (a) Illustration of a uniform distribution and (b) corresponding function of expected distance

Lemma 6. *Let $C \sim U(s, t)$ be an uncertain point. Then the function $E[d(p, C)]$ consists of a parabola in the interval $[s, t]$ and two straight lines of slope $+1$ and -1 in interval (t, ∞) and $(-\infty, s)$, respectively. Furthermore the minimum value of $E[d(p, C)]$ is $\frac{l}{4}$ and the value of $E[d(p, C)]$ at s, t is $\frac{l}{2}$.*

Proof. We use the Eq. 1 to compute the function $E[d(p, C)]$.

$$E[d(p, C)] = \begin{cases} \mu - p & \text{if } p < s \\ \frac{1}{l}(p - \mu)^2 + \frac{l}{4} & \text{if } s \le p \le t \\ -\mu + p & \text{if } p > t \end{cases} \tag{2}$$

At $p = s$ we get $E[d(s, C)] = \frac{1}{t-s}\left(s - \frac{s+t}{2}\right)^2 + \frac{t-s}{4} = \frac{t-s}{2} = \frac{l}{2}$. Similarly, $E[d(t, C)] = \frac{l}{2}$. Now for $p < s$ and $p > t$, $E[d(p, C)] \ge \frac{t-s}{2}$. The minimum value of the parabola $\frac{1}{t-s}\left(p - \frac{s+t}{2}\right)^2 + \frac{t-s}{4}$ is $\frac{l}{4}$ at $p = \frac{s+t}{2}$. $\square$

We have the following lemma.

Lemma 7. *Let $C \sim U(s, t)$ be an uncertain point and b be a number. Then the (C, b)-interval can be computed in $O(1)$ time.*

Proof. To find the (C, b)-interval, we first compute the inverse of the Eq. 2. For $E[d(p, C)] = b > \frac{l}{2}$, we have $p < s$ or $p > t$. Thus we get, $p = \mu \pm b$. For $\frac{l}{4} \le E[d(p, C)] = b \le \frac{l}{2}$, we have $s \le p \le t$. Thus we get $p = \mu \pm \sqrt{l(b - \frac{l}{4})}$. Finally there is no p for which $E[d(p, C)] < \frac{l}{4}$. Hence the (C, b)-interval for $b < \frac{l}{4}$ is empty. Thus the (C, b)-interval I can be written as following.

$$I = \begin{cases} [\mu - b, \mu + b] & \text{if } b > \frac{l}{2} \\ [\mu - \sqrt{l(b - \frac{l}{4})}, \mu + \sqrt{l(b - \frac{l}{4})}] & \text{if } \frac{l}{4} \le b \le \frac{l}{2} \\ \emptyset & \text{if } b < \frac{l}{4} \end{cases} \tag{3}$$

By Eq. 3 we can compute (C, b)-interval in $O(1)$ time. $\square$

Let $C_i \sim U(s_i, t_i), C_j \sim U(s_j, t_j)$ be two uncertain points. Let $l_{max} = \max\{l_i, l_j\}$ and $l_{min} = \min\{l_i, l_j\}$. We call C_i, C_j *well-separated* if none of the intervals $[s_i, t_i]$ and $[s_j, t_j]$ is contained within the other and $|\mu_i - \mu_j| \geq \frac{1}{2}\sqrt{l_{min}(l_{max} - l_{min})}$.

Lemma 8. *Let $C_i \sim U(s_i, t_i), C_j \sim U(s_j, t_j)$ be two uncertain well-separated points and b be a number. Let I_i, I_j be the (C_i, b)-interval and (C_j, b)-interval respectively. Then none of I_i and I_j is contained in the other.*

Proof. Omitted.   □

If the customer locations are specified by well-separated uniform distributions, we can solve the decision version of uncertain r-gathering problem by dynamic programming as follows. A subproblem asks to determine whether there is an r-gathering with cost at most b for the set of customers $C_1, C_2, \cdots, C_i$. Thus we have at most n distinct subproblems, and to solve a subproblem we need to check n smaller subproblems, so we can design an $O(m + n^2)$ time algorithm.

We can improve the running time as follows. A subproblem $P(i)$ asks to find a set of customers $\mathcal{C}_i$ and an interval r-gathering A of customers $\mathcal{C}_i \subseteq \mathcal{C}$ to $F_i = \{f_1, f_2, \cdots, f_i\}$ such that (1) $\mathcal{C}_i$ contains every customer C_i with $t(C_i) \leq f_i$ (possibly with more customers), (2) f_i serves at least r customers, and (3) $\max_{C \in \mathcal{C}_i}\{t(C)\}$ is minimum. Let $C_{z(i)}$ be the customer with $\max_{C \in \mathcal{C}_i}\{t(C)\}$. We can observe that there is a proper interval r-gathering of $\mathcal{C}$ to F if and only if some $P(i)$ with $f_i \geq s(C_n)$ has a solution.

Lemma 9. *If $P(i)$ has a solution, then there is an interval r-gathering where customers assigned to each open facility have consecutive right end-points.*

Proof. Omitted.   □

We now have the following lemma.

Lemma 10. *If $P(i)$ and $P(j)$ have solutions and $i < j$, then $t(C_{z(i)}) \leq t(C_{z(j)})$.*

Proof. For a contradiction assume $t(C_{z(i)}) > t(C_{z(j)})$. Let A_j be an interval r-gathering corresponding to $P(j)$. Since all the intervals are proper, we have $s(C_{z(i)}) > s(C_{z(j)})$, and $s(C_{z(j)}) \leq f_i$. Let $\mathcal{C}'_j$ be the set of customers assigned to any facility between f_i to f_j (including f_i, f_j) in A_j. For any customer $C_k \in \mathcal{C}'_j$, we have $s(C_k) \leq f_i$ and $t(C_k) \geq f_i$. We now derive a new interval r-gathering A'_j by reassigning the leftmost r customers $\mathcal{C}'_j$ to f_i. Clearly, $\max_{C \in \mathcal{C}'_j}\{t(C)\} < t(C_{z(i)})$ and thus A'_j is a solution of $P(i)$, a contradiction.   □

Using Lemmas 9 and 10, we can determine whether $P(i)$ has solution or not. We have two cases. If $f_i \leq t(C_1)$, then $P(i)$ may have a solution with exactly one open facility f_i, and the solution exists if and only if f_i is contained within at least r intervals. Otherwise $f_i > t(C_1)$, then $P(i)$ may have a solution with two or more open facilities. In this case $P(i)$ has a solution if and only if for some $j < i$ $P(j)$ has a solution, there is no customer C with $f_j < s(C) \leq t(C) < f_i$, and there are at least r customers in $\mathcal{C} \setminus \mathcal{C}_j$ containing f_i. Intuitively f_j is a possible second rightmost open facility in a solution of $P(i)$.

We fix the $P(j)$ with minimum j, if $P(i)$ has a solution, and we say f_j the mate of f_i, and denoted as $mate(f_i)$. We have the following lemma.

Lemma 11. *If $P(i)$ and $P(i+1)$ have solutions, then $mate(f_i) \le mate(f_{i+1})$.*

Proof. For a contradiction assume $mate(f_i) > mate(f_{i+1})$. Let $f_j = mate(f_i)$ and $f_{j'} = mate(f_{i+1})$. By Lemma 10 we have $t(C_{z(j)}) \ge t(C_{z(j')})$. Since $f_{j'}$ is mate of f_{i+1}, there is no customer C such that $f_{j'} < s(C) \le t(C) < f_{i+1}$. If $t(C_{z(j)}) < f_i$, then $f_{j'}$ is also a mate of f_j, a contradiction. Now if $t(C_{z(j)}) \ge f_j$, then $f_{j'}$ is a mate of f_j since $t(C_{z(j')}) \le t(C_{z(j)})$, a contradiction. □

We now have the following lemma.

Lemma 12. *Let f_i be a facility with $f_i > t(C_1)$ and for some $j < i$, $P(j)$ has a solution, and $\mathcal{C} \setminus \mathcal{C}_j$ contains no customer C with $f_j < s(C)$ and $t(C) < f_i$. Fix the $P(j)$ with minimum j. Then the following holds.*

(a) If $\mathcal{C} \setminus \mathcal{C}_j$ has less than r customers containing f_i, then no facility $f_{j'}$ with $f_{j'} \ge f_j$ is a mate of f_i, and $P(i)$ has no solution.
(b) If $P(i+1)$ has a solution, then $mate(f_{i+1}) \ge f_j$.

Proof. (a) By Lemma 10 for any facility $f_{j'} \ge f_j$, if $P(j')$ has a solution, then $t(C_{z(j')}) \ge t(C_{z(j)})$. Thus the number of customers in $\mathcal{C} \setminus \mathcal{C}_{j'}$ containing f_i in their interval is less than r.
(b) Assume for a contradiction that $mate(f_{i+1}) \le f_j$. Let $f_{i'} = mate(f_{i+1})$. Thus there is no customer C with $f_{i'} < s(C)$ and $t(C) < f_{i+1}$. Since $f_{i'} \le f_i \le f_{i+1}$, there is no customer C such that $f_{i'} < s(C)$ and $t(C) < f_i$. Hence, $f_{i'}$ is the leftmost facility such that $P(i')$ has a solution and there is no customer C with $f_{i'} < s(C)$ and $t(C) < f_i$, a contradiction. □

By Lemmas 11 and 12, we observe that we can search for $mate(f_{i+1})$ from where the search for mate of $mate(f_i)$ ends. We now give the following Algorithm called **Proper-interval-r-gather**.

If the intervals are sorted according to their right end-points and the facilities are ordered from left to right, then we can preprocess the set of customers containing each facility in linear time. Each customer and each facility have to be processed for a constant number of times. Hence the algorithm runs in $O(n + m)$ time. We thus have the following theorem.

Theorem 3. *Let $F = \{f_1, f_2, \cdots, f_m\}$ be a set of facilities on a line and $\mathcal{C} = \{C_1, C_2, \cdots, C_n\}$ be a set of customers where each customer C_i has an interval $I_i = [s(C_i), t(C_i)]$ and no interval is contained within any other interval. The algorithm **Proper-interval-r-gather** decides whether there is an interval r-gathering of $\mathcal{C}$ to F, and constructs one if exists in $O(n + m)$ time.*

Algorithm 2. Proper-interval-r-gather($\mathcal{C}, F$)

Input : A set $\mathcal{C}$ of customers each having an interval where no interval is contained within other, a set of F of facilities on the line
Output: An interval r-gathering if exists
if $|\mathcal{C}| < r$ *or* $F = \emptyset$ **then**
  | **return** $\emptyset$;
endif
$i \leftarrow 1$;
/* One open facility */
while $f_i \leq t(C_1)$ **do**
  | **if** $f_i \geq s(C_r)$ **then**
  | | $z(i) \leftarrow r$;
  | **endif**
  | $i \leftarrow i + 1$;
end
$j \leftarrow 1$;
/* Two or more open facilities */
while $i \leq m$ **do**
  | $\mathcal{C}_i \leftarrow \{C_1, C_2, \cdots, C_{z(i)}\}$;
  | **while** $j \leq i$ **do**
  | | **if** $\mathcal{C} \setminus \mathcal{C}_j$ *has at least* r *customers containing* f_i *and* $\mathcal{C} \setminus \mathcal{C}_j$ *has no customer* C *with* $f_j < s(C)$ *and* $t(C) < f_i$ **then**
  | | | $z(i) \leftarrow$ index of the r-th customer in $\mathcal{C} \setminus \mathcal{C}_j$ containing f_i; /* $P(i)$
  | | | has a solution */
  | | | $mate(i) \leftarrow j$;
  | | | break;
  | | **endif**
  | | **if** *There is no customer between* f_j *and* f_i, *and* $\mathcal{C} \setminus \mathcal{C}_j$ *has less than* r *customers containing* f_i **then**
  | | | break; /* $P(i)$ has no solution, Lemma 12(a) */
  | | **endif**
  | | $j \leftarrow j + 1$;
  | **end**
  | $i \leftarrow i + 1$;
end
if *Some* $P(i)$ *with* $f_i \geq s(C_n)$ *has a solution* **then**
  | Compute an interval r-gathering A of $\mathcal{C}$ to F;
  | **return** A;
endif
return $\emptyset$;

We now give outline of the algorithm to solve uncertain r-gathering problem on a line where the customer locations are specified by well-separated uniform distributions. Computing the function $E[d(p, C_i)]$ for all the customers takes $O(n)$ time. We can compute the expected distances between customer C_i and all the facilities in $O(m)$ time. Since the function $E[d(p, C_i)]$ is unimodal, the expected distances between C_i and all the facilities can be sorted in $O(m)$ time.

Computing the expected distances between each pair of customers and facilities takes $O(mn)$ time and we can merge the of n sorted list of expected distances in $O(mn \log n)$ time using heap. We do binary search on the ordered list of expected distances to find the optimal r-gathering. Given b we can compute the (C, b)-intervals for all customers in $O(n)$ time. The (C, b)-intervals can be sorted in $O(n \log n)$ time. Solving each decision instance takes $O(m + n)$ time. Thus to find the optimal solution by binary search we need to solve the decision instances $\log mn$ times, so $O((n \log n + m + n) \log mn)$ in total. Hence the running time is $O(mn \log n + (n \log n + m) \log mn)$. Thus we have the following theorem.

Theorem 4. *Let $F = \{f_1, f_2, \cdots, f_m\}$ be a set of facilities on a line and $C = \{C_1, C_2, \cdots, C_n\}$ be a set of customers where each customer C_i has a well-separated uniform distribution. Then an optimal r-gathering of C to F can be constructed in $O(mn \log n + (n \log n + m) \log mn)$ time.*

4   Conclusion

In this paper we presented an $O(nk + mn \log n + (m + n \log k + n \log n + nr^{\frac{n}{r}}) \log mn)$ time algorithm for the one-dimensional uncertain r-gathering problem when the customers are given by piecewise uniform functions. We also gave an $O(mn \log n + (n \log n + m) \log mn)$ time algorithm when the customers are given by well-separated uniform distributions.

References

1. Agarwal, P.K., Cheng, S., Tao, Y., Yi, K.: Indexing uncertain data. In: Proceedings of the Twenty-Eighth ACM SIGMOD-SIGACT-SIGART Symposium on Principles of Database Systems, PODS 2009, pp. 137–146 (2009)
2. Agarwal, P.K., Efrat, A., Sankararaman, S., Zhang, W.: Nearest-neighbor searching under uncertainty I. Discrete Comput. Geom. **58**(3), 705–745 (2017)
3. Agarwal, P.K., Har-Peled, S., Suri, S., Yildiz, H., Zhang, W.: Convex hulls under uncertainty. Algorithmica **79**(2), 340–367 (2017)
4. Ahmed, S., Nakano, S., Rahman, M.S.: r-gatherings on a star. In: Das, G.K., Mandal, P.S., Mukhopadhyaya, K., Nakano, S. (eds.) WALCOM 2019. LNCS, vol. 11355, pp. 31–42. Springer, Cham (2019). https://doi.org/10.1007/978-3-030-10564-8_3
5. Akagi, T., Nakano, S.: On r-gatherings on the line. In: Wang, J., Yap, C. (eds.) FAW 2015. LNCS, vol. 9130, pp. 25–32. Springer, Cham (2015). https://doi.org/10.1007/978-3-319-19647-3_3
6. Armon, A.: On min-max r-gatherings. Theoret. Comput. Sci. **412**(7), 573–582 (2011)
7. Drezner, Z., Hamacher, H.W.: Facility Location: Applications and Theory. Springer, New York (2004)
8. Guha, S., Meyerson, A., Munagala, K.: Hierarchical placement and network design problems. In: Proceedings 41st Annual Symposium on Foundations of Computer Science, pp. 603–612 (2000)

9. Han, Y., Nakano, S.: On r-gatherings on the line. In: Proceedings of FCS 2016, pp. 99–104 (2016)

10. Kamousi, P., Chan, T.M., Suri, S.: Closest pair and the post office problem for stochastic points. Comput. Geom. **47**(2), 214–223 (2014)

11. Karget, D.R., Minkoff, M.: Building steiner trees with incomplete global knowledge. In: Proceedings 41st Annual Symposium on Foundations of Computer Science, pp. 613–623 (2000)

12. Nakano, S.: A simple algorithm for r-gatherings on the line. In: Rahman, M.S., Sung, W.-K., Uehara, R. (eds.) WALCOM 2018. LNCS, vol. 10755, pp. 1–7. Springer, Cham (2018). https://doi.org/10.1007/978-3-319-75172-6_1

13. Sarker, A., Sung, W., Rahman, M.S.: A linear time algorithm for the r-gathering problem on the line (extended abstract). In: Das, G.K., Mandal, P.S., Mukhopadhyaya, K., Nakano, S. (eds.) WALCOM 2019. LNCS, vol. 11355, pp. 56–66. Springer, Cham (2019). https://doi.org/10.1007/978-3-030-10564-8_5

14. Snyder, L.V.: Facility location under uncertainty: a review. IIE Trans. **38**(7), 547–564 (2006)

15. Suri, S., Verbeek, K.: On the most likely voronoi diagram and nearest neighbor searching. Int. J. Comput. Geom. Appl. **26**(3–4), 151–166 (2016)

16. Tao, Y., Xiao, X., Cheng, R.: Range search on multidimensional uncertain data. ACM Trans. Database Syst. **32**(3), 15 (2007)

17. Wang, H., Zhang, J.: One-dimensional k-center on uncertain data. Theoret. Comput. Sci. **602**, 114–124 (2015)

18. Yiu, M.L., Mamoulis, N., Dai, X., Tao, Y., Vaitis, M.: Efficient evaluation of probabilistic advanced spatial queries on existentially uncertain data. IEEE Trans. Knowl. Data Eng. **21**(1), 108–122 (2009)

Improved Algorithms for Ranking and Unranking (k, m)-Ary Trees

Yu-Hsuan Chang[1], Ro-Yu Wu[2], Ruay-Shiung Chang[1],
and Jou-Ming Chang[1(✉)]

[1] Institute of Information and Decision Sciences,
National Taipei University of Business, Taipei, Taiwan
{10766004,rschang,spade}@ntub.edu.tw
[2] Department of Industrial Management,
Lunghwa University of Science and Technology, Taoyuan, Taiwan
eric@mail.lhu.edu.tw

Abstract. Du and Liu (2007) introduced (k, m)-ary trees as a generalization of k-ary trees. In a (k, m)-ary tree, every node on even level has degree k (i.e., has k children), and every node on odd level has degree m (which is called a crucial node) or is a leaf. In particular, a (k, m)-ary tree of order n has exactly n crucial nodes. Recently, Amani and Nowzari-Dalini (2019) presented a generation algorithm to produce all (k, m)-ary trees of order n in B-order using Zaks' encoding, and show that the generated ordering of this encoding results in a reverse-lexicographical ordering. They also proposed the corresponding ranking and unranking algorithms for (k, m)-ary trees according to such a generated ordering. These algorithms take $\mathcal{O}(kmn^2)$ time and space for building a precomputed table in which (k, m)-Catalan numbers (i.e., a kind of generalized Catalan numbers) are stored in advance. In this paper, we revisit the ranking and unranking problems. With the help of an encoding scheme called "right-distance" introduced by Wu et al. (2011), we propose new ranking and unranking algorithms for (k, m)-ary trees of order n in B-order using Zaks' encoding. We show that both algorithms can be improved in $\mathcal{O}(kmn)$ time and $\mathcal{O}(n)$ space without building the precomputed table.

Keywords: (k, m)-ary trees · Ranking/Unranking algorithms ·
Zaks' sequences · RD-sequences ·
Lexicographic/Reverse-lexicographic order · Amortized cost

1 Introduction

In computer science, many practical applications dealing with a huge amount of combinatorial objects require the help of generation, ranking and unranking algorithms to accomplish. In general, combinatorial objects in a certain family are encoded by using integer sequences so that all objects (or their corresponding sequences) are generated in a particular order. For a specific order of objects, a *ranking algorithm* is a function that determines the rank of a given object in the

© Springer Nature Switzerland AG 2019
D.-Z. Du et al. (Eds.): AAIM 2019, LNCS 11640, pp. 16–28, 2019.
https://doi.org/10.1007/978-3-030-27195-4_2

generated list, and an *unranking algorithm* is one that produces the object (or sequence) corresponding to a given rank. Efficient ranking and unranking are important and useful for storing and retrieving elements in a class of combinatorial objects.

Trees are one of the most fundamental combinatorial objects and the problem of generating trees and related works have been widely studied in the literature. For example, many generation, ranking and unranking algorithms have been developed for binary trees [8,12,15], k-ary trees [14,17,18], AVL trees [7], non-regular trees [13,16], trees with n nodes and m leaves [9,10], ordered trees with bounded degree [4], neuronal trees [3,5], Fibonacci-isomorphic trees [1], and (k, m)-ary trees [2].

As to the (k, m)-ary trees, the family of trees is first introduced by Du and Liu [6] when they studied hook length polynomials for plane trees. A (k, m)-ary tree is a generalization of k-ary tree such that the degree of each internal node is determined based on the level of the node resided (formally defined later in Sect. 2.2). After that, it seems to be no literature addressing on the discussion of (k, m)-ary trees until recently, Amani and Nowzari-Dalini [2] presented a generation algorithm to produce all (k, m)-ary trees of order n in B-order (see Definition 3). To design this generation algorithm, they adopted Z-sequences defined by Zaks [18] to encode the trees. As a result, all Z-sequences corresponding to trees are generated in reverse-lexicographical ordering. In particular, each sequence can be generated in a constant amortized time and $\mathcal{O}(n)$ time complexity in the worst case. Moreover, based on this generated ordering, they proposed ranking and unranking algorithms. The main technique of these two algorithms is to use a pre-computation that builds a table for storing a kind of generalized Catalan numbers called (k, m)-Catalan numbers in advance. The pre-computation requires a total of $\mathcal{O}(kmn^2)$ time and space. Then, each ranking and unranking algorithm can be performed by accessing the table in $\mathcal{O}(n)$ and $\mathcal{O}(n \log n)$ time to accomplish, respectively.

In this paper, we are interested in making more explorations on the study of (k, m)-ary trees. We first try to encode a (k, m)-ary tree T by using another coding scheme called "right-distance" introduced by Wu et al. [17] for measuring the distance between nodes and the right arm (i.e., a path from the root to its rightmost leaf) of T. A sequence obtained by using the above measure for internal nodes on odd levels is called a right-distance sequence (or RD-sequence) of T. Then, we are easy to derive a complementary relation between Z-sequences and RD-sequences so that the transformation between the two kinds of sequences can be done in $\mathcal{O}(n)$ time. According to this relation and four extending formulae obtained from (k, m)-Catalan numbers, we redesign ranking and unranking algorithms for (k, m)-ary trees of order n in B-order using Z-sequences. As a consequence, we show that algorithms we proposed are more efficient and can be run in $\mathcal{O}(kmn)$ time and $\mathcal{O}(n)$ space without building the precomputed table.

The rest of this paper is organized as follows. In Sect. 2, we establish all necessary background knowledge. In Sects. 3 and 4, we present our ranking and unranking algorithms, respectively, and show the correctness and efficiency. Finally, concluding remarks are given in the last section.

2   Preliminaries

A rooted tree is a k-*ary tree* if it is an ordered k-regular tree (i.e., every internal node has exactly k-ordered children). A k-*ary tree of order* n is a k-ary tree with exactly n internal nodes. Let T be a tree rooted at r. A node v in T is said to be *on the level* ℓ if the unique path from r to v has length ℓ, and the root r is on the level 0. For notational convenience, we write $i \in T$ to mean that i is an internal node of T. Let $\mathcal{T}_k(n)$ denote the set of k-ary trees of order n. It is well known that the number of k-ary trees of order n is counted by generalized Catalan number [11], i.e., $|\mathcal{T}_k(n)| = C_k(n) = 1/(kn + 1)\binom{kn+1}{n}$.

2.1   Zaks' Sequences vs. Right-Distance Sequences

In most generation algorithms, trees are encoded as integer sequences and these sequences are generated in specific order. In the following, we will introduce two types of integer sequences for representing k-ary trees of order n.

Zaks [18] gave the following representation of k-ary trees. Let T be a k-ary tree of order n in which all internal nodes are numbered from 1 to n in preorder (i.e., visit the root and then recursively the subtrees of T from left to right). Henceforth, we will not distinguish the terms between an internal node and its preorder number. For each internal node $i \in T$, we denote by z_i the visited order of node i when both internal nodes and leaves of T are traversed in preorder. Then, the resulting sequence $z(T) = (z_1, z_2, \ldots, z_n)$ is called the *Zaks' sequence* (or *Z-sequence* for short) of T.

Wu et al. [17] used another measure called "right-distance" to represent k-ary trees. Let T be a k-ary tree defined as above. A *right-distance sequence* (or *RD-sequence* for short) of T, denoted by $d(T) = (d_1, d_2, \ldots, d_n)$, is an integer sequence in which the term d_i for each internal node $i \in T$ is recursively defined as follows:

$$d_i = \begin{cases} 0 & \text{if } i \text{ is the root of } T \text{ (i.e., } i = 1); \\ d_{p(i)} + k - h & \text{otherwise,} \end{cases} \tag{2.1}$$

where $p(i)$ stands for the parent of node i in T, and h is the order in which the node i occurs among all sons of $p(i)$ from left to right. A fact has been pointed out in [17] that the use of RD-sequences is more concise than that of Z-sequences when we consider the range of integers to be used for encoding. Also, it has shown in [17] the following complementary relation between Z-sequences and RD-sequences.

Theorem 1 ([17]). *Let T be a k-ary tree of order n encoded by Z-sequence $(z_1, z_2, \ldots, z_n)$ and RD-sequence $(d_1, d_2, \ldots, d_n)$, respectively. Then, for each $i = 1, 2, \ldots, n$,*

$$z_i + d_i = k(i - 1) + 1.$$

2.2 (k, m)-Ary Trees and (k, m)-Catalan Number

As we have mentioned earlier, a (k, m)-ary tree is a generalization of k-ary tree such that there are two different degrees of nodes in the tree. The following is the formal definition.

Definition 1 ([6]). For $k, m \geqslant 1$ and $n \geqslant 0$, a (k, m)-*ary tree of order* n is an ordered tree, such that

(1) All nodes on even levels have degree k (where the root is on the level 0).
(2) All node on odd levels have degree m or 0, and there are exactly n nodes of degree m in all odd levels.

Let $\mathcal{T}_{k,m}(n)$ denote the set of (k, m)-ary trees of order n. Note that there are a total of $(mn + 1)(k + 1)$ nodes for each tree $T \in \mathcal{T}_{k,m}(n)$, with $mn + 1$ nodes on even levels and $(mn + 1)k$ nodes on odd levels (in which there are n internal nodes and $(mn+1)k-n$ leaves). In particular, the n internal nodes on odd levels are called *crucial nodes* in [6].

Du and Liu [6] also defined (k, m)-*Catalan number of order* n as $C_{k,m}(n)$, and proved that (k, m)-ary trees can be counted by (k, m)-Catalan numbers as follows.

Theorem 2 ([6]). *The number of* (k, m)-*ary trees of order* n *is equal to*

$$|\mathcal{T}_{k,m}(n)| = C_{k,m}(n) = \frac{1}{mn + 1} \binom{(mn + 1)k}{n}. \tag{2.2}$$

According to Eq. (2.2), the number of $(2, 3)$-ary trees of order 4 is equal to $\frac{1}{13} \binom{26}{4} = 1150$.

Amani and Nowzari-Dalini [2] adopted Z-sequences to encode (k, m)-ary trees in the following way. For any (k, m)-ary tree of order n, nodes on even levels (including the root) are ignored, and nodes on odd levels are labeled by 1 if they are crucial nodes and labeled by 0 otherwise. Then, by a preorder traversal of T, we can obtain a binary string called codeword of T, denoted by $x(T)$. The Z-sequence of T is $z(T) = (z_1, z_2, \ldots, z_n)$ such that the term z_i for each crucial node $i \in T$ is the position of the ith "1" in $x(T)$. For example, Fig. 1(a) shows a $(2, 3)$-ary tree of order 4, where white nodes on even levels are internal nodes with degree 2 and the shaded nodes on odd levels indicate crucial nodes with degree 3.

Inspired by the above encoding, we now show how to encode (k, m)-ary trees by using RD-sequences. Let T be a given (k, m)-ary tree of order n. We first imagine that T is extended by adding a dummy node numbered by 0 as the root of the extended tree and let $d_0 = 0$. Also, we assume that T is the rightmost subtree of the dummy root. With a notion similar to Eq. (2.1), the term d_i for each crucial node $i \in T$ is defined as follows:

$$d_i = d_{g(i)} + km - h. \tag{2.3}$$

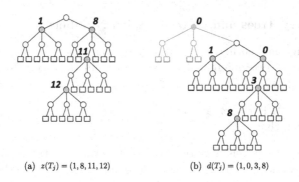

(a) $z(T_j) = (1, 8, 11, 12)$ (b) $d(T_j) = (1, 0, 3, 8)$

Fig. 1. A $(2, 3)$-ary tree of order 4 encoded by (a) Z-sequence and (b) RD-sequence.

where $g(i)$ stands for the grandparent of node i in the extended tree, and h is the order in which the node i occurs among all grandchildren of $g(i)$ from left to right. For example, if we consider the $(2, 3)$-ary tree shown in Fig. 1(a), then the corresponding extended tree is shown in Fig. 1(b). Clearly, for nodes 1 and 2 in T, since they are the fifth and sixth grandchild (from left to right) of the dummy root 0, we have $d_1 = d_0 + 6 - 5 = 1$ and $d_2 = d_0 + 6 - 6 = 0$. Also, as node 3 is the third grandchild of node 2, we have $d_3 = d_2 + 6 - 3 = 3$. Similarly, as node 4 is the first grandchild of node 3, we have $d_4 = d_3 + 6 - 1 = 8$. Thus, $d(T_j) = (1, 0, 3, 8)$.

By using the same proof technique of Theorem 1, we can acquire the following property (here we omit the proof for the sake of brevity).

Theorem 3. *Let T be a (k, m)-ary tree encoded by Z-sequence $(z_1, z_2, \ldots, z_n)$ and RD-sequence $(d_1, d_2, \ldots, d_n)$, respectively. Then, for each $i = 1, 2, \ldots, n$,*

$$z_i + d_i = (i - 1)km + k. \tag{2.4}$$

Corollary 1. *For a (k, m)-ary tree T of order n, transformations between the Z-sequence and RD-sequence of T can be done in $\mathcal{O}(n)$ time.*

2.3   B-order and Reverse-Lexicographical Ordering

In what follows, we establish the related background of the ordering on the family of (k, m)-ary trees generated in [2]. We first give the definitions of two orderings, one is for sequences and the other is for trees.

Definition 2. *Two sequences $x = (x_1, x_2, \ldots, x_n)$ and $y = (y_1, y_2, \ldots, y_m)$ are in lexicographical order (denote as $x \prec_{\text{lex}} y$) if there exists $i \in [1, \min\{n, m\}]$ such that (i) $x_j = y_j$ for all $j = 1, 2, \ldots, i - 1$, and (ii) $x_i < y_i$.*

Definition 3 ([18]). *Let T and T' be two trees and $k = \max\{\deg(T), \deg(T')\}$. We say that T is less than T' in B-order (denote as $T \prec_{\text{B}} T'$) if (i) $\deg(T) < \deg(T')$, or (ii) $\deg(T) = \deg(T')$ and there exists $i \in [1, k]$ such that $T_i \prec_{\text{B}} T'_i$*

and $T_j =_B T'_j$ for all $j = 1, 2, \ldots, i-1$, where $\deg(T)$ denote the degree (i.e., the number of children) of the root of T, and T_i (resp. T'_i) is the ith subtree of T (resp. T').

For instance, Fig. 2 shows a list of $(2, 3)$-ary trees of order 2 in B-order. To generate (k, m)-ary trees of order n in B-order, Amani and Nowzari-Dalini [2] proved that the corresponding Z-sequences of such trees should be generated in reverse-lexicographical ordering. Because k and m are treated as two fixed values in the generation algorithm, by the complementary relation stated in Theorem 3, the generated ordering corresponding to RD-sequences results in a lexicographical ordering.

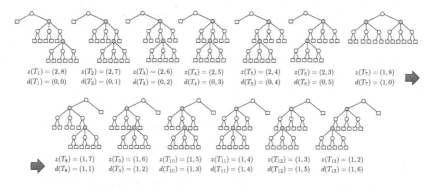

Fig. 2. A list of $(2, 3)$-ary trees of order 2 in B-order.

3 An Improved Ranking Algorithm

In this section, we first introduce the ranking algorithm proposed in [2], and then present our improvement. To calculate the number of (k, m)-ary trees, Amani and Nowzari-Dalini accomplished this by counting the number of *extended-(k, m, d)-ary trees of order* n, where such a tree is defined by preserving the root and nodes on odd levels (and ignoring nodes on even levels) in a (k, m)-ary tree and changing the degree of each non-leaf node with d. Let $B_{n,d}^{k,m}$ be the number of extended-(k, m, d)-ary trees of order n. Since k and m are fixed as constant values in designing ranking and unranking algorithms, for the sake of simplicity, they used $B_{n,d}$ instead of $B_{n,d}^{k,m}$ and provided the following recursive formula of $B_{n,d}$.

$$B_{n,d} = \begin{cases} 1 & \text{if } n = 0 \\ d & \text{if } n = 1 \\ B_{n-1,km} & \text{if } d = 1 \\ B_{n,d-1} + B_{n-1,km+d-1} & \text{if } n, d > 1. \end{cases} \tag{3.1}$$

In fact, it needs an amendment to the above formula by adding $B_{n,0} = 0$, otherwise, we will find a flaw (see Example 1). Also, we note that $B_{n,k} = C_{k,m}(n)$. Amani and Nowzari-Dalini [2] presented a function namely ZseqBorderRank to compute the rank of a given (k, m)-ary tree T with Z-sequence encoding (see Fig. 3). In this function, we observe that the term $B_{n+1-i,k+km(i-1)-z_i}$ can be simplify to write as B_{n+1-i,d_i} by Eq. (2.4). Thus, the variable $NegR$ is used to store the number of trees whose corresponding Z-sequences have rank preceding that of T in lexicographic ordering. Since the equivalence between the rank of a tree in B-order and the rank of the corresponding Z-sequence in reverse-lexicographical ordering, we can obtain the result by subtracting $NegR$ from the total number of trees in $T_{k,m}(n)$.

Function ZseqBorderRank($z_1, z_2, \ldots z_n$: **integer sequence**): **integer**;

   begin
       $NegR \leftarrow 0$;
       for $i \leftarrow 1$ **to** n **do**
          $NegR \leftarrow NegR + B_{n+1-i,k+km(i-1)-z_i}$;
          $Rank \leftarrow B_{n,k} - NegR$;
   return $Rank$;

Fig. 3. The ranking function proposed in [2].

Example 1. Consider a $(2,3)$-tree T_j with $z(T_j) = (1, 8, 11, 12)$ shown in Fig. 1. Since $n = 4$ and $(n-1)km + k = 20$, the algorithm first builds the table $B_{\ell,d}$ where $\ell \in [0, 4]$ and $d \in [0, 20]$ by Eq. (3.1). Note that an amendment is needed for $B_{\ell,0} = 0$. For the sake of saving space, Fig. 4 only shows part of the table. From the function ZseqBorderRank, we can easily check that the value of $NegR$ is computed as follows:

$$NegR = \sum_{i=1}^{4} B_{5-i,2+6(i-1)-z_i} = B_{4,1} + B_{3,0} + B_{2,3} + B_{1,8} = 506 + 0 + 21 + 8 = 535.$$

Thus, the rank of T_j is equal to $B_{n,k} - NegR = 1150 - 535 = 615$.   □

$\ell \backslash d$	0	1	2	3	4	5	6	7	8	9	10
0	0	1	1	1	1	1	1	1	1	1	1
1	0	1	2	3	4	5	6	7	8	9	10
2	0	6	13	21	30	40	51	63	76	90	105
3	0	51	114	190	280	385	506	644	800	975	1170
4	0	506	1150	1950	2925	4095	5481	7105	8729	10353	11977

Fig. 4. An illustration of Example 1.

According to Eq. (3.1), the following theorem is easily proved by induction.

Theorem 4. $B_{\ell,d}$ can also be computed by the following closed form:

$$B_{\ell,d} = \frac{d}{km\ell + d}\binom{km\ell + d}{\ell},\tag{3.2}$$

where $0 \leqslant \ell \leqslant n$ and $0 \leqslant d < (n-1)km + k$.

For convenience, we called a table built by Eq. (3.2) as a *precomputed table*. From this equation, we can also obtain the following four formulae. Indeed, if any entry in the precomputed table is given, we can immediately compute the four related entries in a constant time, which are useful for designing efficient ranking and unranking algorithms.

$$B_{\ell,d+1} = B_{\ell,d} \cdot \frac{(d+1)(km\ell + d)}{d(\ell(km-1) + d + 1)}\tag{3.3}$$

$$B_{\ell,d-1} = B_{\ell,d} \cdot \frac{(d-1)(\ell(km-1) + d)}{d(km\ell + d - 1)}\tag{3.4}$$

$$B_{\ell+1,d-(km-1)} = B_{\ell,d} \cdot \frac{(d - km + 1)(km\ell + d)}{d(\ell + 1)}\tag{3.5}$$

$$B_{\ell-1,d+(km-1)} = B_{\ell,d} \cdot \frac{\ell(d + km - 1)}{d(km\ell + d - 1)}\tag{3.6}$$

We now describe how to design an efficient ranking algorithm. We may visualize that there is a built precomputed table, but indeed it is unnecessary to build such a table. Also, to facilitate the description of the entire process of calculation, we assume that the table is upside down when we compare with the table in Fig. 4. Conceptually, for calculating the rank of a tree, we imagine that the precomputed table is a chess board and only elements along a certain path where a chess B moves on the chess board are accumulated in the rank. Technically, we do not employ Eq. (3.2) to extract elements in the chess board, and alternatively, the movement of B can be done by using Eqs. (3.3) and (3.5). The details of the ranking function are shown in Fig. 5.

In the first step, the input Z-sequence of a (k,m)-ary tree is converted into the corresponding RD-sequence. For convenience, we assume that there exists at least one nonzero item in the sequence $d_1, d_2, \ldots, d_n$ (otherwise, the rank of the tree equals to 0). Let d_g and d_s be the largest nonzero item and the smallest nonzero item, respectively. Step 2 determines the initial location of the chess B corresponding to d_g. Step 3 is the main loop of the function for calculating the rank by a sequence of nonzero entries B_{ℓ,d_ℓ} for $s \leqslant \ell \leqslant g$. For each iteration, we first calculate the index of the next nonzero item d_w in Step 3.1. According to d_w and d_f where $f = n + 1 - \ell$, we can determine the number of steps, say h, of the right movement in Step 3.2. Then, in Step 3.3, we move B to the right h cells in the current row by repeatedly updating B using Eq. (3.3). Further, in Step 3.4, we move B to an upper-left cell (the cell is precisely located in the upper row and to the left $km-1$ cells) by using Eq. (3.5) and acquire a candidate term B_{ℓ,d_ℓ}. Note that this step may be repeated if $d_\ell = 0$. After this step, we

---

Function Ranking($z_1, z_2, \ldots z_n$: **integer sequence**): **integer**;

> **begin**
>
> Step 1 │ // Convert Z-sequence to RD-sequence
> │ **for** $i \leftarrow 1$ **to** n **do** $d_i \leftarrow (i-1)km + k - z_i$;
>
> Step 2 │ // Find the initial term of B
> │ $g \leftarrow \max\{i : d_i \neq 0 \text{ and } 1 \leqslant i \leqslant n\}$;
> │ $s \leftarrow \min\{i : d_i \neq 0 \text{ and } 1 \leqslant i \leqslant n\}$;
> │ $\ell \leftarrow 1$; $B \leftarrow d \leftarrow d_g$;
> │ **for** $i \leftarrow 1$ **to** $n - g$ **do**
> │ │ **for** $r \leftarrow 1$ **to** $km - 1$ **do**
> │ │ │ $B \leftarrow B \cdot \frac{(d+1)(km\ell+d)}{d(\ell(km-1)+d+1)}$;   // Refer to Eq.(3.3), move right
> │ │ └ $d \leftarrow d + 1$;
> │ │ $B \leftarrow B \cdot \frac{(d-km+1)(km\ell+d)}{d(\ell+1)}$;   // Refer to Eq.(3.5), move upper-left
> │ └ $\ell \leftarrow \ell + 1$; $d \leftarrow d - (km - 1)$;
> │ $Rank \leftarrow B$
>
> Step 3 │ // The main loop for accumulating nonzero terms B_{ℓ, d_ℓ}
> │ **while** $g > s$ **do**
> 3.1 │ │ $w \leftarrow \max\{i : d_i \neq 0 \text{ and } s \leqslant i < g\}$;
> │ │ $f \leftarrow n + 1 - \ell$;
> 3.2 │ │ $h \leftarrow (n - (w - 1) - \ell) \cdot (km - 1) + d_w - d_f$;
> 3.3 │ │ **for** $d \leftarrow d_f$ **to** $d_f + h - 1$ **do**
> │ │ └ $B \leftarrow B \cdot \frac{(d+1)(km\ell+d)}{d(\ell(km-1)+d+1)}$;   // Refer to Eq.(3.3), move right
> │ │ $d \leftarrow d_f + h$;
> 3.4 │ │ **for** $i \leftarrow 1$ **to** $n - (w - 1) - \ell$ **do**
> │ │ │ $B \leftarrow B \cdot \frac{(d+1)(km\ell+d)}{d(\ell(km-1)+d+1)}$;   // Refer to Eq.(3.5), move upper-left
> │ │ └ $d \leftarrow d - (km - 1)$
> 3.5 │ └ $\ell \leftarrow n - (w - 1)$; $Rank \leftarrow Rank + B$; $g \leftarrow w$;
> └ **return** $Rank$;

Fig. 5. An improved ranking function.

obtain the desired term in which the current B resides and then add the term into the rank in Step 3.5. We repeat the loop until all subsequent nonzero terms are added into the rank.

Example 2. Consider a $(2, 3)$-ary tree with Z-sequence $(1, 5, 14, 17, 25, 30)$ as the input. Clearly, $km - 1 = 5$. In Step 1, we convert the input into the corresponding RD-sequence $(1, 3, 0, 3, 1, 2)$. In Step 2, we determine $g = 6$, $s = 1$ and $\ell = 1$. Initially, we have $B = 2$ which is located in $B_{1,2}$ and is added into the rank. Figure 6 shows the track of the chess B moving on the chess board when the function Ranking() is performed, where an entry B_{ℓ, d_ℓ} marked by a circle means the term which will be added into the rank. Also, since $B_{6,2} = C_{2,3}(6) = 145299$, we only provide the entries no more than $B_{6,2}$ in the table. In particular, entries with bold face number are cells on which B moves. There are four rounds in Step 3 as follows: In the first round, we have $w = 5$, $f = 6$, and $h = (6 - 4 - 1) \cdot 5 + d_w - d_f = 5 + 1 - 2 = 4$. After B moves from $B_{1,2}$ to $B_{1,6}$, it then moves to $B_{1+1,6-5} = B_{2,1}$. Set $\ell = 2$, $g = 5$, and add the term $B_{2,1} = 6$ into the rank. In the second round, we have $w = 4$, $f = 5$, and $h = (6 - 3 - 2) \cdot 5 + d_w - d_f = 5 + 3 - 1 = 7$. After B moves from $B_{2,1}$ to $B_{2,8}$, it then moves to $B_{2+1,8-5} = B_{3,3}$. Set $\ell = 3$, $g = 4$, and add the term $B_{3,3} = 190$ into the rank. In the third round, we have $w = 2$, $f = 4$, and $h = (6 - 1 - 3) \cdot 5 + d_w - d_f = 10 + 3 - 3 = 10$. After B moves from $B_{3,3}$ to $B_{3,13}$, since $d_3 = 0$, it then moves to $B_{3+1,13-5} = B_{4,8}$ and to $B_{4+1,8-5} = B_{5,3}$. Set $\ell = 5$, $g = 2$, and add the term $B_{5,3} = 21576$ into the

$\ell \backslash d$	0	1	2	3	4	5	6	7	8	9	10	11	12	13
6	0	62832	145299											
5	0	5481	12586	21576	32736	46376	62832	82467						
4	0	506	1150	1950	2925	4095	5481	7150	8990	11160	13640	16456	19635	
3	0	51	114	190	280	385	506	644	800	975	1170	1386	1624	1885
2	0	6	13	21	30	40	51	63	76	90	105	121	138	156
1	0	1	2	3	4	5	6	7	8	9	10	11	12	13
0	0	1	1	1	1	1	1	1	1	1	1	1	1	1

Fig. 6. An illustration of Example 2.

rank. In the last round, we have $w = 1$, $f = 2$, and $h = (6-0-5)\cdot 5 + d_w - d_f = 5+1-3 = 3$. After B moves from $B_{5,3}$ to $B_{5,6}$, it then moves to $B_{5+1,6-5} = B_{6,1}$. Set $\ell = 6$, $g = 1$, and add the term $B_{6,1} = 62832$ into the rank.

Now, since $\ell = s = 1$, the loop terminates. As consequence, we have the rank $B_{1,2} + B_{2,1} + B_{3,3} + B_{5,3} + B_{6,1} = 2 + 6 + 190 + 21576 + 62832 = 84606$.   □

We are now ready to give our main result for ranking algorithm.

Theorem 5. *The Function* Ranking() *can correctly determine the rank of a (k,m)-ary tree of order n in B-order using Z-sequence encoding in $\mathcal{O}(kmn)$ time and $\mathcal{O}(n)$ space.*

4   An Improved Unranking Algorithm

In what follows, we provide a reverse procedure, called Unranking(), to convert a positive integer N to its corresponding Z-sequence of a (k,m)-ary tree of order n in B-order. As before, the corresponding RD-sequence has been chosen purposely to assist the conversion. The basic ideal inside the function is to decompose N into a sequence of fragment values. From these fragmentations, the corresponding RD-sequence can be determined and then be converted into the desired Z-sequence. The details of the unranking function are shown in Fig. 7.

Initially, we assume that $d_i = 0$ for all $i = 1, 2, \ldots, n$. In Step 1, we begin to put B on the initial location by using Eq. (3.2). For example, if we take a $(2,3)$-ary tree of order 6 as an instance, the initial location of B is at the cell $B_{6,2} = 145299$. In Step 2, we first set $i = 1$, $\ell = n$ and $d = k$, where i is the index for which the current item d_i in the RD-sequence will be set up, and ℓ and d indicate the row and column in which the current B resides, respectively. Then, we carry out the main loop till the condition that N has been decreased to 0. For each round in the outer loop, we first search for a possible fragmentation from right to left in a row by the inner loop, where the percolation is made by Eq. (3.4). After this step, we test the condition $B \leqslant N$ to determine if there exists a fragmentation that can be subtracted from N. If it is so in this case, then we update N by subtracting B from N. In the meanwhile, the term d_i is determined. Otherwise, we set $d_i = 0$. Finally, we move B to a bottom-right cell

```
Function Unranking(N: integer): integer sequence;
   begin
Step 1    // Calculate the number of (k,m)-ary trees
          B ← B_{n,k};
Step 2    // Decompose N into a sequence of fragment values
          i ← 1;   d ← k;   ℓ ← n;
          while N > 0 do
             while B > N and d > 1 do
                B ← B · (d-1)(ℓ(km-1)+d)/d(kmℓ+d-1);     // Refer to Eq.(3.4), move left
                d ← d - 1;
             if B ⩽ N then
                N ← N - B;   d_i ← d;
             else
                d_i ← 0;
             i ← i + 1;
             B ← B · ℓ(d+km-1)/d(kmℓ+d-1);     // Refer to Eq.(3.6), move bottom-right
             ℓ ← ℓ - 1;   d ← d + (km - 1);
Step 3    // Convert RD-sequence to Z-sequence
          for i ← 1 to n do   z_i ← (i-1)km + k - d_i;
          return z_1, z_2, ... z_n;
```

Fig. 7. An improved unranking function.

(the cell is precisely located in the lower row and to the right $km - 1$ cells) by using Eq. (3.6) to finish the round. We repeat the loop until all fragmentations are decomposed from N, and thus obtain the resulting RD-sequence. In the final step, the RD-sequence is converted into the corresponding Z-sequence.

Suppose that we are given an integer $N = 84606$. Figure 8 shows the track of performing Unranking(N) for $(2, 3)$-ary trees of order 6. Similarly, a cell marked by a circle means a fragmentation which is decomposed from N. Then, the corresponding RD-sequence can be obtained from the column indices of these cells with fragmentations, i.e., $(1, 3, 0, 3, 1, 2)$ in this case. As a result, referring to Example 2, we have the desired Z-sequence $(1, 5, 14, 17, 25, 30)$.

ℓ\d	0	1	2	3	4	5	6	7	8	9	10	11	12	13
6	0	62832	145299											
5	0	5481	12586	21576	32736	46376	62832	82467						
4	0	506	1150	1950	2925	4095	5481	7150	8990	11160	13640	16456	19635	
3	0	51	114	190	280	385	506	644	800	975	1170	1386	1624	1885
2	0	6	13	21	30	40	51	63	76	90	105	121	138	156
1	0	1	2	3	4	5	6	7	8	9	10	11	12	13
0	0	1	1	1	1	1	1	1	1	1	1	1	1	1

Fig. 8. An illustration of Example 2.

Since Unranking is the reverse procedure of Ranking, the correctness directly follows from Theorem 5. Also, an argument similar to the above one that uses aggregation method can analyze the time complexity and space requirement, and thus we have the following theorem.

Theorem 6. *Given a positive integer N, the function* Unranking() *can correctly determine the Z-sequence $(z_1, z_2, \ldots, z_n)$ of a (k, m)-ary tree of order n such that* Ranking$(z_1, z_2, \ldots, z_n) = N$ *in $\mathcal{O}(kmn)$ time and $\mathcal{O}(n)$ space.*

5 Concluding Remarks

In this paper, we propose improved algorithms of ranking and unranking a (k, m)-ary tree of order n in B-order using Zak's encoding. With the help of the RD-sequence representation and four expedient formulae derived from a closed form related to (k, m)-Catalan numbers, we show that the newly proposed algorithms can be run with time complexity of $\mathcal{O}(kmn)$ and space requirement of $\mathcal{O}(n)$. This improves a previous result that requires $\mathcal{O}(kmn^2)$ time and space for building a precomputed table. As aforementioned, so far little results are known for (k, m)-ary trees. As future research, we expect to find more structural properties of (k, m)-ary trees and applications that can be dealt with by this kind of trees.

References

1. Amani, M.: Gap terminology and related combinatorial properties for AVL trees and Fibonacci-isomorphic trees. AKCE Int. J. Graphs Comb. **15**, 14–21 (2018)
2. Amani, M., Nowzari-Dalini, A.: Efficient generation, ranking, and unranking of (k, m)-ary trees in B-order. Bull. Iranian Math. Soc. (2019). https://doi.org/10.1007/s41980-018-0190-y
3. Amani, M., Nowzari-Dalini, A.: Ranking and unranking algorithm for neuronal trees in B-order. J. Phys. Sci. **20**, 19–34 (2015)
4. Amani, M., Nowzari-Dalini, A.: Generation, ranking and unranking of ordered trees with degree bounds. In: Proceedings of DCM 2015. Electronic Proceedings in Theoretical Computer Science, vol. 204, pp. 31–45 (2015)
5. Amani, M., Nowzari-Dalini, A., Ahrabian, H.: Generation of neuronal trees by a new three letters encoding. Comput. Inform. J. **33**, 1428–1450 (2014)
6. Du, R.R.X., Liu, F.: (k, m)-Catalan numbers and hook length polynomials for plane trees. Euro. J. Combin. **28**, 1312–1321 (2007)
7. Li, L.: Ranking and unranking AVL trees. SIAM J. Comput. **15**, 1025–1035 (1986)
8. Pai, K.-J., Chang, J.-M., Wu, R.-Y., Chang, S.-C.: Amortized efficiency of generation, ranking and unranking left-child sequences in lexicographic order. Discrete Appl. Math. (2018). https://doi.org/10.1016/j.dam.2018.09.035
9. Pallo, J.: Generating trees with n nodes and m leaves. Int. J. Comput. Math. **21**, 133–144 (1987)
10. Seyedi-Tabari, E., Ahrabian, H., Nowzari-Dalini, A.: A new algorithm for generation of different types of RNA. Int. J. Comput. Math. **87**, 1197–1207 (2010)
11. Stanley, R.P.: Enumerative Combinatorics, vol. 2. Cambridge University Press, Cambridge (1999)
12. Wu, R.-Y., Chang, J.-M., Chan, H.-C., Pai, K.-J.: A loopless algorithm for generating multiple binary tree sequences simultaneously. Theor. Comput. Sci. **556**, 25–33 (2014)

13. Wu, R.-Y., Chang, J.-M., Chang, C.-H.: Ranking and unranking of non-regular trees with a prescribed branching sequence. Math. Comput. Model. **53**, 1331–1335 (2011)
14. Wu, R.-Y., Chang, J.-M., Chen, A.-H., Liu, C.-L.: Ranking and unranking t-ary trees in a Gray-code order. Comput. J. **56**, 1388–1395 (2013)
15. Wu, R.-Y., Chang, J.-M., Wang, Y.-L.: A linear time algorithm for binary tree sequences transformation using left-arm and right-arm rotations. Theor. Comput. Sci. **355**, 303–314 (2006)
16. Wu, R.-Y., Chang, J.-M., Wang, Y.-L.: Loopless generation of non-regular trees with a prescribed branching sequence. Comput. J. **53**, 661–666 (2010)
17. Wu, R.-Y., Chang, J.-M., Wang, Y.-L.: Ranking and unranking of t-ary trees using RD-sequences. IEICE Trans. Inform. Syst. **E94–D**, 226–232 (2011)
18. Zaks, S.: Lexicographic generation of ordered trees. Theor. Comput. Sci. **10**, 63–82 (1980)

A Probabilistic Algorithm for Verification of Geometric Theorems

Mingyan Chen and Zhenbing Zeng[✉]

Department of Mathematics, Shanghai University, Shanghai 200444, China
yetibetan@sina.com, zbzeng@shu.edu.cn

Abstract. In this paper we combine the Schwartz-Zippel theorem with statistical inference theory and develop a new probabilistic algorithm instead of deterministic algorithms for geometry theorem proving. Our work includes an improved algorithm for estimating the upper bounds in the pseudo-remainder, and three selection criteria for statistical populations.

Keywords: Geometric theorems proving · Probabilistic algorithm · Selection criterion for statistical population · Statistical inference

1 Introduction

The common approaches of automated geometric theorem proving can be divided into three categories: algebraic methods, vector methods and search methods based on deductive database. Algebraic methods can be classified into two types, symbolic computation type which includes Wu's method [1,2], Gröbner bases method [3,4], resultant elimination method [5], etc.), and numerical computation type which includes the single-instance numerical verification method [6], and parallel numerical verification method [7,8,21,22]. All methods mentioned above belong to deterministic algorithms which will always get deterministic results if calculated by correct steps. However, when it comes to complicated problems, the complexity of deterministic algorithms can be very high which will seriously affect the efficiency of problem-solving. Thus, "probabilistic" algorithms (also called "non-deterministic" algorithms) are proposed which perform efficiently within a short time. Probabilistic algorithms have a wide range of applications in the field of computer algebra, such as prime number judgment and solving the largest invariant factor, etc. Probabilistic algorithms have two significant features: the algorithms are executed within the specified time and returning the computational results; the computational results may be incorrect but can control them within a small scope of 0. So the key question is that can we adopt a fast probabilistic algorithm instead of deterministic algorithms to improve the efficiency of geometric theorem proving?

This work is supported by the Project 11471209 of the National Natural Science Foundation of China.

© Springer Nature Switzerland AG 2019
D.-Z. Du et al. (Eds.): AAIM 2019, LNCS 11640, pp. 29–41, 2019.
https://doi.org/10.1007/978-3-030-27195-4_3

The answer is affirmative. In 1997, Carrá et al. [10] developed a probabilistic algorithm based on Schwartz-Zippel theorem [9] and Wu's method to prove constructive geometric theorems (see [1,2] for the definition) by verifying a number of random instances, the probability of incorrect result is also provided. They combined the bounds of the exponent of a polynomial in the radical of an ideal given by Brownawell and Kollar [11,12] with the bounds of the degree of Wu-Ritt's characteristic sets given by Gallo and Mishra [13,14] to estimate the upper bound about the total degree of the pseudo-remainder. The following is their research result. If a constructive geometric theorem (see Sect. 2) is constructed by C points and P circles or straight lines, then the bound calculated by their algorithm is $D = c \cdot 2^{C^3} 3^{C^3} C^{C^2}$ where c is a constant. Select N instances randomly from J where the J is a set of $2D$ different real number, then the probability that the result is correct is larger than $1 - 2^{-N}$ if the N instances all satisfy the geometric theorem. Unfortunately, this enormous bound D led Carrá et al. to fail to implement their algorithm on a computer. Besides, the extended characteristic sets and the pseudo-remainder are needed to be calculated if instance meets the degenerative conditions, which will increase the complexity of their algorithm inevitably.

Tulone et al. [15] proposed a probabilistic test for the vanishing of radical expression, and soon they developed the Core Library which was designed as a general C++ package and support the Exact Geometric Computation approach to robust algorithms. In 2001, they developed a geometry theorem prover based on probabilistic algorithm and the Core Library [16]. Their result can be summarized as follows.

Theorem 1. *Suppose $g(u, x)$ is the polynomial about the conclusion of a constructive geometric theorem with $\deg(g) = d$, and $G(u)$ be any of the 2^r radical expressions derived from $g(u, x)$ after eliminating dependent variables. If the theorem is constructed by k steps of ruler and compass constructions and $g(u, x)$ contains t terms, then $r \deg_u(G) \leq td2^r 85^k$ holds. Select the independent variables from J randomly, where J is a set formed by $td2^{c+2r} 85^k (c \geq 1)$ real numbers, if the geometric theorem is false, then the probability that and the instance satisfies the theorem is at most 2^{-c}.* □

The above bound is much better, but when it comes to the class of non-linear constructive geometric theorems, the efficiency is still not satisfied. To refine the probabilistic algorithm of geometric theorems proving, we have improved the algorithm in this paper for estimating the upper bounds about the degrees of the pseudo-remainder in each independent variable, proposed three selection criteria for statistical population and apply two checking methods to verify instances, and designed a combined probabilistic checking model for mechanical geometry theorem proving on the basis of statistical error analysis and significance test.

This paper is organized as follows. In Sect. 2 we introduce the methods for representation of geometric theorems and related concepts of irreducible ascending sets. In Sect. 3 we give an improved algorithm to estimate the upper bounds of the pseudo-remainder. In Sect. 4 we introduce two checking methods for instances verification and propose a new probabilistic algorithm for geometry theorem proving based on Schwartz-Zippel Theorem with three selection criteria for statistical population. In Sect. 5 we discuss the statistical error analysis and significance test.

2 Algebraic Representation of Geometry Theorems

It is well known that geometric theorems can be expressed as certain relations of algebraic equations using coordinates. For a large class of elementary geometric theorems, we can translate them into simple quadratic algebraic equations by adopting an appropriate coordinate system. A theorem is called a constructive geometric theorem if it is constructed according to some construction rules (e.g. ruler and compass constructions).

For a constructive geometric theorem, we can translate its hypotheses into a set of multivariate polynomial equations $H : \{f_i(u_1, u_2, \ldots, u_m, x_1, x_2, \ldots, x_n) = 0, i = 1, 2, \ldots, n\}$ and its conclusion is also a multivariate polynomial equation G: $g(u_1, u_2, \ldots, u_m, x_1, x_2, \ldots, x_n) = 0$ where $u_1, u_2, \ldots, u_m$ are independent variables (or parameters) and $x_1, x_2, \ldots, x_n$ are dependent variables, $f_1, f_2, \ldots, f_n$ and g are quadratic equations (i.e., the degree of each variable in each polynomial is not bigger than 2) in $Q(u_1, u_2, \ldots, u_m)[x_1, x_2, ..., x_n]$.

In what follows we use abbreviation $u = u_1, u_2, \ldots, u_m$, $x = x_1, x_2, \ldots, x_n$, $Q[u]$ denotes $Q[u_1, u_2, \ldots, u_m]$ and $Q(u)[x]$ the polynomial ring of $x_1, x_2, \ldots, x_n$ over the field of rational expressions $Q(u_1, u_2, \ldots, u_m)$.

A Maple package (EPSILON) developed by Wang in [17] can be used to translate a geometric theorem into algebraic form automatically by invoking the commands of **Load** and **Algebraic** in its submodule GEOTHER. The preparatory work is to formalize a geometric theorem as Theorem (H, G, X) where H is the hypotheses, G is the conclusion, and X is the set of dependent variables.

In general, the hypotheses H can be simplified into an equivalent ascending set (triangular form) by applying Wu-Ritt's algorithm, see [18–20]. Furthermore, based on the ascending set, following we will give the definition of irreducible ascending set as it plays an important role in this paper.

Definition 1. *If each polynomial f_i in an ascending set is irreducible in the ring $Q(u)[x_1, x_2, \ldots, x_i]/(f_1, f_2, \ldots, f_{i-1})$, then we'll call it an irreducible ascending set (IAS):*

$$IAS \quad \begin{cases} f_1(u, x_1) = 0; \\ f_2(u, x_1, x_2) = 0; \\ \ldots \\ f_n(u, x_1, x_2, \ldots, x_n) = 0. \end{cases} \tag{1}$$

In order to describe the algebraic feature of the class of constructive geometric theorems as well as the linear class clearly, here we use the i-th element $D_f{}^i$ in the $n + m$-length array D_f to denote the degree of f in the i-th variable where $f \in Q_n(u)[x]$, and $m_f = \max\{D_f^i, 1 \le i \le m + n\}$. Define $Q_j(D)$ as a set of specific polynomials and D^i denotes the i-th element in the n+m-length array D, for any $f \in Q_j(u)[x_1, x_2, \ldots, x_j]$, if f satisfies formula (2), then $f \in Q_j(D)$.

$$Q_j(D) = \{f \in Q_j | D_f^i \leq D^i, j = 1, 2, \ldots, m + j\} \qquad (2)$$

For any constructive geometric theorem, each f_i in formula (1) satisfies either $D_{f_i}^{m+i} = 1, m_{f_i} \leq 2$ (suppose that there are l polynomials satisfy this condition, then $\frac{n}{2} \leq l \leq n$ will always hold) or $D_{f_i}^{m+i} = 2, m_{f_i} \leq 4$. If every polynomial f_i satisfies $D_{f_i}^{m+i} = 1$, then we call it a constructive geometric theorem of linear class.

3   Estimating the Degree Bounds for the Pseudo-remainder

Our goal in this section is to establish an algorithm of estimating the upper bounds of the degrees of the pseudo-remainder. We need the following result.

Theorem 2. *Let the i-th element $D_{g_j}^i$ in the (m+j)-length array D_{g_j} denote the degree of $g_j(u)[x_1, x_2, \ldots, x_j]$, in the i-th variable, then for any j-stage irreducible branch $\Re_j$ of formula (1), there exists a non-zero polynomial I on $\Re_j$ and*

$$g_{j-1}(u)[x_1, x_2, \ldots, x_{j-1}] \in Q_{j-1}(D_{f_j}^{m+j} D_{g_j} + D_{f_j}^{m+j} D_{g_j}^{m+j} D_{f_j}),$$

such that

$$I(u)[x_1, x_2, \ldots, x_j] \cdot g_j(u)[x_1, x_2, \ldots, x_j] = g_{j-1}(u)[x_1, x_2, \ldots, x_{j-1}] \qquad (3)$$

holds on $\Re_j$, and $g_j \equiv 0$ on $\Re_j$ if and only if $g_{j-1} \equiv 0$.   □

We refer the reader to the original paper [5] for the proof of this theorem. According to Theorem 2 and the n triangular polynomials in (1), we adopt inductive reasoning to deduce the following theorem.

Theorem 3. *Let the i-th element D_g^i in the (m+n)-length array D_g denote the degree of the i-th variable in the conclusion of a geometric theorem $g(u)[x]$, then for any n-stage irreducible branch $\Re$ of IAS (i.e., formula (1)) derived from its hypotheses, there exists a non-zero polynomial I on $\Re$ and $R \in Q_0(D_0)$, such that*

$$I(u)[x] \cdot g(u)[x] = R[u] \qquad (4)$$

holds on $\Re$, and $g \equiv 0$ on $\Re$ if and only if $R \equiv 0$.

Proof. According to Theorem 2, for any n-stage irreducible branch $\Re_n$ of IAS (hypotheses) and conclusion $g(u)[x]$ of a given constructive geometric theorem, there exist a non-zero polynomial I_n on $\Re_n$ and $g_{n-1}(u)[x_1, x_2, ..., x_{n-1}] \in Q_{n-1}(D_{f_n}^{m+n} D_g + D_{f_n}^{m+n} D_g^{m+n} D_{f_n})$, such that

$$I_n(u)[x] \cdots g(u)[x] = g_{n-1}(u)[x_1, x_2, \ldots, x_{n-1}]$$

holds on $\Re_n$, and $g \equiv 0$ on $\Re_n$ if and only if $g_{n-1} \equiv 0$. Similarly, repeat applying Theorem 2 to eliminate the last i dependent variables, for any $(n - i + 1)$-stage irreducible branch $\Re_{n-i+1}$ of IAS, there exists I_{n-i} and

$$g_{n-i}(u)[x_1, \ldots, x_{n-i}] \in Q_{n-i}(D_{f_{n-i+1}}^{m+n-i+1} D_{g_{n-i+1}} + D_{f_{n-i+1}}^{m+n-i+1} D_{g_{n-i+1}}^{m+n-i+1} D_{f_{n-i+1}}),$$

such that

$$g_{n-i} = I_{n-i}g_{n-i+1} = I_{n-i}I_{n-i+1}g_{n-i+1} = \cdots$$

$$= I_{n-i}I_{n-i+1} \cdots I_{n-1}g_{n-1} = I_{n-i}I_{n-i+1} \cdots I_n g$$

holds on $\Re_{n-i+1}$, and $g_{n-i+1} \equiv 0$ on $\Re_{n-i+1}$ if and only if $g_{n-i} \equiv 0$. Let $i = n$, i.e., have finished eliminating all the dependent variables, then for any 1-stage irreducible branch $\Re_1$ of IAS, there exists a non-zero polynomial I_1 and

$$g_0[u] \in Q_0(D_{f_1}^{m+1} D_{g_1} + D_{f_1}^{m+1} D_{g_1}^{m+1} D_{f_1}) = Q_0(D_0),$$

such that

$$g_0 = I_1 g_1 = I_1 I_2 g_2 = \ldots = I_1 I_2 \cdots I_n g$$

holds on $\Re_1$, and $g_1 \equiv 0$ on $\Re_1$ if and only if $g_0 \equiv 0$. Set $R = g_0$ and $I = I_1 I_2 \cdots I_n$, then $Ig = R$ holds. According to Theorem 2 and the recursive process of I and R, $g \equiv 0$ on $\Re$ if and only if $R \equiv 0$ holds.   □

By Theorem 3 and the whole process of its proof, we can design the following algorithm to estimate the degrees of the polynomial R in every independent variables.

Algorithm 1. Estimate the upper bounds of the degrees of R in every independent variables.

Input: g, IAS, $ux = [u_1, u_2, \ldots, u_m, x_1, x_2, \ldots, x_n]$.

Output: An array D where the first m elements are the upper bounds of the degrees of R in every independent variables.

```
degreebounds:=proc(g, IAS, ux)
 n:=nops(IAS): m:=nops(ux)-n:
 D:=[seq(degree(G, op(i, ux)), i=1.. m+n)]:
   for i from n to 1 by -1 do
   FI:= IAS[i]: dFI:=[seq(degree(FI,op(j, ux)), j=1...m+n)]:
   D:=dFI[m+i]*D + dFI[m+i]*D[m+i]*dFI:
   end do:
 return D:
end proc:
```

By Theorem 3 and Algorithm 1, we have the following corollary.

Corollary 1. *If there are $l(\frac{n}{2} \leq l \leq n)$ polynomials in the irreducible ascending set IAS of a constructive geometric theorem satisfy $\deg(f_i, x_i) = 1$, $m_{f_i} \leq 2$ and $n-l$ polynomials satisfy $\deg(f_i, x_i) = 2$, $m_{f_i} \leq 4$, then $\deg(R, u_i) \leq 2 \cdot 3^l 10^{n-l} \leq 2 \cdot 30^{n/2}(i = 1, 2, ..., m)$ and $T \deg(R) \leq 2m \cdot 3^l \cdot 10^{n-l} \leq 2m \cdot 30^{n/2}$ hold, i.e., the upper bound of the degree of R in each independent variable is not bigger than $2 \cdot 30^{n/2}$ and the upper bound of the total degree of R is not bigger than $2m \cdot 30^{n/2}$. Furthermore, if limited to the linear class, then the bounds can still be improved to $T \deg(R) \leq 2m \cdot 3^n$ and $\deg(R, u_i) \leq 2 \cdot 3^n (i = 1, 2, \ldots, m)$.*

Proof. We first prove the conclusion that $\deg(g_{n-t}, u_i) \leq 2 \cdot 3^{l'} 10^{t-l'}$ $(i = 1, 2, \ldots, m)$ holds for $1 \leq t \wedge t \leq n$ where g_{n-t} is derived by eliminating the last t dependent variables in g and l' denotes the number of polynomials that satisfy $D_{f_i}^{m+i} = 1$ in the last t polynomials of *IAS*. We use mathematical induction to prove the above conclusion.

When $h = 1$, i.e., we can eliminate the last independent variable x_n with f_n, then $g_{n-1} \in Q_{n-1}(D_{f_n}^{m+n} D_g + D_{f_n}^{m+n} D_g^{m+n} D_{f_n})$ holds according to Theorem 2. If $D_{f_n}^{m+n} = 1$, then $l' = 1$, $D_{f_n}^j \leq 2(j = 1, 2, \ldots, m+n)$ holds. Since $D_g^m \leq 2$, then $D_{g_{n-1}}^i \leq D_{f_n}^{m+n} \cdot \max_{1 \leq j \leq m}(D_g^j) + D_{f_n}^{m+n} \cdot \max_{1 \leq j \leq m}(D_g^j) \cdot \max_{1 \leq j \leq m}(D_{f_1}^j) = 1 \cdot 2 + 1 \cdot 2 \cdot 2 = 2 \cdot 3^{l'} \cdot 10^{1-l'}$ holds where $1 \leq i \leq m$, i.e., the conclusion holds. On the other hand, if $D_{f_n}^{m+n} = 2$, then $l' = 0$, $D_{f_n}^j \leq 4(j = 1, 2, \ldots, m)$ holds. Similarly, the conclusion holds. Suppose that when $h = t(1 \leq t < n)$ the conclusion holds, following we will prove when $h = t + 1$ the conclusion also holds. According to Theorem 3, after eliminating x_{n-t} with f_{n-t}, $g_{n-t-1} \in Q_{n-t-1}(D_{f_{n-t}}^{m+n-t} D_{g_{n-t}} + D_{f_{n-t}}^{m+n-t} D_{g_{n-t}}^{m+n-t} D_{f_{n-t}})$ holds obviously. If $D_{f_{n-t}}^{m+n-t} = 1$, then $l' = l'+1$, $D_{f_{n-t}}^j \leq 2(j = 1, 2, \cdot, m)$ holds. Since $h = t$ the conclusion holds, then we have $D_{g_{n-t}}^i \leq 2 \cdot 3^{l'-1} 10^{t-l'-1}(i = 1, 2, \cdot, m)$ holds, now we can derive the upper bound of g_{n-t-1} as follows, $D_{g_{n-t-1}}^i \leq D_{f_{n-t}}^{m+n-t} \cdot \max_{1 \leq j \leq m}(D_{g_{n-t}}^j) + D_{f_{n-t}}^{m+n-t} \cdot \max_{1 \leq j \leq m}(D_{g_{n-t}}^j) \cdot \max_{1 \leq j \leq m}(D_{f_{n-t}}^j) = 1 \cdot 2 \cdot 3^{l'-1} 10^{t-l'-1} + 1 \cdot 3^{l'-1} 10^{t-l'-1} \cdot 2 = 2 \cdot 3^{l'} \cdot 10^{t+1-l'}$, by this the conclusion holds obviously. If $\deg(f_{n-t}, x_{n-t}) = 2$, then $l' = l'$, $D_{f_{n-t}}^{m+n-t} = 2$, $D_{f_{n-t}}^j \leq 4(j = 1, 2, \cdot, m)$ holds, similarly, the conclusion also holds. That is, we have proven that the conclusion will also hold for $h = t + 1$ if $h = t$ the conclusion holds. By mathematical induction, the above conclusion holds for $1 \leq t \wedge t \leq n$. By Theorem 3, after eliminating all the dependent variables we will obtain a polynomial g_0 where $g_0 = R$. If there are l polynomials in *IAS* satisfy $\deg(f_i, x_i) = 1$, then claims that $\deg(R, u_i) \leq 2 \cdot 3^l 10^{n-l}(i = 1, 2, \cdot, m)$ holds obviously by the above conclusion. Since there are m independent variables in R and their degrees are not bigger than $2 \cdot 3^l 10^{n-l}$, so $T \deg(R) \leq 2m \cdot 3^l 10^{n-l}$ holds. Moreover, if limited to the linear class, i.e., every polynomials in *IAS* satisfy $\deg(f_i, x_i) = 1$, so $l = n$. Substitute $l = n$ into $\deg(R, u_i) \leq 2 \cdot 3^l 10^{n-l}(i = 1, 2, \ldots, m)$, the conclusion $\deg(R, u_i) \leq 2 \cdot 3^n (i = 1, 2, ..., m)$ holds immediately, similarly, $T \deg(R) \leq 2m \cdot 3^n$ also holds, i.e., Corollary 1 holds. □

Corollary 1 shows that, if the *IAS* of a constructive geometric theorem contains n polynomials, then the upper bound of the degree of R in each independent variable is at most $B = 2 \cdot 3^l 10^{n-l}$. The bound can be improved to $2 \cdot 3^n$ if limited to the linear class. For constructive geometric theorems, l always satisfies $\frac{n}{2} \leq l \leq n$, so the bound can generalized as $\deg(R, u_i) \leq 2 \cdot 30^{n/2}$. That is, for a constrictive geometric theorem, its bounds B_1 satisfy $B_1 \leq 2 \cdot 30^{n/2}$, and for the linear class, the bounds B_2 can be improved to $B_2 \leq 2 \cdot 3^n$.

4   Probabilistic Estimates of Truth and Selection Criteria For statistical Population

Many geometric theorems can be transformed into the following form of conjunct logic relationship:

$$(\forall u, x)[(f_1 = 0 \wedge f_2 = 0 \wedge \ldots \wedge f_n = 0) \Rightarrow (g = 0)] \tag{5}$$

Thus, if a geometric theorem is true, then we can claim that remainder R (obtained in Theorem 3) is identically zero. It is certain that we can prove a theorem by calculating R from *IAS* and g according to Theorem 3 and then checking whether $R \equiv 0$ holds. We can also avoid calculating R directly via estimating the degrees of R by using Corollary 1. Then the key question is how to judge whether an instance satisfies the geometric theorem. Two kinds of checking methods are given in [6].

Checking Method 1: Numerical checking method, which uses a specific numerical instance to check whether $R \equiv 0$ holds. By Theorem 3, $g \equiv 0$ on $\Re$ if and only if $R \equiv 0$, i.e., for any $\tilde{u} = \tilde{u}_1, \tilde{u}_2, \ldots, \tilde{u}_m$ and $\tilde{x} = \tilde{x}_1, \tilde{x}_2, \ldots, \tilde{x}_n$ satisfy formula (1), if $I(\tilde{u})[\tilde{x}] \neq 0$, then $g(\tilde{u})[\tilde{x}] = 0$ holds generically. The following steps show the checking process: (1)Select a numerical instance $\tilde{u} = \tilde{u}_1, \tilde{u}_2, \ldots, \tilde{u}_m$ randomly from the statistical population; (2) Substitute $\tilde{u}$ into the *IAS*, solve all the numerical solutions of the dependent variables $\tilde{x} = \tilde{x}_1, \tilde{x}_2, \ldots, \tilde{x}_n$; (3) Substitute $\tilde{u}$ and $\tilde{x}$ into the conclusion polynomial g, if $g(\tilde{u})[\tilde{x}] = 0$, then claims $R[\tilde{u}] = 0$, i.e., this instance satisfies the geometric theorem; if the dependent variables cannot be determined, which indicates that $I(\tilde{u})[\tilde{x}] = 0$ holds, according to formula (4), $R[\tilde{u}] = 0$ still holds. (4) If the instance does not satisfy $g(\tilde{u})[\tilde{x}] = 0$, i.e., $g(\tilde{u})[\tilde{x}] = 0$ does not hold, then claims that R is not identically zero, i.e., the geometric theorem is not true absolutely.

Checking Method 2: Successive pseudo division checking method which means that do not solve the numerical solutions about the dependent variables after substituting $\tilde{u}$ into the *IAS* but calculates R according to the instantiated *IAS* and g by formula (4) by using successive pseudo division algorithm. If $R = 0$, then claims that this instance satisfies the geometric theorem and the geometric theorem is true generically, else claims that the geometric theorem is false absolutely.

Remark. More rigorously, a geometric theorem may be expressed as the following conjunct logic relationship with non-degenerate conditions.

$$(\forall u, x)[(f_1 = 0 \wedge f_2 = 0 \wedge \ldots \wedge f_n = 0, \varDelta_1, \varDelta_2, \ldots, \varDelta_k) \Rightarrow (g = 0)], \quad (6)$$

where $\varDelta_1, \varDelta_2, \ldots, \varDelta_k$ denote the algebraic form of non-degenerate cases. Non-degenerate cases can be divided into two kinds: $\varDelta \neq 0$ and $\varDelta > 0$. Almost all prover now available can only deal with the first kind non-degenerate cases, following we also discuss the first kind of non-degenerate cases only.

As the results obtained by probabilistic algorithms are not always true, it is very essential to make probabilistic algorithms more reliable by providing the probability that the result is true (or false).

Randomization procedure is an essential part of our probabilistic algorithm, which means that instances should be randomly selected from the statistical population while checking. If a random instance does not match with the theorem, then the program will terminate and return the running result that the theorem is false, and the probability that the result is incorrect is 0. If N instances all match with the theorem, then the program will return a running result that the theorem is true. If R is not identically zero and all the N random instances are the zeros of R, then the running result will be incorrect. Can we control and obtain the upper bound of the probability that the result is incorrect? To solve this problem, we first introduce the famous Schwartz-Zippel Theorem proposed by Schwartz in 1980 [9] by which the upper bound of the number of the zeros of a nonzero polynomial in a specific sets can be determined.

Theorem 4. *Suppose that $G \in F[x_1, x_2, ..., x_n]$ and G is not identically zero. Let G_1 be the standard simplified form of G and d_1 be the degree of G_1 in x_1, G_2 be the coefficient of x_1^d in G_1. Then, inductively, let d_i be the degree of G_i in x_i and G_{i+1} be the coefficient of $x_i^{d_i}$ in G_i where $1 \le i \le n$. For any $x_i(1 \le i \le n)$, if $x_i \in I_i$ (here, $I_i \subset F$ and $|I_i| < d_i$), then in the set $I_1 \times I_2 \times \ldots \times I_n$, G has at most*

$$|I_1 \times I_2 \times \ldots \times I_n| \left(\frac{d_1}{|I_1|} + \frac{d_2}{|I_2|} + \ldots + \frac{d_n}{|I_n|} \right) \quad (7)$$

zeros. □

The following corollary is obtained immediately from Schwartz-Zippel Theorem.

Corollary 2. *Suppose that $G \in Q[x_1, x_2, \ldots, x_n]$ and G is not identically zero. If $x_1, x_2, \ldots, x_n$ are picked randomly from I where $I \subset Q$, then the probability that G is not identically zero in at most $\frac{d}{|I|}$, here d denotes the total degree of G and $|I| < d$.* □

Together with Algorithm 1, Schwartz-Zippel Theorem enables us to estimate the upper bound of the probability that the result is incorrect. In the rest of this section, we will propose three selection criteria for statistical population according to the Schwartz-Zippel Theorem and Corollary 2.

The First Selection Criterion. Select $c \times m \times d_i$ distinct positive integers to form a finite set U_i, e.g., $U_i = (1, 2, \ldots, cmd_i)$ where c is a positive integer and $i = 1, 2, \ldots, m$. Then an instance (called individual in statistical terminology) is formed by selecting an element from each U_i. Statistical population $S1$ is made up of all the instances, i.e., $S1 = U_1 \times U_2 \times \ldots \times U_m$, then the statistical population size $\# S1$ equal to the number of instances (individuals), i.e., $\# S1 = c^m m^m \prod_{i=1}^{m} d_i$. If R is not identically zero, then according to Theorem 4 we have:

$$|U_1 \times U_2 \times \ldots \times U_m| \left(\frac{d_1}{|U_1|} + \frac{d_2}{|U_2|} + \ldots + \frac{d_m}{|U_m|} \right)$$

$$= (cm)^m \prod_{i=1}^{m} d_i \cdot \left(\frac{d_1}{cmd_1} + \frac{d_2}{cmd_2} + \ldots + \frac{d_m}{cmd_m} \right) = c^{m-1} m^m \prod_{i=1}^{m} d_i \quad (8)$$

By formula (8), we can claim that the number of instances in $S1$ that are zeros of R is at most $c^{m-1} m^m \prod_{i=1}^{m} d_i$. Therefore, the probability that the instance $\tilde{u} = \tilde{u}_1, \tilde{u}_2, \ldots, \tilde{u}_m$ selected randomly from $S1$ satisfies $R[\tilde{u}] = 0$ can be deduced as follows:

$$\text{Prob1}(R[\tilde{u}] = 0 | R \neq 0) \leq \frac{c^{m-1} m^m \prod_{i=1}^{m} d_i}{\# S1} = \frac{c^{m-1} m^m \prod_{i=1}^{m} d_i}{c^m m^m \prod_{i=1}^{m} d_i} = c^{-1} \quad (9)$$

That is, if a geometric theorem is not true, then the probability that $R[\tilde{u}] = 0$ is at most c^{-1}, where $\tilde{u} = \tilde{u}_1, \tilde{u}_2, \ldots, \tilde{u}_m$ is an instance selected randomly from the statistical population $S1$.

The Second Selection Criterion. Let $D = \sum_{i=1}^{m} d_i$ where $d_1, d_2, \ldots, d_m$ are calculated by Algorithm, it is easy to see that the total degree of R in all the independent variables is at most D. Select cD distinct positive integers to form a finite set U, e.g., $U = (1, 2, \ldots, cD)$ where c is also a positive integer. Similarly, we obtain the statistical population: $S2 = \underbrace{U \times U \times \ldots \times U}_{\text{the number of } U \text{ is } m}$, and its size $\# S2$ as following:

$$\# S2 = c^m D^m = c^m (\sum_{i=1}^{m} d_i)^m = c^m m^m (\sum_{i=1}^{m} d_i / m)^m = c^m m^m (\bar{d})^m \quad (10)$$

Repeat m times that select an element from U randomly can form an random instance $\tilde{u} = \tilde{u}_1, \tilde{u}_2, \ldots, \tilde{u}_m$, if R is not identically zero, then the probability that $R[\tilde{u}] = 0$ satisfies the following formula:

$$\text{Prob2}(R[\tilde{u}] = 0 | R \neq 0) \leq \frac{D}{\# U} = \frac{D}{cD} = c^{-1} \quad (11)$$

according to Corollary 2, which implies that, if a geometric theorem is not true, then the probability that $R[\tilde{u}] = 0$ is at most c^{-1}, where $\tilde{u}$ is an instance selected randomly from the statistical population $S2$.

The Third Selection Criterion. As indicated in Corollary 1, if the *IAS* of a geometric theorem contains n polynomials, then the upper bounds satisfy $\deg(R, u_i) \leq 2 \cdot 3^l 10^{n-l} \leq 2 \cdot 30^{n/2} (i = 1, 2, \ldots, m)$. In other words, if the *IAS* is too difficult to calculate, then the upper bounds can still be estimated quickly in accordance with Corollary 1. Let $d_i = 2 \cdot 30^{n/2}$ and $D = \sum_{i=1}^{m} d_i = 2m \cdot 30^{n/2}$, then the statistical population $S3$ can be obtained in accordance with Corollary 2 and its size as following:

$$\# S3 = c^m D^m = c^m (2m \cdot 30^{n/2})^m = 2^m c^m m^m 30^{nm/2} \tag{12}$$

Similarly, the probability that $R[\tilde{u}] = 0$ is at most c^{-1} if the geometric theorem is false and $\tilde{u}$ is selected randomly from $S3$, i.e.,

$$\mathrm{Prob3}(R[\tilde{u}] = 0 | R \neq 0) \leq c^{-1}.$$

5   Statistical Error Analysis and Significance Test

In this section, we compare the three selection criteria for statistical population, and then discuss statistical error analysis and significance test of our method. We have seen that all three selection criteria satisfy $\mathrm{Prob}(R[\tilde{u}] = 0 | R \neq 0) \leq c^{-1}$, which means that, if a geometric theorem is false, then the probability of the checking result that theorem is true is at most c^{-1}. Their main differences lie in the value ranges of the instances and the statistical populations sizes.

If the statistical population is collected by the first selection criterion, then the statistical population size is $\# S1 = c^m \cdot m^m \prod_{i=1}^{m} d_i$. The second selection criterion can determine $S2$ and $\# S2 = c^m m^m (\overline{d})^m$. And therefore,

$$P = \frac{\# S1}{\# S2} = \frac{c^m m^m \prod_{i=1}^{m} d_i}{c^m m^m (\overline{d})^m} = \frac{d_1 d_2 \ldots d_m}{(\sum_{i=1}^{m} d_i / m)^m} \leq 1 \tag{13}$$

where m, c, $d_i (1 \leq i \leq m)$ are all positive integers. By formula (13)shows that adopting the first selection criterion to collect statistical population will more precise than the second one. Moreover, the complexity of the algorithm will decrease with the refined statistical population and more compact value ranges of instances, thus to avoid data-overrun error caused by the limited precision of computation and achieve our goal that to prove geometric theorems fast and accurately.

After simplifying H into *IAS*, the upper bounds of the degrees of R in the independent variables can be estimated by Algorithm 1. However, for some high-complexity geometric theorems, such as the Five-Circles Theorem and Miquel's Theorem, H will be very complicated which will inevitably lead to wasting lots of time and consuming large amounts of memory in the process of simplifying H into *IAS*, and this will contrary to our original intention that design a probabilistic algorithm with high efficiency to prove geometric theorems. Can we avoid calculating the IAS before instantiation if the geometric theorem is very complicated? To achieve this is easy by the Corollary 1 and it is also the reason why we propose the third selection criterion to collect the statistical population. That is, for some high-complexity geometric theorems, in the circumstance that data-overrun error will not occur during the whole actual operation, we can avoid calculating the irreducible ascending set before instantiation, and estimate the upper bounds crudely by Corollary 1 instead of Algorithm 1, thus to avoid failing to get the running result within the specified time or program interruption for out of memory.

The core content of mathematical statistics is the study of the relationship between statistical population and sample, and statistical inference is to infer statistical population in accordance with sample. In general, statistical inference can be divided into two categories: parameter estimation and significance test. The main researched in this paper belongs to inferring statistical population by sample which involves significance test only. We refer the reader to [23] for the general concept and terminology of significant test.

Statistical significance test is a common method of statistical inference whose principle judge whether there exists significant difference between the statistical population and the null hypothesis $H0$ by the sample information. Its essence is the "small probability theory" and logic approach of the "reductio ad absurdum". First of all, define the null hypothesis $H0$, and then calculate the probability that $H0$ holds base on the sample information by the corresponding statistical methods. If the probability is small enough (i.e., less than the significance level $a = 0.01$), then judge that $H0$ does not hold and reject the null hypothesis. Otherwise, accept the null hypothesis.

Statistical significance test involves two types of errors: Type I error and type II error. Type I error is the incorrect rejection of a true null hypothesis and type II error is the failure to reject a false null hypothesis. Unlike many statistical problem, the major research problem in this paper involves Type I error only and Type II error will not exist. If a geometric theorem is false and a counter example is found successfully, then the program will terminate and return the running result that the theorem is false absolutely, and the probability that the result is incorrect is 0. If N random instances all match the theorem, then the program will reject the null hypothesis and return the running result that the theorem is true. In this case, Type I error will occur unluckily. In view of this, we use the statistical significance test which has no relationship with type II error to control the probability that the occurrence of Type I error.

6   Conclusions

In this paper we presented a new probabilistic algorithm for automated geometry theorem proving which combined the Schwartz-Zippel theorem with statistical inference theory. Our main work includes an improved algorithm for estimating the upper bounds of the pseudo-remainder and three selection criteria for statistical populations. We have implemented the prover with Maple and verified the performance with experiments. Due to the page limit of this paper, more results on the experiment results and the prover implementation detail will be published in forthcoming papers.

References

1. Chou, S.C.: Proving elementary geometry theorem using Wu's Algorithm. Department of Mathematics, University of Texas at Austin, Ph.D. thesis (1985)
2. Wu, W.T.: Basic principles of mechanical theorem proving in elementary geometries. J. Symb. Comput. **2**(4), 221–25 (1986)
3. Kapur, D.: Using Grobner bases to reason about geometry problems. J. Symb. Comput. **2**, 399–408 (1986)
4. Kutzler, B., Stifter, S.: Automated geometry theorem proving using Buchberger's algorithm. In: On Symbolic and Algebraic Computation, pp. 209–214. ACM Press (1986)
5. Kapur, D., Saxena, T., Yang, L.: Algebraic and geometric reasoning using Dixon resultant. In: Proceedings of ISSAC 1994, vol. 7, pp. 97–107 (1994)
6. Hong, J.W.: Can we prove geometry theorem by computing an example? Sci. China Math. (Ser. A) **16**(3), 234–243 (1986)
7. Zhang, J.Z., Yang, L., Deng, M.K.: The parallel numerival methods in mechanical theorem proving. Theoret. Comput. Sci. **74**, 253–271 (1990)
8. Bellman, R.E.: On Proving Theorems in Plane Geometry via Digital Computer. RAND Corporation, Santa Monica (1965)
9. Schwartz, J.T.: Fast probabilistic algorithms for verification of polynomical identities. J. ACM **27**, 701–717 (1980)
10. Carrá Ferro, G., Gallo, G., Gennaro, R.: Probabilistic verification of elementary geometry statements. In: Wang, D. (ed.) ADG 1996. LNCS, vol. 1360, pp. 87–101. Springer, Heidelberg (1997). https://doi.org/10.1007/BFb0022721
11. Brownawell, W.D.: Bounds for the degrees in the Nullstellensatz. Ann. Math. **126**, 577–591 (1987)
12. Kollar, J.: Sharp effective Nullstellensatz. J. Am. Math. Soc. **1**, 963–975 (1988)
13. Gallo, G., Mishra, B.: Efficient algorithm and bounds for Wu-Ritt characteristic sets. In: Mora, T., Traverso, C. (eds.) Effective Methods in Algebraic Geometry. Progress in Mathematics, vol. 94, pp. 119–142. Birkhauser, Boston (1990). https://doi.org/10.1007/978-1-4612-0441-1_8
14. Gallo, G., Mishra, B.: Wu-Ritt characteristic sets and their complexity. DIMACS Ser. **6**, 111–136 (1991)
15. Tulone, D., Yap, C., Li, C.: Randomized xero testing of radical expressions and elementary geometry theorem proving. In: Richter-Gebert, J., Wang, D. (eds.) ADG 2000. LNCS (LNAI), vol. 2061, pp. 58–82. Springer, Heidelberg (2001). https://doi.org/10.1007/3-540-45410-1_5
16. Tulone, D., Yap, C., Li, C.: Core Library. http://cs.nye.edu/exact/cpre/
17. Wang, D.M.: EPSILON. http://www-calfor.lip6.fr/wang/epsilon/
18. Chou, S.C.: An introduction to Wu's method for mechanical theorem proving in geometry. J. Autom. Reason. **4**, 237–267 (1988)
19. Wang, D.M.: A new theorem discovered by computer prover. J. Geom. **36**, 173–182 (1989)
20. Gao, X.-S., Lin, Q.: MMP/Geometer – a software package for automated geometric reasoning. In: Winkler, F. (ed.) ADG 2002. LNCS (LNAI), vol. 2930, pp. 44–66. Springer, Heidelberg (2004). https://doi.org/10.1007/978-3-540-24616-9_4

21. Deng, M.K.: The parallel numerical method of proving the construction geometric theorem. Chin. Sci. **34**, 1066–1070 (1989)
22. Yang, L., Zhang, J.Z., Li, C.Z.: A prover for papallel numerical verification to a class of constructive geometirc theorem. J. Guangzhou Univ. (Nat. Sci. Ed.) **1**(3), 29–34 (2002)
23. Casella, G., Berger, R.L.: Statistical Inference, 2nd edn. Duxbury Press, Duxbury (2001)

Approximating Closest Vector Problem in ℓ_∞ Norm Revisited

Wenbin Chen[1(✉)] and Jianer Chen[1,2]

[1] School of Computer Science and Cyber Engineering, Guangzhou University,
Guangzhou, People's Republic of China
cwb2011@gzhu.edu.cn
[2] Department of Computer Science and Engineering, Texas A&M University,
College Station, TX 77843, USA

Abstract. The security of most lattice-based cryptography schemes are based on two computational hard problems which are the Short Integer Solution (SIS) and Learning With Errors (LWE) problems. The computational complexity of SIS and LWE problems are related to approximating Shortest Vector Problem (SVP) and Bounded Distance Decoding Problem (BDD). Approximating BDD is a special case of approximating Closest Vector Problem (CVP).

In this paper, we revisit the study for approximating Closest Vector Problem. We give one proof that approximating the Closest Vector Problem over ℓ_∞ norm (CVP$_\infty$) within any constant factor is NP-hard. The result is obtained by the gap-preserving reduction from Min Total Label Cover problem in ℓ_1 norm to CVP$_\infty$. This proof is simpler than known proofs.

Keywords: Closest vector problem · Computational complexity · NP-hardness · Min total label cover problem · Probabilistically checkable proofs

1 Introduction

Let $\mathbf{B} = \{\mathbf{v}_1, \ldots, \mathbf{v}_n\}$ be a set of linearly independent vectors in $\mathbf{R}^m$. The n-dimensional lattice L generated by B is the set of vectors $\{\sum_{i=1}^{n} a_i \mathbf{v}_i | a_i \in \mathbf{Z}\}$ where $\mathbf{B}$ is called the basis for the lattice L. The lattice L is also an additive group. The same lattice could be generated by many different bases. Given a basis for an n-dimensional lattice L and an arbitrary vector $\mathbf{t}$, the Closest Vector Problem (CVP) is to find a vector in L closest to $\mathbf{t}$ in a certain norm. The Shortest Vector Problem (SVP) is a homogeneous analog of CVP, and is defined to be the problem of finding the shortest non-zero vector in L. These lattice problems have a long history and we present some of the results below. The more comprehensive list of references can be found in [18] and [25].

These lattice problems have been studied since they were introduced in the 19th century. Gauss gave an algorithm that works for 2-dimensional lattices

© Springer Nature Switzerland AG 2019
D.-Z. Du et al. (Eds.): AAIM 2019, LNCS 11640, pp. 42–50, 2019.
https://doi.org/10.1007/978-3-030-27195-4_4

([11], 1801). In 1842, Dirichlet formulated the general problem for arbitrary dimensions. The existence of short non-zero vectors in lattices was dealt with by Minkowski in a field called Geometry of Numbers [26]. It was much later that Lenstra, Lenstra, Lovász [22] presented a polynomial time algorithm approximating the Shortest Lattice Vector to within the exponential factor $2^{n/2}$, where n is the dimension of the lattice. The LLL algorithm can be applied to many other important optimization and combinatorial problems, such as factoring rational polynomials [22], low density subset-sum [19], solvability of radicals [20], integer programming [13,21,22]. Applying LLL's method, Babai gave an algorithm that approximates CVP to within a similar factor [6]. Schnorr [28] improved the factor of approximation to $2^{O(n(\log \log n)^2/\log n)}$ for both CVP and SVP. These approximation results are quite weak, achieving only large exponential factors. It is an important open problem whether there exists a polynomial time algorithm for approximating SVP within polynomial factor. The exponential time algorithms to compute SVP were also presented in [2,14].

The first intractability results for lattice problems date back to 1981. In [29], van Emde Boas proved that CVP in any ℓ_p norm and SVP in ℓ_∞ norm are NP-hard. He also conjured that SVP in any ℓ_p norm is NP-hard. However proving the conjured NP-hard was an open problem for a long time. Until 1998, Ajtai [1] finally proved that SVP is NP-hard under randomized reductions. In the same paper it was also shown that approximating SVP within a factor $1 + \frac{1}{2^{n^c}}$ is NP-hard for some constant c. The non-approximability factor is improved to $1 + \frac{1}{n^\epsilon}$ by Cai and Nerurkar [8]. An breakthrough was made by Micciancio in [24] where he proved that approximating SVP in any ℓ_p norm within any constant less than $2^{1/p}$. Khot [16] improved the non-approximation factor to $p^{1-\epsilon}$ for high ℓ_p norms. Khot also made another great breakthrough in [17] where he showed that it is hard to approximate SVP within any constant factor and $2^{(\log n)^{1/2-\epsilon}}$ factor for any $\ell_p(p > 1)$ norm.

CVP was proved to be NP-hard by van Emde Boas [29]. Arora et al. [4] used the PCP characterization of NP to show that approximating CVP within factor $2^{(\log n)^{1-\epsilon}}$ is NP-hard unless $NP \subseteq DTIME(2^{poly(\log n)})$ and approximating CVP in l_p norm within any constant factor is NP-hard unless $P = NP$. Assuming $P \neq NP$, Dinur et al. [10] proved that it is hard to approximate CVP within factor $n^{c/\log \log n}$ for some constant $c > 0$. All above results for CVP work in all ℓ_p norms. It was showed by Dinur et al. that both CVP in ℓ_∞ norm and SVP in ℓ_∞ norm are hard to approximate within factor $n^{c/\log \log n}$ for some constant $c > 0$ [9].

Our Result

In this paper, we give one proof that assuming $P \neq NP$, it is NP-hard to approximate CVP_∞ within any constant factor. This proof is simpler than that of [9].

Technique

In order to prove our result, we give a reduction from the Min Total Label Cover problem in ℓ_1 norm to CVP_∞.

Structure of the Paper

In Sect. 2, we introduce some definitions. In Sect. 3 we prove that it is NP-hard to approximate CVP_∞ within any constant factor by reducing Min Total Label Cover problem to it. Finally, in Sect. 4 we present some conclusions and some open problems.

2   Preliminaries

In this section we present formal description of the problems which are used in our reductions.

In the following $G = (V_1, V_2, E)$ denotes a bipartite graph, $\mathcal{B}$ a set of labels for the vertices in $V_1 \cup V_2$, and for every $e \in E$ there exists a partial function $\rho_e : \mathcal{B} \rightarrow \mathcal{B}$ describing the admissible pairs of labels. We adapts the notations of [4, 5].

Definition 1. *A labelling of $G = (V_1, V_2, E)$ is a pair $(\mathcal{P}_1, \mathcal{P}_2)$ of functions $\mathcal{P}_i : V_i \rightarrow 2^\mathcal{B}, i = 1, 2$, which assign to each vertex in $V_1 \cup V_2$ a possibly empty set of labels.*

Definition 2. *Let $(\mathcal{P}_1, \mathcal{P}_2)$ be a labelling of $G = (V_1, V_2, E)$ and let $e = (v_1, v_2), v_1 \in V_1, v_2 \in V_2$, be an edge of G. We call $e = (v_1, v_2)$ covered iff $\mathcal{P}_1(v_1) \neq \emptyset, \mathcal{P}_2(v_2) \neq \emptyset$ and for all labels $b_2 \in \mathcal{P}_2(v_2)$ there exists a label $b_1 \in \mathcal{P}_1(v_1)$ such that $\rho_e(b_1) = b_2$. A labelling $(\mathcal{P}_1, \mathcal{P}_2)$ of $G = (V_1, V_2, E)$ is called a total-cover of G iff every edge of G is covered by the labelling $(\mathcal{P}_1, \mathcal{P}_2)$.*

Definition 3. *The ℓ_1-cost of a labelling of $(\mathcal{P}_1, \mathcal{P}_2)$ for a graph $G = (V_1, V_2, E)$ is defined as $cost(\mathcal{P}_1, \mathcal{P}_2) = \sum_{v_j \in V_i, i=1,2} |\mathcal{P}_i(v_j)|$.*

Definition 4. *Min Total Label Cover in ℓ_1-norm ($MinTLC_1$)*

INSTANCE: A d-regular bipartite graph $G = (V_1, V_2, E)$, a set of labels $\mathcal{B} = \{1, \ldots, \mathcal{N}\}, \mathcal{N} \in \mathbf{N}_+$, and for every edge of $e \in E$ a partial function $\rho_e : \mathcal{B} \rightarrow \mathcal{B}$ such that $\rho_e^{-1}(1) \neq \emptyset$ for the distinguished label $1 \in \mathcal{B}$

SOLUTION: A total-cover $(\mathcal{P}_1, \mathcal{P}_2)$ of G

MEASURE: The ℓ_1-cost $cost(\mathcal{P}_1, \mathcal{P}_2)$ of the total cover $(\mathcal{P}_1, \mathcal{P}_2)$

Remark 1. We can always ensure the existence of a total-cover with ℓ_1-cost at most $(|V_1| + 1)\mathcal{N}$; we simply let $\mathcal{P}_1(v_1) = \mathcal{B}$ for all $v_1 \in V_1$ and $\mathcal{P}_2(v_2) = \{1\}$ for all $v_2 \in V_2$.

The Min Label Cover in ℓ_1-norm is explicitly due to Khanna, Sudan and Trevisan [15] and a similar form of the following Lemma is implicitly proved in Lund and Yannakakis [23].

Lemma 1. *For every constant $g \geq 1$ there exists a polynomial time transformation τ from 3-SAT to Min Total Label Cover such that, for all instances I:*

$$I \in 3\text{-}SAT \Rightarrow \exists total\text{-}cover\ (\mathcal{P}_1, \mathcal{P}_2)\ of\ \tau(I): cost(\mathcal{P}_1, \mathcal{P}_2) = 1 \cdot (V_1 + V_2)$$
$$I \notin 3\text{-}SAT \Rightarrow \forall total\text{-}cover\ (\mathcal{P}_1, \mathcal{P}_2)\ of\ \tau(I): cost(\mathcal{P}_1, \mathcal{P}_2) > g \cdot (V_1 + V_2)$$

For studying the hardness of approximation problems we introduce the following reduction due to Arora [3].

Definition 5. *Let Π and Π' be two minimization optimization problems and $\rho, \rho' \geq 1$. A gap-preserving reduction from Π to Π' with parameters $((c, \rho), (c', \rho'))$ is a polynomial transformation τ mapping every instance I of Π to an instance $I' = \tau(I)$ of Π' such that for the optima $opt_\Pi(I)$ and $opt_{\Pi'}(I')$ of I and I', respectively, the following hold:*

$$opt_\Pi(I) \leq c \Longrightarrow opt_{\Pi'}(I') \leq c'$$
$$opt_\Pi(I) \geq c \cdot \rho \Longrightarrow opt_{\Pi'}(I') \geq c' \cdot \rho',$$

where c, ρ and c', ρ' depend on the instance sizes I and I', respectively.
In the following, we give the definition of CVP_∞.

Definition 6. *The Closest Vector Problem over ℓ_∞ norm (CVP_∞) is the problem in which one is given a lattice basis $\mathbf{B}$ and a target vector $\mathbf{y}$ and must find a lattice vector $\mathbf{Bx}$ $(\mathbf{x} \in \mathbf{Z}^n)$ such that $\|\mathbf{Bx} - \mathbf{y}\|_\infty$ is minimum. In the decisional version of CVP_∞ one is also given a real number t, and must decide whether there exist an integer vector $\mathbf{x}$ such that $\|\mathbf{Bx} - \mathbf{y}\|_\infty \leq t$.*
Approximating CVP_∞ to within factor $g = g(n)$ means finding a lattice vector $\mathbf{Bx}$ $(\mathbf{x} \in \mathbf{Z}^n)$ such that $\|\mathbf{Bx} - \mathbf{y}\|_\infty$ is no more than g times the minimum of all $\|\mathbf{Bx} - \mathbf{y}\|_\infty$. The gap version of CVP_∞ is a decision problem as follows, g-CVP_∞. Given $(\mathbf{B}, \mathbf{y}, d)$ for a lattice $\mathbf{B}$, a vector $\mathbf{y}$ and a number d, distinguish between the following two cases:

Yes: There exists a lattice vector $\mathbf{Bx}$ for which $\|\mathbf{Bx} - \mathbf{y}\|_\infty \leq d$.
No: For every lattice vector $\mathbf{Bx}$, $\|\mathbf{Bx} - \mathbf{y}\|_\infty > g \cdot d$.

Proving that g-CVP_∞ is NP-hard means that having an approximation algorithm to within factor g would imply $P = NP$.

3   Hardness of Approximating CVP_∞

In the section, we shall prove that approximating CVP_∞ is NP-hard within any constant factor by reducing MinTLC_1 to it. This reduction follows the same lines of the reduction from $MinLC_\infty$ to MinLS_∞ in [27] and [12].

Let $g(\geq 1)$ be a constant. From a given Minimum Total Label Cover instance $I = (V_1, V_2, E, \rho, \mathcal{B}, \mathcal{N})$, we (efficiently) construct an instance $\tau(I)$ of CVP_∞, $(\mathbf{A}, \mathbf{b})$ with $\mathbf{A}$ an $m \times n$ matrix of entries $\{-1, 0, 1, g\}$, $\mathbf{b}$ an m-dimensional vector of entries $\{0, g\}$, $m = |E|(\mathcal{N} + 1) + |V_1|\mathcal{N} + |V_2|\mathcal{N}$ and $n = (|V_1|\mathcal{N} + |V_2|\mathcal{N})$. We then show that the 'yes' instances of MinTLC_1 are mapped to 'yes' instances of CVP_∞ and 'no' instances to 'no' instances.

For every pair (v, b) with $v \in V_1 \cup V_2$ and $b \in \mathcal{B}$ we define a column vector $\mathbf{a}_{v,b} \in \{-1, 0, 1, g\}^m$ of $\mathbf{A}$ as follows. The first $|E|(\mathcal{N} + 1)$ coordinates of $\mathbf{a}_{v,b}$

are split into $|E|$ blocks of *e-projections* $\mathbf{u}_e(\mathbf{a}_{v,b})-$ one $(\mathcal{N}+1)$-length block for every edge $e \in E$. In particular, we define for every $(v_2, b_2) \in V_2 \times \mathcal{B}$

$$\mathbf{u}_e(\mathbf{a}_{v_2,b_2}) = \begin{cases} g \cdot \mathbf{e}_{b_2} \; iff \; e \; incident \; to \; v_2 \\ \mathbf{0} \qquad otherwise \end{cases}$$

and for every $(v_1, b_1) \in V_1 \times \mathcal{B}$

$$\mathbf{u}_e(\mathbf{a}_{v_1,b_1}) = \begin{cases} g \cdot (1 - \mathbf{e}_{\rho_e(b_1)}) \; iff \; e \; incident \; to \; v_1 \; and \; \rho_e(b_1) \neq \emptyset \\ \mathbf{0} \qquad\qquad\qquad otherwise \end{cases}$$

where $\mathbf{e}_j, j = 1, \ldots, \mathcal{N}$ denotes the jth-unit vector and $\mathbf{0}, \mathbf{1}$ the all-zero, all-one vector in $\mathbf{R}^{\mathcal{N}+1}$, respectively.

The definition of the remaining $|V_1|\mathcal{N} + |V_2|\mathcal{N}$ coordinates of $\mathbf{a}_{v,b}$ uses the properties of Hadamard matrices. A *Hadamard matrix of order* ℓ, denoted by $\mathbf{H}_\ell$, is an $\ell \times \ell$ matrix with ± 1 entries such that $\mathbf{H}_\ell \mathbf{H}_\ell^\top = \ell \mathbf{I}_\ell$. The $\frac{1}{\sqrt{\ell}}\mathbf{H}_\ell$ is clearly an orthonormal matrix. So $\|\frac{1}{\sqrt{\ell}}\mathbf{H}_\ell \mathbf{z}\|_2 = \|\mathbf{z}\|_2$ for every $\mathbf{z} \in \mathbf{Z}^\ell$. Because for every $\mathbf{z} \in \mathbf{R}^n$, $\|\mathbf{z}\|_\infty \geq \|\mathbf{z}\|_2/\sqrt{n}$, we obtain $\|\mathbf{H}_\ell \mathbf{z}\|_\infty \geq \frac{\|\mathbf{H}_\ell \mathbf{z}\|_2}{\sqrt{\ell}} = \|\frac{1}{\sqrt{\ell}}\mathbf{H}_\ell \mathbf{z}\|_2 = \|\mathbf{z}\|_2$. Hadamard matrices can be constructed in time linear in the size of matrix if ℓ is a power of 2 ([7], p. 74). Otherwise we use the matrix $\mathbf{H}_\ell$ consisting in the first ℓ columns of the Hadamard matrix of order $2^{\lceil \log \ell \rceil}$. Clearly, $\|\mathbf{H}_\ell \mathbf{z}\|_\infty \geq \|\mathbf{z}\|_2$ remains valid.

We may assume that for $\ell = \mathcal{N}$ there exists a Hadamard matrix $[\mathbf{H}_\ell] = [\mathbf{h}_1, \ldots, \mathbf{h}_\ell]$ with column vectors $\mathbf{h}_b$ of $\mathbf{H}_\ell$, each of them uniquely identified with a label $b \in \mathcal{B}$. We now split the last $(|V_1| + |V_2|)\mathcal{N}$ coordinates of $\mathbf{a}_{v,b}$ into $|V_1|+|V_2|$ blocks of *v'-projections* $\mathbf{u}_{v'}(\mathbf{a}_{v,b})-$ one $\mathcal{N}$-length block for every vertex $v' \in V_1 \cup V_2-$ where the v'-projections for every vertex $v' \in V_1 \cup V_2-$ and $b \in \mathcal{B}$ are defined as follows

$$\mathbf{u}_{v'}(\mathbf{a}_{v,b}) = \begin{cases} \mathbf{h}_b \; iff \; v = v' \\ \mathbf{0} \; otherwise \end{cases}$$

and $\mathbf{0}$ denotes the all-zero vector in $\mathbf{R}^\mathcal{N}$.

Eventually, we define the right side hand of our linear system— the vector $\mathbf{b}-$ as the vector having g in each of the first $|E|(\mathcal{N}+1)$ coordinates and 0 in the remaining ones. Thus the instance of CVP_∞ is $(\mathbf{A}, \mathbf{b})$, which are showed in the Fig. 1.

Proving Correctness. Let us now show that 'yes' instances of the MinTLC_1 map to 'yes' instances of the CVP_∞.

Proposition 1 (Completeness). *If there is a total label cover with cost $(|V_1|+|V_2|)$ to I (i.e. $opt_{MinTLC_1}(I) = (|V_1| + |V_2|)$), then there is a vector $\mathbf{x}$ such that $\|\mathbf{Ax} - \mathbf{b}\|_\infty = 1$ (i.e. $opt_{CVP_\infty}(\tau(I)) = 1$).*

Proof. Let $(\mathcal{P}_1, \mathcal{P}_2)$ be a total cover with cost $(|V_1| + |V_2|)$. So it assigns a label to exactly one vertex. We define the $(|V_1|\mathcal{N} + |V_2|\mathcal{N})$-length vector $\mathbf{x}$ as follows:

$$x_{v_j, \mathcal{P}_i(v_j)} = 1 \quad \forall v_j \in V_i, i = 1, 2$$
$$x_{v_j, b} = 0 \quad \forall v_j \in V_i, \forall b \in \mathcal{B} \backslash \mathcal{P}_i(v_j), i = 1, 2$$

$$\mathbf{u}_e(\mathbf{a}_{v_2,b_2}) = \begin{pmatrix} 0 \\ \vdots \\ 0 \\ g \\ 0 \\ \vdots \\ 0 \end{pmatrix} b_2^{th} - entry, \quad \mathbf{u}_e(\mathbf{a}_{v_1,b_1}) = \begin{pmatrix} g \\ \vdots \\ g \\ 0 \\ g \\ \vdots \\ g \end{pmatrix} \rho_e(b_1)^{th} - entry,$$

$|V_1|\mathcal{N}$ columns $|V_2|\mathcal{N}$ columns

$$\mathbf{A} = \begin{pmatrix} [\mathbf{u}_e(\mathbf{a}_{v,b_1}),\ldots,\mathbf{u}_e(\mathbf{a}_{v,b_\mathcal{N}})]_{e\in E, v\in V_1} & [\mathbf{u}_e(\mathbf{a}_{v,b_1}),\ldots,\mathbf{u}_e(\mathbf{a}_{v,b_\mathcal{N}})]_{e\in E, v\in V_2} \\ [\mathbf{u}_{v_1}(\mathbf{a}_{v,b_1}),\ldots,\mathbf{u}_{v_1}(\mathbf{a}_{v,b_\mathcal{N}})]_{v_1\in V_1, v\in V_1} & [\mathbf{u}_{v_2}(\mathbf{a}_{v,b_1}),\ldots,\mathbf{u}_{v_2}(\mathbf{a}_{v,b_\mathcal{N}})]_{v_2\in V_2, v\in V_2} \end{pmatrix}$$

$$\mathbf{b} = \begin{pmatrix} g \\ \vdots \\ g \\ 0 \\ \vdots \\ 0 \end{pmatrix}, |E|(\mathcal{N}+1) \ \ rows \ \ are \ \ g, (|V_1|+|V_2|)\mathcal{N} \ \ rows \ \ are \ \ 0.$$

Fig. 1. Lattice $\mathbf{A}$ and the target vector $\mathbf{b}$

Restricting $\mathbf{Ax}$ to an arbitrary row of the first $|E|(\mathcal{N}+1)$ rows, gives g, because Let $(\mathcal{P}_1,\mathcal{P}_2)$ be a total cover and assigns a label to exactly one vertex. Subtracting this from $\mathbf{b}$ gives zero in these rows.

In the remaining $|V_1|\mathcal{N}+|V_2|\mathcal{N}$ rows, since every vertex is assigned one label by $(\mathcal{P}_1,\mathcal{P}_2)$, $\mathbf{Ax}$-$\mathbf{b}$ restricted to these rows equals some column of the Hadamard matrix which is a ±1 matrix. Thus $\|\mathbf{Ax}-\mathbf{b}\|_\infty = 1$.

We will now show that 'no' instances of the MinTLC$_1$ map to 'no' instances of the CVP$_\infty$.

Proposition 2 (Soundness). If $opt_{MinTLC_1(I)} > g\cdot(|V_1|+|V_2|)$, then opt_{CVP_∞} $(\tau(I)) > \sqrt{g}$.

Proof. Given a vector $\mathbf{y} \in \mathbf{R}^{|E|(\mathcal{N}+1)+(V_1+V_2)\mathcal{N}}$, let $\mathbf{u}_E(\mathbf{y})$ denote the vector $\mathbf{y}$ *restricted to* its first $|E|(\mathcal{N}+1)$ coordinates. Given a matrix $\mathbf{A} \in R^{m\times n}$ where $m = |E|(\mathcal{N}+1)+|V_1|\mathcal{N}+|V_2|\mathcal{N}$ and $n = (|V_1|\mathcal{N}+|V_2|\mathcal{N})$, let $\mathbf{u}_E(\mathbf{A})$ denote the matrix $\mathbf{A}$ *restricted to* its first $|E|(\mathcal{N}+1)$ rows. Let $\mathbf{x} = \sum x_{v,b}\mathbf{u}_E(\mathbf{a}_{v,b})$ be an integral linear combination of the 'restricted' column vectors $\mathbf{u}_E(\mathbf{a}_{v,b})$. Then, assigning every vertex v a label b iff $x_{v,b} \neq 0$ defines a labelling $(\mathcal{P}_1^\mathbf{x}, \mathcal{P}_2^\mathbf{x})$ induced by the vector $\mathbf{x}$. From corollary 12 in [4], it follows that any such $\mathbf{x}$ with $\mathbf{x} = \mathbf{u}_E(\mathbf{b})$, induces a total-cover of (V_1, V_2, E).

Suppose $\mathbf{x} \in \mathbf{Z}^{(|V_1|\mathcal{N}+|V_2|\mathcal{N})}$ is any one vector. If $\mathbf{u}_E(\mathbf{A})\mathbf{x} - \mathbf{u}_E(\mathbf{b}) \neq 0$, then $\|\mathbf{u}_E(\mathbf{A})\mathbf{x} - \mathbf{u}_E(\mathbf{b})\|_\infty > g \geq \sqrt{g}$. So $\|\mathbf{A}\mathbf{x} - \mathbf{b}\|_\infty \geq \|\mathbf{u}_E(\mathbf{A})\mathbf{x} - \mathbf{u}_E(\mathbf{b})\|_\infty > \sqrt{g}$. If $\mathbf{u}_E(\mathbf{A})\mathbf{x} - \mathbf{u}_E(\mathbf{b}) = 0$, then $\mathbf{u}_E(\mathbf{b})$ induces a total cover $(\mathcal{P}_1^\mathbf{x}, \mathcal{P}_2^\mathbf{x})$ of (V_1, V_2, E). Thus $\|\mathbf{x}\|_1 \geq cost(\mathcal{P}_1^\mathbf{x}, \mathcal{P}_2^\mathbf{x}) \geq opt_{MinTLC_1(I)} > g \cdot (|V_1| + |V_2|)$.

Since $\mathbf{x}$ is a $(|V_1|\mathcal{N}+|V_2|\mathcal{N})$-length vector, we can split it to (V_1+V_2) vectors with $\mathcal{N}$-length. For every vertex $v \in V_1 \cup V_2$, we have a $\mathcal{N}$-length vector $\mathbf{x}_v = (x_{v,1}, \ldots, x_{v,\mathcal{N}})$. Thus there is some v' such that $\|\mathbf{x}_{v'}\|_1 = Max_{v \in V_1 \cup V_2}\|\mathbf{x}_v\|_1$. So $\|\mathbf{x}_{v'}\|_1 \cdot (V_1+V_2) \geq \sum_{v \in V_1 \cup V_2} \|\mathbf{x}_v\|_1 = \|\mathbf{u}_E(\mathbf{x})\|_1 > g \cdot (|V_1|+|V_2|)$. Therefore, $\|\mathbf{x}_{v'}\|_1 > g$. Thus $\|\mathbf{H}_\ell \mathbf{x}_{v'}\|_\infty \geq \|\mathbf{x}_{v'}\|_2 \geq \sqrt{\|\mathbf{x}_{v'}\|_1} > \sqrt{g}$.

Hence, $\|\mathbf{A}\mathbf{x} - \mathbf{b}\|_\infty \geq \|\mathbf{H}_\ell \mathbf{x}_{v'}\|_\infty > \sqrt{g}$. Since $\mathbf{x}$ is any one vector, $opt_{CVP_\infty}(\tau(I)) > \sqrt{g}$.

By Lemma 1, we come to the following conclusion:

Theorem 3. *Approximating CVP_∞ is NP-hard within any constant factor.*

4   Conclusion

In this paper, we give one proof that it is NP-hard to approximate CVP_∞ within any constant factor. This proof is simpler than known proofs. The result is obtained via the reduction from the Min Total Label Cover problem in ℓ_1 norm.

One open problem is to achieve hardness of approximation polynomial factors for CVP_∞, i.e. the factor n^ϵ for some $\epsilon > 0$. Current techniques seems unlikely to attack on the problem. Such a result is not known even for $CVP_p(1 \leq p < \infty)$. It seems to require new techniques in order to attack on the open problem.

Acknowledgments. We would like to thank the anonymous referees for their careful readings of the manuscripts and many useful suggestions.

Wenbin Chen's research has been supported by the National Natural Science Foundation of China (NSFC) under Grant No. 11271097., and by the Program for Innovative Research Team in Education Department of Guangdong Province Under No. 2016KCXTD017. Jianer Chen has been supported by the National Natural Science Foundation of China (NSFC) under Grant No. 61872097.

References

1. Ajtai, M.: The shortest vector problem in L_2 is NP-hard for randomized reductions. In: Proceedings of the 30th Annual ACM Symposium on Theory of Computing, pp. 10–19 (1998)
2. Ajtai, M., Kumar, R., Sivakumar, D.: A sieve algorithm for the shortest lattice vector problem. In: Proceedings of the 33rd Annual ACM Symposium on Theory of Computing, pp. 601–610 (2001)
3. Arora, S.: Probabilistic checking of proofs and the hardness of approximation problems. Ph.D. thesis, UC Berkeley (1994)
4. Arora, S., Babai, L., Stern, J., Sweedyk, E.Z.: The hardness of approximate optima in lattices, codes, and systems of linear equations. J. Comput. Syst. Sci. **54**, 317–331 (1997)

5. Arora, S., Lund, C.: Hardness of approximations. In: Approximation Algorithms for NP-Hard Problems. PWS Publishing (1996)
6. Babai, L.: On Lovász's lattice reduction and the nearest lattice point problem. Combinatorica **6**, 1–14 (1986)
7. Bollobás, B.: Combinatorics. Cambridge University Press, Cambridge (1986)
8. Cai, J.Y., Nerurkar, A.: Approxiamting the SVP to within a factor $(1 + 1/dim^\epsilon)$ is NP-hard under randomizied reductions. In: Proceedings of the 13th Annual IEEE Conference on Computational Complexity, pp. 151–158 (1998)
9. Dinur, I.: Approximating SVP_∞ to within almost-polynomial factors is NP-hard. In: Bongiovanni, G., Petreschi, R., Gambosi, G. (eds.) CIAC 2000. LNCS, vol. 1767, pp. 263–276. Springer, Heidelberg (2000). https://doi.org/10.1007/3-540-46521-9_22
10. Dinur, I., Kindler, G., Safra, S.: Approximating CVP to within almost-polynomial factors is NP-hard. In: Proceedings of the 39th IEEE Symposium on Foundations of Computer Science (1998)
11. Gauss, C.F.: Disquisitiones arithmeticae. (leipzig 1801), art. 171. Yale University Press. English Translation by A.A. Clarke (1966)
12. Havas, G., Seifert, J.-P.: The complexity of the extended GCD problem. In: Kutyłowski, M., Pacholski, L., Wierzbicki, T. (eds.) MFCS 1999. LNCS, vol. 1672, pp. 103–113. Springer, Heidelberg (1999). https://doi.org/10.1007/3-540-48340-3_10
13. Kannan, R.: Improved algorithm for integer programming and related lattice problems. In: Proceedings of the 15 Annual ACM Symposium on Theory of Computing, pp. 193–206 (1983)
14. Kannan, R.: Minkowski's convex body theorem and integer programming. Math. Oper. Res. **12**, 415–440 (1987)
15. Kannan, R., Sudun, M., Trevisan, L.: Constraint satisfaction: the approximability of minimization problems. In: Proceedings of the 12th IEEE Conference of Computational Complexity, pp. 282–296 (1997)
16. Khot, S.: Hardness of approximating the shortest vector problem in high L_p norms. In: Proceedings of the 44th IEEE Symposium on Foundations of Computer Science (2003)
17. Khot, S.: Hardness of approximating the shortest vector problem in lattices. In: Proceedings of the 45th IEEE Symposium on Foundations of Computer Science (2004)
18. Kumar, R., Sivakumar, D.: Complexity of SVP-A reader's digest. SIGACT News **32**(3), 40–52 (2001). Complexity Theory Column (ed. L.Hemaspaandra)
19. Lagarias, J.C., Odlyzko, A.M.: Solving low-density subset sum problems. J. ACM **32**(1), 229–246 (1985)
20. Landau, S., Miller, G.L.: Solvability of radicals in polynomial time. J. Comput. Syst. Sci. **30**(2), 179–208 (1985)
21. Lenstra, H.W.: Integer programming with a fixed number of variables. Technical report 81-03, University of Amsterdam, Amsterdam (1981)
22. Lenstra, A.K., Lenstra, H.W., Lovász, L.: Factoring polynomials with rational coefficients. Math. Ann. **261**, 513–534 (1982)
23. Lund, C., Yannakakis, M.: On the hardness of minimizaiton problems. J. ACM **41**, 960–981 (1994)
24. Micciancio, D.: The shortest vector problem is NP-hard to approximate to within some constant. In: Proceedings of the 39th IEEE Symposium on Foundations of Computer Science (1998)

25. Micciancio, D., Goldwasser, S.: Complexity of Lattice Problems, A Cryptographic Perspective. Klumer Academic Publishers (2002)
26. Minkowski, H.: Geometrie der zahlen. Leizpig, Tuebner (1910)
27. Rössner, C., Seifert, J.-P.: The complexity of approximate optima for greatest common divisor computations. In: Cohen, H. (ed.) ANTS 1996. LNCS, vol. 1122, pp. 307–322. Springer, Heidelberg (1996). https://doi.org/10.1007/3-540-61581-4_64
28. Schnorr, C.P.: A hierarchy of polynomial-time basis reduction algorithms. In: Proceedings of Conference on Algorithms, Péecs (Hungary), pp. 375–386 (1985)
29. van Emde Boas, P.: Another NP-complete problem and the complexity of computing short vectors in a lattice. Technical report 81-04, Mathematische Instiut, University of Amsterdam (1981)

Low-Dimensional Vectors with Density Bounded by 5/6 Are Pinwheel Schedulable

Wei Ding[(✉)]

Zhejiang University of Water Resources and Electric Power,
Hangzhou 310018, Zhejiang, China
dingweicumt@163.com

Abstract. Given an n-dimensional integer vector $v = (v_1, v_2, \ldots, v_n)$ with $2 \leq v_1 \leq v_2 \leq \cdots \leq v_n$, a *pinwheel schedule* for v is referred to as an infinite symbol sequence $S_1 S_2 S_3 \cdots$, which satisfies that $S_j \in \{1, 2, \ldots, n\}, \forall j \in \mathbb{Z}$ and every $i \in \{1, 2, \ldots, n\}$ occurs at least once in every v_i consecutive symbols $S_{j+1} S_{j+2} \cdots S_{j+v_i}, \forall j \in \mathbb{Z}$. If v has a pinwheel schedule then v is called *(pinwheel) schedulable*. The *density* of v is defined as $d(v) = \sum_{i=1}^{n} \frac{1}{v_i}$. Chan and Chin [4] made a conjecture that every vector v with $d(v) \leq \frac{5}{6}$ is schedulable.

In this paper, we examine the conjecture from the point of view of *low-dimensional* vectors, including 3-, 4- and 5-dimensional ones. We first discover some simple but important properties of schedulable vectors, and then apply these properties to test whether or not a vector is schedulable. As a result, we prove that the maximum density guarantee for low-dimensional vectors is $\frac{5}{6}$, which partially support this conjecture.

Keywords: Pinwheel schedule · Low-dimensional · Density guarantee

1 Introduction

Let $v = (v_1, v_2, \ldots, v_n), v_i \in \mathbb{Z}^+, 1 \leq i \leq n$ be an n-dimensional positive integer vector. W.l.o.g, we always suppose that $v_1 \leq v_2 \leq \cdots \leq v_n$. If an infinite symbol sequence, $S_1 S_2 S_3 \cdots$, satisfies that $S_j \in \{1, 2, \ldots, n\}, \forall j \in \mathbb{Z}$ and each $i \in \{1, 2, \ldots, n\}$ appears at least once in every v_i consecutive symbols $S_{j+1} S_{j+2} \cdots S_{j+v_i}, \forall j \in \mathbb{Z}$, it is called a *pinwheel schedule*. Furthermore, if v has a pinwheel schedule then it is called *(pinwheel) schedulable*. For example, $(2, 3)$ is schedulable as both $\cdots |12| \cdots$ and $\cdots |121| \cdots$ are a pinwheel schedule for it. However, any $(2, 3, v_3), v_3 \geq 3$ is unschedulable since it has no pinwheel schedule. The *density* of v is defined as

$$d(v) = \sum_{i=1}^{n} \frac{1}{v_i}. \tag{1}$$

The **pinwheel scheduling problem** asks to find a pinwheel schedule of v if it is schedulable and answers "NO" otherwise, for a given positive integer vector

© Springer Nature Switzerland AG 2019
D.-Z. Du et al. (Eds.): AAIM 2019, LNCS 11640, pp. 51–61, 2019.
https://doi.org/10.1007/978-3-030-27195-4_5

v. This problem is a special case of real-time scheduling problems [1,7,10], and has important applications in the scheduling of satellite communication with a ground station [7]. Refer readers to [2,5,11] for more related problems and applications.

Given a vector set V, if each $v \in V$ with $d(v) \leq \mathbf{D}$ is schedulable, then $\mathbf{D}$ is called the *density guarantee* for V. In [8], Holte *et al.* pointed out that 1 is a density guarantee for the vectors with at most two distinct integers. Later, Lin and Lin [9] considered the vectors with at most three distinct integers, and proved that $\frac{5}{6}$ is the maximum density guarantee for such vectors.

As we know, all $(2, 3, v_3)$ with $v_3 \geq 3$ are unschedulable [6]. Therefore, the density guarantee for all vectors can not be greater than $\frac{5}{6}$. In [3], Chan and Chin proved that $\frac{7}{10}$ is a density guarantee. In [4], furthermore, they presented a variety of pinwheel scheduling algorithms with different density guarantees, and conjectured that $\frac{5}{6}$ is the maximum density guarantee for all vectors. Later, Fishburn and Lagarias [6] proved that each v with $v_1 = 2$ and $d(v) \leq \frac{5}{6}$ is schedulable, and each v with $d(v) \leq \frac{3}{4}$ is surely schedulable. Moreover, they also considered the m-**pinwheel scheduling problem**, where each $i \in \{1, 2, \cdots, n\}$ occurs at least m times in every mv_i consecutive symbols $S_{j+1}S_{j+2} \cdots S_{j+mv_i}, \forall j \in \mathbb{Z}$, and proved that there always exist unschedulable vectors v with $d(v) = 1 - \frac{1}{(m+1)(m+2)} + \epsilon$, for any $\epsilon > 0$. In fact, Chan and Chin's conjecture was partially supported by Lin and Lin's theorem [9] and Fishburn and Lagarias's theorem [6].

In past nearly two decades, however, almost no new theoretical results came out after Fishburn and Lagarias [6]. This paper revisits the pinwheel scheduling problem and studies it from the prospective of *low-dimensional* vectors (i.e., 3-, 4- and 5-dimensional vectors). We first discover some simple but important properties of schedulable vectors and then use these properties to test whether or not a vector is schedulable. By testing every vector v with dimension at most five and $d(v) \leq \frac{5}{6}$, we claim that $\frac{5}{6}$ is the maximum density guarantee for all these low-dimensional vectors. In others words, our results partially support Chan and Chin's conjecture. It is noted that our results impose no additional requirement on the values of integers in vectors and the number of distinct integers appearing in vectors reaches up to five.

The rest of this paper is organized as follows. In Sect. 2, we define some notations and show several fundamental properties. In Sect. 3, we prove that the 3-, 4- and 5-dimensional vectors with density bounded by $\frac{5}{6}$ are pinwheel schedulable. In Sect. 4, we conclude this paper.

2 Preliminaries

2.1 Notations

Let $v^{(c)} = (v_1, v_2, \ldots, v_c)$ denote a c-dimensional positive integer vector with $v_1 \leq v_2 \leq \cdots \leq v_c$, for any $c \geq 2$, and let $V^{(c)}$ be the set of all c-dimensional vectors. Furthermore, we let $v^{(c)}(f) = (f, v_2, \ldots, v_c)$ denote a c-dimensional vector with $v_1 = f$ and $f \leq v_2 \leq \cdots \leq v_c$, and let $V^{(c)}(f)$ be the set of all such c-dimensional vectors with $v_1 = f$. Besides, any c-dimensional vector that satisfies $v_i \in \mathbb{Z}^+$ and $a \leq v_i \leq b$, for any $1 \leq i \leq c$, is collectively denoted by

$$(v_1, \ldots, v_{i-1}, [a, b], v_{i+1}, \ldots, v_c), \quad \forall 1 \leq i \leq c. \tag{2}$$

Similarly, we let $v^{(c)}([a, b])$ denote any c-dimensional vector with $v_1 \in \mathbb{Z}^+$ and $a \leq v_1 \leq b$, and let $V^{(c)}([a, b])$ be the set of all such c-dimensional vectors. Also, $v^{(c)}([a, +\infty))$ contains $v_1 \geq a$.

Example. Both $(2, 5, 5, 8, 9, 9)$ and $(5, 5, 6, 9, 10, 10)$ are 6-dimensional vectors, and both $(3, 3, 5, 6)$ and $(3, 4, 5, 5)$ are 4-dimensional vectors with $v_1 = 3$. Also, $(3, 5, 5, 8, 9) \in V^{(5)}([2, 4])$ while $(5, 5, 6, 6, 9) \notin V^{(5)}([2, 4])$, and $(6, 6, 6, 8, 8) \in V^{(5)}([5, +\infty))$ while $(3, 4, 5, 6, 8) \notin V^{(5)}([5, +\infty))$. In addition, $(3, 3, [5, 7], 7, 9)$ includes

$$(3, 3, 5, 7, 9), \ (3, 3, 6, 7, 9), \ (3, 3, 7, 7, 9).$$

Every $v^{(1)} = (v_1), v_1 \geq 1$ is schedulable since it has a pinwheel schedule $\cdots |1| \cdots$. Clearly, every $v^{(c)}(1) = (1, v_2, \ldots, v_c)$ is unschedulable, for any $c \geq 2$. In the rest of this paper, therefore, we focus on n-dimensional vectors, $v = (v_1, v_2, \ldots, v_n), n \geq 2$, with $2 \leq v_1 \leq v_2 \leq \cdots \leq v_n$.

2.2 Fundamental Properties

In this section, we show several simple but important properties of schedulable vectors, which will play an important role in testing whether a vector is schedulable.

Lemma 1. *Let $v^{(c)}$ and $u^{(c)}, c \geq 2$ be two c-dimensional vectors satisfying that $u_{i^*} \geq v_{i^*}$ and $u_i = v_i, i \neq i^*$, for any given $1 \leq i^* \leq c$. If $v^{(c)}$ is schedulable and $d(v^{(c)}) = D$, then $u^{(c)}$ is schedulable and has $d(u^{(c)}) \leq D$.*

Proof. If $v^{(c)} = (v_1, v_2, \cdots, v_c)$ is schedulable, then there exists a pinwheel schedule $\mathcal{S}_0$ of $v^{(c)}$ where every $v_i, 1 \leq i \leq c$ consecutive symbols contain at least one i. If $u^{(c)} = (u_1, u_2, \cdots, u_c)$ satisfies that $u_{i^*} \geq v_{i^*}, \forall 1 \leq i^* \leq c$ and $u_i = v_i, i \neq i^*$, then $\mathcal{S}_0$ surely satisfies that every $u_i, 1 \leq i \leq c$ consecutive symbols of $\mathcal{S}_0$ also contain at least one i. This implies that $\mathcal{S}_0$ is also a pinwheel schedule of $u^{(c)}$. We have

$$\begin{aligned} d(u^{(c)}) &= \sum_{i=1}^{c} \frac{1}{u_i} \\ &= \frac{1}{u_{i^*}} + \sum_{i \neq i^*} \frac{1}{v_i} \\ &\leq \frac{1}{v_{i^*}} + \sum_{i \neq i^*} \frac{1}{v_i} \\ &= d(v^{(c)}) \\ &= D. \end{aligned}$$

$\square$

Furthermore, we extend Lemma 1 to the following more general and fundamental lemma.

Lemma 2. *Let $v^{(c)}$ and $u^{(c)}, c \geq 2$ be two c-dimensional vectors satisfying that $u_i \geq v_i, i^* \leq i \leq j^*$ and $u_i = v_i, 1 \leq i < i^*, j^* < i \leq n$, for any given $1 \leq i^* < j^* \leq c$. If $v^{(c)}$ is schedulable and has $d(v^{(c)}) = D$ then $u^{(c)}$ is schedulable and has $d(u^{(c)}) \leq D$, as well as if $u^{(c)}$ is unschedulable and has $d(u^{(c)}) = D'$ then $v^{(c)}$ is unschedulable and has $d(v^{(c)}) \geq D'$.*

Lemma 3. *For any $v^{(c)}, c \geq 2$, if there are at least one index $1 \leq i^* \leq c$ such that $v_{i^*} < i^*$, then it is unschedulable.*

Lemma 4. *Let $v^{(c)} = (f, f, \ldots, f)_{1 \times c}$ be a c-dimensional vector where $c \geq 2$ and $f \geq 1$. If $f \geq c$ then $(f, f, \cdots, f)_{1 \times c}$ is schedulable, and if $f < c$ then it is unschedulable.*

Proof. Since $\cdots |123 \cdots c| \cdots$ is a pinwheel schedule for $(c, c, \ldots, c)_{1 \times c}$, we claim that $(c, c, \cdots, c)_{1 \times c}$ is schedulable. If $f \geq c$, we conclude by Lemma 2 that $(f, f, \cdots, f)_{1 \times c}$ is schedulable. If $f < c$, we have $v_c = f < c$ and then claim by Lemma 3 that $(f, f, \cdots, f)_{1 \times c}$ is unschedulable. □

Lemma 5. *Every $v^{(c)}([c, +\infty)), c \geq 2$ is schedulable.*

Proof. Let $v^{(c)} = (v_1, v_2, \ldots, v_c)$. From $v_1 \geq c$, it follows that $v_c \geq v_{c-1} \geq \cdots \geq v_1 \geq c$. Lemma 4 implies that $(c, c, \cdots, c)_{1 \times c}$ is schedulable. By Lemma 2, we claim that every $v^{(c)}([c, +\infty))$ is schedulable. □

3   Main Results

In this section, we prove that the maximum density guarantee for 3-, 4-, 5-dimensional vectors is $\frac{5}{6}$, respectively.

Theorem 1. *Every $v^{(2)} \in V^{(2)}$ is schedulable, and $v^{(3)} \in V^{(3)}$ with $d(v^{(3)}) \leq 1$ except $(2, 3, m)$ is schedulable.*

Theorem 1 and its proof imply that every $v^{(\ell)}, \ell = 2, 3$ with $d(v^{(\ell)}) \leq \frac{5}{6}$ is schedulable. In fact, any $v^{(2)}$ has at most two distinct integers and any $v^{(3)}$ has at most three distinct integers. So, Theorem 1 accords with the theorems in [8,9]. Furthermore, it extends the density guarantee from $\frac{5}{6}$ to 1 for 3-dimensional vectors except $(2, 3, m)$.

Obviously, any $v^{(4)}$ has up to four distinct integers and any $v^{(5)}$ has up to five distinct integers. Although it remains as an open problem whether the maximum density guarantee for all vectors is $\frac{5}{6}$, we show in Theorem 2 that any $v^{(4)}$ with $d(v^{(4)}) \leq \frac{5}{6}$ is schedulable and in Theorem 3 that any $v^{(5)}$ with $d(v^{(5)}) \leq \frac{5}{6}$ is schedulable.

Theorem 2. *Every $v^{(4)} \in V^{(4)}$ with $d(v^{(4)}) \leq \frac{5}{6}$ is schedulable.*

Proof. In the following, we discuss the cases of $v_1 = 2, 3$ and $v_1 \geq 4$.

Case 1 : $v_1 = 2$. In [6], Fishburn *et al.* point out that every v with $v_1 = 2$ and $d(v) \leq \frac{5}{6}$ is schedulable. So, it follows immediately that every $v^{(4)}(2)$ with $d(v^{(4)}(2)) \leq \frac{5}{6}$ is schedulable.

Case 2 : $v_1 = 3$. We need to discuss the subcases of $v_2 = 3, v_2 = 4$ and $v_2 \geq 5$, respectively.

Subcase 2.1 : $v_2 = 3$. Clearly, $v_3 \geq 3$. It is well-known that every v with $d(v) > 1$ is unschedulable. We have

$$1 = \frac{1}{3} \times 3 < d((3,3,3,v_4)) \leq \frac{1}{3} \times 4 = \frac{4}{3}.$$

Thus, $(3,3,3,v_4)$ is unschedulable. Since $(3,3,6,6)$ has a pinwheel schedule $\cdots |123124| \cdots$, we claim that it is schedulable. Further, we conclude by Lemma 2 that every $(3,3,[6,+\infty),v_4)$ is schedulable. We have

$$\frac{2}{3} = \frac{1}{3} \times 2 < d((3,3,[6,+\infty),v_4)) \leq \frac{1}{3} \times 2 + \frac{1}{6} \times 2 = 1.$$

The subcases of $(3,3,4,v_4)$ and $(3,3,5,v_4)$ will be analyzed at the end of the proof.

Subcase 2.2 : $v_2 = 4$. Clearly, $v_3 \geq 4$. By using the Chan and Chin's algorithm [3], we get a pinwheel schedule $\cdots |21312314| \cdots$ for $(3,4,5,8)$. So, we claim that $(3,4,5,8)$ is schedulable. Furthermore, we conclude from Lemma 2 that every $(3,4,5,[8,+\infty))$ is schedulable. We have

$$\frac{47}{60} = \frac{1}{3} + \frac{1}{4} + \frac{1}{5} < d((3,4,5,[8,+\infty))) \leq \frac{1}{3} + \frac{1}{4} + \frac{1}{5} + \frac{1}{8} = \frac{109}{120}.$$

Since $(3,4,6,6)$ has a pinwheel schedule $\cdots |123124| \cdots$, we claim that it is schedulable. By Lemma 2, we conclude that every $(3,4,[6,+\infty),v_4)$ is schedulable. We have

$$\frac{7}{12} = \frac{1}{3} + \frac{1}{4} < d((3,4,[6,+\infty),v_4)) \leq \frac{1}{3} + \frac{1}{4} + \frac{1}{6} + \frac{1}{6} = \frac{11}{12}.$$

The analysis on the subcases of $(3,4,4,v_4)$ and $(3,4,5,[5,7])$ are left at the end of the proof. Then, we claim that every $(3,4,v_3,v_4)$ except $(3,4,4,v_4)$ and $(3,4,5,[5,7])$ is schedulable and its density is at most $\min\{\frac{109}{120}, \frac{11}{12}\} = \frac{109}{120}$.

Subcase 2.3 : $v_2 \geq 5$. We claim that $(3,5,5,5)$ is schedulable since it has a pinwheel schedule $\cdots |12134| \cdots$. By Lemma 2, we conclude that every $(3,[5,+\infty),v_3,v_4)$ is schedulable. We have

$$d((3,[5,+\infty),v_3,v_4)) \leq \frac{1}{3} + \frac{1}{5} \times 3 = \frac{14}{15}.$$

Case 3 : $v_1 \geq 4$. It follows from Lemma 5 that every $v^{(4)}$ with $v_1 \geq 4$ is schedulable. We have

$$d(([4,+\infty),v_2,v_3,v_4)) \leq \frac{1}{4} \times 4 = 1.$$

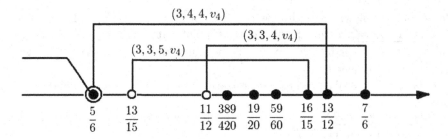

Fig. 1. Illustration for the proof of Theorem 2.

By Case 1, we claim that every $v^{(4)}(2)$ with $d(v^{(4)}(2)) \leq \frac{5}{6}$ is schedulable. By Case 2, we claim every $v^{(4)}(3)$, except the subcases left, with $d(v^{(4)}(3)) \leq \min\{1, \frac{109}{120}, \frac{14}{15}\} = \frac{109}{120}$ is schedulable. By Case 3, we claim each $v^{(4)}([4, +\infty))$ with $d(v^{(4)}([4, +\infty))) \leq 1$ is schedulable. So, we conclude each $v^{(4)}$, except the subcases left, with $d(v^{(4)}) \leq \min\{\frac{5}{6}, \frac{109}{120}, 1\} = \frac{5}{6}$ is schedulable (Fig. 1).

Now, we consider the subcases left. We obtain

$$\frac{11}{12} = \frac{1}{3} + \frac{1}{3} + \frac{1}{4} < d((3,3,4,v_4)) \leq \frac{1}{3} + \frac{1}{3} + \frac{1}{4} \times 2 = \frac{7}{6},$$
$$\frac{13}{15} = \frac{1}{3} + \frac{1}{3} + \frac{1}{5} < d((3,3,5,v_4)) \leq \frac{1}{3} + \frac{1}{3} + \frac{1}{5} \times 2 = \frac{16}{15},$$
$$\frac{5}{6} = \frac{1}{3} + \frac{1}{4} + \frac{1}{4} < d((3,4,4,v_4)) \leq \frac{1}{3} + \frac{1}{4} \times 3 = \frac{13}{12},$$
$$d((3,4,5,5)) = \frac{1}{3} + \frac{1}{4} + \frac{1}{5} + \frac{1}{5} = \frac{59}{60},$$
$$d((3,4,5,6)) = \frac{1}{3} + \frac{1}{4} + \frac{1}{5} + \frac{1}{6} = \frac{19}{20},$$
$$d((3,4,5,7)) = \frac{1}{3} + \frac{1}{4} + \frac{1}{5} + \frac{1}{7} = \frac{389}{420}.$$

Since

$$(\frac{11}{12}, \frac{7}{6}] \cup (\frac{13}{15}, \frac{16}{15}] \cup (\frac{5}{6}, \frac{13}{12}] \cup \{\frac{59}{60}, \frac{19}{20}, \frac{389}{420}\} \cap (0, \frac{5}{6}] = \emptyset,$$

we claim that every $v^{(4)}$ with $d(v^{(4)}) \leq \frac{5}{6}$ is schedulable no matter whether the subcases left are schedulable or not. The proof is completed. □

Lemma 6. Every $v^{(5)}(3) \in V^{(5)}(3)$ with $d(v^{(5)}(3)) \leq \frac{5}{6}$ is schedulable.

Proof. In the following, we discuss the subcases of $v_2 = 3, 4, 5$ and $v_2 \geq 6$, respectively.

Case 1: $v_2 = 3$. Clearly, $v_3 \geq 3$. As $(3, 3, 9, 9, 9)$ has a pinwheel schedule $\cdots |123124125| \cdots$, we claim that it is schedulable. Then, we conclude by Lemma 2 that each $(3, 3, [9, +\infty), v_4, v_5)$ is schedulable. Hence, we only need to consider the subcases of $v_3 = 3, 4, 5, 6, 7, 8$. We have

$$1 = \frac{1}{3} \times 3 < d((3,3,3,v_4,v_5)) \leq \frac{1}{3} \times 5 = \frac{5}{3},$$
$$\frac{11}{12} = \frac{1}{3} \times 2 + \frac{1}{4} < d((3,3,4,v_4,v_5)) \leq \frac{1}{3} \times 2 + \frac{1}{4} \times 3 = \frac{17}{12},$$
$$\frac{13}{15} = \frac{1}{3} \times 2 + \frac{1}{5} < d((3,3,5,v_4,v_5)) \leq \frac{1}{3} \times 2 + \frac{1}{5} \times 3 = \frac{19}{15},$$
$$\frac{5}{6} = \frac{1}{3} \times 2 + \frac{1}{6} < d((3,3,6,v_4,v_5)) \leq \frac{1}{3} \times 2 + \frac{1}{6} \times 3 = \frac{7}{6}.$$

Since

$$(1, \frac{5}{3}] \cup (\frac{11}{12}, \frac{17}{12}] \cup (\frac{13}{15}, \frac{19}{15}] \cup (\frac{5}{6}, \frac{7}{6}] \cap (0, \frac{5}{6}] = \emptyset \, ,$$

we need no considering the subcases of $v_3 = 3, 4, 5, 6$. So, the work left is to analyze the subcases of $v_3 = 7, 8$.

Subcase 1.1 : $v_3 = 7$. So, $v_4 \geq 7$. Since $(3, 3, 7, 12, 12)$ has a pinwheel schedule $\cdots |124123125123| \cdots$, we claim that it is schedulable. Furthermore, we conclude by Lemma 2 that any $(3, 3, 7, [12, +\infty), v_5)$ is schedulable. Therefore, we only need to consider the subcases of $v_4 = 7, 8, 9, 10, 11$. We have

$$\frac{20}{21} = \frac{1}{3} \times 2 + \frac{1}{7} \times 2 < d((3, 3, 7, 7, v_5)) \leq \frac{1}{3} \times 2 + \frac{1}{7} \times 3 = \frac{23}{21},$$
$$\frac{157}{168} = \frac{1}{3} \times 2 + \frac{1}{7} + \frac{1}{8} < d((3, 3, 7, 8, v_5)) \leq \frac{1}{3} \times 2 + \frac{1}{7} + \frac{1}{8} \times 2 = \frac{89}{84},$$
$$\frac{58}{63} = \frac{1}{3} \times 2 + \frac{1}{7} + \frac{1}{9} < d((3, 3, 7, 9, v_5)) \leq \frac{1}{3} \times 2 + \frac{1}{7} + \frac{1}{9} \times 2 = \frac{65}{63},$$
$$\frac{191}{210} = \frac{1}{3} \times 2 + \frac{1}{7} + \frac{1}{10} < d((3, 3, 7, 10, v_5)) \leq \frac{1}{3} \times 2 + \frac{1}{7} + \frac{1}{10} \times 2 = \frac{106}{105},$$
$$\frac{208}{231} = \frac{1}{3} \times 2 + \frac{1}{7} + \frac{1}{11} < d((3, 3, 7, 11, v_5)) \leq \frac{1}{3} \times 2 + \frac{1}{7} + \frac{1}{11} \times 2 = \frac{229}{231}.$$

Since

$$(\frac{20}{21}, \frac{23}{21}] \cup (\frac{157}{168}, \frac{89}{84}] \cup (\frac{58}{63}, \frac{65}{63}] \cup (\frac{191}{210}, \frac{106}{105}] \cup (\frac{208}{231}, \frac{229}{231}] \cap (0, \frac{5}{6}] = \emptyset \, ,$$

we need no considering the subcases of $v_4 = 7, 8, 9, 10, 11$. So, we conclude that every $(3, 3, 7, v_4, v_5)$ with $d((3, 3, 7, v_4, v_5)) \leq \frac{5}{6}$ is schedulable.

Subcase 1.2 : $v_3 = 8$. Clearly, $v_4 \geq 8$. Recall that $(3, 3, 7, 12, 12)$ given in Subcase 1.1 is schedulable. By Lemma 1, we claim that $(3, 3, 8, 12, 12)$ is also schedulable. Hence, we only need to consider the subcases of $v_4 = 8, 9, 10, 11$. We have

$$\frac{11}{12} = \frac{1}{3} \times 2 + \frac{1}{8} \times 2 < d((3, 3, 8, 8, v_5)) \leq \frac{1}{3} \times 2 + \frac{1}{8} \times 3 = \frac{25}{24},$$
$$\frac{65}{72} = \frac{1}{3} \times 2 + \frac{1}{8} + \frac{1}{9} < d((3, 3, 8, 9, v_5)) \leq \frac{1}{3} \times 2 + \frac{1}{8} + \frac{1}{9} \times 2 = \frac{73}{72},$$
$$\frac{107}{120} = \frac{1}{3} \times 2 + \frac{1}{8} + \frac{1}{10} < d((3, 3, 8, 10, v_5)) \leq \frac{1}{3} \times 2 + \frac{1}{8} + \frac{1}{10} \times 2 = \frac{119}{120},$$
$$\frac{233}{264} = \frac{1}{3} \times 2 + \frac{1}{8} + \frac{1}{11} < d((3, 3, 8, 11, v_5)) \leq \frac{1}{3} \times 2 + \frac{1}{8} + \frac{1}{11} \times 2 = \frac{257}{264}.$$

Since

$$(\frac{11}{12}, \frac{25}{24}] \cup (\frac{65}{72}, \frac{73}{72}] \cup (\frac{107}{120}, \frac{119}{120}] \cup (\frac{233}{264}, \frac{257}{264}] \cap (0, \frac{5}{6}] = \emptyset \, ,$$

we need no considering the subcases of $v_4 = 8, 9, 10, 11$. So, we conclude that every $(3, 3, 8, v_4, v_5)$ with $d((3, 3, 8, v_4, v_5)) \leq \frac{5}{6}$ is schedulable.

Case 2 : $v_2 = 4$. Clearly, $v_3 \geq 4$. As $(3, 4, 8, 8, 8)$ has a pinwheel schedule $\cdots |14213125| \cdots$, we claim that it is schedulable. Then, we conclude by Lemma 2 that each $(3, 4, [8, +\infty), v_4, v_5)$ is schedulable. We have

$$d((3, 4, 4, v_4, v_5)) > \frac{1}{3} + \frac{1}{4} + \frac{1}{4} = \frac{5}{6},$$
$$d((3, 4, 4, v_4, v_5)) \leq \frac{1}{3} + \frac{1}{4} \times 4 = \frac{4}{3}.$$

Since $(\frac{5}{6}, \frac{4}{3}] \cap (0, \frac{5}{6}] = \emptyset$, we only need to analyze the subcases of $v_3 = 5, 6, 7$ in the following.

Subcase 2.1 : $v_3 = 5$. Clearly, $v_4 \geq 5$. As $(3, 4, 5, 16, 16)$ has a pinwheel schedule $\cdots |1231421312513213| \cdots$, we claim that it is schedulable. Then, we conclude by Lemma 2 that each $(3, 4, 5, [16, +\infty), v_5)$ is schedulable. We have

$$d((3, 4, 5, [5, 15], v_5)) > \tfrac{1}{3} + \tfrac{1}{4} + \tfrac{1}{5} + \tfrac{1}{15} = \tfrac{17}{20} > \tfrac{5}{6},$$
$$d((3, 4, 5, [5, 15], v_5)) \leq \tfrac{1}{3} + \tfrac{1}{4} + \tfrac{1}{5} \times 3 = \tfrac{71}{60}.$$

Since $(\tfrac{17}{20}, \tfrac{71}{60}] \cap (0, \tfrac{5}{6}] = \emptyset$, we need no discussing the subcases of $v_4 = 5, 6, \ldots, 15$. So, we conclude every $(3, 4, 5, v_4, v_5)$ with $d((3, 4, 5, v_4, v_5)) \leq \tfrac{5}{6}$ is schedulable.

Subcase 2.2 : $v_3 = 6$. So, $v_4 \geq 6$. Since $(3, 4, 6, 12, 12)$ has a pinwheel schedule $\cdots |124132152132| \cdots$, we claim that it is schedulable. By Lemma 2, we conclude each $(3, 4, 6, [12, +\infty), v_5)$ is schedulable. We have

$$d((3, 4, 6, [6, 11], v_5)) > \tfrac{1}{3} + \tfrac{1}{4} + \tfrac{1}{6} + \tfrac{1}{11} = \tfrac{37}{44},$$
$$d((3, 4, 6, [6, 11], v_5)) \leq \tfrac{1}{3} + \tfrac{1}{4} + \tfrac{1}{6} \times 3 = \tfrac{13}{12}.$$

Because of $(\tfrac{37}{44}, \tfrac{13}{12}] \cap (0, \tfrac{5}{6}] = \emptyset$, we need no analyzing the subcases of $v_4 = 6, 7, 8, 9, 10, 11$. Therefore, every $(3, 4, 6, v_4, v_5)$ with $d((3, 4, 6, v_4, v_5)) \leq \tfrac{5}{6}$ is schedulable.

Subcase 2.3 : $v_3 = 7$. Clearly, $v_4 \geq 7$. As $(3, 4, 7, 11, 11)$ has a pinwheel schedule $\cdots |14213125132| \cdots$, we claim that it is schedulable. Further, we conclude by Lemma 2 that each $(3, 4, 7, [11, +\infty), v_5)$ is schedulable. So, we only need to consider the subcases of $v_4 = 7, 8, 9, 10$. We have

$$\tfrac{73}{84} = \tfrac{1}{3} + \tfrac{1}{4} + \tfrac{1}{7} \times 2 < d((3, 4, 7, 7, v_5)) \leq \tfrac{1}{3} + \tfrac{1}{4} + \tfrac{1}{7} \times 3 = \tfrac{85}{84},$$
$$\tfrac{143}{168} = \tfrac{1}{3} + \tfrac{1}{4} + \tfrac{1}{7} + \tfrac{1}{8} < d((3, 4, 7, 8, v_5)) \leq \tfrac{1}{3} + \tfrac{1}{4} + \tfrac{1}{7} + \tfrac{1}{8} \times 2 = \tfrac{41}{42},$$
$$\tfrac{211}{252} = \tfrac{1}{3} + \tfrac{1}{4} + \tfrac{1}{7} + \tfrac{1}{9} < d((3, 4, 7, 9, v_5)) \leq \tfrac{1}{3} + \tfrac{1}{4} + \tfrac{1}{7} + \tfrac{1}{9} \times 2 = \tfrac{239}{252}.$$

Since

$$(\frac{73}{84}, \frac{85}{84}] \cup (\frac{143}{168}, \frac{41}{42}] \cup (\frac{211}{252}, \frac{239}{252}] \cap (0, \frac{5}{6}] = \emptyset,$$

we need no considering the subcases of $v_4 = 7, 8, 9$. Hence, we only need to analyze the subcase of $v_4 = 10$.

As $(3, 4, 7, 10, 16)$ has a pinwheel schedule $\cdots |1231421312413215| \cdots$, we claim that it is schedulable, and then conclude by Lemma 1 that every $(3, 4, 7, 10, [16, +\infty))$ is schedulable. We have

$$d((3, 4, 7, 10, [10, 15])) \geq \tfrac{1}{3} + \tfrac{1}{4} + \tfrac{1}{7} + \tfrac{1}{10} + \tfrac{1}{15} = \tfrac{25}{28},$$
$$d((3, 4, 7, 10, [10, 15])) \leq \tfrac{1}{3} + \tfrac{1}{4} + \tfrac{1}{7} + \tfrac{1}{10} \times 2 = \tfrac{389}{420}.$$

Since $[\tfrac{25}{28}, \tfrac{389}{420}] \cap (0, \tfrac{5}{6}] = \emptyset$, we need no discussing the subcases of $v_5 = 10, 11, 12, 13, 14, 15$. So, each $(3, 4, 7, 10, v_5)$ with $d((3, 4, 7, 10, v_5)) \leq \tfrac{5}{6}$ is schedulable.

Case 3 : $v_2 = 5$. Clearly, $v_3 \geq 5$. Since $(3, 3, 9, 9, 9)$ is schedulable, we claim by Lemma 1 that $(3, 5, 9, 9, 9)$ is also schedulable. Furthermore, we conclude from Lemma 2 that each $(3, 5, [9, +\infty), v_4, v_5)$ is schedulable. Next, we analyze the subcases of $v_3 = 5, 6, 7, 8$.

Subcase 3.1 : $v_3 = 5$. Clearly, $v_4 \geq 5$. As $(3, 5, 5, 10, 10)$ has a pinwheel schedule $\cdots |1231412315| \cdots$, we claim that it is schedulable. Furthermore, we conclude by Lemma 2 that each $(3, 5, 5, [10, +\infty), v_5)$ is schedulable. We have

$$d((3, 5, 5, [5, 9], v_5)) > \tfrac{1}{3} + \tfrac{1}{5} + \tfrac{1}{5} + \tfrac{1}{9} = \tfrac{38}{45},$$
$$d((3, 5, 5, [5, 9], v_5)) \leq \tfrac{1}{3} + \tfrac{1}{5} \times 4 = \tfrac{17}{15}.$$

Since $(\tfrac{38}{45}, \tfrac{17}{15}] \cap (0, \tfrac{5}{6}] = \emptyset$, we need no discussing the subcases of $v_4 = 5, 6, 7, 8, 9$. Therefore, every $(3, 5, 5, v_4, v_5)$ with $d((3, 5, 5, v_4, v_5)) \leq \tfrac{5}{6}$ is schedulable.

Subcase 3.2 : $v_3 = 6$. Clearly, $v_4 \geq 6$. As $(3, 5, 5, 10, 10)$ given in Subcase 3.1 is schedulable, we claim by Lemma 1 that $(3, 5, 6, 10, 10)$ is also schedulable. So, we conclude by Lemma 2 that each $(3, 5, 6, [10, +\infty), v_5)$ is schedulable. We have

$$d((3, 5, 6, [6, 7], v_5)) > \tfrac{1}{3} + \tfrac{1}{5} + \tfrac{1}{6} + \tfrac{1}{7} = \tfrac{59}{70},$$
$$d((3, 5, 6, [6, 7], v_5)) \leq \tfrac{1}{3} + \tfrac{1}{5} + \tfrac{1}{6} \times 3 = \tfrac{31}{30}.$$

Since $(\tfrac{59}{70}, \tfrac{31}{30}] \cap (0, \tfrac{5}{6}] = \emptyset$, we only need to consider the subcase of $v_4 = 8, 9$.

As $(3, 5, 6, 8, 12)$ has a pinwheel schedule $\cdots |123142153124| \cdots$, we claim that it is schedulable and then conclude by Lemma 1 that every $(3, 5, 6, 8, [12, +\infty))$ is schedulable. We have

$$d((3, 5, 6, 8, [8, 11])) \geq \tfrac{1}{3} + \tfrac{1}{5} + \tfrac{1}{6} + \tfrac{1}{8} + \tfrac{1}{11} = \tfrac{403}{440},$$
$$d((3, 5, 6, 8, [8, 11])) \leq \tfrac{1}{3} + \tfrac{1}{5} + \tfrac{1}{6} + \tfrac{1}{8} \times 2 = \tfrac{19}{20}.$$

Since $[\tfrac{403}{440}, \tfrac{19}{20}] \cap (0, \tfrac{5}{6}] = \emptyset$, we do not need to consider the subcases of $(3, 5, 6, 8, v_5)$ with $v_5 = 8, 9, 10, 11$. So, we conclude that every $(3, 5, 6, 8, v_5)$ with $d((3, 5, 6, 8, v_5)) \leq \tfrac{5}{6}$ is schedulable. As $(3, 5, 6, 8, 12)$ is schedulable, we claim by Lemma 1 that $(3, 5, 6, 9, 12)$ is also schedulable. Further, we conclude that every $(3, 5, 6, 9, [12, +\infty))$ is schedulable. We have

$$d((3, 5, 6, 9, [9, 11])) \geq \tfrac{1}{3} + \tfrac{1}{5} + \tfrac{1}{6} + \tfrac{1}{9} + \tfrac{1}{11} = \tfrac{893}{990},$$
$$d((3, 5, 6, 9, [9, 11])) \leq \tfrac{1}{3} + \tfrac{1}{5} + \tfrac{1}{6} + \tfrac{1}{9} \times 2 = \tfrac{83}{90}.$$

Since $[\tfrac{893}{990}, \tfrac{83}{90}] \cap (0, \tfrac{5}{6}] = \emptyset$, we do not need to consider the subcases of $(3, 5, 6, 9, v_5)$ with $v_5 = 9, 10, 11$. So, we conclude that every $(3, 5, 6, 9, v_5)$ with $d((3, 5, 6, 9, v_5)) \leq \tfrac{5}{6}$ is schedulable.

Subcase 3.3 : $v_3 = 7$. Clearly, $v_4 \geq 7$. As $(3, 5, 7, 9, 9)$ has a pinwheel schedule $\cdots |123142153| \cdots$, we claim that it is schedulable. Furthermore, we conclude by Lemma 2 that every $(3, 5, 7, [9, +\infty), v_5)$ is schedulable. Next, we only need to analyze the subcases of $v_4 = 7, 8$.

Since $(3, 5, 7, 7, 12)$ has a pinwheel schedule $\cdots |123142153124| \cdots$, we claim that it is schedulable. Then, we conclude by Lemma 1 that every $(3, 5, 7, 7, [12, +\infty))$ is schedulable. We have

$$d((3, 5, 7, 7, [7, 11])) \geq \tfrac{1}{3} + \tfrac{1}{5} + \tfrac{1}{7} + \tfrac{1}{7} + \tfrac{1}{11} = \tfrac{1051}{1155},$$
$$d((3, 5, 7, 7, [7, 11])) \leq \tfrac{1}{3} + \tfrac{1}{5} + \tfrac{1}{7} \times 3 = \tfrac{101}{105}.$$

Since $[\frac{1051}{1155}, \frac{101}{105}] \cap (0, \frac{5}{6}] = \emptyset$, we do not need to consider the subcases of $(3, 5, 7, 7, v_5)$ with $v_5 = 7, 8, 9, 10, 11$. Thus, we conclude each $(3, 5, 7, 7, v_5)$ with $d((3, 5, 7, 7, v_5)) \le \frac{5}{6}$ is schedulable. As $(3, 5, 7, 7, 12)$ is schedulable, we claim by Lemma 1 that $(3, 5, 7, 8, 12)$ is also schedulable. So, we conclude that every $(3, 5, 7, 8, [12, +\infty))$ is schedulable. We have

$$d((3, 5, 7, 8, [8, 11])) \ge \tfrac{1}{3} + \tfrac{1}{5} + \tfrac{1}{7} + \tfrac{1}{8} + \tfrac{1}{11} = \tfrac{8243}{9240},$$
$$d((3, 5, 7, 8, [8, 11])) \le \tfrac{1}{3} + \tfrac{1}{5} + \tfrac{1}{7} + \tfrac{1}{8} \times 2 = \tfrac{389}{420}.$$

Since $[\frac{8243}{9240}, \frac{389}{420}] \cap (0, \frac{5}{6}] = \emptyset$, we do not need to consider the subcases of $(3, 5, 7, 8, v_5)$ with $v_5 = 8, 9, 10, 11$. Therefore, every $(3, 5, 7, 8, v_5)$ with $d((3, 5, 7, 8, v_5)) \le \frac{5}{6}$ is schedulable.

Subcase 3.4 : $v_3 = 8$. Clearly, $v_4 \ge 8$. Since $(3, 5, 8, 8, 8)$ has a pinwheel schedule $\ldots |12314125| \cdots$, we claim that it is schedulable. Further, we conclude by Lemma 2 that every $(3, 5, 8, v_4, v_5)$ with $d((3, 5, 8, v_4, v_5)) \le \frac{5}{6}$ is schedulable.

Case 4 : $v_2 \ge 6$. We claim $(3, 6, 6, 6, 6)$ is schedulable since it has a pinwheel schedule $\cdots |123145| \cdots$. By Lemma 2, we conclude that every $(3, [6, +\infty), v_3, v_4, v_5)$ is schedulable.

Based on above discussions, we conclude that every $(3, v_2, v_3, v_4, v_5)$ with $d((3, v_2, v_3, v_4, v_5)) \le \frac{5}{6}$ is schedulable. □

Lemma 7. *Every $v^{(5)}(4) \in V^{(5)}(4)$ with $d(v^{(5)}(4)) \le \frac{5}{6}$ is schedulable.*

Theorem 3. *Every $v^{(5)} \in V^{(5)}$ with $d(v^{(5)}) \le \frac{5}{6}$ is schedulable.*

Since all $(2, 3, v_3)$ with $v_3 \ge 3$ are unschedulable [6], we claim that the maximum density guarantee for all low-dimensional vectors can not be greater than $\frac{5}{6}$. Combining Theorems 1, 2 and 3, we immediately obtain the following corollary.

Corollary 1. *The maximum density guarantee for all low-dimensional (i.e., 3-, 4-, 5-dimensional) vectors is $\frac{5}{6}$.*

4   Concluding Remarks

In this paper, we examine Chan and Chin's conjecture [4] from the angle of low-dimensional (i.e., 3-, 4- and 5-dimensional) vectors. We first show several fundamental properties of schedulable vectors, and then use these properties to test whether or not a vector is schedulable. By testing all the low-dimensional vectors, we prove that the maximum density guarantee for low-dimensional vectors is $\frac{5}{6}$, which partially supports this conjecture.

It is also of great interest to prove this conjecture for general vectors. Although the way used in this paper can not be directly applied to general vectors (with arbitrary dimension), it indeed points out a potential technical line to prove the conjecture.

References

1. Baruah, S., Bestavros, A.: Pinwheel scheduling for fault-tolerant broadcast disks in real-time database systems. In: Proceedings of ICDE 1997, pp. 543–551 (1997)
2. Baruah, S.K., Lin, S.S.: Pfair scheduling of generalized pinwheel task systems. IEEE Trans. Comput. **47**(7), 812–816 (1998)
3. Chan, M.Y., Chin, F.: General schedulers for the pinwheel problem based on double-integer reduction. IEEE Trans. Comput. **41**(6), 755–768 (1992)
4. Chan, M.Y., Chin, F.: Schedulers for larger classes of pinwheel instances. Algorithmica **9**(5), 425–462 (1993)
5. Dhall, S.K., Liu, C.L.: On a real-time scheduling problem. Oper. Res. **26**(1), 127–140 (1978)
6. Fishburn, P.C., Lagarias, J.C.: Pinwheel scheduling: achievable densities. Algorithmica **34**(1), 14–38 (2002)
7. Holte, R., Mok, A., Rosier, L., Tulchinsky, I., Varvel, D.: The pinwheel: a real-time scheduling problem. In: Proceedings of the 22nd Hawaii International Conference on System Science, pp. 693–702 (1989)
8. Holte, R., Rosier, L., Tulchinsky, I., Varvel, D.: Pinwheel scheduling with two distinct numbers. Theor. Comput. Sci. **100**(1), 105–135 (1992)
9. Lin, S.S., Lin, K.J.: A pinwheel scheduler for three distinct numbers with a tight schedulability bound. Algorithmica **19**(4), 411–426 (1997)
10. Liu, C.L., Layland, J.W.: Scheduling algorithms for multiprogramming in a hard real-time environment. J. ACM **20**(1), 46–61 (1973)
11. Romer, T.H., Rosier, L.E.: An algorithm reminiscent of euclidean-gcd for computing a function related to pinwheel scheduling. Algorithmica **17**, 1–10 (1997)

Constant-Factor Greedy Algorithms for the Asymmetric p-Center Problem in Parameterized Complete Digraphs

Wei Ding[1]($\boxtimes$) and Ke Qiu[2]

[1] Zhejiang University of Water Resources and Electric Power,
Hangzhou 310018, Zhejiang, China
dingweicumt@163.com
[2] Department of Computer Science, Brock University,
St. Catharines, Canada
kqiu@brocku.ca

Abstract. This paper studies the **asymmetric p-center problem** (**ApCP**) and the **vertex-weighted asymmetric p-center problem** (**WApCP**) in complete digraphs (CD) satisfying the triangle inequality. First, we propose two classes of *parameterized complete digraphs*, α-CD and $\langle \alpha, \beta \rangle$-CD from the angle of the parameterized upper bound on the ratio of two asymmetric edge-weights between two different vertices and on the ratio of two vertex-weights, respectively. Using the greedy method, we design a $(1 + \alpha)$-approximation algorithm for the ApCP in α-CD's and a $(1 + \alpha\beta)$-approximation algorithm for the WApCP in $\langle \alpha, \beta \rangle$-CD's, respectively.

Keywords: Asymmetric p-center · Greedy Algorithm · Parameterized graph

1 Introduction

Let $G = (V, E, w)$ be an undirected graph, where V is the set of n vertices, E is the set of m edges, and $w : E \to \mathbb{R}^+$ is an edge-weight function. Given a subset $X \subset V$ and a vertex $v \in V$, the *distance* from X to v, $w(X, v)$, is the distance from the closest vertex in X to v. The **discrete (vertex) p-center problem** (**DpCP**) asks for a subset $X \subset V$ with $|X| \leq p$ such that the maximum distance from X to all vertices is minimized. The vertices in X are called the *facilities*. Furthermore, if the facilities can be on the edges, the DpCP is generalized to the **continuous p-center problem** (**CpCP**). Both of them were proposed by Hakimi [9,10], and were proved to be NP-hard [15]. Without the triangle inequality, they are both NP-hard to approximate within any factor. They have many applications in real-world problems, e.g., establishing p emergency facilities (hospitals or fire stations), and have been extensively studied [1,2,4,7,8,11–15,17–19].

D.-Z. Du et al. (Eds.): AAIM 2019, LNCS 11640, pp. 62–71, 2019.
https://doi.org/10.1007/978-3-030-27195-4_6

1.1   Exact Algorithms

The DpCP admits an $O(n^{p+1}/(p-1)!)$-time exhaustive algorithm. For the CpCP in vertex-weighted and vertex-unweighted graphs, Kariv and Hakimi [15] gave an $O(m^p n^{2p-1} \log n/(p-1)!)$- and an $O(m^p n^{2p-1}/(p-1)!)$-time exact algorithm, respectively. For $p = 1$, they gave an $O(mn \log n)$- and an $O(mn + n^2 \log n)$-time algorithm, respectively, when the distance matrix is available. In [5], they considered a *variant* of the C1CP with a subset $S \subset V$ of s terminals in vertex-unweighted graphs and obtained an $O(s(n-s)^2 + s^2(n-s))$-time exact algorithm. For the CpCP in vertex-weighted graphs, Tamir [19] presented an $O(m^p n^p \alpha(n) \log n)$-time exact algorithm, where $\alpha(n)$ is the inverse Ackermann function. Recently, Bhattacharya and Shi [2] developed a significantly improved algorithm with time complexity of $O(m^p n^{p/2} 2^{\log^* n} \log n)$, where $\log^* n$ denotes the iterated logarithm of n, based on the observation that the CpCP can be transformed to the well-known Klee's measure problem [3]. Specifically, its time complexity is $O(m^2 n \log^2 n)$ when $p = 2$.

1.2   Approximation Algorithms

It is known that the DpCP in a connected undirected graph is equivalent to that in the corresponding complete graph, $G_c = (V, E_c, d)$, where E_c is the set of $\frac{1}{2}n(n-1)$ edges and the edge-weight function $d : V \times V \to \mathbb{R}^+ \cup \{0\}$ satisfies the *triangle inequality*. For the DpCP in vertex-unweighted G_c, Hsu and Nemhauser [14] proved that it is NP-hard to find $(2-o(1))$-approximation unless P = NP. In [12], Hochbaum and Shmoys gave a (best possible) 2-approximation algorithm with time complexity of $O(m \log m)$. They also studied the *weighted* version of DpCP [13], where each vertex has a weight representing the opening cost and an upper bound on the total budget for opening facilities is given, and designed a 3-approximation algorithm, which was shown to be tight [4]. In [16], Liang considered the *restricted* version of DpCP, where only one part of vertices admit facilities, and devised a 2-approximation algorithm. For the DpCP in vertex-weighted G_c, Dyer and Frieze [7] presented an $O(np)$-time approximation algorithm with a factor of $\min\{3, 1+\alpha\}$, where α is the maximum ratio between vertex weights. In [18], Plesník designed an $O(n^2 \log n)$-time 2-approximation algorithm.

For the CpCP in vertex-unweighted undirected graphs, it is proved by Hsu and Nemhauser [14] to be NP-hard to find a $(\frac{3}{2} - \epsilon)$-approximation, for any $\epsilon > 0$, unless P = NP. In [18], Plesník questioned whether or not it is NP-hard to find a $(2 - \epsilon)$-approximation. More recently, Ding and Qiu [6] considered the *restricted* version of CpCP with only a subset of p terminals, and presented an $O(mp^2 \log m)$-time 2-approximation algorithm. For the CpCP in vertex-weighted undirected graphs, Plesník [18] showed that the $O(n^2 \log n)$-time 2-approximation algorithm for DpCP is also a 4-approximation algorithm for CpCP, and further developed an $O(mn^2 \log n)$-time 2-approximation algorithm.

If a complete digraph allows the two edge-weights between two vertices to be different, it is called *asymmetric*. The **asymmetric p-center problem (ApCP)** is actually the DpCP in asymmetric complete digraphs satisfying the (directed) triangle inequality. For the ApCP, Panigrahy and Vishwanathan [17] gave the first approximation algorithm with a factor of $O(\log^* n)$. In [1], Archer designed two elegant $O(\log^* p)$-approximation algorithms using many ideas in [17]. The ApCP was proved to be NP-hard to approximate within a factor of $\log^* n - \Theta(1)$ unless NP $\subseteq$ DTIME($n^{\log \log n}$) in [4,11]. Later, Gørtz and Wirth [8] considered the *weighted* version of ApCP with the total budget of opening facilities bounded, and designed an $O(\log^* n)$-approximation algorithm.

1.3   Our Works

In this paper, we propose two classes of parameterized complete digraphs. If a complete digraph satisfies that the ratio of two asymmetric edge-weights between two different vertices is always bounded by a parameter $\alpha \geq 1$, it is called an *α-complete digraph* (α-CD). For the ApCP in α-CD's, we design a greedy approximation algorithm with a factor of $1 + \alpha$. If a vertex-weighted complete digraph is an α-CD and satisfies that the ratio of two vertex-weights is always bounded by a parameter $\beta \geq 1$, it is called an *$\langle \alpha, \beta \rangle$-complete digraph* ($\langle \alpha, \beta \rangle$-CD). This paper also studies the **vertex-weighted asymmetric p-center problem (WApCP)** in $\langle \alpha, \beta \rangle$-CD's, and develops a $(1 + \alpha\beta)$-approximation greedy algorithm. Both of them are constant-factor approximation algorithms when α and β are constants, which break the barrier $\log^* n$ on the approximability of the ApCP in complete digraphs satisfying the triangle inequality [4].

Organization. The rest of this paper is organized as follows. In Sect. 2, we define the ApCP and WApCP formally, provide some observations and define parameterized complete digraphs. In Sect. 3, we present a greedy approximation algorithm for the ApCP in α-CD's. In Sect. 4, we develop a greedy approximation algorithm for the WApCP in $\langle \alpha, \beta \rangle$-CD's. In Sect. 5, we conclude this paper.

2   Preliminaries

2.1   Definitions and Notations

Let $D_c = (V, A_c, d)$ be a complete digraph, where $V = \{v_1, v_2, \ldots, v_n\}$ is the set of n vertices, A_c is the set of $n(n-1)$ directed edges, and the edge-weight function $d : V \times V \to \mathbb{R}^+ \cup \{0\}$ must be *metric* (satisfying the *triangle inequality*), i.e., $d(v_i, v_j) \leq d(v_i, v_k) + d(v_k, v_j), \forall i, j, k$, and $d(v_i, v_i) = 0$. Let $D_c^\rho = (V, A_c, d, \rho)$ be a vertex-weighted complete digraph, where $\rho : V \to \mathbb{R}^+ \cup \{0\}$ is a vertex-weight function. Unless otherwise specified, D_c and D_c^ρ always denote such complete digraphs, respectively, in the rest of this paper.

A set of cardinality of q is called a *q-set*. Let $\mathcal{C} = \{c_1, c_2, \ldots, c_q\} \subset V$ be a q-set of V with no duplicate. The *distance* from $\mathcal{C}$ to v_i, $d(\mathcal{C}, v_i)$, is referred to as the shortest distance from one vertex in $\mathcal{C}$ to v_i, for any $1 \leq i \leq n$.

So, $d(\mathcal{C}, v_i) = \min_{1 \leq j \leq q} d(c_j, v_i)$. The maximum distance from $\mathcal{C}$ to V is called the *radius* of $\mathcal{C}$, denoted by $r(\mathcal{C})$. Given a complete digraph, $D_c = (V, A_c, d)$, the **asymmetric p-center problem (ApCP)** asks for a subset of V with cardinality at most p such that the radius is minimized. Let $\mathcal{C}^*$ be an optimal solution to ApCP in D_c. We have

$$r(\mathcal{C}) = \max_{1 \leq i \leq n} d(\mathcal{C}, v_i), \tag{1}$$

and

$$r(\mathcal{C}^*) = \min_{\mathcal{C} \subseteq V, |\mathcal{C}| \leq p} r(\mathcal{C}). \tag{2}$$

The *weighted distance* from $\mathcal{C}$ to v_i is defined as $\rho(v_i) d(\mathcal{C}, v_i)$, and the maximum weight distance from $\mathcal{C}$ to V is called *weighted radius* of $\mathcal{C}$, denoted by $wr(\mathcal{C})$. Given a vertex-weighted complete digraph, $D_c^\rho = (V, A_c, d, \rho)$, the **weighted asymmetric p-center problem (WApCP)** asks for a subset of V such that the weighted radius is minimized. Let $\mathcal{C}_\rho^*$ be an optimum to WApCP in D_c^ρ. We have

$$wr(\mathcal{C}) = \max_{1 \leq i \leq n} \rho(v_i) d(\mathcal{C}, v_i), \tag{3}$$

and

$$wr(\mathcal{C}_\rho^*) = \min_{\mathcal{C} \subseteq V, |\mathcal{C}| \leq p} wr(\mathcal{C}). \tag{4}$$

Obviously, the classic ApCP is just a special case of WApCP where all the vertices have a uniform weight. The focus of this paper is to study the WApCP in a vertex-weighted complete digraph.

2.2   Observations

Let $D = (V, A, w)$ be a digraph, where $V = \{v_1, v_2, \ldots, v_n\}$ is the set of n vertices, A is the set of m directed edges, and each directed edge $(v, u) \in A$ has a nonnegative weight $w(a) \geq 0$. Let $d(v_i, v_j)$ denote the v_i-to-v_j *shortest path distance* (SPD) in D, for any pair of vertices, v_i and v_j. In general, $d(v_i, v_j) \neq d(v_j, v_i)$. Therefore, the *shortest path graph* (SPG) induced by D is in general an *asymmetric* complete digraph. In rare cases where $d(v_i, v_j) = d(v_j, v_i), \forall i, j$, the corresponding SPG is called *symmetric* and is essentially an undirected complete graph. Of course, the SPG induced by an undirected graph is also an undirected complete graph. Furthermore, we let the weight of directed edge (v, u) in SPG be ∞ if v is not connected to u when D is not strongly connected.

The following two observations point out a way of solving the DpCP in general (vertex-weighted) digraphs.

Observation 1. *Let $D = (V, A, w)$ be a digraph and $D_c = (V, A_c, d)$ be the corresponding SPG induced by D. If D_c is asymmetric then the DpCP in D is equivalent to the ApCP in D_c, and the DpCP in G_c otherwise. Here, G_c is the corresponding undirected complete graph.*

Observation 2. *Let $D^\rho = (V, A, w, \rho)$ be a vertex-weighted digraph and $D_c^\rho = (V, A_c, d, \rho)$ be the corresponding SPG induced by D^ρ. If D_c^ρ is asymmetric then the vertex-weighted version of DpCP in D^ρ is equivalent to the WApCP in D_c^ρ.*

2.3   Parameterized Complete Digraphs

In Sect. 2.2, it is shown in observations that the DpCP in a general digraph can be reduced to ApCP in the corresponding complete digraph in general. In this section, we consider several classes of *parameterized* complete digraphs, both of which arise from the real-world problems.

In a general strongly connected digraph, there are cases where the ratio between the v-to-u distance and the u-to-v distance is bounded by a parameter, for any pair of vertices, v and u. This inspires us to consider a single-parameterized complete digraph as follows.

Definition 1. *Given a complete digraph $D_c = (V, A_c, d)$ and a real number $\alpha \geq 1$, if $d : V \times V \to \mathbb{R}^+ \cup \{0\}$ is metric and satisfies that $d(v_i, v_j) \leq \alpha \cdot d(v_j, v_i), \forall i, j$, then D_c is called an α-**complete digraph**, abbreviated as α-CD and denoted by D_α.*

Furthermore, when the ratio between the weight of vertex v and the weight of vertex u is also bounded by a parameter, we propose the following double-parameterized complete digraph based on Definition 1.

Definition 2. *Given a vertex-weighted complete digraph $D_c^\rho = (V, A_c, d, \rho)$ and two real numbers, $\alpha \geq 1$ and $\beta \geq 1$, if D_c^ρ is an α-CD and satisfies that $\frac{\rho(v_i)}{\rho(v_j)} \leq \beta, \forall i, j$, then D_c^ρ is called an $\langle \alpha, \beta \rangle$-**complete digraph**, abbreviated as $\langle \alpha, \beta \rangle$-CD and denoted by D_α^β.*

3   A $(1 + \alpha)$-Approximation to ApCP in α-CD

In this section, we study ApCP in α-CD's with $\alpha \geq 1$, and design a $(1 + \alpha)$-approximation greedy algorithm, called ApCP-CDA, based on the framework in [12,18]. Let $\mathcal{C}_1^*$ be an optimal solution to ApCP, and opt_1 be the optimal value. Also, we let $\mathcal{C}_1^A$ be an algorithm solution of ApCP-CDA.

3.1   A Test Procedure Using Greedy Method

In this subsection, we use greedy method to design a test procedure, called TEST_1, which will play an important role in ApCP-CDA. The input of TEST_1 consists of an α-CD with $\alpha \geq 1$, $D_\alpha = (V, A_c, d)$, and a real number $\lambda > 0$. So, TEST_1 has two input parameters, $\alpha \geq 1$ and $\lambda > 0$. $\mathsf{TEST}_1(\alpha, \lambda)$ picks out unlabelled vertices successively to obtain a solution, denoted by $\mathcal{C}_1(\lambda)$. During the process of obtaining $\mathcal{C}_1(\lambda)$, we let $\mathcal{U}$ denote the set of unlabelled vertices. Initially, set $\mathcal{C}_1(\lambda) = \emptyset$ and $\mathcal{U} = V$. TEST_1 uses greedy method to select an unlabelled vertex into $\mathcal{C}_1(\lambda)$ and then label vertices. Specifically, every time an unlabelled vertex $\widehat{u} \in \mathcal{U}$ is selected arbitrarily into $\mathcal{C}_1(\lambda)$, each unlabelled vertex $u \in \mathcal{U}$ satisfying that $d(\widehat{u}, u) \leq (1 + \alpha)\lambda$ is labelled. Repeat above operation until $\mathcal{U} = \emptyset$ or $|\mathcal{C}_1(\lambda)| = p$. If TEST$_1$ terminates with $\mathcal{U} = \emptyset$, then TEST$_1$ finds a subset $\mathcal{C}_1(\lambda) \subset V$ with cardinality at most p, satisfying that $r(\mathcal{C}_1(\lambda)) \leq (1+\alpha)\lambda$, and so returns YES. Otherwise, TEST$_1$ fails and returns NO.

Theorem 1. *Given $D_\alpha = (V, A_c, d)$ with $\alpha \geq 1$, TEST$_1(\alpha, \lambda)$ can find a subset $\mathcal{C}_1(\lambda) \subset V$ with $|\mathcal{C}_1(\lambda)| \leq p$ satisfying that $r(\mathcal{C}_1(\lambda)) \leq (1 + \alpha)\lambda$ in $O(pn)$ time for the ApCP in D_α if there is a subset $\mathcal{F}_1 \subset V$ with $|\mathcal{F}_1| \leq p$ satisfying that $r(\mathcal{F}_1) \leq \lambda$, for any real number $\lambda > 0$.*

TEST$_1(\alpha, \lambda)$: Test Procedure.
Input: $D_\alpha = (V, A_c, d)$ with $\alpha \geq 1$, and $\lambda \in \mathbb{R}^+$.
Output: NO or YES with $\mathcal{C}_1(\lambda) \subset V$.
01: $\mathcal{C}_1(\lambda) \leftarrow \emptyset; \mathcal{U} \leftarrow V;$
02: **while** $\mathcal{U} \neq \emptyset$ and $\|\mathcal{C}_1(\lambda)\| < p$ **do**
03:    Select a vertex $\widehat{u} \in \mathcal{U}$ arbitrarily;
04:    $\mathcal{C}_1(\lambda) \leftarrow \mathcal{C}_1(\lambda) \cup \{\widehat{u}\};$
05:    $N(\widehat{u}; (1 + \alpha)\lambda) \leftarrow \{u \in \mathcal{U} \| d(\widehat{u}, u) \leq (1 + \alpha)\lambda\};$
06:    $\mathcal{U} \leftarrow \mathcal{U} \setminus N(\widehat{u}; (1 + \alpha)\lambda);$
07: **end**
08: **if** $\mathcal{U} \neq \emptyset$ **then** Return NO;
09: **else** Return YES and $\mathcal{C}_1(\lambda);$ **endif**

By letting $\mathcal{F}_1 = \mathcal{C}_1^*$ and $\lambda = opt_1$ in Theorem 1, we conclude from $r(\mathcal{C}_1^*) \leq opt_1$ that TEST$_1(\alpha, opt_1)$ can find a subset $\mathcal{C}_1' \subset V$ with $|\mathcal{C}_1'| \leq p$ satisfying that $r(\mathcal{C}_1') \leq (1 + \alpha)opt_1$. Immediately, we obtain the following corollary. TEST$_1$ surely can find a $(1 + \alpha)$-approximation to the ApCP in α-CD's.

Corollary 1. *Given $D_\alpha = (V, A_c, d)$ with $\alpha \geq 1$, TEST$_1(\alpha, opt_1)$ can produce a $(1 + \alpha)$-approximation to the ApCP in D_α.*

3.2 A $(1 + \alpha)$-Approximation Algorithm

Based on the test procedure, we design a constant-factor approximation algorithm, called ApCP-CDA, for the ApCP in α-CD's. Given an α-CD with n vertices, there are $n(n - 1)$ possible values of radius, i.e., $d(v_i, v_j), \forall i, j, i \neq j$. We delete the duplicates in all these values and arrange the remaining s values into an increasing sequence, $g_1 < g_2 < \cdots < g_s$. Clearly, $s \leq n(n - 1)$.

In Step 1 of ApCP-CDA, the above sequence is obtained by sorting all the remaining values. Let α' be the *tight* upper bound on the ratio of two asymmetric edge-weights between two different vertices in the given α-CD, i.e.,

$$\alpha' = \max_{1 \leq i, j \leq n, i \neq j} \frac{d(v_i, v_j)}{d(v_j, v_i)}. \tag{5}$$

Clearly, $\alpha' \leq \alpha$. In Step 2, the value of α' is determined. The key task is to find the smallest one, g_{k^*}, in $g_1, g_2, \ldots, g_s$, which makes $\mathsf{TEST}_1(\alpha', g_k)$ return YES. Of course, we can use a straightforward *sequential search* to find g_{k^*}. Every search calls TEST_1 once and as a result the whole search calls TEST_1 $O(s)$ times. Technically, we can apply a *binary search* to $g_1, g_2, \ldots, g_s$ to find g_{k^*}, as described in Step 3 of ApCP-CDA. It is sufficient to apply TEST_1 $O(\log s)$ times, which reduces $O(s)$ times greatly.

The binary search is described as follows. If $\mathsf{TEST}_1(\alpha', g_1)$ returns YES, then we know that $g_{k^*} = g_1$ and the output is $\mathcal{C}_1(g_1)$, i.e., $\mathcal{C}_1^A = \mathcal{C}_1(g_1)$. Otherwise, we use a binary search to find g_{k^*}. Let LB and UB be the lower bound and upper bound on k^*, respectively. Initially, set LB $= 1$ and UB $= s$. Let B $= \lfloor \frac{\text{LB}+\text{UB}}{2} \rfloor$. If $\mathsf{TEST}_1(\alpha', g_B)$ returns YES, then B becomes a new upper bound and LB remains the lower bound. If $\mathsf{TEST}_1(\alpha', g_B)$ returns NO, then B becomes a new lower bound and UB remains the upper bound. Repeat above operation until UB and LB become two consecutive integers. In the whole operation, $\mathsf{TEST}_1(\alpha', g_{\text{UB}})$ always returns YES while $\mathsf{TEST}_1(\alpha', g_{\text{LB}})$ always returns NO. Hence, the final UB is just k^*, and the output is $\mathcal{C}_1(g_{\text{UB}})$, i.e., $\mathcal{C}_1^A = \mathcal{C}_1(g_{\text{UB}})$.

ApCP-CDA: Algorithm for ApCP in D_α with $\alpha \geq 1$.

Input: $D_\alpha = (V, A_c, d)$ with $\alpha \geq 1$.

Output: $\mathcal{C}_1^A \subset V$ with $|\mathcal{C}_1^A| \leq p$.

Step 1: Delete the duplicates in $d(v_i, v_j), 1 \leq i, j \leq n, i \neq j$, and arrange the remaining values into an increasing sequence, $g_1 < g_2 < \cdots < g_s$.

Step 2: $\alpha' \leftarrow \max_{1 \leq i,j \leq n, i \neq j} \frac{d(v_i, v_j)}{d(v_j, v_i)}$;

Step 3: if $\mathsf{TEST}_1(\alpha', g_1) = $ YES **then**
    Return $\mathcal{C}_1(g_1)$;
else
    LB $\leftarrow 1$; UB $\leftarrow s$;
    while UB $-$ LB > 1 **do**
      B $\leftarrow \lfloor \frac{\text{LB}+\text{UB}}{2} \rfloor$;
      if $\mathsf{TEST}_1(\alpha', g_B) = $ NO **then** LB $\leftarrow$ B;
      else $\mathsf{TEST}_1(\alpha', g_B) = $ YES **then** UB $\leftarrow$ B;
      endif
    end
    Return $\mathcal{C}_1(g_{\text{UB}})$;
endif

Theorem 2. *Given $D_\alpha = (V, A_c, d), \alpha \geq 1$ with n vertices, ApCP-CDA can produce a $(1 + \alpha)$-approximation to the ApCP in D_α within $O(n^2 \log n)$ time.*

Moreover, we can obtain that the performance factor of ApCP-CDA is actually $1 + \alpha'$ for the ApCP in the given α-CD.

4   WApCP in $\langle \alpha, \beta \rangle$-CD

In this section, we study the WApCP in $\langle \alpha, \beta \rangle$-CD's with $\alpha \geq 1$ and $\beta \geq 1$. Let opt_2 and $\mathcal{C}_2^*$ be the optimal value and the optimum, respectively. By adjusting ApCP-CDA to the WApCP in $\langle \alpha, \beta \rangle$-CD's, we derive a constant-factor approximation algorithm, called ApCP-CAG. Let $\mathcal{C}_2^A$ be an algorithm solution of ApCP-CAG.

The framework of ApCP-CAG is the same as ApCP-CDA. Let TEST_2 denote the test procedure of ApCP-CAG. The key difference between TEST_2 and TEST_1 is that every unlabelled vertex $u \in \mathcal{U}$ satisfying that $\rho(u)d(\widehat{u}, u) \leq (1 + \alpha\beta)\lambda$ is labelled when $\widehat{u} \in \mathcal{U}$ is selected as a facility. Let $N(\widehat{u}; (1 + \alpha\beta)\lambda)$ be the subset of such vertices to be labelled. By replacing $N(\widehat{u}; (1+\alpha)\lambda)$ in TEST_1 with $N(\widehat{u}; (1+\alpha\beta)\lambda)$, we obtain TEST_2. Obviously, TEST_2 has three parameters (real numbers), $\alpha \geq 1$, $\beta \geq 1$ and $\lambda > 0$. Let $\mathcal{C}_2(\lambda)$ be the output of $\mathsf{TEST}_2(\alpha, \beta, \lambda)$. Accordingly, Theorem 1 is adjusted to the following one.

Theorem 3. *Given $D_\alpha^\beta = (V, A_c, d, \rho)$ with $\alpha \geq 1, \beta \geq 1$, $\mathsf{TEST}_2(\alpha, \beta, \lambda)$ can find a subset $\mathcal{C}_2(\lambda) \subset V$ with $|\mathcal{C}_2(\lambda)| \leq p$ satisfying that $wr(\mathcal{C}_2(\lambda)) \leq (1 + \alpha\beta)\lambda$ in $O(pn)$ time for the WApCP in D_α^β if there is a subset $\mathcal{F}_2 \subset V$ with $|\mathcal{F}_2| \leq p$ satisfying that $wr(\mathcal{F}_2) \leq \lambda$, for any real number $\lambda > 0$.*

Let β' be the tight upper bound on the ratio between any two difference vertex-weights in the given $\langle \alpha, \beta \rangle$-CD, i.e.,

$$\beta' = \max_{1 \leq i, j \leq n, i \neq j} \frac{\rho(v_i)}{\rho(v_j)}. \tag{6}$$

Clearly, $\beta' \leq \beta$. In Step 2 of ApCP-CAG, the value of β' is computed, additionally, which requires $O(n^2)$ time.

Given an $\langle \alpha, \beta \rangle$-CD with n vertices, there are $n(n-1)$ possible values of weighted radius, i.e., $\rho(v_j)d(v_i, v_j), \forall i, j, i \neq j$. We delete the duplicates in all the possible values and arrange the remaining t values into an increasing sequence, $h_1 < h_2 < \cdots < h_t$. Clearly, $t \leq n(n-1)$. Step 1 of ApCP-CAG computes this sequence instead of $g_1, g_2, \ldots, g_s$. Let $h_{k_\triangle}$ be the smallest value in $h_1, h_2, \ldots, h_t$ such that $\mathsf{TEST}_2(\alpha', \beta', h_k)$ returns YES. Similarly, we can prove that $h_{k_\triangle} \leq opt_2$. Combining with Theorem 3, we conclude that

$$wr(\mathcal{C}_2^A) = wr(\mathcal{C}_2(h_{k_\triangle})) \leq (1 + \alpha'\beta')h_{k_\triangle} \leq (1 + \alpha'\beta')opt_2 \leq (1 + \alpha\beta)opt_2. \tag{7}$$

Theorem 4. *Given $D_\alpha^\beta = (V, A_c, d, \rho)$ with $\alpha \geq 1$ and $\beta \geq 1$, ApCP-CAG can produce a $(1+\alpha\beta)$-approximation to the WApCP in D_α^β within $O(n^2 \log n)$ time.*

Furthermore, we can obtain that the performance factor of ApCP-CAG is actually $1 + \alpha'\beta'$ for the WApCP in the given $\langle \alpha, \beta \rangle$-CD.

5   Conclusions

This paper studied the classic ApCP and the vertex-weighted version of ApCP (WApCP). First, we proposed two classes of parameterized complete digraphs, α-CD and $\langle \alpha, \beta \rangle$-CD, from the prospective of the parameterized upper bound on the ratio of two asymmetric edge-weights between two different vertices and on the ratio of two vertex-weights, respectively. For the classic ApCP in α-CD's, we present a $(1 + \alpha)$-approximation greedy algorithm. For the WApCP in $\langle \alpha, \beta \rangle$-CD's, we develop a $(1 + \alpha\beta)$-approximation greedy algorithm.

In [4], the ApCP in general complete digraphs satisfying the triangle inequality is proved to be NP-hard to approximate within a factor of $\log^* n - \Theta(1)$. It is also of great interest to study the inapproximability of the WApCP in general vertex-weighted complete digraphs satisfying the triangle inequality.

References

1. Archer, A.: Two $O(\log^* k)$-approximation algorithms for the asymmetric k-center problem. In: Aardal, K., Gerards, B. (eds.) IPCO 2001. LNCS, vol. 2081, pp. 1–14. Springer, Heidelberg (2001). https://doi.org/10.1007/3-540-45535-3_1
2. Bhattacharya, B., Shi, Q.S.: Improved algorithms to network p-center location problems. Comput. Geom. **47**, 307–315 (2014)
3. Chan, T.M.: A (slightly) faster algorithm for Klees measure problem. In: Proceedings of the 24th ACM SoCG, pp. 94–100 (2008)
4. Chuzhoy, J., Guha, S., Halperin, E., Kortsarz, G., Khanna, S., Naor, S.: Asymmetric k-center is $\log^* n$-hard to approximate. In: Proceedings of the 36th STOC, pp. 21–27 (2004)
5. Ding, W., Qiu, K.: Algorithms for the minimum diameter terminal steiner tree problem. J. Comb. Optim. **28**(4), 837–853 (2014)
6. Ding, W., Qiu, K.: Minimum diameter k-steiner forest. In: Proceedings of the 12th AAIM, pp. 1–11 (2018)
7. Dyer, M.E., Frieze, A.M.: A simple heuristic for the p-centre problem. Oper. Res. Lett. **3**, 285–288 (1985)
8. Gørtz, I.L., Wirth, A.: Asymmetry in k-center variants. Theor. Comput. Sci. **361**, 188–199 (2006)
9. Hakimi, S.L.: Optimum locations of switching centers and the absolute centers and medians of a graph. Oper. Res. **12**(3), 450–459 (1964)
10. Hakimi, S.L.: Optimal distribution of switching centers in a communications network and some related graph-theoretic problems. Oper. Res. **13**, 462–475 (1965)
11. Halperin, E., Kortsarz, G., Krauthgamer, R.: Tight lower bounds for the asymmetric k-center problem. Technical report, 03-035, Electronic Colloquium on Computational Complexity (2003)
12. Hochbaum, D.S., Shmoys, D.B.: A best possible heuristic for the k-center problem. Math. Oper. Res. **10**(2), 180–184 (1985)
13. Hochbaum, D.S., Shmoys, D.B.: A unified approach to approximation algorithms for bottleneck problems. J. ACM **33**, 533–550 (1986)
14. Hsu, W.L., Nemhauser, G.L.: Easy and hard bottleneck location problems. Discrete Appl. Math. **1**, 209–215 (1979)

15. Kariv, O., Hakimi, S.L.: An algorithmic approach to network location problems. I: the p-centers. SIAM J. Appl. Math. **37**(3), 513–538 (1979)
16. Liang, H.Y.: The hardness and approximation of the star p-hub center problem. Oper. Res. Lett. **41**, 138–141 (2013)
17. Panigrahy, R., Vishwanathan, S.: An $O(\log^* n)$ approximation algorithm for the asymmetric p-center problem. J. Algorithms **27**, 259–268 (1998)
18. Plesník, J.: A heuristic for the p-center problem in graphs. Discrete Appl. Math. **17**, 263–268 (1987)
19. Tamir, A.: Improved complexity bounds for center location problems on networks by using dynamic data structures. SIAM J. Discrete Math. **1**(3), 377–396 (1988)

Updating Matrix Polynomials

Wei Ding[1(✉)] and Ke Qiu[2]

[1] Zhejiang University of Water Resources and Electric Power,
Hangzhou 310018, Zhejiang, China
`dingweicumt@163.com`
[2] Department of Computer Science, Brock University, St. Catharines, Canada
`kqiu@brocku.ca`

Abstract. Given a square matrix $M = (u_{ij})_{n \times n}$ and an m-order matrix polynomial $f_m(M) = \sum_{k=0}^{m} a_k M^k = a_0 I + a_1 M + a_2 M^2 + \cdots + a_m M^m$, if M is a dense matrix and is perturbed to become M' at a single entry, say u_{pq}, a straightforward re-calculation of $f_m(M')$ would require $O(n^\omega \cdot \alpha(m))$ arithmetic operations, where $\omega < 2.3728639$ and $\alpha(m)$ depends on the strategy of computing $M'^k, 1 \leq k \leq m$ appearing in $f_m(M')$, using the fastest square matrix multiplication algorithm by François Le Gall (ISSAC'14). In this paper, we assume that M is a *dense* matrix and that $f_m(M)$ is known while no other additional information is available. From the perspective of the naive (a.k.a., standard *row-by-column*) matrix multiplication, we discuss the update of matrix polynomials. First, we present $O(n)$-, $O(n^2)$- and $O(n^2)$-operations update algorithms for 2-order, 3-order and 4-order matrix polynomials, respectively. Furthermore, we discuss the update of high-order matrix polynomials with a *sparse* coefficient vector and as a result, propose a *combinatorial* heuristic updating method based on *directed Steiner tree* in a directed acyclic graph.

Keywords: Matrix polynomials · Update · Directed Steiner tree

1 Introduction

1.1 Motivation

A frequently occurring problem in control theory and some other application areas is that of computing a function $f(M)$ of an n by n matrix M [7]. A rich class of such natural matrix functions includes polynomial functions of a matrix [9]. Let $f_m(x) = \sum_{k=0}^{m} a_k x^k = a_0 + a_1 x + a_2 x^2 + \cdots + a_m x^m$ be an m-order polynomial, where $m \geq 0$, and $M = (u_{ij})_{n \times n}$ be an n by n square matrix, the *matrix polynomial* [1,7,9] is defined as

$$f_m(M) = \sum_{k=0}^{m} a_k M^k = a_0 I + a_1 M + a_2 M^2 + \cdots + a_m M^m.$$

© Springer Nature Switzerland AG 2019
D.-Z. Du et al. (Eds.): AAIM 2019, LNCS 11640, pp. 72–82, 2019.
https://doi.org/10.1007/978-3-030-27195-4_7

The minimal and characteristic polynomials are examples of matrix polynomials. Also, the approximation of transcendental matrix functions often involves the evaluation of matrix polynomials [7]. Due to many applications of matrix polynomials, it is important to compute matrix polynomials efficiently. A special case of matrix polynomials is power of matrices. It is also important to compute power of matrices as matrices can be used to represent (weighted) graphs and the power of matrices (and their closures) are related to weights of all walks of certain lengths [9] and lengths of shortest paths between two vertices using at most a certain number of edges [8].

Suppose, for a square matrix M and a polynomial $f(x)$, we have evaluated $f(M)$. If M is changed to M', it is necessary to evaluate $f(M')$. One could certainly re-compute the matrix polynomial at M' from scratch (static re-evaluation). However, when there exist only a bit of perturbations from M to M', it is costly to re-evaluate f at M'. So, it becomes very important to develop a *dynamic* algorithm for updating $f(M)$ to $f(M')$ which runs faster than re-evaluation. The problem of updating $f(M)$ to $f(M')$ is named **updating matrix polynomials problem (UMPP)**.

For instance, when one system involves the computation of a collection of polynomials with high-order square matrices as input and the matrices are perturbed regularly, the dynamic update of matrix polynomials will reduce the time cost greatly than static re-evaluation. In the rest of this paper, M is always a *dense* matrix without otherwise specified, and $f(M)$ is known while no other additional information is available.

1.2   Related Works

For matrix computations, a good example is the Sherman-Morrison formula with one of its main applications being to compute the inverse of a matrix A (assuming that A^{-1} exists and is known) when A is corrected (perturbed) to $A+uv^T$, where u and v are column vectors, cheaply, without having to compute $(A + uv^T)^{-1}$ from scratch. In [13], Sankowski studied many dynamic matrix problems. Reif and Tate considered the problem of incrementally evaluating algebraic functions which include *discrete Fourier transform, multipoint polynomial evaluation, and matrix-matrix product*, among many others [12]. The incremental algorithm is to quickly process on-line requests such as "change the input value x to x'". General methods are developed which result in many incremental algorithms for solving a host of problems. In addition, they also gave lower bounds on time costs of incremental algorithms for the problems studied there. Specifically, for the matrix-matrix product problem, their general methods imply a time cost of $O(\sqrt{M(n)})$ where $M(n)$ is the best matrix multiplication time, which now stands at $O(n^\omega)$ where $\omega < 2.3728639$ [11], as well a lower bound for the problem. This lower bound has been subsequently improved to $\Omega(n)$ [5]. We refer readers to [3–6] for more other dynamic matrix or algebraic problems.

1.3   Our Results

The focus of this paper is to deal with the case of UMPP where a single entry of a given matrix changes while all the other entries stay unchanged. Let M be an n by n dense matrix and $f_m(x)$ be an m-order polynomial. When M is changed to M' at a single entry, say u_{pq}, a straightforward re-evaluation of $f_m(M')$ would require $O(n^\omega \cdot \alpha(m))$ operations, where $\omega < 2.3728639$ and $\alpha(m)$ depends on the strategy of computing $M'^k, 1 \le k \le m$ appearing in $f_m(M')$, using the fastest square matrix multiplication algorithm by François Le Gall [11]. In this paper, we discuss how to update $f_m(M)$ to $f_m(M')$ from the perspective of standard *row-by-column* matrix multiplication.

For the update of *low-order* matrix polynomials, we establish update formulas and as a result present seemingly direct but nontrivial update algorithms, i.e., $O(n)$-, $O(n^2)$- and $O(n^2)$-operations algorithms for 2-order, 3-order and 4-order matrix polynomials, respectively. Our algorithms utilize only $f_m(M)$ and need no other additional information. It is shown by computational results that update algorithms run substantially faster than the static re-evaluation. For the update of *high-order* matrix polynomials with a *sparse* coefficient vector, we propose a heuristic updating method based on *directed Steiner tree* in a directed acyclic graph.

Organization. The rest of this paper is organized as follows. We show some preliminary works in Sect. 2, and present updating algorithms for 2-, 3- and 4-order matrix polynomials in Sects. 3, 4 and 5, respectively. In Sect. 6, we offer some concluding remarks. Due to page limit, the updating method for high-order matrix polynomials and the experimental results are omitted in this paper.

2   Preliminaries

Let $f_m(x)$ be an m-order polynomial,

$$f_m(x) = \sum_{k=0}^{m} a_k x^k = a_0 + a_1 x + a_2 x^2 + \cdots + a_m x^m, \quad m \ge 0, \qquad (1)$$

and $M = (u_{ij})_{n \times n}$ be an n by n dense matrix,

$$M = \begin{pmatrix} u_{11} & \cdots & u_{1j} & \cdots & u_{1n} \\ \vdots & \ddots & \vdots & \ddots & \vdots \\ u_{i1} & \cdots & u_{ij} & \cdots & u_{in} \\ \vdots & \ddots & \vdots & \ddots & \vdots \\ u_{n1} & \cdots & u_{nj} & \cdots & u_{nn} \end{pmatrix} \qquad (2)$$

where u_{ij} is the element at the intersection of the i-th row and the j-th column. By taking M into Eq. (1), we get a matrix polynomial as follows,

$$f_m(M) = \sum_{k=0}^{m} a_k M^k = a_0 I_n + a_1 M + a_2 M^2 + \cdots + a_m M^m, \quad m \ge 0, \qquad (3)$$

where I_n is an n-order identity matrix.

Let M' be a new matrix derived from replacing u_{pq} with u'_{pq} and all the other elements staying unchanged. Set $\delta = u'_{pq} - u_{pq}$. Let $\mathbf{E}_n(p,q) = (e_{ij})_{n \times n}$ be an $n \times n$ matrix with $e_{pq} = 1$ and all the other elements equal to zero. We have

$$M' = M + \delta \mathbf{E}_n(p,q). \tag{4}$$

Let $\nabla_{pq} f_k, 1 \le k \le m$ be a function in M and δ based on $f_k(x)$,

$$\nabla_{pq} f_k(M, \delta) = f_k(M + \delta \mathbf{E}_n(p,q)) - f_k(M), \quad 1 \le p, q \le n. \tag{5}$$

Specifically, $\nabla_{pq} f_0(M, \delta) = \mathbf{0}_n$, where $\mathbf{0}_n$ is an $n \times n$ zero matrix. Lemma 1 shows the recursive equation of $\nabla_{pq} f_k(M, \delta), 1 \le k \le m$.

Lemma 1. *Given an $n \times n$ matrix M and $1 \le p, q \le n$, for any real number δ,*

$$\nabla_{pq} f_k(M, \delta) = \nabla_{pq} f_{k-1}(M, \delta) + a_k(M + \delta \mathbf{E}_n(p,q))^k - a_k M^k, \ 1 \le k \le m. \tag{6}$$

3 Update of 2-Order Matrix Polynomials

In this section, we discuss the update of 2-order matrix polynomials. We consider 2-order polynomials as follows and let $\mathcal{A}_2 = (a_0, a_1, a_2)$,

$$f_2(x) = a_0 + a_1 x + a_2 x^2.$$

Lemma 2. *Given an $n \times n$ matrix M and $1 \le p, q \le n$, for any real number δ,*

$$\nabla_{pq} f_2(M, \delta) = a_1 \delta \mathbf{E}_n(p,q) + a_2 \delta[M \mathbf{E}_n(p,q) + \mathbf{E}_n(p,q)M + \delta \mathbf{E}_n^2(p,q)]. \tag{7}$$

Lemma 3. *For any $n \times n$ matrix M and $1 \le p, q \le n$, we have*

1. $M\mathbf{E}_n(p,q) = (\alpha_{ij}^{(1)})_{n \times n} = \begin{pmatrix} 0 & \cdots & \overset{q\text{-th}}{u_{1p}} & \cdots & 0 \\ \vdots & \ddots & \vdots & \ddots & \vdots \\ 0 & \cdots & u_{np} & \cdots & 0 \end{pmatrix}$;

2. $\mathbf{E}_n(p,q)M = (\alpha_{ij}^{(2)})_{n \times n} = \begin{pmatrix} 0 & \cdots & 0 \\ \vdots & \ddots & \vdots \\ u_{q1} & \cdots & u_{qn} \\ \vdots & \ddots & \vdots \\ 0 & \cdots & 0 \end{pmatrix}$ $p\text{-th}$;

3. $\mathbf{E}_n^2(p,q) = (\varepsilon_{ij})_{n \times n} = \begin{cases} \mathbf{0}_n \ if \ p \ne q, \\ \mathbf{E}_n(p,p) \ if \ p = q. \end{cases}$

Theorem 1. *Let* $\nabla_{pq} f_2(M, \delta) = (b_{ij})_{n \times n}$. *We have*

$$b_{ij} = \begin{cases} 0, & if\ i \neq p, j \neq q, \\ \delta a_2 u_{qj}, & if\ i = p, j \neq q, \\ \delta a_2 u_{ip}, & if\ i \neq p, j = q, \end{cases} \tag{8}$$

and

$$b_{pq} = \begin{cases} \delta(a_1 + a_2(u_{pp} + u_{qq})), & if\ p \neq q, \\ \delta(a_1 + 2a_2 u_{pp}) + \delta^2 a_2, & if\ p = q. \end{cases} \tag{9}$$

Combining Eqs. (8), (9) and (5), we obtain our update algorithm, called Update-2OMP. Its output is the desired matrix polynomial $f_2(M')$.

Theorem 2. *Given an* $n \times n$ *matrix* M *and a real number* δ, *when* M *changes to* M' *by* u_{pq} *of* M *increasing by* δ *and all the other elements remaining unchanged,* Update-2OMP *updates* $f_2(M)$ *to* $f_2(M')$ *in* $O(n)$ *arithmetic operations.*

4   Update of 3-Order Matrix Polynomials

In this section, we discuss the update of 3-order matrix polynomials. We consider 3-order polynomials as follows and let $\mathcal{A}_3 = (a_0, a_1, a_2, a_3)$,

$$f_3(x) = a_0 + a_1 x + a_2 x^2 + a_3 x^3.$$

Update-2OMP($M, f_2(M), \mathcal{A}_2, (p, q), \delta$):

Let $f_2(M) = (f_{ij})_{n \times n}$;
if $p = q$ **then** $f_{pq} \leftarrow f_{pq} + \delta(a_1 + 2a_2 u_{pp}) + \delta^2 a_2$;
else $f_{pq} \leftarrow f_{pq} + \delta(a_1 + a_2(u_{pp} + u_{qq}))$; **endif**
for $j := 1$ to n **do**
    if $j \neq q$ **then** $f_{pj} \leftarrow f_{pj} + \delta a_2 u_{qj}$; **endif**
endfor
for $i := 1$ to n **do**
    if $i \neq p$ **then** $f_{iq} \leftarrow f_{iq} + \delta a_2 u_{ip}$; **endif**
endfor
Return $f_2(M)$;

Lemma 4. *Given an* $n \times n$ *matrix* M *and* $1 \leq p, q \leq n$, *for any real number* δ,

$$\nabla_{pq} f_3(M, \delta) = \nabla_{pq} f_2(M, \delta) + a_3 \delta [M^2 \mathbf{E}_n(p, q) + M \mathbf{E}_n(p, q) M$$
$$+ \mathbf{E}_n(p, q) M^2 + \delta(M \mathbf{E}_n^2(p, q) + \mathbf{E}_n(p, q) M \mathbf{E}_n(p, q) \tag{10}$$
$$+ \mathbf{E}_n^2(p, q) M) + \delta^2 \mathbf{E}_n^3(p, q)].$$

Lemma 5. *For any $n \times n$ matrix M and $1 \leq p, q \leq n$, we have*

1. $M^2 \mathbf{E}_n(p,q) = (\beta_{ij}^{(1)})_{n \times n} = \begin{pmatrix} 0 \cdots \overset{q\text{-}th}{\sum_{s=1}^{n} u_{1s}u_{sp}} \cdots 0 \\ \vdots \ddots \vdots \ddots \vdots \\ 0 \cdots \sum_{s=1}^{n} u_{ps}u_{sp} \cdots 0 \\ \vdots \ddots \vdots \ddots \vdots \\ 0 \cdots \sum_{s=1}^{n} u_{ns}u_{sp} \cdots 0 \end{pmatrix} \begin{matrix} \\ \\ p\text{-}th; \\ \\ \end{matrix}$

2. $M\mathbf{E}_n(p,q)M = (\beta_{ij}^{(2)})_{n \times n} = \begin{pmatrix} u_{1p}u_{q1} & \cdots & \overset{q\text{-}th}{u_{1p}u_{qq}} & \cdots & u_{1p}u_{qn} \\ \vdots & \ddots & \vdots & \ddots & \vdots \\ u_{pp}u_{q1} & \cdots & u_{pp}u_{qq} & \cdots & u_{pp}u_{qn} \\ \vdots & \ddots & \vdots & \ddots & \vdots \\ u_{np}u_{q1} & \cdots & u_{np}u_{qq} & \cdots & u_{np}u_{qn} \end{pmatrix} \begin{matrix} \\ \\ p\text{-}th; \\ \\ \end{matrix}$

3. $\mathbf{E}_n(p,q)M^2 = (\beta_{ij}^{(3)})_{n \times n} = \begin{pmatrix} 0 & \cdots & \overset{q\text{-}th}{0} & \cdots & 0 \\ \vdots & \ddots & \vdots & \ddots & \vdots \\ \sum_{s=1}^{n} u_{qs}u_{s1} & \cdots & \sum_{s=1}^{n} u_{qs}u_{sq} & \cdots & \sum_{s=1}^{n} u_{qs}u_{sn} \\ \vdots & \ddots & \vdots & \ddots & \vdots \\ 0 & \cdots & 0 & \cdots & 0 \end{pmatrix} \begin{matrix} \\ \\ p\text{-}th; \\ \\ \end{matrix}$

4. $M\mathbf{E}_n^2(p,q) = \begin{cases} \mathbf{0}_n & \text{if } p \neq q, \\ M\mathbf{E}_n(p,p) & \text{if } p = q. \end{cases}$

5. $\mathbf{E}_n(p,q)M\mathbf{E}_n(p,q) = (\beta_{ij}^{(4)})_{n \times n} = \begin{pmatrix} 0 \cdots \overset{q\text{-}th}{0} \cdots 0 \\ \vdots \ddots \vdots \ddots \vdots \\ 0 \cdots u_{qp} \cdots 0 \\ \vdots \ddots \vdots \ddots \vdots \\ 0 \cdots 0 \cdots 0 \end{pmatrix} \begin{matrix} \\ \\ p\text{-}th; \\ \\ \end{matrix}$

6. $\mathbf{E}_n^2(p,q)M = \begin{cases} \mathbf{0}_n & \text{if } p \neq q, \\ \mathbf{E}_n(p,p)M & \text{if } p = q. \end{cases}$

7. $\mathbf{E}_n^3(p,q) = \begin{cases} \mathbf{0}_n & \text{if } p \neq q, \\ \mathbf{E}_n(p,p) & \text{if } p = q. \end{cases}$

Theorem 3. *Let* $\nabla_{pq} f_3(M, \delta) = (c_{ij})_{n \times n}$. *We have*

$$
c_{ij} = \begin{cases} \delta a_3 u_{ip} u_{qj}, & \text{if } i \neq p, j \neq q, \\ \delta(a_2 u_{qj} + a_3(u_{pp} u_{qj} + \sum_{s=1}^{n} u_{qs} u_{sj})), & \text{if } i = p, j \neq q, \\ \delta(a_2 u_{ip} + a_3(u_{ip} u_{qq} + \sum_{s=1}^{n} u_{is} u_{sp})), & \text{if } i \neq p, j = q, \end{cases} \tag{11}
$$

and

$$
c_{pq} = \begin{cases} \delta(a_1 + a_2(u_{pp} + u_{qq}) + a_3(u_{pp} u_{qq} & \text{if } p \neq q. \\ \quad + \sum_{s=1}^{n} (u_{ps} u_{sp} + u_{qs} u_{sq}))) + \delta^2 a_3 u_{qp}, \\ \delta(a_1 + 2a_2 u_{pp} + a_3(u_{pp}^2 + 2 \sum_{s=1}^{n} u_{ps} u_{sp})) & \text{if } p = q. \\ \quad + \delta^2(a_2 + 3a_3 u_{pp}) + \delta^3 a_3. \end{cases} \tag{12}
$$

Based on Eqs. (11), (12) and (5), we design our update algorithm, called Update-3OMP. For simplicity, we use $c_{ij}^{(1)}, c_{pj}^{(2)}, c_{iq}^{(3)}$ to represent the three items in Eq. (11), and $c_{pq}^{(1)}$ and $c_{pq}^{(2)}$ to represent the two items in Eq. (12), respectively.

Update-3OMP$(M, f_3(M), \mathcal{A}_3, (p, q), \delta)$:

Let $f_3(M) = (g_{ij})_{n \times n}$;

if $p = q$ **then** $g_{pq} \leftarrow g_{pq} + c_{pq}^{(2)}$;

else $g_{pq} \leftarrow g_{pq} + c_{pq}^{(1)}$; **endif**

for $\{i := 1 \text{ to } n; j := 1 \text{ to } n\}$ **do**

    if $\{i \neq p, j \neq q\}$ **then** $g_{ij} \leftarrow g_{ij} + c_{ij}^{(1)}$;

    if $\{i = p, j \neq q\}$ **then** $g_{pj} \leftarrow g_{pj} + c_{pj}^{(2)}$;

    if $\{i \neq p, j = q\}$ **then** $g_{iq} \leftarrow g_{iq} + c_{iq}^{(3)}$;

endfor

Return $f_3(M)$;

Theorem 4. *Given an* $n \times n$ *matrix* M *and a real number* δ, *when* M *changes to* M' *by* u_{pq} *of* M *increasing by* δ *and all the other elements remaining unchanged,* Update-3OMP *updates* $f_3(M)$ *to* $f_3(M')$ *in* $O(n^2)$ *arithmetic operations.*

5   Update of 4-Order Matrix Polynomials

In this section, we discuss the update of 4-order matrix polynomials. We consider 4-order polynomials as follows and let $\mathcal{A}_4 = (a_0, a_1, a_2, a_3, a_4)$,

$$
f_4(x) = a_0 + a_1 x + a_2 x^2 + a_3 x^3 + a_4 x^4.
$$

Lemma 6. *Given an $n \times n$ matrix M and $1 \leq p, q \leq n$, for any real number δ,*

$$
\begin{aligned}
&\nabla_{pq} f_4(M, \delta) \\
&= \nabla_{pq} f_3(M, \delta) + a_4 \delta [M^3 \mathbf{E}_n(p, q) + M^2 \mathbf{E}_n(p, q) M + M \mathbf{E}_n(p, q) M^2 \\
&\quad + \mathbf{E}_n(p, q) M^3 + \delta (M^2 \mathbf{E}_n^2(p, q) + M \mathbf{E}_n(p, q) M \mathbf{E}_n(p, q) \\
&\quad + M \mathbf{E}_n^2(p, q) M + \mathbf{E}_n(p, q) M^2 \mathbf{E}_n(p, q) + \mathbf{E}_n(p, q) M \mathbf{E}_n(p, q) M \\
&\quad + \mathbf{E}_n^2(p, q) M^2) + \delta^2 (M \mathbf{E}_n^3(p, q) + \mathbf{E}_n(p, q) M \mathbf{E}_n^2(p, q) \\
&\quad + \mathbf{E}_n^2(p, q) M \mathbf{E}_n(p, q) + \mathbf{E}_n^3(p, q) M) + \delta^3 \mathbf{E}_n^4(p, q)].
\end{aligned}
\tag{13}
$$

Lemma 7. *For any $n \times n$ matrix M and $1 \leq p, q \leq n$, we have*

$$
1.\ M^3 \mathbf{E}_n(p, q) = (\theta_{ij}^{(1)})_{n \times n} =
\begin{pmatrix}
0 & \cdots & \overset{q\text{-th}}{\sum_{s=1}^{n} u_{1s}\beta_{sq}^{(1)}} & \cdots & 0 \\
\vdots & \ddots & \vdots & \ddots & \vdots \\
0 & \cdots & \sum_{s=1}^{n} u_{ps}\beta_{sq}^{(1)} & \cdots & 0 \\
\vdots & \ddots & \vdots & \ddots & \vdots \\
0 & \cdots & \sum_{s=1}^{n} u_{ns}\beta_{sq}^{(1)} & \cdots & 0
\end{pmatrix}
\begin{matrix} \\ \\ p\text{-th} \\ \\ \\ \end{matrix};
$$

$$
2.\ M^2 \mathbf{E}_n(p, q) M = (\theta_{ij}^{(2)})_{n \times n} =
\begin{pmatrix}
\beta_{1q}^{(1)} u_{q1} & \cdots & \overset{q\text{-th}}{\beta_{1q}^{(1)} u_{qq}} & \cdots & \beta_{1q}^{(1)} u_{qn} \\
\vdots & \ddots & \vdots & \ddots & \vdots \\
\beta_{pq}^{(1)} u_{q1} & \cdots & \beta_{pq}^{(1)} u_{qq} & \cdots & \beta_{pq}^{(1)} u_{qn} \\
\vdots & \ddots & \vdots & \ddots & \vdots \\
\beta_{nq}^{(1)} u_{q1} & \cdots & \beta_{nq}^{(1)} u_{qq} & \cdots & \beta_{nq}^{(1)} u_{qn}
\end{pmatrix}
\begin{matrix} \\ \\ p\text{-th} \\ \\ \\ \end{matrix};
$$

$$
3.\ M \mathbf{E}_n(p, q) M^2 = (\theta_{ij}^{(3)})_{n \times n} =
\begin{pmatrix}
u_{1p}\beta_{p1}^{(3)} & \cdots & \overset{q\text{-th}}{u_{1p}\beta_{pq}^{(3)}} & \cdots & u_{1p}\beta_{pn}^{(3)} \\
\vdots & \ddots & \vdots & \ddots & \vdots \\
u_{pp}\beta_{p1}^{(3)} & \cdots & u_{pp}\beta_{pq}^{(3)} & \cdots & u_{pp}\beta_{pn}^{(3)} \\
\vdots & \ddots & \vdots & \ddots & \vdots \\
u_{np}\beta_{p1}^{(3)} & \cdots & u_{np}\beta_{pq}^{(3)} & \cdots & u_{np}\beta_{pn}^{(3)}
\end{pmatrix}
\begin{matrix} \\ \\ p\text{-th} \\ \\ \\ \end{matrix};
$$

$$
4.\ \mathbf{E}_n(p, q) M^3 = (\theta_{ij}^{(4)})_{n \times n} =
\begin{pmatrix}
0 & \cdots & \overset{q\text{-th}}{0} & \cdots & 0 \\
\vdots & \ddots & \vdots & \ddots & \vdots \\
\sum_{s=1}^{n} \beta_{ps}^{(3)} u_{s1} & \cdots & \sum_{s=1}^{n} \beta_{ps}^{(3)} u_{sq} & \cdots & \sum_{s=1}^{n} \beta_{ps}^{(3)} u_{sn} \\
\vdots & \ddots & \vdots & \ddots & \vdots \\
0 & \cdots & 0 & \cdots & 0
\end{pmatrix}
\begin{matrix} \\ \\ p\text{-th} \\ \\ \\ \end{matrix}.
$$

Lemma 8. *For any* $n \times n$ *matrix* M *and* $1 \le p, q \le n$, *we have*

1. $M^2 \mathbf{E}_n^2(p,q) = \begin{cases} \mathbf{0}_n \ if \ p \ne q, \\ M^2 \mathbf{E}_n(p,p) \ if \ p = q. \end{cases}$

2. $ME_n(p,q)ME_n(p,q) = (\lambda_{ij}^{(1)})_{n \times n} = \begin{pmatrix} 0 & \cdots & \beta_{1p}^{(2)} & \cdots & 0 \\ \vdots & \ddots & \vdots & \ddots & \vdots \\ 0 & \cdots & \beta_{pp}^{(2)} & \cdots & 0 \\ \vdots & \ddots & \vdots & \ddots & \vdots \\ 0 & \cdots & \beta_{np}^{(2)} & \cdots & 0 \end{pmatrix} \begin{matrix} \\ \\ p\text{-}th \\ \\ \\ \end{matrix}$; (with q-th above the β column)

3. $ME_n^2(p,q)M = \begin{cases} \mathbf{0}_n \ if \ p \ne q, \\ ME_n(p,p)M \ if \ p = q. \end{cases}$

4. $\mathbf{E}_n(p,q)M^2\mathbf{E}_n(p,q) = (\lambda_{ij}^{(2)})_{n \times n} = \begin{pmatrix} 0 & \cdots & 0 & \cdots & 0 \\ \vdots & \ddots & \vdots & \ddots & \vdots \\ 0 & \cdots & \beta_{qq}^{(1)} & \cdots & 0 \\ \vdots & \ddots & \vdots & \ddots & \vdots \\ 0 & \cdots & 0 & \cdots & 0 \end{pmatrix} \begin{matrix} \\ \\ p\text{-}th \\ \\ \\ \end{matrix}$; (with q-th above)

5. $\mathbf{E}_n(p,q)M\mathbf{E}_n(p,q)M = (\lambda_{ij}^{(3)})_{n \times n} = \begin{pmatrix} 0 & \cdots & 0 & \cdots & 0 \\ \vdots & \ddots & \vdots & \ddots & \vdots \\ \beta_{q1}^{(2)} & \cdots & \beta_{qq}^{(2)} & \cdots & \beta_{qn}^{(2)} \\ \vdots & \ddots & \vdots & \ddots & \vdots \\ 0 & \cdots & 0 & \cdots & 0 \end{pmatrix} \begin{matrix} \\ \\ p\text{-}th \\ \\ \\ \end{matrix}$; (with q-th above)

6. $\mathbf{E}_n^2(p,q)M^2 = \begin{cases} \mathbf{0}_n \ if \ p \ne q, \\ \mathbf{E}_n(p,p)M^2 \ if \ p = q. \end{cases}$

Lemma 9. *For any* $n \times n$ *matrix* M *and* $1 \le p, q \le n$, *we have*

1. $M\mathbf{E}_n^3(p,q) = \begin{cases} \mathbf{0}_n \ if \ p \ne q, \\ M\mathbf{E}_n(p,p) \ if \ p = q. \end{cases}$

2. $\mathbf{E}_n(p,q)M\mathbf{E}_n^2(p,q) = \begin{cases} \mathbf{0}_n \ if \ p \ne q, \\ \mathbf{E}_n(p,p)M\mathbf{E}_n(p,p) \ if \ p = q. \end{cases}$

3. $\mathbf{E}_n^2(p,q)M\mathbf{E}_n(p,q) = \begin{cases} \mathbf{0}_n \ if \ p \ne q, \\ \mathbf{E}_n(p,p)M\mathbf{E}_n(p,p) \ if \ p = q. \end{cases}$

4. $\mathbf{E}_n^3(p,q)M = \begin{cases} \mathbf{0}_n \ if \ p \ne q, \\ \mathbf{E}_n(p,p)M \ if \ p = q. \end{cases}$

5. $\mathbf{E}_n^4(p,q) = \begin{cases} \mathbf{0}_n \ if \ p \ne q, \\ \mathbf{E}_n(p,p) \ if \ p = q. \end{cases}$

Theorem 5. *Let* $\nabla_{pq} f_4(M, \delta) = (d_{ij})_{n \times n}$. *If* $p \neq q$ *then*

$$d_{ij} = \begin{cases} c_{ij} + a_4\delta(\beta_{iq}^{(1)}u_{qj} + u_{ip}\beta_{pj}^{(3)}), & \text{if } i \neq p, j \neq q, \\ c_{pj} + a_4\delta(\beta_{pq}^{(1)}u_{qj} + u_{pp}\beta_{pj}^{(3)} + \sum_{s=1}^{n} \beta_{ps}^{(3)}u_{sj}) + a_4\delta^2\beta_{qj}^{(2)}, & \text{if } i = p, j \neq q, \\ c_{iq} + a_4\delta(\beta_{iq}^{(1)}u_{qq} + u_{ip}\beta_{pq}^{(3)} + \sum_{s=1}^{n} u_{is}\beta_{sq}^{(1)}) + a_4\delta^2\beta_{ip}^{(2)}, & \text{if } i \neq p, j = q, \\ c_{pq} + a_4\delta(\beta_{pq}^{(1)}u_{qq} + u_{pp}\beta_{pq}^{(3)} + \sum_{s=1}^{n}(u_{ps}\beta_{sq}^{(1)} + \beta_{ps}^{(3)}u_{sq})) & \text{if } i = p, j = q. \\ \quad + a_4\delta^2(\beta_{pp}^{(2)} + \beta_{qq}^{(2)} + \beta_{qq}^{(1)}), \end{cases} \quad (14)$$

and if $p = q$ *then*

$$d_{ij} = \begin{cases} c_{ij} + a_4\delta(\beta_{ip}^{(1)}u_{pj} + u_{ip}\beta_{pj}^{(3)}) + a_4\delta^2\beta_{ij}^{(2)}, & \text{if } i \neq p, j \neq p, \\ c_{pj} + a_4\delta(\beta_{pp}^{(1)}u_{pj} + u_{pp}\beta_{pj}^{(3)} + \sum_{s=1}^{n} \beta_{ps}^{(3)}u_{sj}) & \text{if } i = p, j \neq p, \\ \quad + a_4\delta^2(2\beta_{pj}^{(2)} + \beta_{pj}^{(3)}) + a_4\delta^3 u_{pj}, \\ c_{ip} + a_4\delta(\beta_{ip}^{(1)}u_{pp} + u_{ip}\beta_{pp}^{(3)} + \sum_{s=1}^{n} u_{is}\beta_{sp}^{(1)}) & \text{if } i \neq p, j = p, \\ \quad + a_4\delta^2(\beta_{ip}^{(1)} + 2\beta_{ip}^{(2)}) + a_4\delta^3 u_{ip}, \\ c_{pp} + a_4\delta(\beta_{pp}^{(1)}u_{pp} + u_{pp}\beta_{pp}^{(3)} + \sum_{s=1}^{n}(u_{ps}\beta_{sp}^{(1)} + \beta_{ps}^{(3)}u_{sp})) & \text{if } i = j = p. \\ \quad + a_4\delta^2(2\beta_{pp}^{(1)} + 3\beta_{pp}^{(2)} + \beta_{pp}^{(3)}) + 4a_4\delta^3 u_{pp}, \end{cases} \quad (15)$$

Combining Eqs. (14), (15) and (5), we obtain our update algorithm, called Update-4OMP. For simplicity, we let $d_{ij}^{(0,1)}, d_{pj}^{(0,2)}, d_{iq}^{(0,3)}, d_{pq}^{(0,4)}$ represent the four items in Eq. (14), and $d_{ij}^{(1,1)}, d_{pj}^{(1,2)}, d_{iq}^{(1,3)}, d_{pq}^{(1,4)}$ represent the four items in Eq. (15), respectively.

Update-4OMP$(M, f_4(M), \mathcal{A}_4, (p, q), \delta)$:
Let $f_4(M) = (h_{ij})_{n \times n}$;
for $\{i := 1 \text{ to } n;\ j := 1 \text{ to } n\}$ **do**
if $\{p \neq q, i \neq p, j \neq q\}$ **then** $h_{ij} \leftarrow h_{ij} + d_{ij}^{(0,1)}$;
if $\{p = q, i \neq p, j \neq q\}$ **then** $h_{ij} \leftarrow h_{ij} + d_{ij}^{(1,1)}$;
if $\{p \neq q, i = p, j \neq q\}$ **then** $h_{pj} \leftarrow h_{pj} + d_{pj}^{(0,2)}$;
if $\{p = q, i = p, j \neq q\}$ **then** $h_{pj} \leftarrow h_{pj} + d_{pj}^{(1,2)}$;
if $\{p \neq q, i \neq p, j = q\}$ **then** $h_{iq} \leftarrow h_{iq} + d_{iq}^{(0,3)}$;
if $\{p = q, i \neq p, j = q\}$ **then** $h_{iq} \leftarrow h_{iq} + d_{iq}^{(1,3)}$;
if $\{p \neq q, i = p, j = q\}$ **then** $h_{pq} \leftarrow h_{pq} + d_{pq}^{(0,4)}$;
if $\{p = q, i = p, j = q\}$ **then** $h_{pq} \leftarrow h_{pq} + d_{pq}^{(1,4)}$;
endfor
Return $f_4(M)$;

Theorem 6. *Given an* $n \times n$ *matrix* M *and a real number* δ, *when* M *changes to* M' *by* u_{pq} *of* M *increasing by* δ *and all the other elements remaining unchanged,* Update-4OMP *updates* $f_4(M)$ *to* $f_4(M')$ *in* $O(n^2)$ *arithmetic operations.*

6 Conclusions

In this paper, we studied the update of low-order matrix polynomials, and designed $O(n)$-, $O(n^2)$-, and $O(n^2)$-operations update algorithms for 2-, 3- and 4-order matrix polynomials, respectively, based on the naive matrix multiplication. For high-order matrix polynomials with a sparse coefficient vector, we proposed a combinatorial heuristic updating method.

It is of great interest to obtain faster algorithms for the update of high-order matrix polynomials. Although it seems very hard to design update algorithms based on faster square matrix multiplication algorithms [2,10,11,14], it also remains as a future research topic.

References

1. Brualdi, R.A., Cvetković, D.: A Combinatorial Approach to Matrix Theory and Its Applications. CRC Press, Boca Raton (2009)
2. Coppersmith, D., Winograd, S.: Matrix multiplication via arithmetic progressions. J. Symb. Comput. **9**, 251–280 (1990)
3. Frandsen, G.S.: Dynamic matrix algorithms, manuscript, BRICS, University of Aarhus, Denmark, 11 April 2011
4. Frandsen, G.S., Frandsen, P.F.: Dynamic matrix rank. Theor. Comput. Sci. **410**, 4085–4093 (2009)
5. Frandsen, G.S., Hansen, J.P., Miltersen, P.B.: Lower bounds for dynamic algebraic problems. Inf. Comput. **171**, 333–349 (2001)
6. Frandsen, G.S., Sankowski, P.: Dynamic normal forms and dynamic characteristic polynomial. Theor. Comput. Sci. **412**, 1470–1483 (2011)
7. Golub, G.H., Van Loan, C.F.: Matrix Computations, 3rd edn. The Johns Hopkins University Press, Baltimore and London (1996)
8. Goodrich, M.T., Tamassia, R.: Algorithm Design, Foundations, Analysis, and Internet Examples. Wiley, Hoboken (2001)
9. Horn, R.A., Johnson, C.R.: Matrix Analysis, 2nd edn. Cambridge University Press, Cambridge (2012)
10. Le Gall, F.: Faster algorithms for rectangular matrix multiplication. In: Proceedings of 53rd FOCS, pp. 514–523 (2012)
11. Le Gall, F.: Powers of tensors and fast matrix multiplication. In: Proceedings of 39th ISSAC, pp. 296–303 (2014)
12. Reif, J.H., Tate, S.R.: On dynamic algorithms for algebraic problems. J. Algorithms **22**, 347–371 (1997)
13. Sankowski, P.: Dynamic transitive closure via dynamic matrix inverse. In: Proceedings of 45th FOCS, pp. 509–517 (2004)
14. Vassilevska Williams, V.: Multiplying matrices faster than coppersmith-winograd. In: Proceedings of 44th STOC, pp. 887–898 (2012)

On the Structure of Discrete Metric Spaces Isometric to Circles

Andreas W. M. Dress[1,2], Hiroshi Maehara[3], Sabrina Xing Mei Pang[4], and Zhenbing Zeng[1,5(✉)] (iD)

[1] CAS-MPG Partner Institute for Computational Biology,
Shanghai Institutes for Biological Sciences, Chinese Academy of Sciences,
320 Yue Yang Road, Shanghai 200031, People's Republic of China
{andreas,zbzeng}@picb.ac.cn
[2] MPI for Mathematics in the Sciences, 04103 Leipzig, Germany
[3] College of Education, Ryukyu University, Nishihara, Okinawa 903-0213, Japan
hmaehara@edu.u-ryukyu.ac.jp
[4] Center for Combinatorics, LPMC, Nankai University, Tianjin 300071, China
pangxingmei@mail.nankai.edu.cn
[5] Department of Mathematics, Shanghai University, Shanghai 200444, China
zbzeng@shu.edu.cn

Abstract. A metric space (X, D) is called circular if it is isometric to a subspace of a *metric circle*, that is, a circle in which distances are measured by the length of the shorter arc connecting them. We show that the following three conditions are equivalent: (1) X is circular, (2) every 4-point subset of X can be labeled as a, b, c, d so that $D(a, b) + D(b, c) = D(a, c)$, $D(b, c) + D(c, d) = D(b, d)$ holds, and (3) every 4-point subset of X is circular.

Keywords: Metric space · Circular metrics · Distance

1 Introduction

A metric space $X = (X, D)$ with *point space* X and metric $D : X \times X \to \mathbb{R}$ is called *Euclidean* if it is isometric to a subspace of a Euclidean space $\mathbb{R}^n$, and *linear* if it is isometric to a subspace of the real line $\mathbb{R}$. In this note, a circle Γ is always meant to be a *metric circle*, that is, a metric space (Γ, δ) whose point set Γ is a circle of some positive radius ρ contained in $\mathbb{R}^2$ in which, however, the *circle distance* $\delta(u, v) = \delta_\Gamma(u, v)$ between any two points $u, v \in \Gamma$ is the *arc length* between u and v in Γ, i.e., it is the minimum of the lengths of the two arcs in Γ with endpoints u, v. In other words, δ_Γ it is the canonical *geodesic* (or *shortest path* or *intrinsic*) metric induced on Γ by the standard Euclidean metric of $\mathbb{R}^2$.

This paper was supported in part by the National Natural Science Foundation of China under Grant No. 11471209.

© Springer Nature Switzerland AG 2019
D.-Z. Du et al. (Eds.): AAIM 2019, LNCS 11640, pp. 83–94, 2019.
https://doi.org/10.1007/978-3-030-27195-4_8

A circle of circumference ρ is denoted by Γ_ρ in which case we also write δ_ρ rather than δ or δ_Γ for the metric on Γ_ρ. Note that the following holds for all $u, v, w \in \Gamma_\rho$:

(ρ 1) $\rho \geq \delta_\rho(u, v) + \delta_\rho(v, w) + \delta_\rho(w, u) \geq 2\delta_\rho(u, v)$.

(ρ 2) One has $\delta_\rho(u, w) = \delta_\rho(u, v) + \delta_\rho(v, w)$ if and only if v is contained in an arc of minimal length with endpoints u, w.

(ρ 3) At least one of the following assertions $v \in [u, w]$, $u \in [w, v]$, $w \in [v, u]$, or $\rho = \delta_\rho(u, v) + \delta_\rho(v, w) + \delta_\rho(w, u)$ always holds[1].

(ρ 4) Denoting more specifically, for every $x \in \Gamma_\rho$, the unique *antipodal* point of x in Γ_ρ (i.e., the unique point in Γ_ρ with δ_ρ-distance $\rho/2$ to x) by x^*, we have

$$w \in [u, v] \iff u \in [w, v^*] \iff v \in [w, u^*] \iff w^* \in [u^*, v^*],$$

and we have $\delta_\rho(x, y) + \delta_\rho(y, x^*) = \delta_\rho(x, x^*) = \rho/2$ and

$$\Gamma_\rho = [x, y] \cup [y, x^*] \cup [x^*, y^*] \cup [y^*, x]$$

for all $x, y \in \Gamma_\rho$.

(ρ 5) In particular, the four alternatives mentioned in (ρ 3) correspond exactly to the four alternatives $w \in [v, u^*]$, $w \in [v^*, u]$, $w \in [v, u]$, and $w \in [u^*, v^*]$, i.e., we have $w \in [v, u^*] \iff \delta_\rho(v, u^*) = \delta_\rho(v, w) + \delta_\rho(w, u^*) \iff \rho/2 - \delta_\rho(v, u) = \delta_\rho(v, w) + \rho/2 - \delta_\rho(w, u) \iff v \in [u, w]$, and so on.

(ρ 6) If $\rho = \delta_\rho(u, v) + \delta_\rho(v, w) + \delta_\rho(w, u)$ holds, one has

$$\delta_\rho(x, w) = \min\left(\delta_\rho(x, u) + \delta_\rho(u, w), \delta_\rho(x, v) + \delta_\rho(v, w)\right)$$

for all $x \in [u, v]$.

(ρ 7) If $v \in [u, x] \cap [u, y]$ holds for some $x, y \in \Gamma_\rho$, either $x \in [v, y]$ or $y \in [v, x]$ holds.

(ρ 8) If $\delta_\rho(u, v) < \rho/2$ holds, the map $\Gamma_\rho \to \mathbb{R}^2 : x \mapsto (\delta_\rho(x, v), \delta_\rho(x, u))$ is injective: Indeed, we can infer from the distances $\delta_\rho(x, v)$ and $\delta_\rho(x, u)$ whether x is contained in $[u, v]$, $[v, u^*]$, $[u^*, v^*]$ or in $[v^*, u]$ in which case the same must hold for any $y \in \Gamma_\rho$ with $\delta_\rho(x, v) = \delta_\rho(y, v)$ and $\delta_\rho(x, u) = \delta_\rho(y, u)$. So, it suffices to note that $a, b \in \Gamma_\rho$, $\delta_\rho(a, b) < \rho/2$, $x, y \in [a, b]$, $\delta_\rho(x, a) = \delta_\rho(y, a)$, and $\delta_\rho(x, a) = \delta_\rho(y, b)$ together implies $x = y$ (as the interval $[a, b]$ is obviously a linear metric space).

A metric space X is called *circular* (resp. *ρ-circular*) if it is isometric to a subspace of a circle Γ (resp. Γ_ρ). For convenience, we regard a linear metric space as a circular metric space with *infinitely large* circumference. Thus, linear metric spaces are special circular spaces.

---

[1] As always, given any metric space $X = (X, D)$ and any two points $x, y \in X$, we denote by $[x, y]$ the *interval* $[x, y] = [x, y]_D := \{z \in X : D(x, y) = D(x, z) + D(z, y)\}$ spanned by x and y relative to D. Recall that $x, y, z, u \in X, y \in [x, z]$, and $z \in [x, u]$ always implies $y \in [x, u]$ and $z \in [y, u]$.

A metric space is called *strictly circular* if it is circular and not linear. And it is called *strictly ρ-circular* for some real number $\rho > 0$ if it is ρ-circular, but neither linear nor ρ'-circular for any $\rho' \neq \rho$.

Apparently, the following holds

Lemma 1. (*i*) *Given a positive real number ρ, a metric space is strictly ρ-circular for some $\rho > 0$ if and only if it is strictly circular and ρ-circular.*

(*ii*) *In particular, every strictly circular space is strictly ρ-circular for that number ρ for which it is ρ-circular.*

(*iii*) *A linear metric space (X, D) is ρ-circular for some positive real number ρ if and only if $\mathrm{diam}(X) \leq \rho/2$ holds for its diameter*

$$\mathrm{diam}(X) := \sup(D(x, y) : x, y \in X).$$

In particular, a linear metric space (X, D) is not 'properly circular' (i.e., it is not ρ-circular for any positive real number ρ) if and only if $\mathrm{diam}(X) = \infty$ holds.

(*iv*) *Every circular space that contains a strictly circular (strictly ρ-circular) subspace must, of course, be strictly circular (strictly ρ-circular), too.*

In this note, we present some results on linear, and circular metric spaces derived by perfectly elementary and direct arguments. Note that this problem can also be investigated from the point of view of isometric embedding of finite metric spaces into particular metric spaces like Euclidean spaces, spherical caps and hyperspheres, as initially done by Blumenthal in [3]. Many work of this kind can be found in Blumenthal's book [4], Tóth [9], Robinson [8], and the book [7] of Deza and Laurent. The motivation of our paper is to exhibit our investigation focusing on circular metrics from a rather elementary and intrinsic point of view, without considering the embedding into other types of "symmetric spaces", as the results we will show are quite useful in analysis of real distance data, in particular data coming from biology representing similarities between species or genes.

Let x, y, z be any three points in a metric space (X, D). Then, the largest one of the three terms $D(x, y), D(y, z), D(x, z)$ of X is equal to the sum of the other two if and only if the subspace $\{x, y, z\}$ is linear and, hence, circular.

Otherwise, each of the three terms $D(x, y), D(y, z), D(z, x)$ is smaller than the sum of the other two, and we can find three points $\hat{x}, \hat{y}, \hat{z}$ on a circle Γ with circumference $D(x, y) + D(y, z) + D(z, x)$, so that $\delta(\hat{x}, \hat{y}) = D(x, y), \delta(\hat{y}, \hat{z}) = D(y, z), \delta(\hat{z}, \hat{x}) = D(z, x)$ holds. Hence, $\{x, y, z\}$ is strictly circular. Thus, all 3-point subspaces of any metric space are circular, but not necessarily linear.

A metric space (X, D) is called *3-point linear* (3PL) if every 3-point subspace of X is linear, and (X, D) is called *4-point circular* (4PC) if every 4-point subspace is circular.

It is well known (cf. Blumenthal [4] or Section 5 below) that a 3PL metric space (X, D) is linear if the cardinality $|X|$ of X is not equal to 4, and circular (possibly linear) if $|X| = 4$.

We prove that, given a metric space (X, D), the following three assertions are equivalent: (1) (X, D) is circular. (2) Every 4-point subset of (X, D) can be labeled as x_1, x_2, x_3, x_4 so that $D(x_{i-1}, x_{i+1}) = D(x_{i-1}, x_i) + D(x_i, x_{i+1})$ holds for $i = 2, 3$. (3) (X, D) is 4PC. We also prove that if (X, D) is strictly circular, then (X, D) is never Euclidean. In the final section, we consider metric spaces that are isometric to a subspace of a sphere (with intrinsic metric). We show that every open subset of a (2-dimensional) sphere contains a 4-point subset that is not isometrically embeddable into any other sphere of different radius.

2   Some Basic Definitions and Facts

For a metric space (X, D) and two points $a, b \in X$, we define $[a, b]$ by

$$[a, b] := \{x \in X \mid D(a, x) + D(x, b) = D(a, b)\}.$$

Note that $v \in [u, w]$ holds for some $u, v, w \in \Gamma_\rho$ if and only if v is contained in an arc of minimal length with endpoints u, w while $\Gamma_\rho = [u, v] \cup [v, w] \cup [w, u]$ holds for all $u, v, w \in \Gamma_\rho$ for which $\rho = \delta_\rho(u, v) + \delta_\rho(v, w) + \delta_\rho(w, u)$ holds.

An ordered sequence $(x_1, x_2, \ldots, x_k)$ of k points, $k \geq 3$, in (X, D) is called a (discrete) *geodesic* if $x_i \in [x_{i-1}, x_{i+1}]$ holds for all $i = 2, 3, \ldots, k-1$ in which case we will also say that $[x_1 - x_2 - \cdots - x_k]$ holds. And if, in addition, $D(x_1, x_k) = D(x_1, x_2) + D(x_2, x_3) + \cdots + D(x_{k-1}, x_k)$ holds, the sequence is called a *shortest (discrete) geodesic* in which case we will also say that $[x_1 = x_2 = \cdots = x_k]$ holds. In particular, we have

$$[a - b - c] \iff [a = b = c] \iff D(a, b) + D(b, c) = D(a, c)$$

for all $a, b, c \in X$. And our footnote on Page 2 can now be rephrased as saying that $[x - y - z]$ and $[x - z - w]$ holds for some points x, y, z, w in the point set X of some metric space (X, D) if and only if $[x = y = z = w]$ holds.

The following lemma follows easily from repeated applications of the triangle inequality:

Lemma 2. (i) *Any shortest geodesic* $(x_1, x_2, \ldots, x_k)$ *is, in particular, a geodesic.* (ii) *A finite subspace* Y *of* (X, D) *is linear if and only if its points can be labeled so as to form a shortest geodesic.*

The next lemma is more interesting (see also Theorem 1 and its corollary for a more general assertion):

Lemma 3. *A 4-point space* (X, D) *is circular if and only if the four points of* X *can be labeled so as to form a geodesic.*

Proof. First, suppose that X is ρ-circular. We may regard X as a 4-point subspace of some circle $\Gamma = \Gamma_\rho$. Suppose that the four points of X appear on Γ_ρ in cyclic order as a, b, c, d. Then,

$$\delta_\rho(a, c) = \min\{\delta_\rho(a, b) + \delta_\rho(b, c), \delta_\rho(a, d) + \delta_\rho(d, c)\}$$

must hold because one of the two arcs connecting a and c must contain b and the other one d, and (at least) one of those two arcs must have minimum length. And

$$\delta_\rho(b,d) = \min\{\delta_\rho(b,c) + \delta_\rho(c,d),\ \delta_\rho(b,a) + \delta_\rho(a,d)\}$$

must hold for similar reasons. So, without loss of generality, we may suppose that $\delta_\rho(a,c) = \delta_\rho(a,b) + \delta_\rho(b,c)$ holds. If also $\delta_\rho(b,d) = \delta_\rho(b,c) + \delta_\rho(c,d)$ holds, then (a,b,c,d) is a geodesic. Otherwise, $\delta_\rho(b,d) = \delta_\rho(b,a) + (a,d)$ holds and (d,a,b,c) is a geodesic.

Conversely, suppose that the four points in X are indexed as x_1, x_2, x_3, x_4 so that (x_1, x_2, x_3, x_4) forms a geodesic and put

$$\rho := D(x_1, x_2) + D(x_2, x_3) + D(x_3, x_4) + D(x_4, x_1).$$

Then, by applying the triangle inequality, we can deduce that $D(x_i, x_j) \le \rho/2$ holds for all $i, j \in \{1, 2, 3, 4\}$.

Take four points $\hat{x}_i$, $i = 1, 2, 3, 4$ on Γ_ρ in cyclic order relative to some fixed orientation of Γ_ρ so that, in that orientation, the arcs from $\hat{x}_i$ to $\hat{x}_{i+1}$ have length $D(x_i, x_{i+1}) \le \rho/2$ for every $i = 1, 2, 3$ implying that also $\delta(\hat{x}_i, \hat{x}_{i+1}) = D(x_i, x_{i+1})$ must hold for $i = 1, 2, 3$ and that, in view of $D(x_1, x_2) + D(x_2, x_3) + D(x_3, x_4) \le \rho$, the arc from $\hat{x}_1$ to $\hat{x}_4$ has length $D(x_1, x_2) + D(x_2, x_3) + D(x_3, x_4)$. Then, since

$$\rho/2 \le D(x_1, x_2) + D(x_2, x_3) + D(x_3, x_4) = \delta_\rho(\hat{x}_1, \hat{x}_2) + \delta_\rho(\hat{x}_2, \hat{x}_3) + \delta_\rho(\hat{x}_3, \hat{x}_4),$$

we have $\delta_\rho(\hat{x}_4, \hat{x}_1) = \rho - \delta_\rho(\hat{x}_1, \hat{x}_2) - \delta_\rho(\hat{x}_2, \hat{x}_3) - \delta_\rho(\hat{x}_3, \hat{x}_4) = D(x_1, x_4)$. Further, since $D(x_1, x_2) + D(x_2, x_3) = D(x_1, x_3) \le \rho/2$, we have $\delta_\rho(\hat{x}_1, \hat{x}_2) + \delta_\rho(\hat{x}_2, \hat{x}_3) \le \rho/2$ and, hence, $\delta_\rho(\hat{x}_1, \hat{x}_3) = \delta_\rho(\hat{x}_1, \hat{x}_2) + \delta_\rho(\hat{x}_2, \hat{x}_3) = D(x_1, x_2) + D(x_2, x_3) = D(x_1, x_3)$. Similarly, we have $\delta_\rho(\hat{x}_2, \hat{x}_4) = D(x_2, x_4)$. Therefore,

$$X \to \Gamma_\rho : x_i \mapsto \hat{x}_i \quad (i = 1, 2, 3, 4)$$

is an isometric embedding, and X is circular. $\qquad\square$

A space (X, D) of cardinality $|X| = 4$ is called *antipodally circular* if we can label the four points of X as a, b, c, d so that (d, a, b, c, d, a) is a geodesic.

Notice that every 3-point subspace of an antipodally circular 4-point space X is linear, but X itself is strictly circular. Note also that the diameter of X must coincide with $D(a, c)$ as well as with $D(b, d)$ as

$$D(a, c) = D(a, b) + D(b, c) = D(c, d) + D(d, a)$$

and

$$D(b, d) = D(b, c) + D(c, d) = D(d, a) + D(a, b)$$

and, hence, also

$$2D(a, c) = D(a, b) + D(b, c) + D(c, d) + D(d, a) = 2D(b, d)$$

must hold.

In particular, if X' and X'' are two antipodally circular 4-point subspaces of a metric space (X, D) with $|X' \cap X''| = 3$, we must $D(x, y') = D(x, y'')$ for all $x \in X \cap X'$ and the two unique elements $y' \in X' - X''$ and $y'' \in X'' - X'$.

Recall also that, more generally, a metric space (X, D) is called antipodal if there exists an involutory map $\tau_D : X \to X : x \mapsto \bar{x}$ from X onto itself such that $X = [x, \bar{x}]$ holds for every $x \in X$ in which case this map must be an isometry in view of the above discussion of antipodally circular spaces of cardinality 4, and note that a metric space (X, D) with $|X| = 4$ is antipodal if and only if it is antipodally circular.

The first four assertions in the following lemma are obvious.

Lemma 4. (1) *Let X be a ρ-circular metric space, and $x, y \in X$ be two distinct points such that $D(x, y) < \rho/2$. Then every isometric embedding $f : \{x, y\} \to \Gamma_\rho$ can be uniquely extended to an isometric embedding $\hat{f} : X \to \Gamma_\rho$.*

(2) *If the 4-point space $\{a, b, c, d\}$ is linear with diameter at most $\rho/2$, then $\{a, b, c, d\}$ is isometrically embeddable in Γ_ρ.*

(3) *A 3-point space $\{a, b, c\}$ is strictly circular if and only if the sum of any two of the three terms $D(a, b), D(b, c), D(a, c)$ is greater than the third one, and it is strictly ρ-circular if and only if, in addition, $D(a, b) + D(b, c) + D(c, a) = \rho$ holds.*

(4) *If the 3-point space $\{a, b, c\}$ as well as its one-point extension $X = \{a, b, c, x\}$ are strictly circular and $x \in [a, b]$ holds, we have $\rho - D(c, a) - D(a, x) = D(a, b) + D(b, c) - D(a, x) = D(c, b) + D(b, x)$ and, therefore, also*

$$D(c, x) = \min\{D(c, a) + D(a, x), D(c, b) + D(b, x)\})$$
$$= \min\{D(c, a) + D(a, x), \rho - D(c, a) - D(a, x)\}.$$

(5) *If the 4-point space $X = \{a, b, c, d\}$ is strictly, but not antipodally circular, then it contains a strictly circular 3-point subspace.*

Proof. To establish (5), suppose that $\{a, b, c, d\}$ is strictly ρ-circular, but not antipodally circular. We may suppose that (a, b, c, d), but not (a, b, c, d, a) is a geodesic. Then, $D(c, a) < D(c, d) + D(d, a)$ must hold. Further, we must also have $D(a, d) < D(a, c) + D(c, d)$ as $D(a, d) = D(a, c) + D(c, d)$ would imply $D(a, d) = D(a, c) + D(c, d) = D(a, b) + D(b, c) + D(c, d)$ and, hence, that (a, b, c, d) would be a shortest geodesic and $\{a, b, c, d\}$, therefore, linear (cf. Lemma 2). Finally, we must also have $D(c, d) < D(c, a) + D(a, d)$ in view of $D(c, a) + D(a, d) = D(c, b) + D(b, a) + D(a, d) \geq D(c, b) + D(b, d) = D(c, b) + D(b, c) + D(c, d) > D(c, d)$. So, $\{a, c, d\}$ is strictly circular by (3). □

3   4-point Circularity

Lemma 5. *Let (X, D) be a 4PC metric space and suppose that $\{a, b, c\} \subset X$ is strictly ρ-circular. Then, every embedding $\{a, b, c\} \to \Gamma_\rho$ can be uniquely extended to an embedding $X \to \Gamma_\rho$. In particular, X is strictly ρ-circular.*

Remark 1. It is interesting that the 4PC condition eventually restricts the cardinality of X so that it is less than, or at most equal to, the cardinality of continuum.

Proof. Let us first note that

(i) $X = [a, b] \cup [b, c] \cup [c, a]$ must hold,
(ii) $x \in [a, b]$ implies $D(c, x) = \min\{D(c, a) + D(a, x), D(c, b) + D(b, x)\}$ and $D(a, x) + D(x, b) + D(b, c) + D(c, a) = \rho$,
(iii) $x \in [a, b]$ and $D(c, a) + D(a, x) \le D(c, b) + D(b, x)$ implies
- $D(c, x) = D(c, a) + D(a, x) \le D(c, b) + D(b, x) = \rho - D(c, a) - D(a, x)$,
- $D(b, c) + D(c, x) + D(x, b) = D(b, c) + D(c, a) + D(a, x) + D(x, b) = D(b, c) + D(c, a) + D(a, b) = \rho$,
- $D(x, b) < D(b, c) + D(c, x)$ (in view of $D(x, b) \le D(b, a) + D(a, x) < D(b, c) + D(c, a) + D(a, x) = D(b, c) + D(c, x)$),
- and $D(b, c) < D(c, x) + D(x, b)$ (in view of $D(b, c) < D(c, a) + D(a, b) = D(c, a) + D(a, x) + D(x, b) = D(c, x) + D(x, b)$).

So, either $\{b, c, x\}$ is strictly ρ-circular in this case (that is, it is strictly ρ-circular if $D(x, b) + D(b, c) > D(c, x)$ holds), or $D(x, b) + D(b, c) = D(c, x) = D(c, a) + D(a, x)$ holds and, therefore, also $D(c, x) = \rho/2$.

Now, fix an isometric embedding $f : \{a, b, c\} \to \Gamma_\rho; a, b, c \mapsto \hat{a}, \hat{b}, \hat{c} \in \Gamma_\rho$. For each $x \in X - \{a, b, c\}$, since $\{a, b, c, x\}$ is ρ-circular, this embedding f can be uniquely extended to an embedding $\tilde{f}_x : \{a, b, c, x\} \to \Gamma_\rho$. Together, this yields an extension

$$\hat{f} : X \to \Gamma_\rho : x \mapsto \begin{cases} f(x) & \text{if } x \in \{a, b, c\} \\ \hat{x} := \tilde{f}_x(x) & \text{else .} \end{cases}$$

It suffices to show that $D(x, y) = \delta_\rho(\hat{x}, \hat{y})$ holds for all $x, y \in X - \{a, b, c\}$. To this end, we may further suppose that

$$x \in [a, b], D(c, a) + D(a, x) \le D(c, b) + D(b, x), \text{ and } y \in [a, b] \cup [b, c]$$

holds. If $\{b, c, x\}$ is strictly ρ-circular, $\{b, c, x, y\}$ must be strictly ρ-circular, too, as it is circular. Furthermore, our assumption that $D(b, c) < \rho/2$ holds, implies that the embedding $f|_{\{b,c\}} : \{b, c\} \to \Gamma_\rho$ can be uniquely extended to an embedding $\{b, c, x, y\} \to \Gamma_\rho$, i.e., we can find points $\bar{x}, \bar{y} \in \Gamma_\rho$ with $D(x, y) = \delta_\rho(\bar{x}, \bar{y})$ so that also

$$D(b, x) = \delta_\rho(\hat{b}, \bar{x}), \ D(c, x) = \delta_\rho(\hat{c}, \bar{x}), \ D(b, y) = \delta_\rho(\hat{b}, \bar{y}), \text{ and } D(c, y) = \delta_\rho(\hat{c}, \bar{y})$$

holds. Since also

$$D(b, x) = \delta_\rho(\hat{b}, \hat{x}), \ D(c, x) = \delta_\rho(\hat{c}, \hat{x}), \ D(b, y) = \delta_\rho(\hat{b}, \hat{y}), \text{ and} D(c, y) = \delta_\rho(\hat{c}, \hat{y})$$

holds, we must have

$$\delta_\rho(\hat{b}, \bar{x}) = \delta_\rho(\hat{b}, \hat{x}), \ \delta_\rho(\hat{c}, \bar{x}) = \delta_\rho(\hat{c}, \hat{x}), \ \delta_\rho(\hat{b}, \bar{y}) = \delta_\rho(\hat{b}, \hat{y}), \text{ and } \delta_\rho(\hat{c}, \bar{y}) = \delta_\rho(\hat{c}, \hat{y}).$$

So, $D(b,c) = \delta_\rho(\hat{b}, \hat{c}) < \rho/2$ together with $(\rho\, 8)$ above implies $\bar{x} = \hat{x}$ and $\bar{y} = \hat{y}$ and, therefore, also $D(x,y) = \delta_\rho(\bar{x}, \bar{y}) = \delta_\rho(\hat{x}, \hat{y})$ as claimed.

The same argument can be applied in case $y \in [a,b]$ holds and $\{b, c, y\}$ is strictly ρ-circular.

Otherwise, however,

$$D(x,b) + D(b,c) = D(c,x) = D(c,a) + D(a,x) = \rho/2$$

and either $y \in [a,b]$ and

$$D(y,b) + D(b,c) = D(c,y) = D(c,a) + D(a,y) = \rho/2$$

or $y \in [b,c]$ must hold.

In the first case, we get $D(x,u) = D(y,u)$ for all $u \in \{a,b,c\}$ and $\hat{x} = \hat{y} = \hat{c}^*$ and while $b \in [c,x] \cap [c,y]$ and therefore, as $\{b,c,x,y\}$ is circular, also $x \in [b,y]$ or $y \in [b,x]$ implies that $D(x,y) = |D(b,x) - D(b,y)|$ must hold. So, we get $D(x,y) = 0 = \delta_\rho(\hat{x}, \hat{y})$, as claimed.

In the second case, we get $b \in [c,x]$ and $y \in [c,b]$ and, therefore, also $\hat{b} \in [\hat{c}, \hat{x}]$ and $\hat{y} \in [\hat{c}, \hat{b}]$ (as $\hat{f}$ is an isometry on $\{a,b,c,x\}$ and on $\{a,b,c,y\}$). So, we must have $b \in [y,x]$ as well as $\hat{b} \in [\hat{y}, \hat{x}]$ and therefore $D(x,y) = D(x,b) + D(b,y) = \delta_\rho(\hat{x}, \hat{b}) + \delta_\rho(\hat{x}, \hat{b}) = \delta_\rho(\hat{x}, \hat{y})$, once again as claimed.   □

Lemma 6. *Let X be a metric space with $|X| \geq 4$, and suppose that every 4-point subspace of X is linear. Then X is linear.*

Proof. First note that every 3-point subspace of X is linear. Let $a, b \in X$ be two distinct points, and let $f : \{a,b\} \to \mathbb{R}$ be an isometric embedding. For every $x \in X$, since $\{a,b,x\}$ is linear, there is a unique element $\hat{x}$ for which $D(a,x) = |f(a) - \hat{x}|$, $D(b,x) = |f(b) - \hat{x}|$ holds, giving rise to a map $\hat{f} : X \to \mathbb{R} : x \mapsto \hat{x}$. For any $x, y \in X - \{a,b\}$, since $\{a,b,x,y\}$ is linear, there is an isometric embedding $g : \{a,b,x,y\} \to \mathbb{R}$. Since $|g(a) - g(b)| = D(a,b) = |\hat{f}(a) - \hat{f}(b)|$, there is an isometry $\sigma : \mathbb{R} \to \mathbb{R}$ such that $\sigma(g(a)) = \hat{a}$, $\sigma(g(b)) = \hat{b}$. Let $h = \sigma \circ g$. Since $h : \{a,b,x,y\} \to \mathbb{R}$ is an isometric embedding, we have automatically $h(x) = \hat{x}$ and $h(y) = \hat{y}$ and $D(x,y) = |h(x) - h(y)| = |\hat{x} - \hat{y}|$. Therefore, $\hat{f} : X \to \mathbb{R}$ is an isometric embedding, and X is linear.   □

Notice that Lemma 6 is a special case of Menger's Theorem, here we have shown that it also follows in a straightforward way from some rather simple observations.

4   Circular Metric Spaces

Lemma 7. *Let $|X| \geq 5$ and suppose that X is 4PC. If X is not linear, then X has a strictly circular 3-point subspace.*

Proof. Since X is not linear and $|X| \geq 5$, X contains a strictly circular 4-point subspace $Y = \{a, b, c, d\}$ by Lemma 6. Then (cf. Lemma 4), Y either contains a strictly circular 3-point subspace and we are done, or it is antipodally circular in which case (see the discussion of antipodally circular 4-point spaces preceding Lemma 4) we may assume that

$$D(a, c) = D(a, b) + D(b, c) = D(c, d) + D(d, a)$$

and

$$D(b, d) = D(b, c) + D(c, d) = D(d, a) + D(a, b)$$

and, hence, also

$$2D(a, c) = D(a, b) + D(b, c) + D(c, d) + D(d, a) = 2D(b, d)$$

holds. So, assume that Y is antipodally circular, choose an arbitrary element $x \in X - Y$ and put $Y_y := \{x\} \cup Y - \{y\}$ for every $y \in Y$. If they were all antipodally circular, we would get $D(x, y) = D(y', y)$ for all $y' \in Y$ and $y \in Y - \{y'\}$ (also in view of that discussion) which, however, is impossible. And if one of these subspaces , say Y_d, were linear, we have either $a, b \in [x, c]$ or $x \in [a, b]$ or $x \in [b, c]$ or $b, c \in [a, x]$. In the first and the second case, $\{x, b, c, d\}$ is neither linear nor antipodally circular. And in the other two cases, $\{x, a, b, d\}$ is neither linear nor antipodally circular.

So, in any case, one of the four spaces Y_y $(y \in Y)$ must be neither linear nor antipodally circular and, thus, must contain a strictly circular 3-point subspace, as required. □

Theorem 1. *For a metric space (X, D), the following four conditions are equivalent:*

(1) *(X, D) is circular.*
(2) *Every finite subset of (X, D) forms a geodesic in some order.*
(3) *Every four points of (X, D) form a geodesic in some order.*
(4) *(X, D) is 4PC.*

Remark 2. Theorem 39.2 in Blumenthal [4], p. 97 implies, as a special case, that if every 4-point subspace of a metric space (X, D) is ρ-circular for a prescribed ρ, then (X, D) is ρ-circular. Our condition (4) is slightly, but also significantly weaker, since it uses "circular" instead of "ρ-circular", and that's it what makes establishing it quite a bit more complicated.

Proof. It is enough to consider the case $|X| \geq 5$.

The implications $(1) \Rightarrow (2) \Rightarrow (3) \Rightarrow (4)$ are either obvious or have been established already above. And $(4) \Rightarrow (1)$ holds because, if X is linear, then it is certainly circular, and if X is not linear, then it contains, according Lemma 5, a strictly circular 3-point subspace and is, therefore, circular in view of by Lemma 5. □

Ptolemy's theorem asserts that, for every four points $x, y, z, w \in \mathbb{R}^2$,

$$\|x - y\| \cdot \|z - w\| + \|y - z\| \cdot \|w - x\| \geq \|x - z\| \cdot \|y - w\|$$

holds, with equality only if x, y, z, w lie on a circle or line in this cyclic order, see e.g. Berger [2], p. 308. Apostol [1] pointed out that this inequality can be extended to $\mathbb{R}^3$, and hence to $\mathbb{R}^n$.

A metric space (X, D) is called a *Ptolemaic* (see Deza and Deza [6]) if every four points $x, y, z, w \in X$ satisfy the inequality

$$D(x, y) \cdot D(z, w) + D(y, z) \cdot D(w, x) \geq D(x, z) \cdot D(y, w).$$

Thus, $\mathbb{R}^2$ with its Euclidean metric is Ptolemaic by Ptolemy's theorem.

Theorem 2. *If (X, D) is strictly circular, and $|X| \geq 4$, then (X, D) is not Ptolemaic and, hence, not Euclidean.*

Proof. Since X is not linear, X contains a strictly circular 4-point subspace $Y = \{a, b, c, d\}$. We may suppose that (a, b, c, d) is a geodesic. Put $\rho := D(a, b) + D(b, c) + D(c, d) + D(d, a)$. Then

$$\begin{aligned}
D(a, c) \cdot D(b, d) &= (D(a, b) + D(b, c)) \cdot (D(b, c) + D(c, d)) \\
&= D(a, b) \cdot D(c, d) + D(b, c) \cdot (D(a, b) + D(b, c) + D(c, d)) \\
&\geq D(a, b) \cdot D(c, d) + D(b, c) \cdot D(a, d),
\end{aligned}$$

and the equality holds only when $D(a, d) = D(a, b) + D(b, c) + D(c, d)$, that is, only when (a, b, c, d) is a shortest geodesic, i.e., $\{a, b, c, d\}$ is linear which is not our case. Therefore, (X, D) is not a Ptolemaic metric space.   □

5   The 3PL Condition

We continue to considering an arbitrary metric space (X, D), and we will establish a special case of a beautiful, much more general theorem of Menger's (see Blumenthal [4]):

Theorem 3. *Every 3PL metric space (X, D) is linear for $|X| \neq 4$, and antipodally circular or linear for $|X| = 4$.*

Proof. First, we show that (X, D) is 4-point circular. Let a, b, c, d be four distinct points in X. Since $\{a, b, c\}$ is linear, we may suppose that $[a - b - c]$ holds. Since $\{a, b, d\}$ and $\{a, c, d\}$ are linear, we have either $[d - a - b]$ or $[a - d - b]$ or $[a - b - d]$ as well as either $[a - c - d]$ or $[a - d - c]$ or $[d - a - c]$.

If $[d - a - b]$ holds, we have $[d - a - b - c]$ by definition. And if $[a - d - b]$ holds, we even have $[a = d = b = c]$ in view $[a - b - c]$. So, we may suppose that $[a - b - d]$ holds.

Further, we have $[a = b = c = d]$ in case $[a - c - d]$ holds. And we have $[a = b = d = c]$ in case $[a - d - c]$ in view of our assumption that $[a - b - d]$ holds. And if $[d - a - c]$ holds, we have $[d = a = b = c]$, again in view of $[a - b - c]$.

Thus, (X, D) is 4-point circular and, hence, circular. If $|X| \geq 5$ and X is not linear, then X has a strictly circular 3-point subspace by Lemma, contradicting that (X, D) is 3-point linear. Hence, if $|X| \geq 5$ or $|X| \leq 3$, then X is linear. And if $|X| = 4$, then X is linear or antipodally circular by Lemma 4.   □

6   Finite Circular Metrics Are Kalmanson Metrics

In this section, we consider a finite metric space (X, D). Recall that the metric D is called a *Kalmanson metric* if there exists a 2-regular connected graph $G = G_D$ with vertex set X and edge set $E = E_D \subseteq \binom{X}{2}$ and a map $w = w_D : \binom{E_D}{2} \to \mathbb{R}_{\geq 0}$ such that $D(x, y) = \sum_{\{e,f\} \in \text{Sep}(x,y)} w(\{e, f\})$ holds for all $x, y \in X$ where $\text{Sep}(x, y)$ denotes the set of all 2-subsets $\{e, f\} \in \binom{E}{2}$ for which x and y are contained in distinct connected components of the (necessarily not connected) graph $G^{\{e,f\}} = (X, E - \{e, f\})$ (cf., e.g., Chepoi and Fichet [5] and the literature quoted there).

Note that, given a 2-regular connected graph G with vertex set X and edge set $E \subseteq \binom{X}{2}$, there exists always, for every metric D defined on X, a unique map $w_D : \binom{E}{2} \to \mathbb{R}$ for which $D(x, y) = \sum_{\{e,f\} \in \text{Sep}(x,y)} w(\{e, f\})$ holds for all $x, y \in X$, as observed by Chepoi and Fichet in [5]. So, the point of this definition is that it requires the map w to have non-negative values, only. We claim:

Theorem 4. *The metric D of any finite circular metric space (X, D) is a Kalmanson metric.*

Remark 3. Michel Deza suggested that this should hold, and we are happy to supply here the simple proof.

Proof. This is a simple consequence of the following four observations: □

Lemma 8. *(i) The restriction $D'|_{X \times X}$ of any Kalmanson metric D' defined on a finite set X' to a subset X of X' is a Kalmanson metric.*

(ii) Given a finite circular metric space (X, D), there exists a finite circular antipodal metric space (X', D') with $X \subseteq X'$ and $D = D'_{X \times X}$, i.e., any finite circular metric space (X, D) has a finite circular antipodal extension.

(iii) A metric space (X, D) with a finite point set X is circular if and only if the degree

$$\deg_D = \deg_{G_D}(x) := \#\{y \in X : \#[x, y] = 2\}$$

of every vertex x in the associated simple (and necessarily connected) vicinity graph G_D of D (i.e., the graph $G_D = (X, E_D)$ with vertex set X and edge set $E_D := \{\{x, y\} \subseteq X : \#[x, y] = 2\}$) never exceeds 2 (and is exactly 2 for every $x \in X$ unless (X, D) is linear), i.e., one has

$$\#\{y \in X : \#[x, y] = 2\} \leq 2$$

for every $x \in X$.

(iv) If (X, D) is a finite circular antipodal metric space, one has

$$D(x,y) = \sum_{\{e,f\}\in\mathrm{Sep}(x,y)} w(\{e,f\})$$

for the map

$$w = w_D : \binom{E}{2} \to \mathbb{R}_{\geq 0}$$

that maps a 2-subset $\{e, f\}$ of E onto 0 unless the antipodal involution $\tau_D : X \to X : x \mapsto \bar{x}$ maps e onto f in which case $w_D(\{e, f\})$ is defined to be the distance between the two vertices in e (or, equivalently, between those in f).

Proof. (i) We leave the rather straight-forward and elementary verification of the first two assertions to the reader.

(ii) Since the restrict 12-page limit is too narrow to contain the proof of Assertion (iii), we have included it in an Appendix for the review process.

(iii) And we note that Assertion (iv) follows easily from the fact that

(a) the *shortest-path* metric induced on the point set X of a finite metric space (X, D) by the associated vicinity graph $G_D = (X, E_D$ coincides with D provided we assign the length $\ell(e) := D(x, y)$ to any edge $e = \{x, y\} \in E_D$ in that graph and that

(b) $\mathrm{Sep}(x, y)$ coincides, for all $x, y \in X$ with the set consisting of all pairs of the form $\{e, \tau_D(e)\} \in \binom{E_D}{2}$ for which e is contained in a shortest path from x to y (if there are two such path, it apparently doesn't matter which one you choose). $\qquad\qquad\square$

Acknowledgment. We wish to thank Michel Deza for various interesting and helpful discussions of the topic of this paper.

References

1. Apostol, T.M.: Ptolemy's inequality and the chordal metric. Math. Mag. **40**, 233–235 (1967)
2. Berger, M.: Geometry I. Springer, Heidelberg (1967). 1977. MAA
3. Blumenthal, L.: A new concept in distance geometry with applications to spherical subsets. Bull. Am. Math. Soc. **47**(6), 435–443 (1941)
4. Blumenthal, L.: Theory and Applications of Distance Geometry, 2nd edn. Chelsce Pub. Co., Bronx (1970)
5. Chepoi, V., Fichet, B.: A note on circular decomposable metrics. Geom. Ded. **69**, 237–240 (1998)
6. Deza, E., Deza, M.M.: Dictionary of Distances. Elsevier, Tokyo (2006)
7. Deza, M., Laurent, M.: Geometry of Cuts and Metrics. Algorithms and Combinatorics, vol. 15. Springer, Heidelberg (1997). https://doi.org/10.1007/978-3-642-04295-9
8. Robinson, P.L.: The sphere is not flat. Am. Math. Monthly **113**, 171–172 (2006)
9. Tóth, L.F.: Langerungen in der Ebene auf Kugel unt im Raum. Springer, Heidelberg (1972). https://doi.org/10.1007/978-3-642-65234-9

A 2.57-Approximation Algorithm for Contig-Based Genomic Scaffold Filling

Qilong Feng[✉], Xiangzhong Meng, Guanlan Tan, and Jianxin Wang

School of Information Science and Engineering, Central South University,
Changsha 410083, People's Republic of China
{csufeng,jxwang}@mail.csu.edu.cn

Abstract. Genomic Scaffold Filling problem forms an important class of problems, and has been paid lots of attention in the literature. In this paper, we study one of the Genomic Scaffold Filling problem, called One-sided-GSF-max-BC problem. The previous approximation ratio for the problem is 2. However, as we pointed out in the introduction part, the ratio 2 algorithm in the literature can only deal with special instances of the problem, not really solve the One-sided-GSF-max-BC problem. In this paper, we give an approximation algorithm of ratio 2.57 for the One-sided-GSF-max-BC problem. Our method is based on auxiliary graphs constructed and two applications of finding maximum matching in auxiliary graphs.

Keywords: Approximation algorithm · Genomic Scaffold Filling

1 Introduction

With the development of Next Generation Sequencing, it is no longer a difficult problem to produce the genomes quickly and cheaply. However, it is common that the genomes obtained are draft, called scaffolds for short, which means that they are incomplete with the losses of some genes. Because most of the current studies of biological sciences are based on the assumption that the genomes can provide the complete information, the draft genomes cannot directly be applied in biological scientific researches because of the unpredictable errors. Thus, lots of attention has been paid on, based on a given reference genome, how to fill the draft genome with the missing genes to make it complete, and as close as possible to the reference genome.

Given a set Σ of symbols and a string $S = s_1 s_2 \ldots s_n$ on Σ, we use $c(S)$ to represent the set of all symbols in S. We now give the definition of the problem.

This work is supported by the National Natural Science Foundation of China under Grants (61672536, 61420106009, 61872450, 61828205), Hunan Provincial Science and Technology Program (2018WK4001).

© Springer Nature Switzerland AG 2019
D.-Z. Du et al. (Eds.): AAIM 2019, LNCS 11640, pp. 95–107, 2019.
https://doi.org/10.1007/978-3-030-27195-4_9

One-sided-GSF-max-BC problem:
Given a complete string G and a scaffold $S = \langle C_1 C_2 \dots C_m \rangle$ based on a same set Σ of symbols with $c(S) \setminus c(G) = \emptyset$, construct a new string S' by inserting the symbols in $c(G) \setminus c(S)$ into the slots in S such that the number of matched-pairs between G and S' is maximized.

Genomic Scaffold Filling problem was first introduced by Muñoz et al. [13]. It was initially defined as following: given a complete genome G(without gene repetition), and a scaffold S, the goal is to insert all the genes, which are missed in S corresponding to G, into S to obtain a complete genome S', so that the DCJ distance [14] between G and S' is minimum. The problem was showed to be solvable in polynomial time in [13]. When measured by maximizing common adjacency distance or minimizing breakpoint distance, the problem can also be solved in polynomial time [6].

The Genomic Scaffold Filling problem becomes harder when considering the repeatability of genes. The computations of most similarity measures between two complete genomes are NP-complete, for instance, the exemplar breakpoint distance [4], the exemplar adjacency distance [3], etc. Jiang et al. [7] showed that the problem is NP-hard by maximizing common adjacency distance, and gave an approximation algorithm with ratio 1.33. Subsequently, several approximation results were presented [8,11]. The best approximation ratio for this problem is 1.2 [11]. If one input of the Genomic Scaffold Filling problem is a complete genome, and the other is a scaffold, the problem is called One-sided Genomic Scaffold Filling problem. If the two given genomes are both scaffolds, the problem is called Two-sided Genomic Scaffold Filling problem. For the Two-sided Genomic Scaffold Filling problem, an approximation algorithm with ratio 2 was given in [7]. After that, several improvements were obtained [9,12]. The current best approximation ratio is 1.4 [12]. By considering the number of common adjacencies as parameter, Bulteau et al. [1] gave FPT algorithms for the two problems, respectively.

In many applications, a scaffold is usually defined as a series of sequential contigs $C_1 C_2 \dots C_m$, in which any contig C_i cannot be modified, and the insertion of missing genes can only be executed at the both ends of the contig. Under this constraint, when there is no duplicate gene, the One-sided Genomic Scaffold Filling problem can be solved in polynomial time [10]. When there exist duplicate genes, Jiang et al. [5] proved that the problem is NP-complete by maximizing the number of common adjacencies. An approximation algorithm with ratio 2 and an FPT algorithm with two parameters (k, the number of the common adjacencies, and d, the number of the most duplications of a gene) were proposed [5]. Bulteau et al. [2] gave a polynomial kernel for this problem, and presented an FPT algorithm by using the number of k-Mer as parameter and two FPT algorithms with breakpoint distance as parameter. Zhu [15] gave a comprehensive survey of the related results.

For the approximation algorithm with ratio 2 given in [5], we point out that it can only cope with special instances. There exist many cases that the algorithm in [5] cannot deal with. For example, assume that there exists an instance including 4 symbols needed to insert, $\{a, a, b, b\}$, and 3 2-strings to be matched, $\{aa, ab, bb\}$. The optimal solution is $aabb$ with 3 matched 2-strings. However, for the algorithm in [5], matching $\{ab, ab\}$ may be obtained. It is easy to see that for this case, the ratio is 3. The occurrence of this case is caused by the fact that the appearing number of a 2-string to be matched is less than the appearing number of both symbols of the 2-string. The algorithm can deal with the instances without the above fact.

In this paper, we study One-sided-GSF-max-BC problem. Based on special construction of auxiliary graphs and maximum matching algorithm, an approximation algorithm with ratio 2.57 is given.

2 Preliminaries

Given a set Σ of symbols and a string $S = s_1 s_2 \ldots s_n$ on Σ, since set $c(S)$ is possibly a multi-set, for simplicity, let $c(S) = \{s'_1 : i_1, \ldots, s'_r : i_r\}$, where $1 \le r \le n$, $s'_1, \ldots, s'_r$ are contained in S, $i_1 + \ldots + i_r = n$, and $1 \le i_h \le n$, for $h = 1, \ldots, r$. For any two symbols x, y in Σ, if S contains at least one substring from $\{xy, yx\}$, then it is called that x, y are adjacent in S, and we also call that symbols x, y form a block in S. For the string S, let $p(S)$ be the set of all blocks in the string S. Similarly, set $p(S)$ is also possibly a multi-set, and let $p(S) = \{s''_1 s''_2 : j_1, \ldots, s''_{t-1} s''_t : j_{t-1}\}$, where $2 \le t \le n$, $s''_1 s''_2, \ldots, s''_{t-1} s''_t$ are contained in S, $j_1 + \ldots + j_{t-1} = n - 1$, and $1 \le j_h \le n - 1$, for $h = 1, \ldots, t - 1$. Note that for any two symbols x, y in Σ, if xy, yx are contained in S, and the number of appearances of xy, yx are a, b, respectively, then, either $xy : a + b$ or $yx : a + b$ is contained in $p(S)$. For a set T, we use $s(T, x)$ to represent the number of appearances of element x in set T.

Given two strings $A = a_1 a_2 \ldots a_n$ and $B = b_1 b_2 \ldots b_m$ on Σ, for a block $a_i a_{i+1}$ in $p(A)$ and a block $b_j b_{j+1}$ (or $b_{j+1} b_j$) in $p(B)$, if $a_i a_{i+1} = b_j b_{j+1}$ (or $a_i a_{i+1} = b_{j+1} b_j$), then we say that the block $a_i a_{i+1}$ is matched with the block $b_j b_{j+1}$ (or $b_{j+1} b_j$), and $(a_i a_{i+1}, b_j b_{j+1})$ (or $(a_i a_{i+1}, b_{j+1} b_j)$) is called a pair. For the blocks in $p(A)$ and $p(B)$, let H be a set of pairs between $p(A)$ and $p(B)$ with maximum size, in which each block from $p(A)$ or $p(B)$ can be matched at most one time, then each block in H is called a matched-block, and each pair in H is called a matched-pair. It is obvious that set H is unique. Assume that $H = \{(\alpha_1, \beta_1), \ldots, (\alpha_h, \beta_h)\}$, where α_i is a matched-block in $p(A)$, β_i is a matched-block in $p(B)$, and (α_i, β_i) is a matched-pair, for $i = 1, \ldots, h$. Let $W(A) = \bigcup \{\alpha_i\}$, and $W(B) = \bigcup \{\beta_i\}$, i.e., $W(A)$ is the set of matched-blocks in $p(A)$, and $W(B)$ is the set of matched-blocks in $p(B)$. Obviously, $W(A)$ is equal to $W(B)$. If $p(A) \setminus W(A)$ is not empty, then each block in $p(A) \setminus W(A)$ is called a breakpoint. Similarly, if $p(B) \setminus W(B)$ is not empty, then each block in $p(B) \setminus W(B)$ is also called a breakpoint. Let $\overline{W}(A) = p(A) \setminus W(A)$, and $\overline{W}(B) = p(B) \setminus W(B)$, i.e., $\overline{W}(A)$ is the set of breakpoints in $p(A)$, and $\overline{W}(B)$ is the set of breakpoints in $p(B)$.

Given a complete string G, and an incomplete scaffold S, the Genomic Scaffold Filling problem is to insert the symbols in $c(G) \setminus c(S)$ into S to obtain a new string with certain constraints. In scaffold S, there are several contigs contained, which is a string of length at least 1. For a contig C in S, no symbol can be inserted into the position between any two consecutive symbols in C. Assume that scaffold S contains contigs $C_1, C_2, \ldots, C_m$, and let $S = \langle C_1 C_2 \ldots C_m \rangle$. Since no symbol can be inserted into each contig in S, symbols in $c(G) \setminus c(S)$ can only be inserted into the positions between C_i and C_{i+1} ($i = 1, \ldots, m-1$), position before C_1, and position after C_m. For simplicity, each position in S that symbols in $c(G) \setminus c(S)$ can be inserted into is called a slot, and it is easy to see that there are exactly $m+1$ slots in S. We can also get that $c(S) = c(C_1) \cup c(C_2) \cup \ldots \cup c(C_m)$, $p(S) = p(C_1) \cup p(C_2) \cup \ldots \cup p(C_m)$.

In this paper, we will use string and symbol instead of genome and gene for simplicity. We define an operation $\bigoplus$, where $\bigoplus$ is the symmetric difference defined by $A \bigoplus B = (A \setminus B) \cup (B \setminus A)$. Let K', K'' be two matchings in a graph, and $K = K' \bigoplus K''$, then we can get that each component in K is either a simple path, or a simple cycle.

3 An Approximation Algorithm for One-sided-GSF-max-BC

Given an instance (G, S) of the One-sided-GSF-max-BC problem, and for each contig C_i in S, let f_i and l_i be the first symbol and the last one of C_i, respectively. Assume that $F(S)$ is the set of all first and last symbols of the contigs in S, i.e., $F(S) = \{f_1, l_1, \ldots, f_m, l_m\}$. For two contigs C_i, C_{i+1} in S, the slot between C_i and C_{i+1} is denoted as (l_i, f_{i+1}), $i = 1, \ldots, m-1$. For consistency, we define the left-most slot of S as $(\sharp, f_1)$, and the right-most slot as $(l_m, \sharp)$.

Breakpoint-Calculation(G, S)
Input: a complete string, G, a scaffold, S
Output: the multi-set of breakpoints in G, $\overline{W}(G)$
1. calculate set $p(G)$, $p(S)$; construct a graph Φ;
2. **for** each block α in $p(G)$ or β in $p(S)$ **do**
        add a corresponding vertex in Φ;
3. **for** each block α in $p(G)$ **do**
        for each block β in $p(S)$ **do**
            if α and β form a pair **then** add an edge between the two corresponding vertices;
4. calculate a maximum matching M in Φ; based on M, the set H can be obtained; get $W(G)$, $W(S)$, $\overline{W}(G)$, $\overline{W}(S)$;
5. **return** $\overline{W}(G)$.

Fig. 1. Algorithm for calculating the set of breakpoint $\overline{W}(G)$

For an instance (G, S) of the One-sided-GSF-max-BC problem, based on Algorithm BC(G, S) in Fig. 1, the set of breakpoints $\overline{W}(G)$ can be obtained.

Let $X = c(G) \setminus c(S)$. Based on X, $F(S)$, and $\overline{W}(G)$, the remaining task of the One-sided-GSF-max-BC problem is to find a set Q of blocks satisfying the following properties: (1) the blocks in Q are constructed by the symbols in $X \cup F(S)$ with constraints that each symbol from X can be used at most two times and each symbol from $F(S)$ can be only used one time; (2) for each block α in Q, there exists a breakpoint β in $\overline{W}(G)$ such that (α, β) can form a matched-pair; (3) the size of Q is maximum. Due to the hardness of finding the set Q, we expect to obtain an approximation solution for the One-sided-GSF-max-BC problem.

The general idea of our approximation algorithm for the One-sided-GSF-max-BC problem is as follows. Firstly, an auxiliary graph Γ_1 is constructed in the following way. Each symbol in $X \cup F(S)$ is viewed as a vertex in Γ_1. For two symbols x, y, $x \in X$, either $y \in X$ or $y \in F(S)$, if there exists a breakpoint β in $\overline{W}(G)$ such that β and the block constructed by x, y form a pair, then add an edge xy into Γ_1. For two symbols $x, y \in F(S)$, assume that contig C_1 contains symbol x, and contig C_2 contains symbol y, C_1 is before C_2 and C_1 is neighboring to C_2 in S. If x is the last symbol of C_1, y is the first symbol of C_2, and there exists a breakpoint β in $\overline{W}(G)$ such that β and the block constructed by x, y form a pair, then add an edge xy into Γ_1.

A maximum matching M_1 in Γ_1 can be found in polynomial time. For each edge xy in M_1, it is easy to see that xy is a block constructed by the symbols x, y. Thus, for each block xy in M_1, the number of appearances of block xy in M_1 is called the indicator-number of block xy. By the construction process of graph Γ_1, for each block xy in M_1, there must exist a breakpoint β in $\overline{W}(G)$ such that xy and β can form a matched-pair. For each matched-pair (α, β) constructed by the edges in M_1 and breakpoints in $\overline{W}(G)$, the indicator-number of α in M_1 may be larger than the number of β in $\overline{W}(G)$. Moreover, if the indicator-number of α in M_1 is larger than the number of β in $\overline{W}(G)$, we call that block α are redundant. Let $\{\alpha_1, \ldots, \alpha_r\}$ be the set of blocks in M_1 after deleting the redundant ones, and let $\{t_1, \ldots, t_r\}$ be the corresponding indicator-number of the blocks in $\{\alpha_1, \ldots, \alpha_r\}$. Thus, let H' be the set of matched-pairs constructed by the blocks in $\{\alpha_1, \ldots, \alpha_r\}$ and breakpoints in $\overline{W}(G)$, and the size of H' is $t_1 + \ldots + t_r$.

Based on H', we update S to get S_1. For each block xy in H' constructed by the symbols in $X \cup F(S)$, we update S by the following cases.

(1) x, y are both from X. We skip the block.

(2) One of x, y is from X, and the other one is from $F(S)$. Without loss of generality, assume that x is from X, and y is from $F(S)$. For vertex y, assume that contig C contains y as the last symbol of C. Since xy forms a block in H', symbol x can be inserted at the right of y in contig C. After the insertion of x into S, denote the new contig obtained by C'. It is easy to see that x becomes the last symbol of contig C'. Then, update S and $F(S)$ based on C' and C. Delete the block xy from H'.

(3) Both x, y are from $F(S)$. Assume that C_1 and C_2 are two contigs containing x, y, respectively. Without loss of generality, assume that the position of C_2 is

after the one of C_1 in S. Then, C_1 and C_2 must be the neighboring contigs in S. We can also get that x must be the last symbol of C_1, and y must be the first symbol of C_2. Based on C_1, C_2 and block xy, a new contig C' formed by C_1 and C_2 can be obtained. Then, update S and $F(S)$ based on C'. Delete the block xy from H'.

Then, we obtain the new scaffold S_1, the set $F(S_1)$ and the remaining set H'. Let $X' = c(G) \setminus c(S_1)$ and use Algorithm BC(G, S_1) to get the new $\overline{W}(G)$. Based on $X', F(S_1), H'$ and $\overline{W}(G)$, graph Γ_2 can be constructed as the same process of constructing graph Γ_1. In Γ_2, if there exists an edge between two vertices whose corresponding symbols form a block in H', we delete the edge. In graph Γ_2, a maximum matching M_2 can be found. Our approximation solution of the One-sided-GSF-max-BC problem is constructed based on M_1 and M_2.

Before giving the detailed process of the approximation algorithm for the One-sided-GSF-max-BC problem, we remark that the redundant blocks in M may have impact on the approximation ratio for the problem. Figure 2 gives an example to illustrate this. For the multi-set of symbols $X = \{a : 2, b : 2\}$ and multi-set of breakpoints $B = \{aa : 1, ab : 1, bb : 1\}$ in Fig. 2, the optimal solution is $aabb$ with three matched-pairs. For the graph in the Fig. 2, $\{ab, ab\}$ can form a maximum matching. It is easy to see that there has only one matched-pair and a redundant block ab in the maximum matching $\{ab, ab\}$, which results in a ratio 3. If the maximum matching of the graph in Fig. 2 is $\{aa, bb\}$, then two matched-pairs can be found by the maximum matching $\{aa, bb\}$, and an approximation result with ration 1.5 can be obtained. By the example, we can get that redundant blocks have impact on the approximation ratio of the algorithm solving the One-sided-GSF-max-BC problem.

Based on the above discussion, our approximation algorithm solving the One-sided-GSF-max-BC problem has the following obstacles. (1) How to construct an auxiliary graph Γ such that no maximum matching in Γ has redundant blocks. (2) How to get the approximation solution of the One-sided-GSF-max-BC problem based on M_1 and M_2, where M_1 is the maximum matching in Γ_1, and M_2 is the maximum matching in Γ_2.

Fig. 2. Impact of redundant blocks on approximation ratio. The graph is constructed based on a multi-set of symbols $X = \{a : 2, b : 2\}$ that should be inserted into S, and a multi-set of breakpoints $B = \{aa : 1, ab : 1, bb : 1\}$.

3.1   Constructing Auxiliary Graphs Γ Without Redundant Blocks

For a given instance (G, S), we figure out $X = c(G) \setminus c(S)$, $F(S)$, and use Algorithm BC(G, S) to get $\overline{W}(G)$. Let $D = X \cup F(S)$, $B = \overline{W}(G)$, and we divide the breakpoints in B into the following two types.

(1) fallible breakpoints set $B_1 \subseteq B$. A breakpoint $xy \in B$ belongs to B_1 if and only if the following conditions are satisfied: if $x \neq y$, $s(D, x) > s(B, xy)$ and $s(D, y) > s(B, xy)$. Otherwise, $s(D, x) \geq 2s(B, xx) + 2$.

(2) infallible breakpoints set $B_2 = B \setminus B_1$.

It is easy to see that B_1 is the set of breakpoints in B that are possibly used to construct redundant blocks.

We now give the general idea of constructing auxiliary graph Γ based on B_1, B_2, and D. For each kind of symbols x in D, we calculate $TD(x)$, which represents the number of blocks containing x, and $TB(x)$, which represents the number of breakpoints containing x in B_1, and $TS(x) = TB(x) - TD(x)$. For each kind of symbols x in D, we construct an ordering of $s(D, x)$ new symbols $x_1, \ldots, x_{s(D,x)}$. Let D' be the set of new symbols obtained from the symbols in D. Let $LD(x)$ be an indicator denoting which symbol in $\{x_1, \ldots, x_{s(D,x)}\}$ is handling, and let $LDS(x)$ be an indicator denoting which breakpoint of $TB(x)$ breakpoints is handling, and initialize them to 1. Assume that Θ is the set of each kind of symbols x in D sorted by ascendant order according to the value of $TS(x)$. A list B_1' of breakpoints in B_1 can be obtained in the following way. Initially, $B_1' = \emptyset$. For each breakpoint xy in B_1, construct two breakpoints xy, yx in B_1', and a common shared indicator $LB(x, y) = 1$. Sort the breakpoints in B_1' according to Θ. Based on B_1' and B_2, the auxiliary graph $\Gamma = (V, E)$ can be constructed as follows. Let $V = D$, i.e., each symbol in D is viewed as a vertex in Γ.

For each symbol x in Θ in order, there are two steps:

(1) For each breakpoint xy in B_1' in order with $LB(x, y) = 1$, the following operations are taken:

(1.1) y is before x in Θ, $LD(x) \leq s(D, x)$, $LDS(x) \leq TD(x)$ and $LDS(y) \leq TD(y)$. We add edges between $x_{LD(x)}$ and y_i ($1 \leq i \leq s(D, y)$) into E, and let $LD(x) = LD(x) + 1$, $LDS(x) = LDS(x) + 1$, $LDS(y) = LDS(y) + 1$, $LB(x, y) = 0$;

(1.2) $y = x$, $LD(x) \leq s(D, x)$ and $LDS(x) \leq TD(x) - 1$. We add edges between $x_{LD(x)}$ and x_i ($1 \leq i \leq s(D, x)$, $i \neq LD(x)$) into E, and let $LD(x) = LD(x) + 1$, $LDS(x) = LDS(x) + 2$, $LB(x, y) = 0$;

(2) For each breakpoint xy in B_1' in reserve order with $LB(x, y) = 1$, the following operation is taken:

(2.1) $LD(x) \leq s(D, x)$, $LDS(x) \leq TD(x)$, $LDS(y) \leq TD(y)$. We add edges between $x_{LD(x)}$ and y_i ($1 \leq i \leq s(D, y)$) into E, and let $LD(x) = LD(x) + 1$, $LDS(x) = LDS(x) + 1$, $LDS(y) = LDS(y) + 1$, $LB(x, y) = 0$;

For each breakpoint xy in B_2, the edges in Γ are constructed by the following way:

(3) $x \neq y$. We add edges between x_i and y_j ($1 \leq i \leq s(D, x)$, $1 \leq j \leq s(D, y)$);

Auxiliary-Graph-Construction(X, $F(S)$, B)

Input: a multi-set of symbols, X, a set of the ends of contigs, $F(S)$, and a multi-setof breakpoints, B

Output: an auxiliary graph Γ

1. let $D = X \cup F(S)$, and get subsets B_1, B_2, respectively;
2. $D' = B'_1 = V = E = \emptyset$;
3. **for** each symbol x in D **do**
   construct $s(D, x)$ new symbols $\{x_1, \ldots, x_{s(D,x)}\}$, and $D' = D' \cup \{x_1, \ldots, x_{s(D,x)}\}$; $LD(x) = LDS(x) = 1$, $TD(x) = TB(x) = 0$;
4. **for** each symbol x in D **do**
   if $x \in X$ **then** $TD(x) = TD(x) + 2$; **else** $TD(x) = TD(x) + 1$;
5. **for** each symbol x in D **do**
   for each breakpoint α in B_1 containing x **do**
   if α contains one x **then** $TB(x) = TB(x) + 1$; **else** $TB(x) = TB(x) + 2$; $TS(x) = TB(x) - TD(x)$;
6. let Θ be the set of symbols in D sorted by ascending order according to the value of $TS(x)$ of each kind symbol x in D;
7. **for** each breakpoint xy in B_1 **do**
   construct two breakpoints xy, yx, add them into B'_1, and they share a common indicator $LB(x, y) = 1$;
8. sort the breakpoints in B'_1 according to Θ order;
9. $V = D'$;
10. **for** each symbol x in Θ in order **do**
    for each breakpoint xy in B'_1 in order with $LB(x, y) = 1$ **do**
    if y is before of x in Θ, $LD(x) \leq s(D, x)$, $LDS(x) \leq TD(x)$ and $LDS(y) \leq TD(y)$ **then** add edges between $x_{LD(x)}$ and y_i ($1 \leq i \leq s(D, y)$) into E, $LD(x) = LD(x) + 1$, $LDS(x) = LDS(x) + 1$, $LDS(y) = LDS(y) + 1$, $LB(x, y) = 0$;
    else if $y = x$, $LD(x) \leq s(D, x)$ and $LDS(x) \leq TD(x) - 1$ **then** add edges between $x_{LD(x)}$ and x_i ($1 \leq i \leq s(D, x)$, $i \neq LD(x)$) into E, $LD(x) = LD(x) + 1$, $LDS(x) = LDS(x) + 2$, $LB(x, y) = 0$;
    for each breakpoint xy in B'_1 in reserve order with $LB(x, y) = 1$ **do**
    if $LD(x) \leq s(D, x)$, $LDS(x) \leq TD(x)$ and $LDS(y) \leq TD(y)$ **then** add edges between $x_{LD(x)}$ and y_i ($1 \leq i \leq s(D, y)$) into E, $LD(x) = LD(x) + 1$, $LDS(x) = LDS(x) + 1$, $LDS(y) = LDS(y) + 1$, $LB(x, y) = 0$;
11. **for** each breakpoint xy in B_2 **do**
    if $x \neq y$ **then** add edges between x_i and y_j into E ($1 \leq i \leq s(D, x), 1 \leq j \leq s(D, y)$); **else** add edges between x_i and x_j into E ($1 \leq i \leq s(D, x)$, $1 \leq j \leq s(D, x)$, $i \neq j$);
12. **for** each edge xy in E **do**
    if $x, y \in F(S)$ and x, y are not in a same slot **then** delete the edge xy;
13. delete all isolated vertices;
14. **return** graph Γ.

Fig. 3. Algorithm for constructing auxiliary graph Γ

(4) $x = y$. We add edges between x_i and x_j ($1 \leq i \leq s(D, x)$, $1 \leq j \leq s(D, x)$, $i \neq j$);

The specific process of constructing auxiliary graph Γ is given in Fig. 3.

3.2   The Approximation Algorithm Based on Maximum Matching

Based on auxiliary graph constructed in the above subsection, our approximation algorithm for the One-sided-GSF-max-BC problem contains the following two processes. For a given instance (G, S) of the One-sided-GSF-max-BC problem, by calling algorithm AGC(X, $F(S)$, $\overline{W}(G)$), a graph Γ_1 can be obtained. The first process is to find a maximum matching M_1 on graph Γ_1, and we will analyze the relation between the edges in M_1 and the optimal solution of the One-sided-GSF-max-BC problem. Based on M_1, our second process is to construct a new auxiliary graph Γ_2 by calling algorithm AGC, and find a maximum matching M_2 in the new auxiliary graph. By studying the relations of M_1, M_2 with the optimal solution, we analyze that the approximation ratio of our algorithm is 2.57.

The first process is given in Fig. 4, by which we can obtain a partially filled scaffold, S_1, and a set of matched-blocks, T. In this process, an auxiliary graph Γ_1 can be constructed by the algorithm in Fig. 3 AGC. Then, a maximum matching M_1 is obtained in Γ_1. For each edge xy in M_1, the scaffold S can be updated as follows:

(1) x, y are the two end symbols of a same slot. We combine the two contigs of the slot to make a new contig and remove the edge from M_1;

(2) x (or y) $\in F(S)$. We insert y (or x) into the contig containing x (or y) with x, y adjacent, and remove the edge from M_1.

The second process is given in Fig. 5, by which we can obtain a partially filled scaffold S_2. The inputs of the algorithm in Fig. 5 contain a complete string G, a partially scaffold S, and a matching T. In this process, an auxiliary graph Γ_2 can be constructed by Algorithm AGC. We remove all the edges in T from Γ_2. Then, a maximum matching M_2 can be found in Γ_2. For each component κ in $T \oplus M_2$, the scaffold S can be updated as follows:

(1) κ is a simple path. Let $\kappa = p_1 p_2 \ldots p_{t-1} p_t$. Do the following operations:

(1.1) $p_1, p_t \in F(S)$ and p_1, p_t are not in a same slot. We insert $p_2 \ldots p_{t-1}$ into the contig containing p_1 with p_1, p_2 adjacent;

(1.2) $p_1, p_t \in F(S)$ and p_1, p_t are in a same slot. We insert $p_2 \ldots p_{t-1}$ to combine the two contigs of the slot to get a new contig with p_1, p_2 adjacent and p_{t-1}, p_t adjacent;

(1.3) p_1 (or p_t) $\in F(S)$. We insert $p_2 \ldots p_{t-1} p_t$ (or $p_1 p_2 \ldots p_{t-1}$) into the contig containing p_1 (or p_t) with p_1, p_2 (or p_{t-1}, p_t) adjacent;

(1.4) For other cases, we insert $p_1 p_2 \ldots p_{t-1} p_t$ into the rightmost of S;

(2) κ is a simple cycle. We delete an arbitrary edge in κ to get a path $p_1 p_2 \ldots p_{t-1} p_t$, and insert $p_1 p_2 \ldots p_{t-1} p_t$ to the rightmost of S.

Finally, by the algorithm in Fig. 6, we combine the two processes and insert the remaining symbols to get the fully filled S'.

First-Round(G, S)
Input: a complete string G, and a scaffold S
Output: a scaffold S_1, and a subset of matchings, T
1. let $X = c(G) \setminus c(S)$;
2. call algorithm BC(G, S) to get set B of breakpoints;
3. let $F(S)$ be the set of first symbol and last symbol of each contig in S;
4. call algorithm AGC($X, F(S), B$) to get a graph Γ_1;
5. find a maximum matching M_1 in Γ_1;
6. $T = \emptyset$, $S_1 = S$;
7. **for** each edge xy in M_1 **do**
    if x, y are the two end symbols of a same slot **then** combine the two contigs of the slot to make a new contig in S_1; **else if** x (or y) $\in F(S)$ **then** insert y (or x) into the contig containing x (or y) with x, y adjacent in S_1; **else** $T = T \cup \{xy\}$;
8. **return** S_1 and T.

Fig. 4. Algorithm FR

Second-Round(G, S, T)
Input: a complete string G, a scaffold S, and a matching T
Output: a scaffold S_2
1. let $X = c(G) \setminus c(S)$;
2. call algorithm BC(G, S) to get set B of breakpoints;
3. let $F(S)$ be the set of first symbol and last symbol of each contig in S;
4. **for** each edge in T **do**
    delete the corresponding breakpoint in B;
5. call algorithm AGC($X, F(S), B$) to get a graph Γ_2;
6. **for** each edge xy in T **do**
    if xy is an edge in Γ_2 **then** delete the edge xy in Γ_2;
7. find a maximum matching M_2 in Γ_2;
8. $S_2 = S$, $\Lambda = T \bigoplus M_2$;
9. **for** each component κ in Λ **do**
    if κ is a simple path, let $\kappa = p_1 p_2 \ldots p_{t-1} p_t$ **then**
        if $p_1, p_t \in F(S)$ **then**
            if p_1, p_t are not in a same slot **then** insert $p_2 \ldots p_{t-1}$ into the contig containing p_1 with p_1, p_2 adjacent in S_2; **else** insert path $p_2 \ldots p_{t-1}$ to combine the two contigs of the slot to get a new contig with p_1, p_2 adjacent and p_{t-1}, p_t adjacent in S_2;
        else if p_1 (or p_t) $\in F(S)$ **then** insert $p_2 \ldots p_{t-1} p_t$ (or $p_1 p_2 \ldots p_{t-1}$) into the contig containing p_1 (or p_t) with p_1, p_2 (or p_{t-1}, p_t) adjacent in S_2; **else** insert $p_1 p_2 \ldots p_{t-1} p_t$ to the rightmost of S_2;
    else delete an arbitrary edge in κ to get a path $p_1 p_2 \ldots p_{t-1} p_t$, insert $p_1 p_2 \ldots p_{t-1} p_t$ to the rightmost of S_2;
10. **return** S_2.

Fig. 5. Algorithm SR

> **Ratio-2.57 Approximation Algorithm**
> Input: a complete string, G, a scaffold, S
> Output: a filled scaffold, S'
> 1. call FR(G, S) to get S_1 and T;
> 2. call SR(G, S_1, T) to get S_2;
> 3. $S' = S_2$;
> 4. insert all symbols in $c(G) \setminus c(S_2)$ to the rightmost in S';
> 5. **return** S'.

Fig. 6. Ratio-2.57 approximation algorithm

Lemma 1. *Each block which is corresponding to an edge in the maximum matching obtained from the graph Γ is not redundant.*

Given a complete string G and a scaffold S, assume that OPT is the optimal solution, APP is the approximation solution, Z is the set of matched-pairs between G and S, k is the number of matched-pairs between G and OPT, and k' is the number of matched-pairs between G and APP. We can get that $k - |Z|$ is the number of matched-pairs contained in OPT and not in Z, and $k' - |Z|$ is the number of matched-pairs in APP and not in Z. Let $r_1 = (k - k_0)/(k' - k_0)$. Let $r_2 = k/k'$. It is easy to see that $r_1 \geq r_2$. Therefore, in the following, we only analyze the value of r_1.

Observation 1. *In the graph Γ returned by algorithm $AGC(X, F(S), B)$, for each vertex x and a breakpoint xy, the adjacent edges can be constructed by the one of the following two ways:*
(i) For a x_i, construct edges between x_i and y_j, $1 \leq j \leq s(X \cup F(S), y)$ (for case $x \neq y$), or construct edges between x_i and y_j, $1 \leq j \leq s(X \cup F(S), y)$, $j \neq i$ (for case $x = y$);
(ii) For a y_j, construct edges between x_i and y_j, $1 \leq i \leq s(X \cup F(S), x)$.

For a vertex x, if the adjacent edges of x are all constructed by the ways in (ii), then vertex x is called a *type-2 vertex*. Otherwise, vertex x is called a *type-1 vertex*. A type-1 vertex x must have a unique corresponding breakpoint from B, called *type-1 breakpoint*, which is used in (i) to construct some edges, denoted by E_x. The neighbors of the type-1 vertex x, which are adjacent to the edges in E_x, are called *type-1 breakpoint neighbors*.

Lemma 2. *For a given instance (G, S) of the One-sided-GSF-max-BC problem, let OPT be an optimal solution. For the algorithm in Fig. 4, let A_1 be the set of matched-blocks obtained by the maximum matching M_1. Then, the size of A_1 is at least $\frac{|OPT|}{3}$.*

Lemma 3. *For a given instance (G, S) of the One-sided-GSF-max-BC problem, let OPT be an optimal solution. For the algorithm in Fig. 5, let A_2 be the set of matched-blocks obtained by the maximum matching M_2. Then, the number of matched-blocks in A_2 is at least $\frac{|OPT|}{18}$.*

Theorem 1. *The ratio of our approximation algorithm is 2.57.*

Proof. Let A be the approximation solution of our algorithm. Based on Lemmas 2 and 3, there is no redundant matched-block in $A_1 \cup A_2$. Thus, $|A| = |A_1| + |A_2| \geq \frac{|OPT|}{3} + \frac{|OPT|}{18} = \frac{7}{18}|OPT|$. Therefore, the ratio of our approximation algorithm is 2.57. □

References

1. Bulteau, L., Carrieri, A.P., Dondi, R.: Fixed-parameter algorithms for scaffold filling. Theoret. Comput. Sci. **568**, 72–83 (2015)
2. Bulteau, L., Fertin, G., Komusiewicz, C.: Beyond adjacency maximization: Scaffold filling for new string distances. In: LIPIcs-Leibniz International Proceedings in Informatics, vol. 78. Schloss Dagstuhl-Leibniz-Zentrum fuer Informatik (2017)
3. Chen, Z., Fu, B., Xu, J., Yang, B., Zhao, Z., Zhu, B.: Non-breaking similarity of genomes with gene repetitions. In: Ma, B., Zhang, K. (eds.) CPM 2007. LNCS, vol. 4580, pp. 119–130. Springer, Heidelberg (2007). https://doi.org/10.1007/978-3-540-73437-6_14
4. Chen, Z., Fu, B., Zhu, B.: The approximability of the exemplar breakpoint distance problem. In: Cheng, S.-W., Poon, C.K. (eds.) AAIM 2006. LNCS, vol. 4041, pp. 291–302. Springer, Heidelberg (2006). https://doi.org/10.1007/11775096_27
5. Jiang, H., Fan, C., Yang, B., Zhong, F., Zhu, D., Zhu, B.: Genomic scaffold filling revisited. In: LIPIcs-Leibniz International Proceedings in Informatics, vol. 54. Schloss Dagstuhl-Leibniz-Zentrum fuer Informatik (2016)
6. Jiang, H., Zheng, C., Sankoff, D., Zhu, B.: Scaffold filling under the breakpoint and related distances. IEEE/ACM Trans. Comput. Biol. Bioinform. **9**(4), 1220–1229 (2012)
7. Jiang, H., Zhong, F., Zhu, B.: Filling scaffolds with gene repetitions: maximizing the number of adjacencies. In: Giancarlo, R., Manzini, G. (eds.) CPM 2011. LNCS, vol. 6661, pp. 55–64. Springer, Heidelberg (2011). https://doi.org/10.1007/978-3-642-21458-5_7
8. Liu, N., Jiang, H., Zhu, D., Zhu, B.: An improved approximation algorithm for scaffold filling to maximize the common adjacencies. IEEE/ACM Trans. Comput. Biol. Bioinform. **10**(4), 905–913 (2013)
9. Liu, N., Zhu, D., Jiang, H., Zhu, B.: A 1.5-approximation algorithm for two-sided scaffold filling. Algorithmica **74**(1), 91–116 (2016)
10. Liu, N., Zou, P., Zhu, B.: A polynomial time solution for permutation scaffold filling. In: Chan, T.-H.H., Li, M., Wang, L. (eds.) COCOA 2016. LNCS, vol. 10043, pp. 782–789. Springer, Cham (2016). https://doi.org/10.1007/978-3-319-48749-6_60
11. Ma, J., Jiang, H.: Notes on the $\frac{6}{5}$-approximation algorithm for one-sided scaffold filling. In: Zhu, D., Bereg, S. (eds.) FAW 2016. LNCS, pp. 145–157. Springer, Heidelberg (2016). https://doi.org/10.1007/978-3-319-39817-4_15
12. Ma, J., Jiang, H., Zhu, D., Zhang, S.: A 1.4-approximation algorithm for two-sided scaffold filling. In: Xiao, M., Rosamond, F. (eds.) FAW 2017. LNCS, vol. 10336, pp. 196–208. Springer, Cham (2017). https://doi.org/10.1007/978-3-319-59605-1_18
13. Muñoz, A., Zheng, C., Zhu, Q., Albert, V.A., Rounsley, S., Sankoff, D.: Scaffold filling, contig fusion and comparative gene order inference. BMC Bioinform. **11**(1), 304 (2010)

14. Yancopoulos, S., Attie, O., Friedberg, R.: Efficient sorting of genomic permutations by translocation, inversion and block interchange. Bioinformatics **21**(16), 3340–3346 (2005)

15. Zhu, B.: Genomic scaffold filling: a progress report. In: Zhu, D., Bereg, S. (eds.) FAW 2016. LNCS, vol. 9711, pp. 8–16. Springer, Heidelberg (2016). https://doi.org/10.1007/978-3-319-39817-4_2

Profit Parameterizations of Dominating Set

Henning Fernau[1] and Ulrike Stege[2]($\boxtimes$)

[1] Fachbereich 4 – Abteilung Informatikwissenschaften, CIRT,
Universität Trier, 54286 Trier, Germany
fernau@uni-trier.de
[2] Department of Computer Science, University of Victoria, Victoria, Canada
ustege@uvic.ca

Abstract. Dominating Set is one of the most classical NP-complete combinatorial graph problems. Unfortunately, the natural parameterization (by solution size) does not make this problem feasible in the sense of Parameterized Complexity. We propose two new views to consider Dominating Set, and a new parameterization of this problem (by the profit parameter) and give algorithms for these parameterizations that show the problems to be in FPT. More precisely, we give a linear-size kernel and a search-tree procedure.

Keywords: Profit problems · Parameterized Complexity · Domination

1 Introduction

The concept of domination in graphs is one of the central topics in graph theory; see [12], with a large number of applications. Recall that a set S of vertices in an undirected graph $G = (V, E)$ is called *dominating set* (for G) if for each $x \in V \setminus S$, x has a neighbor in S. This leads to the following decision problem. k-Dominating Set

Input: $G = (V, E)$, integer $k \geq 0$
Question: Does there exist $S \subseteq V$ such that S is a dominating set for G and $|S| \leq k$?

As decision problem, k-Dominating Set is shown to be NP-complete by Karp in his classical paper [13]. The natural parameterization by k is not helpful in terms of Parameterized Complexity, as k-Dominating Set is W[2]-complete—a standard example in any textbook even in Parameterized Algorithmics [8, 16]. Also its corresponding optimization problem Minimum Dominating Set is hard to approximate and can be approximated up to a logarithmic factor only [10].

In the spirit of the Art of Problem Parameterization (see Chapter 5 in [16]), it appears attractive to look for parameterizations different from the natural one

© Springer Nature Switzerland AG 2019
D.-Z. Du et al. (Eds.): AAIM 2019, LNCS 11640, pp. 108–120, 2019.
https://doi.org/10.1007/978-3-030-27195-4_10

such that computing a good (or optimal) solution for an instance of DOMINAT-
ING SET becomes viable. A now classical trick is to consider the so-called *dual
parameter*. The dual problem of DOMINATING SET, also called NONBLOCKER
(see, e.g., [9]), is defined as follows.

Input: $G = (V, E)$, integer $k_d \geq 0$
Question: Does there exist $N \subseteq V$ such that $S = V \setminus N$ is a dominating set
for G and $|N| \geq k_d$?

Its natural parameterization k_d-NONBLOCKER is in FPT [9]. As MAXIMUM
SPANNING STAR FOREST, the corresponding optimization problem was considered in [2,7,15], in particular proving APX membership. An optimum solution
S^* to a DOMINATING SET instance $G = (V, E)$ translates into an optimum solution N^* to NONBLOCKER by complementation as $N^* = V \setminus S^*$. Slater already
observed this in [19] where nonblocker sets are called *enclaveless sets*.

In this paper, we propose a new re-parameterization for DOMINATING SET,
formalizing the following idea: whenever we include a vertex x into a set S (that
might become a dominating set), we *conquer* (or *buy*) a certain number of new
vertices (or edges), and pay for them with vertex x. Hence, the *profit* gained by
this transaction is (in a simplified computation) the number of conquered objects
minus one. Depending on whether we conquer vertices or edges, we obtain the
following two problem variants where the goal is to identify subgraphs that
permit the purchase of vertex set S resulting in a profit of at least p:

PROFIT DOMINATING SET, EDGE VARIANT (EV)
Input: $G = (V, E)$, integer $p \geq 0$
Question: Does there exist $W \subseteq V$ with the following edge profit property?
The subgraph $G' = (W, E')$ of G induced by W has *edge profit* at least p,
that is, there exists a dominating set $S \subseteq W$ for G' with $|E'| - |S| \geq p$.

PROFIT DOMINATING SET, VERTEX VARIANT (VV)
Input: $G = (V, E)$, integer $p \geq 0$
Question: Does there exist $W \subseteq V$ with the following vertex profit property?
The subgraph $G' = (W, E')$ of G induced by W has *vertex profit* at least p,
that is, there exists a dominating set $S \subseteq W$ for G' with $|W| - |S| \geq p$.

Remark 1. Consider the following reformulation of PROFIT DOMINATING SET,
VV: Given $G = (V, E)$ and $p \geq 0$, find $S \subseteq V$ such that $NB(S) := |\bigcup_{v \in D} N[v]| -
|S| \geq p$. Notably, if G is isolate-free and S a dominating set, then $NB(S) = V \setminus S$.
Moreover, if S is not a dominating set for G, then $S' = S \cup (V \setminus \bigcup_{v \in S} N[v])$ is a
dominating set with $NB(S') = NB(S)$. Hence, there is a vertex set of profit $\geq p$ in
G if and only if G has a dominating set of size $\leq |V| - p$. Thus NONBLOCKER and
PROFIT DOMINATING SET, VV are the same problem for isolate-free graphs.

This motivates us to mainly focus on the edge variant, also because k_d-NON-
BLOCKER was well-studied before. A similar re-parameterization was considered
for VERTEX COVER [17,20]. Here, the goal was to define a variant that considers
as gain the solution's closeness to being a vertex cover (i.e., the more edges are

covered by solution C, the better). In [20], PROFIT VERTEX COVER, EV was introduced and studied as PROFIT COVER: we ask for a subgraph $G' = (V', E')$ and a vertex cover $C \subseteq V'$ for G' with profit $|E'| - |C| \geq p$. PROFIT VERTEX COVER, EV is solvable in time $O^*(c^p)$ for some quite small constant c: $c \approx 1.15$. For PROFIT DOMINATING SET, EV, we show a linear kernel and a branching algorithm running in time $O^*(1.47^p)$.

2   Preliminaries

Throughout this paper we deal with undirected simple graphs. For a vertex x, $N(x)$ is the *open neighborhood* of x, i.e., the set of all vertices adjacent to x. Likewise, $N[x] = N(x) \cup \{x\}$ is the *closed neighborhood* of x. If $N(x) = \emptyset$, x is called an *isolate*. To avoid confusion, we might add as subscript the graph for which we consider neighborhoods. For $X \subseteq V$, $G[X]$ is the subgraph induced by X in G. Given graph $G = (V, E)$ and $S \subseteq V$, we refer to a vertex $v \in V$ as *conquered* by S if $v \notin S$, $x \in S$, and $xv \in E$. Likewise, we refer to edge $vx \in E$ as *conquered* by S if vertex $v, x \in S \cup \bigcup_{s \in S} N[s]$.

2.1   Simple Observations

Remark 2. When defining our problem variants, determine a subgraph instead of an induced subgraph is not advantageous: if S is a dominating set for $H' = (V, E')$, then S is also a dominating set for any supergraph $H = (V, E)$, $E \supseteq E'$.

Remark 1 proves that the vertex profit parameter is a valid alternative parameterization for DOMINATING SET: (1) for isolate-free graphs, S^* is a minimum dominating set for $G = (V, E)$ if and only if S^* is a set with maximum vertex profit $|V \setminus S^*|$; (2) if S is a vertex set with maximum vertex profit, say p, then there is a minimum dominating set $S^* \supseteq S$ with profit p. From an algorithmic perspective, this means that known algorithms for MINIMUM DOMINATING SET can be applied to solve the vertex profit maximization problem, because isolates can always be treated beforehand. We will make use this in the next subsection.

Considering the edge profit case, for isolate-free graph $G = (V, E)$ a dominating set S has edge profit $|E| - |S|$; hence, when conquering all edges, a minimum dominating set has maximum profit. Conversely, consider vertex set S with maximum edge profit. If there is an edge xy not conquered by S, we can add x to S without changing the edge profit (by profit maximality of S). Doing this repeatedly, we obtain a superset $S^* \supseteq S$ with maximum edge profit, which is also a dominating set. That is, for isolate free graphs, a vertex set conquers all edges if and only if it is a dominating set. This proves that also the edge profit parameter is a valid alternative parameterization for DOMINATING SET.

2.2 Vertex Variant

We inherit the following results as corollaries from [9].

Corollary 1. k_d-NONBLOCKER *admits a kernel with at most* $1.67k_d$ *vertices and hence can be solved in time* $O^*(d^{1.67k_d})$, *where* $d < 1.5$ *is the basis of the running time of the best exact algorithm for solving* DOMINATING SET.

Currently, $d < 1.497$ appears the record holder for exact exponential-time algorithms for MINIMUM DOMINATING SET [18]. Thus k_d-DUAL DOMINATING SET, VV can be solved in time $O^*(1.96^{k_d})$. Using Remark 1 we deduce:

Corollary 2. p-PROFIT DOMINATING SET, VV *admits a kernel with at most* $1.67p$ *vertices and can hence be solved in time* $O^*(1.96^p)$.

3 Main Results

We consider the far more involved edge variant. We present a kernelization and a search-tree algorithm for p-PROFIT DOMINATING SET, EV, proving that this is indeed a good approach to make use of standard FPT techniques for domination.

3.1 A Linear Kernel

We first describe briefly how to obtain a linear vertex kernel. This also shows FPT membership.

Proposition 1. p-PROFIT DOMINATING SET, EV *has a kernel with at most* $5p$ *vertices.*

Proof. As discusses earlier, neither isolated vertices nor isolated edges contribute to obtaining a profit of p. Therefore, we remove all isolated vertices and isolated edges from the graph. Profit parameter p remains unchanged.

Now consider a graph where none of these two reduction rules applies. Let (G, p) be the resultant (irreducible) instance. We perform the following greedy procedure that determines a (profit) set S for G. For this, initialize $G' = (V', E')$ with $V' = V$ and $E' = E$, and let $S = \emptyset$. As long as there exists a vertex of degree at least two in G' and $|S| < p$, select a vertex x of highest degree and include it into S. Remove x from G', together with all ℓ conquered edges (from $G[N[x]]$). This adds value $|N(x)| - (\ell - 1)$ to the obtained profit for S in G, which is subtracted from the current profit parameter. The procedure stops if $|S| \geq p$ (in this case we have a proof that (G, p) is a YES-instance) or if the remaining graph has maximum degree one.

Consider the case where $|S| < p$ and the procedure stopped for the second reason. Then, there exist less than $2p$ neighbors of S-vertices that are in $V \setminus S$. Let D denote the set of these (by S dominated) neighbors. Next, consider $X = V \setminus (S \cup D)$. By its definition, each vertex in X has only neighbors in $D \cup X$. A vertex in $D \cup X$ cannot have more than one neighbor since otherwise it would have been promoted to S during our greedy procedure (since including such a vertex into S would increase the profit). Thus, each vertex in X has degree at most one, and exactly one neighbor in D, as otherwise the isolated edge rule would have applied to two vertices in X connected by an edge. Also, the isolated vertex rule prohibits any isolated vertices to appear in X. Likewise, each vertex in D has at most one neighbor in X (since otherwise it would be included into S). Also, X contains no isolates. Hence, $|X| \leq |D| \leq 2|S| < 2p$, showing $|V| < 5p$.   □

It is not immediately clear if this kernelization idea can be used to obtain good approximation algorithms for MAXIMUM PROFIT DOMINATING SET, EV, as ventured for other problems similar to NONBLOCKER in [1].

By the observations in Subsect. 2.1, we can use the currently best exact algorithm for MINIMUM DOMINATING SET to solve p-PROFIT DOMINATING SET, EV in time $O^*(7.5211^p)$. Below, we improve on the algorithm implicitly presented in Proposition 1 considerably.

3.2   A Search-Tree Algorithm

We describe a branching algorithm that deals with annotated instances of p-PROFIT DOMINATING SET, EV, defined as follows: vertex set V of input graph $G = (V, E)$ is partitioned into $N \cup D \cup B$, i.e., $N \cap D = \emptyset$, $N \cap B = \emptyset$, and $D \cap B = \emptyset$, using the following semantics:

- $x \in N$ means that during branching, vertex x was determined **n**ot to belong to the solution that is being constructed.
- $x \in D$ means that x has already been **d**ominated (by another vertex that has in fact been deleted in the branching).
- $x \in B$ is a **b**lank vertex; no decision has been made about this vertex so far.

Then, given $G = (N \cup D \cup B, E)$ and integer p, the task of this annotated version of p-PROFIT DOMINATING SET, EV is to select $S \subseteq B \cup N$ such that $|E(G[N[S]])| - |S| \geq p$.

We describe a number of reduction rules that allows us to analyze the sketch listed in Algorithm 1 using some specific measure. These reduction rules either provide a new (annotated) problem instance or they answer directly YES or NO. Also, the branching step itself will be described in more detail below. However, due to space restrictions, several parts of the proof have been omitted. We are going to show the following result:

Theorem 1. p-PROFIT DOMINATING SET, EV *is solvable in time* $O^*(1.47^p)$.

We begin with a non-annotated instance, i.e., $V = B$ and $N = D = \emptyset$. From this non-annotated instance, we delete all vertices I of degree zero, as these isolates are not part of any small solution (Proposition 1). The resulting instance contains no blank vertices that are isolates. We will maintain this invariant in the course of our algorithm, i.e., we ensure none of the reduction and branching rules that we introduce from now on produces blank isolates. Call PDS with these arguments.

Algorithm 1. Branching algorithm *PDS* for p-PROFIT DOMINATING SET, EV

Require: $G = (V, E)$, where V is partitioned into $N \cup D \cup B$, integer p
Ensure: YES if there is $S \subseteq B \cup N$ such that $|E(G[N[S]])| - |S| \geq p$, NO otherwise
  if any of the reduction rules is applicable **then**
    Apply first applicable rule
    if rule answers YES or NO **then**
      Return this answer.
    else
      Let $G' = (V', E')$ with $V' = N' \cup D' \cup B'$ and p' be this answer.
      Return $PDS(G' = (V', E'), V' = N' \cup D' \cup B', p')$.
    end if
  else if there is a vertex x of degree at least three in $B \cup N$ **then**
    if possible **then**
      Pick x with a neighbor in N.
    else if possible **then**
      Pick x from D.
    else
      Pick x from B.
    end if
    Branch by either including or excluding x.
    Return YES if one of the two recursive branches returns YES.
  else
    Solve remaining 2-regular instance and return the answer.
  end if

Due to the observations from Subsect. 2.1, we henceforth assume that any solution $V' \subseteq V$ for the isolate-free non-annotated instance is in fact a dominating set of the (whole) graph. In the following analysis we make use this property repeatedly. During the algorithm, we keep track of a measure μ defined as

$$\mu(G, p) := p - \omega|N| - |E(G[N])| - |E(G[D])| \text{ where } G = (N \cup D \cup B, E).$$

Note that weight ω is not chosen yet. We, however, assume that $0 \leq \omega \leq 1$.[1] Clearly, for a non-annotated instance (G, p) we obtain $\mu(G, p) = p$. We ensure that whenever value $\mu(G, p)$ drops below zero, we know whether this branch yields a YES- or a NO-answer. Then, we (can and will) measure the progress of our recursive algorithm in terms of μ instead of p. This is reminiscent of the measure-and-conquer approach that is well established for exact exponential-time algorithms but rarely observed for parameterized algorithms; see [6] and [11] for the overall methodology. Intuitively, the choice of μ can be justified as follows.

- Edges between dominated vertices (i.e., $xy \in E$, $x, y \in D$) can be removed as they already belong to the edge set (of the graph) induced by the closed neighborhood of the solution built so far.

---

[1] Note that below we also discuss the possibility replacing term $\omega|N|$ by some function $\omega : N \to [0, 1]$ that assigns weights to vertices from N.

- Edges between vertices x, y in N are conquered eventually, because some neighbor of x and some neighbor of y must eventually be included into the solution (irrespective of presence of absence of edge xy).
- Vertices $x \in N$ show the potential of producing profit, which we simply deduct as early as possible. We demonstrate how to do this using the following reduction rules, dealing mostly with vertices of degree one.

Given $G = (N \cup D \cup B, E)$ and p, we describe the effect of two operations that form the basic steps of our algorithm. The resulting instance is described by $G^* = (N^* \cup D^* \cup B^*, E^*)$ with $V^* = N^* \cup D^* \cup B^*$, p^* as arguments of PDS.

- $x \in B$ is *included* into the solution. Then, $B^* := B \setminus N_G[x]$, $D^* := D \cup (N_G(x) \setminus N)$, $N^* := N \setminus N_G(x)$, $G^* := G[V^*]$, $p^* := p - |E \setminus E^*| + 1$.
- $x \in D$ is *included* into the solution. Then, $B^* := B \setminus N_G(x)$, $D^* := (D \setminus \{x\}) \cup (N_G(x) \setminus N)$, $N^* := N \setminus N_G(x)$, $G^* := G[V^*]$, $p^* := p - |E \setminus E^*| + 1$.
- $x \in B$ is *excluded* from the solution. Then, $B^* := B \setminus \{x\}$, $D^* := D$, $N^* := N \cup \{x\}$, $G^* := G[V^*]$, $p^* := p$.
- $x \in D$ is *excluded* from the solution. Then, $B^* := B$, $D^* := D \setminus \{x\}$, $N^* := N$, $G^* := G[V^*]$, $p^* := p - |E \setminus E^*|$.

We explain the first case; the other operations can be explained similarly. If x is included into the solution, then x and all neighbors of x are dominated by x, explaining B^* and D^*. We can delete vertices in $D^* \cap N$ as well as their incident edges from G since, by definition, vertices in N no longer serve to dominate any vertex in G. We, however, ensure that they are (eventually) dominated. Once this is achieved, they can be deleted. To understand why at this time we can already count and remove edges incident to conquered vertices in N, recall that we build a solution dominating the entire graph. Thus, while these incident edges are not yet conquered, they will be conquered eventually, as the other endpoint of each such edge must be dominated by our final solution. This justifies why edges not present in $G[V^*]$ can be deleted to obtain G^*, even if they are not (yet) incident to two dominated vertices, and why the profit is updated accordingly.

We next analyze how μ is affected when executing the four operations described above. Consider instance $G = (N \cup D \cup B, E)$ with profit parameter p. We begin by considering what happens if $x \in B$ is excluded from the solution.

$$
\begin{aligned}
\mu(G, p) - \mu(G^*, p^*) &= p - \omega|N| - |E(G[N])| - |E(G[D])| \\
&\quad - (p^* - \omega|N^*| - |E(G^*[N^*])| - |E(G^*[D^*])|) \\
&= |N_G(x) \cap N| + \omega \geq 0, \tag{1}
\end{aligned}
$$

because $B^* = B \setminus \{x\}$, $N^* = N \cup \{x\}$, $D^* = D$, $p^* = p$, i.e., only x is moved from B to N. Next we discuss the case where $x \in D$ is excluded from the solution.

$$
\begin{aligned}
\mu(G, p) - \mu(G^*, p^*) &= p - \omega|N| - |E(G[N])| - |E(G[D])| \\
&\quad - (p^* - \omega_N|N^*| - |E(G^*[N^*])| - |E(G^*[D^*])|) \\
&= |E \setminus E^*| - |N_G(x) \cap D| \\
&= |N_G(x) \cap (B \cup N)| \geq 0, \tag{2}
\end{aligned}
$$

because $B^* = B$, $N^* = N$, $D^* = D \setminus \{x\}$, and $p^* = p - |E \setminus E^*|$. In other words, x as well as its incident edges are (completely) removed.

Next we consider the (more complicated) situations where vertex x is included into the solution. Let $\beta = |B \cap N_G(x)|$, $\delta = |D \cap N_G(x)|$, and $\nu = |N \cap N_G(x)|$.

First, assume $x \in B$. Thus, from above we obtain $B^* = B \setminus N_G[x]$, $D^* = D \cup (N_G(x) \setminus N)$, $N^* = N \setminus N_G(x)$. Hence, $|B^*| = |B| - \beta - 1$, $|D^*| = |D| + \beta - |N \cap N_G(x)|$, $|N^*| = |N| - \nu$. Therefore,

$$\mu(G,p) - \mu(G^*, p^*) = p - \omega|N| - |E(G[D])| - (p^* - \omega(|N| - \nu) - |E(G^*[D^*])|),$$

as D and N are independent in G, which also holds true for N^* in G^*. Moreover,

$$E(G^*[D^*]) \setminus E(G[D]) = \left(\sum_{y \in N_G(x) \cap B} |N_G(y) \cap D| \right) + |E(G[N_G(x) \cap B])|. \quad (3)$$

Comparing E^* with E, we observe that only edges incident to x, and those incident to $N \cap N_G(x)$ are removed. Hence

$$p^* = p - |N_G(x)| - \left(\sum_{y \in N \cap N_G(x)} (|N_G(y)| - 1) \right) + 1. \quad (4)$$

As $|N^*| = |N| - \nu$, the measure difference $\mu(G,p) - \mu(G^*, p^*)$ equals

$$|N_G(x)| - 1 + \left(\sum_{y \in N \cap N_G(x)} (|N_G(y)| - 1 - \omega) \right) + |E(G^*[D^*]) \setminus E(G[D])|. \quad (5)$$

Finally, we assume $x \in D$. Consider the case where x is included into the solution. Then Eq. (3) becomes

$$E(G^*[D^*]) \setminus E(G[D]) = \left(\sum_{y \in N_G(x) \cap B} (|N_G(y) \cap D| - 1) \right) + |E(G[N_G(x) \cap B])|.$$

Interestingly, the updated profit is computed by the same formula as Eq. (4), and the measure difference as in Eq. (5).

We next describe several reduction rules, to be applied in the given order. We only describe steps in detail that actually change the instance. Our (recursive) algorithm makes use of one peculiarity: it may return NO for two reasons. (a) there is no solution on this branch at all; (b) the algorithm detected that another branch of the algorithm might deliver a solution that is no worse and possibly better. This latter decision, formalized as the concept of *reference search trees* in [5], turns out to be helpful for our purposes. The main idea of a reference search tree is that the search space can be traversed using a directed acyclic graph that is not necessarily a directed tree. In our case, this is conceptually achieved by cutting branches where we find a previous decision to be wrong when x was included into N, and pointing instead to the branch where x is included into the solution. A positive termination is signaled by YES. We give further justification for rules where the justification is less straightforward.

Few edges abort If $|E| < p$ then NO.

High-degree 1 If there is some $x \in D \cup B$ of degree larger than p, then YES.

Delete N-edges If $x, y \in N$ and $xy \in E$, then delete xy and decrement p.

High-degree 2 If there is some $x \in N$ of degree larger than p, then YES.

Delete D-edges If $x, y \in D$ and $xy \in E$, then delete xy and decrement p.

Isolate N If $x \in N$ is an isolate, then NO.

Isolate D If $x \in D$ is an isolate, then exclude x from the solution.

Observe that reduction rules that return a conclusive YES or NO do not change our measure. More precisely, for the edge deletion rules, both profit and count on the number of edges between N-vertices or D-vertices, respectively, are decremented. For the exclusion of isolates from D, confer Eq. (2). After executing these reduction rules exhaustively, N and D form independent sets in graph instance G. Hence, the measure simplifies to

$$\mu(G, p) = p - \omega|N|.$$

Next, we describe how to deal with vertices of degree one, and graphs consisting of several connected components. One special case of degree-1 vertices concerns isolated edges, i.e., a neighbored pair of vertices of degree one each.

Support vertex abort Let $N(x) = \{y\}$. If $y \in N$ and $x \in B$ then NO.

Pendant non-dominated vertex Let $N(x) = \{y\}$. If $x \in N \cup B$ then include y into the solution.

Pendant dominated vertex Let $N(x) = \{y\}$. If $x \in D$ then delete x from G.

Many components If G has at least p components then YES.

N-Encycled blank vertex If there is some $x \in B$ such that $N_G(x) \subseteq N$, then include x into the solution.

Pendant induced cycle $\boxed{C_3}$: If $x_1 - x_2 - x_3 - x_1$, is an induced cycle in G and x_1 is the only vertex of degree larger than two on this cycle, meaning that $|N_G(x_1)| > 2$ but $|N_G(x_i)| = 2$ for all $i > 1$, then do the following: (i) If $x_1 \in N$, then NO. (ii) Otherwise, include x_1 into the solution.

$\boxed{C_4}$: If $x_1 - x_2 - x_3 - x_4 - x_1$ is an induced cycle in G and x_1 is the only vertex of degree larger than two on this cycle, meaning that $|N_G(x_1)| > 2$ but $|N_G(x_i)| = 2$ for all $i > 1$, then do the following: (i) If $x_1 \in N$ and $x_2, x_4 \in B$, then NO. (ii) If $x_1 \in N$ and $x_4 \in D$, then include x_2 into the solution. (The case "$x_2 \in D$" is symmetric.) (iii) If $x_1 \in B$, $x_2 \notin N$ and $x_4 \in D$, then include x_2 into the solution. (The case "$x_2 \in D$" is symmetric.) (iv) If $x_1 \in D \wedge x_3 \in B$, or if $x_1 \in B \wedge \{x_2, x_4\} \cap N \neq \emptyset$, or if $\{x_1, x_2, x_3, x_4\} \subseteq B$, then include x_3. (v) Otherwise, include x_1 into the solution.

We can argue correctness of these rules and explain why the measure never increases. Returning NO in these rules is justified by reference search tree arguments. Hence, it is crucial that these rules are applied within the branching.

From now on, in particular we can assume that our instance is of minimum degree two. Next, we consider a reduction rule that is most crucial for the correctness of our approach; hence we provide a detailed proof.

Measure Below Zero. If $\mu(G,p) \leq 0$ then YES.

Proposition 2. *If $\mu(G,p) \leq 0$, then (G,p) is a YES-instance.*

Although above we set $\omega \leq 0.5$, our proof is valid for any $\omega \leq 1$. In fact we assume $\omega = 1$ as the worst case scenario in our reasoning.

Proof. By our reduction rules, we have an irreducible instance (G,p) with $G = (B \cup D \cup N, E)$ such that $\mu(G,p) = p - |N| \leq 0$. Moreover, we know that each vertex is of degree at least two. We show that in such a situation, (G,p) is a YES-instance. We prove this by showing that $S = D \cup B$ always provides sufficient profit. By rule **Isolate** N, S forms a dominating set of G.

First we assume that $G[D \cup B]$ contains no edges. In that case, G is bipartite, because $G[N]$ contains no edges by reduction. We then claim that $S = D \cup B$ is a solution that provides sufficient profit $p(S)$. By construction

$$p(S) = \sum_{v \in D \cup B} (|N_G(v)| - 1) = \left(\sum_{v \in D \cup B} |N_G(v)| \right) - |D \cup B|.$$

As every vertex v is of degree at least two, $p(S) \geq |B \cup D|$. Hence, if $|B \cup D| \geq |N|$, the claim is proved. Assume that $|B \cup D| < |N|$. By double counting of the edges and by making use once more of the minimum degree condition, we see that

$$p(S) = \left(\sum_{v \in D \cup B} |N_G(v)| \right) - |D \cup B|$$

$$= \left(\sum_{v \in N} |N_G(v)| \right) - |D \cup B|$$

$$\geq 2|N| - |D \cup B| > |N|.$$

Hence, the claim is shown for the case that $G[D \cup B]$ contains no edges.

We next show how to reduce the general case to this bipartite one. Namely, we subdivide all edges e in $G[D \cup B]$ by inserting new vertices $[e]$, collected in N'. This results in a new graph $G^* = (V^*, E^*)$, with $B^* = B$, $D^* = D$, $N^* = N \cup N'$ and $E^* = (E \setminus E(G[D \cup B])) \cup \{u[e] \mid u \in D \cup B, e \in E(G[D \cup B]), u \in e\}$. Moreover, for instance (G^*, p^*), with $p^* = p + |N'|$, we get $\mu(G^*, p^*) = p^* - |N^*| = (p + |N'|) - (|N| + |N'|) \leq 0$, and every vertex in G^* has degree at least two. As $G^*[D \cup B]$ contains no edges, we know that (G^*, p^*) is a YES-instance. We also know that solution $S = D \cup B$ obtained for (G^*, p^*) is a dominating set (even a vertex cover) of G^*. Hence, profit $p^*(S)$ can be written as $p^*(S) = |E^*| - |S| \geq p^* = p + |N'|$. By construction, $|E^*| = (|E| - |E(G[D \cup B])|) + 2(|E(G[D \cup B])|) = |E| + |E(G[D \cup B])| = |E| + |N'|$, because each edge in $E(G[D \cup B])$ is subdivided by a vertex from N'. Therefore, $|E^*| - |S| = |E| + |N'| - |S| \geq p + |N'|$ implies $p(S) = |E| - |S| \geq p$, i.e., S is a solution to (G,p) that provides sufficient profit to show that (G,p) is a YES-instance. □

Branching Analysis. A first branching analysis (omitted due to space restrictions) shows that for $\omega = 0.5$ p-PROFIT DOMINATING SET, EV can be solved in time $O^*(2^p)$, resulting from a branching vector of $[2, .5]$ in two situations.

In this analysis it becomes apparent that branching vectors improve if we had not chosen $\omega = 0.5$. The reason for this choice resulted from the case when finding N-vertices of degree one (**Pendant non-dominated vertex**). This motivates us to no longer consider ω to be a fixed constant, but rather let its value depend on the degree of the vertex. In the following setting, $x \in N$ is assumed.

$$\omega(x) = \begin{cases} 0.5, & \text{if } |N_G(x)| = 1; \\ 1, & \text{if } |N_G(x)| > 1. \end{cases}$$

Note that there is no need to define ω on vertices of degree zero, as NO is returned. There is one problem with this new definition, namely Rule **Delete N-edges**. When executing this rule, the measure will now increase, which should not be the case. However: when is this rule executed? Not in the beginning, or after executing reduction rules only, because N-vertices are not introduced by reduction rules. Rather, N-vertices and hence N-edges are only created when excluding a blank vertex from the solution. If this would create an N-edge, we would immediately delete it and include this in our considerations concerning the change of the measure. Also, we have to be a bit more careful when creating N-vertices of degree one, as this could also affect the measure, because the function value of ω would change. This could influence the case when vertices and hence edges are deleted, e.g., when a dominated vertex is excluded from the solution. Notice that in the case when a vertex is included into the solution, it is deleted, but its neighbors will not belong to N (anymore).

A detailed branching analysis is omitted due to space restrictions. Note that Proposition 2 remains valid for this modified measure. The worst case, leading to a branching vector of $[3, 1]$ shows up when branching on blank vertices of degree four.

4   Conclusions and Prospects

We presented two approaches to prove that p-PROFIT DOMINATING SET, EV belongs to FPT. The branching algorithm has two features rarely observed with parameterized algorithms: (a) it uses a non-standard measure on the running time, (b) it employs techniques borrowed from reference search trees. Both features are more prominent in exact exponential-time algorithms. This might point into a direction of FPT branching-algorithm design for similar problems.

We propose to study a scaled version of the profit of an edge set, i.e., a vertex set S has a profit of $|E'| - \sigma|S|$, where E' refers to the set of conquered edges. In this paper, we studied the case $\sigma = 1$. In general, scaling factor σ indicates our investment value for vertices in order to conquer edges (our return value). Similar problems have been studied in the literature. Let us spell out (p, σ)-PROFIT DOMINATING SET, VV for $\sigma = 2$. Given $G = (V, E)$ and $p \geq 0$, find $D \subseteq V$ such that $\partial(D) := |\bigcup_{v \in D}(N(v) \setminus D)| - |D| \geq p$. The abbreviation $\partial(D)$

is also known as the *differential* of D [14]. Hence, $\partial(G) := \max_{D \subseteq V} \partial(D) \geq p$ iff (G, p) is a YES instance to $(p, 2)$-PROFIT DOMINATING SET, VV. By results from [1,3], this problem is in FPT, allowing for some $O^*(c^p)$ algorithm. Interestingly, $(p, 2)$-PROFIT DOMINATING SET, VV can be also viewed as the parameterized dual of ROMAN DOMINATING SET; see [4].

Finally, we did not address the natural and interesting question of approximability of profit problems in their natural form as maximization problems.

Acknowledgements. Part of this work resulted from the 16th Bellairs Workshop on Comp. Geometry (Feb. 2017). The 1st-author's research visit at the 2nd-author's institution in late 2018 was financed by DFG project overhead money. The 2nd-author is supported by an NSERC DG. We are grateful for discussions with Iris van Rooij.

References

1. Abu-Khzam, F.N., Bazgan, C., Chopin, M., Fernau, H.: Data reductions and combinatorial bounds for improved approximation algorithms. J. Comput. Syst. Sci. **82**(3), 503–520 (2016)
2. Athanassopoulos, S., Caragiannis, I., Kaklamanis, C., Kyropoulou, M.: An improved approximation bound for spanning star forest and color saving. In: Královič, R., Niwiński, D. (eds.) MFCS 2009. LNCS, vol. 5734, pp. 90–101. Springer, Heidelberg (2009). https://doi.org/10.1007/978-3-642-03816-7_9
3. Bermudo, S., Fernau, H.: Computing the differential of a graph: hardness, approximability and exact algorithms. Discrete Appl. Math. **165**, 69–82 (2014)
4. Bermudo, S., Fernau, H., Sigarreta, J.M.: The differential and the Roman domination number of a graph. Applicable Anal. Discrete Math. **8**, 155–171 (2014)
5. Binkele-Raible, D., Fernau, H.: An exact exponential time algorithm for Power Dominating Set. Algorithmica **63**, 323–346 (2012)
6. Binkele-Raible, D., Fernau, H.: Parameterized measure & conquer for problems with no small kernels. Algorithmica **64**, 189–212 (2012)
7. Chen, N., Engelberg, R., Nguyen, C.T., Raghavendra, P., Rudra, A., Singh, G.: Improved approximation algorithms for the spanning star forest problem. Algorithmica **65**(3), 498–516 (2013)
8. Cygan, M., et al.: Parameterized Algorithms. Springer, Heidelberg (2015). https://doi.org/10.1007/978-3-319-21275-3
9. Dehne, F., Fellows, M., Fernau, H., Prieto, E., Rosamond, F.: NONBLOCKER: parameterized algorithmics for MINIMUM DOMINATING SET. In: Wiedermann, J., Tel, G., Pokorný, J., Bieliková, M., Štuller, J. (eds.) SOFSEM 2006. LNCS, vol. 3831, pp. 237–245. Springer, Heidelberg (2006). https://doi.org/10.1007/11611257_21
10. Feige, U.: A threshold of $\ln n$ for approximating set cover. J. ACM **45**, 634–652 (1998)
11. Fomin, F.V., Kratsch, D.: Exact Exponential Algorithms. Texts in Theoretical Computer Science. Springer, Heidelberg (2010). https://doi.org/10.1007/978-3-642-16533-7
12. Haynes, T.W., Hedetniemi, S.T., Slater, P.J.: Fundamentals of Domination in Graphs. Monographs and Textbooks in Pure and Applied Mathematics, vol. 208. Marcel Dekker, New York (1998)

13. Karp, R.M.: Reducibility among combinatorial problems. In: Miller, R.E., Thatcher, J.W. (eds.) Complexity of Computer Computations, pp. 85–103. Plenum Press, New York (1972)

14. Mashburn, J.L., Haynes, T.W., Hedetniemi, S.M., Hedetniemi, S.T., Slater, P.J.: Differentials in graphs. Utilitas Mathematica **69**, 43–54 (2006)

15. Nguyen, C.T., Shen, J., Hou, M., Sheng, L., Miller, W., Zhang, L.: Approximating the spanning star forest problem and its application to genomic sequence alignment. SIAM J. Comput. **38**(3), 946–962 (2008)

16. Niedermeier, R.: Invitation to Fixed-Parameter Algorithms. Oxford University Press, Oxford (2006)

17. van Rooij, I.: Tractable cognition: complexity theory in cognitive psychology. Ph.D. thesis, University of Victoria, Canada (2003)

18. van Rooij, J.M.M., Bodlaender, H.L.: Exact algorithms for dominating set. Discrete Appl. Math. **159**(17), 2147–2164 (2011)

19. Slater, P.J.: Enclaveless sets and MK-systems. J. Res. Natl. Bureau Standards **82**(3), 197–202 (1977)

20. Stege, U., van Rooij, I., Hertel, A., Hertel, P.: An $O(pn + 1.151^p)$-algorithm for p-profit cover and its practical implications for vertex cover. In: Bose, P., Morin, P. (eds.) ISAAC 2002. LNCS, vol. 2518, pp. 249–261. Springer, Heidelberg (2002). https://doi.org/10.1007/3-540-36136-7_23

Exponential Time Approximation Scheme for TSP

Zhixiang Chen[1], Qilong Feng[2], Bin Fu[1(✉)], Mugang Lin[3], and Jianxin Wang[2]

[1] Department of Computer Science, University of Texas Rio Grande Valley,
Edinburg, TX 78539, USA
bin.fu@utrgv.edu
[2] School of Information, Central Southern University, Changsha, China
[3] School of Computer Science and Technology, Hengyang Normal University,
Hengyang, China

Abstract. In this paper, we develop an exponential time approximation scheme for the traveling salesman problem (TSP) on undirected graphs. If the weight of each edge is a nonnegative real number, then there is an algorithm to give an $(1 + \epsilon)$ approximation for the TSP problem in $O(\frac{1}{\epsilon} \cdot 1.66^n)$ and a polynomial space. It is in contrast to Golovnen's approximation scheme for TSP on directed graphs with $O(\frac{1}{\epsilon} \cdot 2^n)$ time. We also show that there is no $2^{o(n)}$ time constant factor approximation for the TSP problem under Exponential Time Hypothesis in complexity theory.

1 Introduction

The traveling salesman problem (TSP) is one of the classical NP-hard problems in the field of computer science. This problem is a generalization of the Hamiltonian path problem. TSP has been widely studied in the computer science. In this paper, we develop an approximation scheme for the TSP problem.

Bellman [1], Held and Karp [2] developed $O(2^n)$ time exact algorithm for TSP. Their algorithm is still the fastest after more than half century has passed.

Another related problem is Hamiltonian path problem, which has a parameterized problem called k-path problem for finding a simple path of length k. An algebraic approach for the two problems are developed in [3,13–15,17]. Björklund [3] showed an algorithm with time complexity $O(1.66^n)$ for the Hamiltonian path problem. A $O(1.66^n)$ time algorithm for the k-path problem was derived by Bjorklund, Husfeldt, Kaski, and Koivisto [13]. Their algorithm was simplified by Abasi and Bshouty [14].

The general TSP cannot be approximated by any polynomial time algorithm unless P=NP (Sahni and Gonzalez [16]). There are reports about approximation algorithm for the metric TSP problems. A special case is that each edge has weight either 1 or 2, which is called (1,2)-TSP. The (1,2)-TSP problem is MAX-SNP hard (see Papadimitriou, Yannakakis [8]). There is a 2-approximation algorithm for Metric TSP by Rosenkrantz, Stearns and Lewis [5]. A 1.5 factor

© Springer Nature Switzerland AG 2019
D.-Z. Du et al. (Eds.): AAIM 2019, LNCS 11640, pp. 121–128, 2019.
https://doi.org/10.1007/978-3-030-27195-4_11

approximation was achieved by Christofides' algorithm [6]. Berman and Karpinski [9] designed a $\frac{8}{7}$-approximation algorithm for (1,2)-TSP. Karpinski, Lampis, and Schmied [7] showed that undirected metric TSP cannot be approximated in polynomial time with a ratio better than $\frac{123}{122}$, unless P=NP.

It has been an open problem if there exists a $O(2^n \text{poly}(n))$ time and polynomial space algorithm for TSP (see [10]). Golovnen [4] showed a simple approximation scheme for TSP on directed graphs with $O(\frac{1}{\epsilon} \cdot 2^n)$ time, and a polynomial $\frac{1}{\epsilon} \cdot \text{poly}(n, \log M)$ space if the weights are integers in $[0, M]$. In this paper we study approximation algorithm for TSP problem, which allows the weight of each edge to be a nonnegative real number. We show that there is an algorithm to give an $(1 + \epsilon)$ approximation for the TSP problem on undirected graphs in $O(\frac{1}{\epsilon} \cdot 1.66^n)$. Our algorithm calls Björklund [3]'s algorithm polynomial times and uses polynomial space based on the polynomial space property of his algorithm. The TSP problem does not need to satisfy metric conditions in our algorithm. We also show that there is no $2^{o(n)}$ time constant factor approximation for the TSP problem under the exponential time hypothesis [11].

2   Definitions

For an undirected graph $G(V, E)$, let $w(.)$ be the weighted function for its edges. We always assume $w(e) \geq 0$ throughout the paper. We use $(V, E, w(.))$ to represent the weighted graph G with weight function for its edges. For a set of edges X in G, define $w(X) = \sum_{e \in X} w(e)$ to be the sum of weights of edges in X.

A Hamiltonian path in graph G is a path that visits every vertex of G exactly once. The Traveling Salesman Problem (TSP) is, given a graph $G = (V, E)$ with nonnegative real edge weights given by a function $w(.) : E \to [0, +\infty)$, to find a Hamiltonian path C of minimal weight $w(C)$. Metric TSP is TSP restricted to graphs with edge costs satisfying the triangle inequality.

3   Outline of the Algorithm

The input graph $G(V, E)$ has weight function $w(.) : E \to [0, +\infty)$. Our algorithm works at the subgraph (V, E_0) for the set of edges $E_0 \subseteq E$ with small weight edges in the first phase 0, and find a temporary TSP path. Additional edges S_{u+1} are added to E_{u+1} from E in Phase $u + 1$. The weights of edges in S_{u+1} have a difference by a factor at most $g(n)$, where $g(n)$ is a fixed function. The weights of edges are transformed into integers by losing a small accuracy. When edges with large weights are added, the edges of small weight will have zero weight. This will let the TSP path lose a small accuracy.

4   Algorithm and Its Analysis

Let G be a complete undirected graph with n vertices and integer edge weights in range $[0, M]$. Theorem 1 shows the existence of an algorithm that can handle TSP with integer weights in the range $[0, M]$ (See [3,4]).

Theorem 1. *[3] Let G be a complete undirected graph with n vertices and integer edge weights in range $[0, M]$. Then there is an algorithm to solve the TSP problem in $O(1.66^n M)$.*

Proof (Sketch). Use the dynamic programming method to deal with the integer weights. Define a polynomial $F_{k,h}(.)$ such that it has multilinear monomial if and only if there exists a path of k-edges with sum of weights equal to h. The total number of cases for h is $2^{o(n)}$. Transform it into the case of Hamiltonian path problem.

Our algorithm transforms an TSP algorithm for small integer range weights graph to an approximation TSP algorithm for graph with arbitrary nonnegative real weights. The time complexity of our approximate algorithm for TSP does not depend on the range of edge weights, but the algorithm given in Theorem 1 has a time depending on the weight range and requires that the edges weight to be integer.

Theorem 2. *Assume that there is a $t(n, m, z)$ time exact algorithm for TSP on undirected graphs that allows edge weights to be integers in the range $[0, z]$ for every integer parameter z. Then there is a $O(mt(n, m, \left(\frac{2n}{\epsilon}\right)^4 + 1))$ time algorithm such that given a graph with nonnegative real edge weights, it gives an $(1 + \epsilon)$-approximation the TSP for any fixed $\epsilon > 0$, where n and m are the number of vertices and edges of an input graph, respectively.*

Proof. The algorithm has multiple stages. Part of the edges are added to the graph according to the nondecreasing order at each phase. At Phase i, their weights are scaled into integers with weight function $w_i(.)$ from their original weight function $w(.)$.

Algorithm
Input: parameter $\epsilon > 0$, and an undirected graph $G(V, E, w(.))$, where $w(.)$ is the weight function for the edges in G.

- Initialization: Select function $g(n) = \frac{2n}{\epsilon}$.
  Sort the edges by increasing order of their weights.
- **Phase 0:** Let S_0 contain all the edges of weight 0. Let weight function $w_0(.)$ be zero for each edge in S_0.
  Find a TSP P_0 in the graph $(V, E_0, w_0(.))$ with $E_0 = S_0$. Let $T_0 = E - S_0$. Enter final step if $T_0 = \emptyset$. Otherwise, enter Phase 1.
- **Phase 1:** let edge $e_1 \in E$ have the least positive weight $w(e_1) = \min\{w(e) : e \in E \text{ and } w(e) > 0\}$. Let S_1 contain all edges of weight at most $w(e_1)g(n)$. Let z_1 be the largest weight in S_1.
  Define $w_1(.)$ be the weight function such that $w_1(e) = \left\lceil \frac{w(e)g(n)}{w(e_1)} \right\rceil$ for each edge e in S_1, and $w_1(e) = 0$ for each $e \in E_0$.
  Find a TSP P_1 with edges in the graph $(V, E_1, w_1(.))$ with $E_1 = E_0 \cup S_1$. Let $T_1 = T_0 - S_1$. Enter final step if $T_1 = \emptyset$. Otherwise, enter Phase $u + 1$ with $u = 1$.

- **Phase** $u + 1$: $(1 \leq u < |E|)$, find an edge $e_{u+1} \in E - E_u$ with the least $w(e_{u+1})$. Let S_{u+1} contain all edges e with $w(e_{u+1}) \leq w(e) \leq w(e_{u+1})g(n)$. We discuss two cases:

  Case 1. $w(e_{u+1}) \leq w(e_u)g(n)^2$. Define $w_{u+1}(.)$ be the weight function such that $w_{u+1}(e) = \left\lceil \frac{w(e)g(n)}{w(e_u)} \right\rceil$ for the edges in $S_u \cup S_{u+1}$, and $w_{u+1}(e) = 0$ for each $e \in E_{u-1}$.

  Case 2. $w(e_{u+1}) > w(e_u)g(n)^2$. Define $w_{u+1}(.)$ be the weight function such that $w_{u+1}(e) = \left\lceil \frac{w(e)g(n)}{w(e_{u+1})} \right\rceil$ for the edges in S_{u+1}, and $w_{u+1}(e) = 0$ for each $e \in E_u$.

  Find a TSP P_{u+1} in the graph $(V, E_{u+1}, w_{u+1}(.))$ with $E_{u+1} = E_u \cup S_{u+1}$. Let $T_{u+1} = T_u - S_{u+1}$. Enter final step if $T_{u+1} = \emptyset$. Otherwise, enter Phase $u + 2$.

- **Final step**: Output the TSP P_i among $P_0, P_1, \cdots, P_k$ such that $w(P_i)$ is the least.

End of Algorithm

The correctness of this algorithm is based on the following Lemmas. Lemma 1 gives some basic properties about the algorithm.

Lemma 1. *We have the following properties about the algorithm:*

1. *If there are k phases that have been executed in the algorithm, we have $E_k = E$.*
2. *For each $e \in S_i$, $w(e_i) \leq w(e)$.*
3. *For each $e \in S_i$, $w(e) \leq w(e_i)g(n)$.*

Proof. The algorithm keeps adding edges by increasing order of their weights to E_i. At the end of the algorithm, E_k contains all the edges in E. The other parts follow from the construction of S_i in each phase of the algorithm.

Lemma 2. *For each $w_i(.)$, we have $w_i(e) \in [0, g(n)^4 + 1]$ for each edge in E_i.*

Proof. We discuss two different cases. It is easy to verify for both $w_0(.)$ and $w_1(.)$.

For Case 1 of Phase $i = u + 1$ with $u \geq 1$, the function $w_{u+1}(.)$ is defined to be $w_i(e) = \left\lceil \frac{w(e)g(n)}{w(e_u)} \right\rceil \leq \frac{w(e)g(n)}{w(e_u)} + 1 \leq \frac{w(e_{u+1})g(n)^2}{w(e_u)} + 1 \leq g(n)^4 + 1$.

For Case 2 of Phase $i = u + 1$, the function $w_i(.)$ is defined to be $w_i(e) = 0$ for each $e \in E_u$, and $w_i(e) = \left\lceil \frac{w(e)g(n)}{w(e_{u+1})} \right\rceil \leq \frac{w(e)g(n)}{w(e_{u+1})} + 1 \leq g(n)^2 + 1$ for each $e \in E_{u+1} - E_u$.

Lemma 3 gives the accuracy of the algorithm. It shows it gives an $(1 + \epsilon)$-approximate TSP path if it exists in G.

Lemma 3. *Assume that the algorithm is executed with k phases in total, and P_i is the TSP path found by the algorithm at Phase i with $w(P_i) = \min(w(P_1), \cdots, w(P_k))$. Then P_i is a $(1 + \epsilon)$-approximation to an optimal TSP path in $G(V, E)$.*

Proof. It is trivial for $i = 0$ since S_0 contains all edges of weight zero. Assume that it is true for $i \leq u$. Consider Phase $u + 1$. No TSP path has been found among phases i with $i \leq u$. Therefore, there is no TSP path in the graph (V, E_u).

Assume that P^* is an optimal TSP path for the input graph $G(V, E)$. Let i be the least number such that all edges in P^* are in E_i, which is constructed in Phase i. Let P_i be the TSP path found by Phase i. Since i is the least number that all edges of P^* are in E_i, P^* contains at least one edge e^* in $S_i = E_i - E_{i-1}$.

Let $H_1 = S_1$. For $i > 1$, let $H_i = S_{i-1} \cup S_i$ if case 1 is used in phase i, and $H_i = S_i$ if case 2 is used in phase i. Let $e'_1 = e_1$, $e'_i = e_{i-1}$ if case 1 is used in phase i, and $e'_i = e_i$ if case 2 is used in phase i.

For $i = 1$, we have $\sum_{e \in E_i - H_i} w(e) = 0$ since $w(e) = 0$ for every $e \in E_0$, $H_1 = S_1$, and $E_1 - H_1 = E_0$.

In the case 1 of phase i, we have $H_i = S_{i-1} \cup S_i$. Thus, $E_i - H_i = E_{i-2}$, and $w(e) \leq w(e_{i-1})$ for each $e \in E_i - H_i$. As $e^* \in S_i$, we have $w(e_{i-1}) \leq \frac{w(e_i)}{g(n)} \leq \frac{w(e^*)}{g(n)}$. Therefore, we have $\sum_{e \in P_i \cap (E_i - H_i)} w(e) \leq nw(e_{i-1}) \leq \frac{nw(e^*)}{g(n)}$.

In the case 2 of phase i, we have $H_i = S_i$. Thus, $E_i - H_i = E_{i-1}$. Since $w(e_{i-1})g(n)^2 < w(e_i)$, and $w(e) \leq w(e_{i-1})g(n)$ for each $e \in S_{i-1}$, we have $w(e) < \frac{w(e_i)}{g(n)}$ for each $e \in S_{i-1}$. We have $w(e) < \frac{w(e_i)}{g(n)}$ for each $e \in E_i - H_i$ as $w(e) \leq w(e_{i-1})$ for each $e \in E_{i-2}$. Since $e^* \in S_i$, we have $w(e_i) \leq w(e^*)$. Therefore, we have $\sum_{e \in P_i \cap (E_i - H_i)} w(e) \leq nw(e_{i-1})g(n) \leq \frac{nw(e_i)}{g(n)} \leq \frac{nw(e^*)}{g(n)}$.

Since P_i is the least TSP path found in Phase i with graph $(V, E_i, w_i(.))$, we have $w_i(P_i) \leq w_i(P^*)$. We have $\sum_{e \in P_i} w_i(e) \leq \sum_{e \in P^*} w_i(e)$.

Thus,

$$\sum_{e \in P_i \cap H_i} \frac{w(e)g(n)}{w(e'_i)} \leq \sum_{e \in P_i \cap H_i} \left\lceil \frac{w(e)g(n)}{w(e'_i)} \right\rceil \leq \sum_{e \in P^*} \left\lceil \frac{w(e)g(n)}{w(e'_i)} \right\rceil \qquad (1)$$

$$\leq \sum_{e \in P^*} \frac{w(e)g(n)}{w(e'_i)} + 1. \qquad (2)$$

By inequality (2), we have the inequality

$$\sum_{e \in P_i \cap H_i} w(e) \qquad (3)$$

$$\leq \sum_{e \in P^*} \left(w(e) + \frac{w(e'_i)}{g(n)} \right) \leq \left(\sum_{e \in P^*} w(e) \right) + \frac{nw(e'_i)}{g(n)} \qquad (4)$$

$$\leq \left(\sum_{e \in P^*} w(e) \right) + \frac{nw(e^*_i)}{g(n)}. \qquad (5)$$

Therefore,

$$\sum_{e \in P_i} w(e) = \sum_{e \in P_i \cap H_i} w(e) + \sum_{e \in P_i \cap (E_i - H_i)} w(e) \tag{6}$$

$$\leq \left(\sum_{e \in P^*} w(e)\right) + \frac{nw(e^*)}{g(n)} + \frac{nw(e^*)}{g(n)} \tag{7}$$

$$\leq \left(\sum_{e \in P^*} w(e)\right) + \frac{2n}{g(n)} \cdot \left(\sum_{e \in P^*} w(e)\right) \tag{8}$$

$$\leq (1 + \epsilon) \left(\sum_{e \in P^*} w(e)\right). \tag{9}$$

Lemma 4. *The time complexity is at most $|E| t(|V|, |E|, f(|V|))$, where $t(.,.,.)$ is the time function for TSP with integer weigths in the range from 0 to $f(n) = g(n)^4 + 1$.*

Proof. In each phase i with $i \geq 1$, at least one new edge is added into E_i. The main loop of the algorithm iterates at most $|E|$ times. It is easy to see from the algorithm. The definition of function $f(.)$ is based on Lemma 2.

The theorem is proved by Lemmas 3, and 4.

Corollary 1. *There is a $O(\frac{1}{\epsilon} \cdot 1.66^n)$ time randomized $(1 + \epsilon)$-approximation algorithm for TSP in graphs with nonnegative real edge weights.*

Proof. It follows from Theorems 2 and 1.

We note that our algorithm uses polynomial space since Björklund [3] with time complexity $O(1.66^n)$ runs in polynomial space.

5   Lower Bound

In this section, we show a lower bound for the TSP approximation under the well known exponential time hypothesis. The hypothesis states that 3-SAT cannot be solved in subexponential time in the worst case [11]. Many problems can be reduced to each other while preserving the their computational time complexity. A subexponential time algorithm for any one of them implies subexponential time algorithms for all the others.

Theorem 3. *For any $c > 0$, if there is a $f(n, c)$ time c-approximation algorithm for the TSP problem that has edge weights to be integers in the range $[0, (c+1)n]$, then there is a $O(f(n, c))$ time algorithm for the Hamiltonian path problem.*

Proof. Let $G(V, E)$ be graph. Build a weighted complete graph $G'(V, E')$, where $E' = E \cup \{\{u, v\} \notin E\}$, $w(e) = 1$ for every $e \in E$, and $w(e) = (c + 1)n$ for every $e \in E' - E$. A c-approximation algorithm gives an path of sum of weights at most cn if and only if there exists a Hamiltonian path.

Corollary 2. *For any $c > 0$, there is no $2^{o(n)}$ time c-approximation for the TSP problem under the assumption of exponential time hypothesis.*

Proof. It is well known that there is no $2^{o(n)}$ time algorithm for the Hamiltonian path problem under the exponential time hypothesis [12]. It follows from Theorem 3.

6   Conclusions

We developed an exponential time approximation scheme for the TSP problem. Our lower bound shows it is impossible for the existence of subexponential time approximation unless the exponential hypothesis fails. It will be interesting if the time complexity can be further improved. There might be some chance to get a less time complexity approximation scheme for TSP without improving the exact algorithm for the Hamiltonian path problem. A long standing open problem for TSP is to find a $(2 - \epsilon)^n$-time exact algorithm.

Acknowledgments. The authors would like to thank the reviewers whose suggestions improve the presentation of this paper. This research is supported by NSFC 61772179, NSFC 61872450 and Hunan Provincial Natural Science Foundation 2019JJ40005.

References

1. Bellman, R.: Dynamic programming treatment of the travelling salesman problem. J. ACM **9**, 61–63 (1962)
2. Held, M., Karp, R.M.: A dynamic programming approach to sequencing problems. J. Soc. Ind. Appl. Math. **10**(1), 196–210 (1962)
3. Björklund, A.: Determinant sums for undirected hamiltonicity. In: Proceedings of the 2010 IEEE 51st Annual Symposium on Foundations of Computer Science. FOCS 2010, pp. 173–182. IEEE Computer Society, Washington, DC (2010)
4. Golovnev, A.: Approximating asymmetric TSP in exponential time. Int. J. Found. Comput. Sci.£. **25**(01), 89–99 (2014)
5. Rosenkrantz, D.J., Stearns, R.E., Lewis, P.M.: An analysis of several heuristics for the traveling salesman problem. SIAM J. Comput. **6**(3), 563–581 (1977)
6. Christofides, N.: Worst-case analysis of a new heuristic for the traveling salesman problem. Technical report 338, Graduate School of Industrial Administration, CMU (1976)
7. Karpinski, M., Lampis, M., Schmied, R.: New inapproximability bounds for TSP. CoRR abs/1303.6437 (2013)
8. Papadimitriou, C.H., Yannakakis, M.: The traveling salesman problem with distances one and two. Math. Oper. Res. **18**, 1–11 (1993)

9. Berman, P., Karpinski, M.: $\frac{8}{7}$-approximation algorithm for (1,2)- TSP. In: Proceedings of the Seventeenth Annual ACM-SIAM Symposium on Discrete Algorithm, SODA 2006, pp. 641–648. ACM, New York (2006)

10. Woeginger, G.J.: Open problems around exact algorithms. Discrete Appl. Math. **156**, 397–405 (2008)

11. Impagliazzo, R., Paturi, R.: The complexity of k-SAT. In: Proceedings of the 14th IEEE Conference on Computational Complexity, pp. 237–240 (1999). 1999.766282. https://doi.org/10.1109/CCC

12. Impagliazzo, R., Paturi, R., Zane, F.: Which problems have strongly exponential complexity? J. Comput. Syst. Sci. **63**(4), 512–530 (2001)

13. Björklund, A., Husfeldt, T., Kaski, P., Koivisto, M.: Narrow sieves for parameterized paths and packings. arXiv:1007.1161v1 (2010)

14. Abasi, H., Bshouty, N.H.: A simple algorithm for undirected hamiltonicity. Electronic Colloquium on Computational Complexity, Report No. 12 (2013)

15. Koutis, I.: Faster algebraic algorithms for path and packing problems. In: Aceto, L., Damgård, I., Goldberg, L.A., Halldórsson, M.M., Ingólfsdóttir, A., Walukiewicz, I. (eds.) ICALP 2008. LNCS, vol. 5125, pp. 575–586. Springer, Heidelberg (2008). https://doi.org/10.1007/978-3-540-70575-8_47

16. Sahni, S., Gonzalez, T.: P-complete approximation problems. J. ACM **23**, 555–565 (1976)

17. Williams, R.: Finding paths of length k in $O^*(2^k)$. Inform. Process Lett. **109**(6), 301–338 (2009)

Interaction-Aware Influence Maximization and Iterated Sandwich Method

Chuangen Gao[1], Shuyang Gu[2], Ruiqi Yang[3], Jiguo Yu[1,4(✉)], Weili Wu[2], and Dachuan Xu[3]

[1] School of Computer Science and Technology,
Qilu University of Technology (Shandong Academy of Sciences),
Jinan 250353, Shandong, People's Republic of China
gaochuangen@gmail.com, jiguoyu@sina.com
[2] Department of Computer Science, University of Texas at Dallas, Dallas, USA
{Shuyang.Gu,weiliwu}@utdallas.edu
[3] Department of Information and Operations Research,
Beijing University of Technology, Beijing 100124, People's Republic of China
yangruiqi@emails.bjut.edu.cn, xudc@bjut.edu.cn
[4] Shandong Computer Science Center (National Supercomputer Center in Jinan),
Jinan 250014, Shandong, People's Republic of China

Abstract. Influence maximization problem has been studied extensively with the development of online social networks. Most of the existing works focus on the maximization of influence spread under the assumption that the number of influenced users determines the success of a product promotion. However, the profit of some products such as online game depends on the interactions among users besides the number of users. In this paper, we take both the number of active users and the user-to-user interactions into account and propose the interaction-aware influence maximization problem. To address this practical issue, we analyze its complexity and modularity, propose the sandwich theory which is based on decomposing the non-submodular objective function into the difference of two submodular functions and design iterated sandwich algorithm which is guaranteed to get data dependent approximation solution.

Keywords: Social networks · Influence maximization · Submodular · DS decomposition · Iterated sandwich algorithm

1 Introduction

Viral marketing has long been acknowledged as an effective marketing strategy. The development of online social networks such as Facebook and Twitter provide opportunities for large-scale online viral marketing in social networks. Under this circumstance, influence maximization [8] becomes a very popular research

© Springer Nature Switzerland AG 2019
D.-Z. Du et al. (Eds.): AAIM 2019, LNCS 11640, pp. 129–141, 2019.
https://doi.org/10.1007/978-3-030-27195-4_12

direction in the past decade, which could be described as the problem of finding a small set of most influential nodes so that the spread of influence in the network is maximized.

Most of the works focus on maximization of the spread of influence, which considers the number of users influenced by "word-of-mouth" effect in online social networks. These works are based on the assumption that the number of influenced users determines the profit of product. However, some types of products earn profit in a continuous way besides the sales of product itself. The online game is a good example. The game company's revenue usually comes from two parts, one is the revenue from selling the game product itself, and the other is from the proceeds of advertising and virtual item products. For the first part of revenue, the value of a single game product itself is fixed. The more game players buy game products, the more they earn. For the second part of the revenue is related to the interaction of the game player. When multiple people enter the same game scene online, the advertisement will be displayed and browsed. The more frequent the interaction between players, the more times an advertisement is presented and viewed, which will lead to more advertising revenue. In addition, when players participate in the game, they will use some props to complete the task. These items are virtual equipment, which can increase the experience and fun of the game players. These virtual products will also bring certain benefits [4].

We analyze such revenue model and define the interaction-aware influence maximization problem selecting a seed set to maximize the revenue dependent on the number of influenced users and the interaction between activated nodes. The interaction-aware influence maximization problem is not submodular, thus the greedy strategy can't be directly applied to our problem to get a guaranteed approximate solution. To solve this problem, we propose sandwich theory which is based on the decomposition strategy that represents objective function as a difference of two submodular functions. And based on sandwich theory and the decomposition we design two iterated sandwich algorithms.

The contributions of this paper are summarized as follows.

- We propose a new problem named interaction-aware influence maximization and we prove it is NP-hard and non-submodular.
- To solve this non-submodular problem, we propose the sandwich theory that for any set function there are a modular upper bound and a modular lower bound respectively. The sanwich theory is mainly based on the fact that any set function can be expressed as a difference between two submodular functions. And we successfully decompose our objective function into the difference of two submodular functions which are monotone nondecreasing.
- Based on the sandwich theory and the decomposed submodular functions mentioned above, we design an iterated sandwich algorithm to solve the interaction-aware influence maximization problem, which can get a data dependent approximate solution.

2   Related Works

Influence maximization was first described as an algorithm problem by Domingos and Richardson [3,12], they model the problem using Markov random fields and propose heuristic solutions. Kempe *et al.* [8] formulated the influence maximization problem from the view of combinatorial optimization and showed that the problem is NP-hard under both the IC and LT models, they propose a simple greedy algorithm with an approximation ratio of $(1 - 1/e)$. However a drawback of their work is the scalability of the greedy algorithm. Since then a number of efficient heuristic algorithms haven been proposed in many works [1,2,5,7,10,13]. In [9], Leskovec *et al.* present a "lazy-forward" optimization in selecting new seeds, in which submodularity is exploited.

Most of the works only consider the number of activated users, and the activities between users is first processed by [14]. However, their work does not maximize the influence spread in the meantime and only count activity strength of the directly connected users. In this paper, we propose the interaction-aware influence maximization problem which take both parts into consideration.

3   Problem Formulation

In this section, we formulate interaction-aware influence maximization problem in IC model formally and prove it is neither submodular nor supermodular by counter examples. For complexity we prove it is NP-hard by a special case of the problem.

3.1   Interaction-Aware Influence Maximization

In this paper, we use the directed graph $G = (V, E)$ to represent a social network, where V is the set of users and E is the set of social relations between users. Each edge $(u, v) \in E$ is assigned with a probability p_{uv} so that when u is active, v is activated by u with probability p_{uv}. And the benefit related to the interaction between nodes is represented by a nonnegative function $b : V \times V \rightarrow \mathbb{R}_{\geq 0}$, in which $b(u, v) = b(v, u)$ for the unordered pair $\{u, v\}$ of node u and v. Our goal is to find a set of initial users to maximize total profit related to both the number of the influenced nodes and the interaction between influenced nodes.

Since the randomness of propagation process in IC model, consider a point in the cascade process when node u has just become active, and it attempts to activate its neighbor v, succeeding with probability $p_{u,v}$. We can view the outcome of this random event as being determined by flipping a coin of bias $p_{u,v}$. With all the coins flipped in advance, the edges in G for which the coin flip indicated a successful activation are declared to be live; the remaining edges are declared to be blocked [8]. We use g to represent the outcome of this process which is called a live graph of G since it consists of all edges declared to be live. We denote as $g \sim D$, where D is the distribution of g. For any seed set S, denote by $I_g(S)$ the set of all active nodes at end of the cascade process in live graph g. It's cardinality is represented by $|I_g(S)|$.

Definition 1. *The total expected benefit would be defined as*

$$f(S) = \mathbb{E}_{g \sim D}[\alpha \cdot |I_g(S)| + \beta \cdot \sum_{\{u,v\} \subseteq I_g(S)} b(u,v)]$$

$$= \sum_g Prob[g] \cdot (\alpha \cdot |I_g(S)| + \beta \cdot \sum_{\{u,v\} \subseteq I_g(S)} b(u,v)) \qquad (1)$$

The benefit consists of two parts, the first part denoted as $\alpha \cdot I_g(S)$ is related to the number of nodes that are finally activated, and the second part $\beta \cdot \sum_{\{u,v\} \subseteq I_g(S)} b(u,v)$ is related to the strength of interaction between the active nodes. The parameters α, β are used to balance the weight of two parts of the profits, and $\{u,v\} \subseteq I(S)$ denotes the all unordered pair in the set $I(S)$. Note that for each unordered pair $\{u,v\}$, since $b(u,v) = b(v,u)$, we only compute once the benefit between them. The expectation is respected to g.

In this paper, we study the following problem.

Definition 2 (Interaction-aware Influence Maximization Problem, IAIM). *Given a social network $G = (V, E)$, a propagation probability p_{uv} for each edge (u,v) under the IC model, a benefit function $b : V \times V \to \mathbb{R}_{\geq 0}$, and a positive integer k, find a set S of k seeds to maximize the expected profit through influence propagation:*

$$\max f(S) \qquad (2)$$

$$s.t. |S| \leq k \qquad (3)$$

3.2 Modularity of Objective Function

We say that $g(\cdot)$ is submodular if it satisfies a natural "diminishing returns" property: the marginal gain from adding an element to a set X is at least as high as the marginal gain from adding the same element to a superset of X. Formally, for every set X, Y such that $X \subseteq Y \subseteq V$ and every $e \in V \setminus Y$, it follows that

$$g(X \cup \{e\}) - g(X) \geq g(Y \cup \{e\}) - g(Y)$$

And it is monotone if $g(X) \leq g(Y)$ whenever $X \subseteq Y$.

Theorem 1. *$f(S)$ is neither submodular nor supermodular under IC model.*

Proof. We prove by the counter example shown in Fig. 1. The first element in the tuple tied on each edge represents the propagation probability, and the second one denote the benefit between its two end nodes. For pairs $\{u,v\}$ between which there is no edge set $b(u,v) = 0$ except pair $\{b,d\}$. In Fig. 1, $(0,1)$ on edge (a,b) means propogation probability $p_{ab} = 0$ and $b(a,b) = 1$, then we have $f(\{a\}) = 1 + 0 = 1$, $f(\{a,b\}) = 2 + 1 = 3$, $f(\{a,d\}) = 2 + 0 = 2$ and $f(\{a,b,d\}) = 3 + 3 = 6$. Thus, $f(\{a,d\}) - f(\{a\}) < f(\{a,b,d\}) - f(\{a,b\})$, which implies $f(S)$ is not submodular. Also, we have $f(\{c\}) = 2 + 2 = 4$, $f(\{d,c\}) = 2 + 2 = 4$, $f(\{d\}) = 1$. Thus, $f(\{c\}) - f(\emptyset) > f(\{d,c\}) - f(\{c\})$ which implies $f(S)$ is not supermodular. $\square$

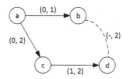

Fig. 1. Counter example

3.3 Hardness Result

Theorem 2. *Interaction-aware influence maximization problem is NP-hard.*

Proof. We prove by showing a special case of Interaction-aware influence maximization problem is NP-hard, where $\beta = 0$, then it become the traditional influence maximization problem which is $NP - hard$. Note that a problem is NP-hard in a special case implies NP-hardness in general case.   □

4 Sandwich Theory

Since interaction-aware influence maximization problem is not submodular, the greedy strategy can't be directly applied to our problem to get a guaranteed approximate solution. To solve this non-submodular problem, we propose the sandwich theory that for any set function there are a modular upper bound and a modular lower bound respectively [15]. The sandwich theory is mainly based on the fact that any set function can be expressed as a difference between two submodular functions [11].

4.1 Preliminary

Before proposing our sandwich theory, let's first introduce a few important conclusions about the submodular function and the set function [6,11].

Lemma 1. *For any submodular set function $g(\cdot)$ on ground set V, we have the following two tight modular upper bounds that are tight at a given set Y [6]:*

$$U_{Y,1}^g(X) \triangleq g(Y) - \sum_{j \in Y \setminus X} g(j \mid Y \setminus j) + \sum_{j \in X \setminus Y} g(j \mid \emptyset) \tag{4}$$

$$U_{Y,2}^g(X) \triangleq g(Y) - \sum_{j \in Y \setminus X} g(j \mid V \setminus j) + \sum_{j \in X \setminus Y} g(j \mid Y) \tag{5}$$

Lemma 2. *For any submodular set function $g(\cdot)$, a modular lower bound of $g(\cdot)$ is tight at a given set Y can be obtained as follows [6]. Let σ be a permutation of V and define $P_i^\sigma = \{\sigma(1), \sigma(2), \ldots, \sigma(i)\}$ as σ's chain containing Y, in which $P_0^\sigma = \emptyset$ and $P_{|Y|}^\sigma = Y$. Define*

$$L^g_{Y,\sigma}(\sigma(i)) = g(P^\sigma_i) - g(P^\sigma_{i-1}).$$ (6)

Then

$$L^g_{Y,\sigma}(X) \triangleq \sum_{v \in X} L^g_{Y,\sigma}(v)$$ (7)

is a tight lower bound of $g(X)$, i.e., $L^g_{Y,\sigma}(X) \le g(X)$, $\forall X \subseteq V$, and $L^g_{Y,\sigma}(Y) = g(Y)$.

Lemma 3. *Every set function $f : 2^X \to R$ can be expressed as the difference of two monotone nondecreasing submodular functions f_1 and f_2, i.e $f = f_1 - f_2$, where X is a finite set [11].*

4.2   Sandwich Theory

Theorem 3. *For any set function $f : 2^X \to R$ and any set $Y \subset X$, there are two modular functions $m^u_f : 2^X \to R$ and $m^l_f : 2^X \to R$ such that $m^u_f(X) \ge f(X) \ge m^l_f(X)$ and $m^u_f(Y) = f(Y) = m^l_f(Y)$.*

Proof. This theorem means that for any set function we can find a modular upper bound and modular lower bound which are exact at some given point. By Lemma 3, there exist submodular function f_1 and f_2 such that $f = f_1 - f_2$. By Lemmas 1 and 2, there exist modular functions $U^{f_1}_Y, L^{f_1}_Y$ such that $U^{f_1}_Y(X) \ge f_1(X) \ge L^{f_1}_Y(X)$, and $U^{f_1}_Y(Y) = f_1(Y) = L^{f_1}_Y(Y)$ for submodular function f_1. By the same reason, there exist modular functions $U^{f_2}_Y, L^{f_2}_Y$ such that $U^{f_2}_Y(X) \ge f_2(X) \ge L^{f_2}_Y(X)$, and $U^{f_2}_Y(Y) = f_2(Y) = L^{f_2}_Y(Y)$ for submodular function f_2. We set $m^u_f = U^{f_1}_Y - L^{f_2}_Y$, and $m^l_f = L^{f_1}_Y - U^{f_2}_Y$, then $m^u_f(X) \ge f(X) \ge m^l_f(X)$ and $m^u_f(Y) = m^l_f(Y) = f(Y) = f_1(Y) - f_2(Y)$. Note that both m^u_f and m^l_f are modular, since the linear combination of modular functions is still modular.   □

4.3   DS Decomposition

Since our sandwich theorem is based on the DS decomposition of a set function that expressing it as a Difference between two Submodular functions. Thus the key point is find such a decomposition. However, it is unknown whether there exists a polynomial-time algorithm for finding such a pair of monotone nondecreasing submodular functions for every given set function. Moreover, the DS decomposition in this paper is nontrivial and two constructed monotone nondecreasing submodular functions are easily computable.

Give a seed set S and a live graph g, we define the $B_1(S)$ as benefit between activated users $I_g(S)$ and all users V, and define $B_2(S)$ as the benefit among all activated users $I_g(S)$ plus the benefit between the activated users $I_g(S)$ and the non-activated users $V \setminus I(S)$, which are formulated as follows:

$$B_1(S) = \sum_{u \in I_g(S)} \sum_{v \in V} b(u, v) \tag{8}$$

$$B_2(S) = \sum_{\{u,v\} \subseteq I_g(S)} b(u, v) + \sum_{u \in I_g(S)} \sum_{v \in V \setminus I_g(S)} b(u, v) \tag{9}$$

And given a seed set S, we define the following functions

$$f_1(S) = \mathbb{E}_{g \sim D}[\alpha \cdot |I_g(S)| + \beta \cdot B_1(S)] \tag{10}$$
$$f_2(S) = \mathbb{E}_{g \sim D}[\beta \cdot B_2(S)] \tag{11}$$

Then we have

$$f(S) = \mathbb{E}_{g \sim D}[\alpha \cdot |I_g(S)| + \beta \cdot \sum_{\{u,v\} \subseteq I(S)} b(u, v)]$$
$$= \mathbb{E}_{g \sim D}[\alpha \cdot |I_g(S)| + \beta \cdot (B_1(S) - B_2(S))]$$
$$= \mathbb{E}_{g \sim D}[\alpha \cdot |I_g(S)| + \beta \cdot B_1(S)] - \mathbb{E}_{g \sim D}[\beta \cdot B_2(S)]$$
$$= f_1(S) - f_2(S) \tag{12}$$

Actually $f(S)$ is decomposed as a difference between function f_1 and f_2, now we prove both of them are submodular.

Lemma 4. $B_1(S)$ *is submodular and monotone under the IC model.*

Proof. According the definition of $B_1(S)$ shown in Eq. 8, we have

$$B_1(S) = \sum_{u \in I_g(S)} \sum_{v \in V} b(u, v)$$

$$= \sum_{\{u,v\} \subseteq I_g(S)} 2 \cdot b(u, v) + \sum_{u \in I(S)} \sum_{v \in V \setminus I_g(S)} b(u, v) \tag{13}$$

$$= \sum_{v \in I_g(S)} w(v) \tag{14}$$

where $I_g(S)$ denotes the set of all active nodes in a live graph g, and $w(v)$ is the weight of the node v which is defined as follows

$$w(v) = \sum_{u \in V} b(v, u) \tag{15}$$

It is actually the sum of benefit between v and the remaining nodes in V. Thus, we can see that the $B_1(S)$ is essentially a weighted version of influence spread. And the submodularity follows immediately [8].   □

Since the profit function $b : V \times V \rightarrow \mathbb{R}_{\geq 0}$ is nonnegative which means the profit of each pair of nodes is non-negative. Thus the weight of every node is non-negative and the monotonicity of $B_1(S)$ follows immediately. For the submodularity, we need prove $B_1(M \cup \{v\}) - B_1(M) \geq B_1(N \cup \{v\}) - B_1(N)$, such that $M \subseteq N \subseteq V$ and $v \in V \setminus N$. The left side of inequality is the weight of nodes which can be activated by v but can not by M. The right side is the weight of nodes which can be activated by v but can not by N. We have $I_g(v) - I_g(M) \supseteq I_g(v) - I_g(N)$, since $M \subseteq N$ and $I_g(M) \subseteq I_g(N)$. And the submodularity follows immediately.   □

Theorem 4. $f_1(S)$ *is submodular and monotone under the IC model.*

Proof. According the definition of $f_1(S)$ shown in Eq. 10, we have

$$f_1(S) = \mathbb{E}_{g \sim D}[\alpha \cdot |I_g(S)| + \beta \cdot B_1(S)]$$
$$= \sum_g Prob[g] \cdot (\alpha \cdot |I_g(S)| + \beta \cdot B_1(S)) \tag{16}$$

The first part $|I_g(S)|$ of the $f_1(S)$ is the traditional influence maximization problem which is submodular [8]. Given $\alpha \geq 0$, $\beta \geq 0$, $Prob[g] \geq 0$ and $B_2(S)$ is submodular prove and monotone proved by Lemma 4, $f_1(S)$ is submodular and monotone follows immediately since the fact that a non-negative linear combination of submodular functions is also submodular.   □

Lemma 5. $B_2(S)$ *is submodular and monotone under the IC model.*

Proof. Let M, N to be any two seed sets such that $M \subseteq N \subseteq V$ and x to be any element such that $x \in V \setminus N$. According the definition of $B_2(S)$ shown in Eq. 9, we have Then we have

$$B_2(M \cup \{x\}) - B_2(M)$$
$$= \sum_{\{u,v\} \subseteq I_g(x) \setminus I_g(M)} b(u,v) + \sum_{u \in I_g(x) \setminus I_g(M)} \sum_{v \in V \setminus I_g(M) \cup I_g(x)} b(u,v) \tag{17}$$

Through the same analysis process, we can get

$$B_2(N \cup \{x\}) - B_2(N)$$
$$= \sum_{\{u,v\} \subseteq I_g(x) \setminus I_g(N)} b(u,v) + \sum_{u \in I_g(x) \setminus I_g(N)} \sum_{v \in V \setminus I_g(N) \cup I_g(x)} b(u,v) \tag{18}$$

Comparing all terms on the right-hand sides of 17 and 18, since $M \subseteq N$, we have $I_g(M) \subseteq I_g(N)$. So $I_g(x) \setminus I_g(M) \supseteq I_g(x) \setminus I_g(N)$ and $V \setminus I_g(M) \cup I_g(x) \supseteq V \setminus I_g(N) \cup I_g(x)$ follows. Thus both the first item and second item of 17 are greater than the first item and second item of 18 respectively. Through above analysis, we obtain $B_2(M \cup \{x\}) - B_2(M) \geq B_1(N \cup \{x\}) - B_2(N)$. Therefore, $B_2(S)$ is submodular.

For monotonicity, we need prove $B_2(M) \leq B_2(N)$, which is non-decreasing. According to Eq. 9, we have

$$B_2(M)$$
$$= \sum_{\{u,v\} \subseteq I_g(M)} b(u,v) + \sum_{u \in I_g(M)} \sum_{v \in V \setminus I_g(M)} b(u,v)$$
$$= \sum_{\{u,v\} \subseteq I_g(M)} b(u,v) + \sum_{u \in I_g(M)} \sum_{v \in I_g(N) \setminus I_g(M)} b(u,v)$$
$$+ \sum_{u \in I_g(M)} \sum_{v \in V \setminus I_g(N)} b(u,v) \qquad (19)$$

$$B_2(N)$$
$$= \sum_{\{u,v\} \subseteq I_g(N)} b(u,v) + \sum_{u \in I_g(N)} \sum_{v \in V \setminus I_g(N)} b(u,v) \qquad (20)$$

Since $I_g(M) \subseteq I_g(N)$, we have $\forall (i,j) \in \{(u,v) \mid u \in I_g(M), v \in I_g(N) \setminus I_g(M)\}$, $i \in I_g(N)$, $j \in I_g(N)$. Thus the sum of first two items of $B_2(M)$ is less than the first item of $B_1(N)$. By the same reason, we have the third item of $B_2(M)$ is less than the second item of $B_2(N)$. Through above analysis, the monotonicity of $B_2(S)$ follows immediately.   □

Theorem 5. $f_2(S)$ *is submodular and monotone under the IC model.*

Proof. According the definition of $f_2(S)$ shown in Eq. 11, we have

$$f_2(S) = \mathbb{E}_{g \sim D}[\beta \cdot B_2(S)]$$
$$= \beta \cdot \sum_g Prob[g] \cdot B_2(S) \qquad (21)$$

Given $\beta \geq 0$, $Prob[g] \geq 0$ and $B_2(S)$ is submodular and monotone proved by Lemma 5, $f_2(S)$ is submodular and monotone follows immediately since the fact that a non-negative linear combination of submodular functions is also submodular.   □

5   Algorithms

According to the sandwich theorem and our DS decomposition, we designed a iterated sandwich algorithm. The algorithm named iterated modular sandwich algorithm is based on the modular upper and lower bounds of our objective function. Our algorithm are guaranteed to gain data dependent approximation solutions.

5.1 Iterated Sandwich Algorithm

Algorithm 1. Iterated Modular Sandwich Algorithm

1: initialize $\epsilon > 0$, an integer k, $t \leftarrow 0$, $S^t \leftarrow$ a random seeds of size k, $S_{\max} = S^0$
2: **repeat**
3:    choose a permutation σ^t whose chain contains S^t
4:    construct a modular upper bound $U^{f_1}_{S^t}(X)$ (and, $U^{f_2}_{S^t}(X)$) and a modular lower
      bound $L^{f_1}_{S^t,\sigma^t}(X)$ (and, $L^{f_2}_{S^t,\sigma^t}(X)$) for f_1 (and, f_2)
5:    $S^t_u \leftarrow \text{argmax}_X m^u_t(X) = U^{f_1}_{S^t}(X) - L^{f_2}_{S^t,\sigma^t}(X)$;
6:    $S^t_l \leftarrow \text{argmax}_X m^l_t(X) = L^{f_1}_{S^t,\sigma^t}(X) - U^{f_2}_{S^t}(X)$;
7:    $S_o \leftarrow \text{argmax}_X f(X)$;
8:    Let $S^{t+1} \leftarrow \text{argmax}_X(f(S^t_u), f(S^t_l), f(S_o))$
9:    **if** $f(S^{t+1}) \geq (1+\epsilon)f(S_{\max})$ **then**
10:       $S_{\max} \leftarrow S^{t+1}$
11:       $t \leftarrow t + 1$
12:    **end if**
13: **until** *converged, i.e.,* $f(S^{t+1}) < (1+\epsilon)f(S_{\max})$
14: **return** $S_{\max}$

For Algorithm 1 named Iterated Modular Sandwich Algorithm, we iteratively find the optimal solutions for three functions: the modular upper bound function $m^u_t(X)$, the modular lower bound function $m^l_t(X)$ and the original objective function $f(X)$, and then choose the best solution from $f(X)$ as the input of next iteration.

5.2 Analysis

We say a set S is local optimum solution of a submodular function f, if for any $T \subseteq S$ or $T \supseteq S$, we have $f(S) \geq f(T)$. Similarly, we say set S is a $(1+\epsilon)$-approximate local optimum solution of a submodular function f, if for any $e \in V$, we have $(1+\epsilon)f(S) \geq f(S \cup \{e\})$ and $(1+\epsilon)f(S) \geq f(S \setminus \{e\})$. Consider iteration t, let $F_t(X) = U^{f_1}_{S^t}(X) - L^{f_2}_{S^t,\sigma^t}(X)$ for any $X \subseteq V$, we firstly give a notation *approximation coefficient* $\eta_t = \max_{X \subseteq V} \frac{F_t(X)}{f(X)}$, which is denoted as how the approximate extent of the replace function $F_t(X)$ to the original function $f(X)$. Let $\eta = \max_t \eta_t$. Now we can bound the value of set returned by the Iterative Modular Sandwich algorithm by the following theorem.

Theorem 6. *Let $S_{\max}$ be the returned set by Algorithm 1, then we have $S_{\max}$ either is a $(1+\epsilon)$-approximate local maximum solution by justly checking $O(n)$ permutations, or is $\frac{1}{\eta(1+\epsilon)}$-approximation solution for the interaction-aware influence maximization problem.*

Proof. For the case that the return set $S_{\max}$ is derived by the modular lower bound and set t as the terminal iteration of Algorithm 1, then we have

$$(1 + \epsilon)f(S^t) \geq f(S^{t+1}) = f_1(S^{t+1}) - f_2(S^{t+1})$$

$$\geq \frac{1}{\eta} \cdot (U_{S^t}^{f_1}(S^{t+1}) - L_{S^t,\sigma^t}^{f_2}(S^{t+1}))$$

$$\geq \frac{1}{\eta} \cdot (U_{S^t}^{f_1}(OPT) - L_{S^{t+1},\sigma^t}^{f_2}(OPT))$$

$$\geq \frac{1}{\eta} \cdot (f_1(OPT) - f_2(OPT))$$

$$= \frac{1}{\eta} \cdot f(OPT)$$

The first inequality is derived by the line 9 of Algorithm 1, the second inequality follows the definition of approximation coefficient, the third inequality is derived by the optimality of S^{t+1} according to $F(\cdot)$ at iteration $t+1$, and last inequality is obtained by the construction of the upper and lower bounds. Thus we have

$$f(S^t) \geq \frac{1}{\eta(1 + \epsilon)} f(OPT).$$

The rest of our proof is to show that the set $S_{\max}$ is obtained from the lower bound is a ϵ-approximate local maximum solution. We follow the ideas presented by Iyer and Bilmes [6], who consider a general DS-decomposition minimization by a iterative modular approximation algorithm. For one subcase, we need to show under the terminate iteration t, if we add a element j, our local solution S^t will not increase a enough amount. By the construction of bounds, and optimality condition, we have

$$(1 + \epsilon)f(S^t) \geq f(S^{t+1}) = f_1(S^{t+1}) - f_2(S^{t+1})$$

$$\geq L_{S^t,\sigma^t}^{f_1}(S^{t+1}) - U_{S^t}^{f_2}(S^{t+1})$$

$$\geq L_{S^t,\sigma^t}^{f_1}(S^t \cup \{j\}) - U_{S^t}^{f_2}(S^t \cup \{j\})$$

$$= f_1(S^t \cup \{j\}) - f_2(S^t \cup \{j\})$$

$$= f(S^t \cup \{j\}).$$

Similarly, we can lower bound $(1 + \epsilon)f(S^t) \geq f(S^t \setminus \{j\})$ under the subcase that of deleting a element j. By the definition of ϵ-approximate local maximum, we know S^t is the ϵ-approximate local maximum solution.

Theorem 7. *The Algorithm 1 terminates at most $O(1/\epsilon \log(OPT/f(S^0)))$ steps and the total time complexity is bounded by $O(C/\epsilon \log(OPT/f(S^0)))$, where C is the upper bound of time of computing optimal solution of modular function.*

Proof. Follows from the repeat process of Algorithm 1, we have $f(S^{i+1}) \geq (1 + \epsilon)f(S^i)$ for any iteration $i(< t)$. It is easy to check out that the number of steps of repeat process is at most $\log_{1+\epsilon} \frac{f(S^t)}{f(S^0)} (\leq O(1/\epsilon \log(OPT/f(S^0))))$. Assume the time of computing the optimal solution of modular function is at most C, with a multiplicative factor C, we can bound the total complexity of Algorithm 1.

6   Conclusion

In this paper, we propose an influence maximization problem that takes the interaction among users into consideration. To solve our non-submodular problem, we propose sandwich theory based on decomposing the original function into the difference between two submodular functions and design an iterated sandwich algorithm.

Acknowledgements. The work is supported by Natural Science Foundation of China (No. 61672321, 61771289, 61832012, 61373027).

References

1. Chen, W., et al.: Influence maximization in social networks when negative opinions may emerge and propagate. In: Proceedings of the 2011 SIAM International Conference on Data Mining, pp. 379–390. SIAM (2011)
2. Chen, W., Wang, Y., Yang, S.: Efficient influence maximization in social networks. In: Proceedings of the 15th ACM SIGKDD International Conference on Knowledge Discovery and Data Mining, pp. 199–208. ACM (2009)
3. Domingos, P., Richardson, M.: Mining the network value of customers. In: Proceedings of the Seventh ACM SIGKDD International Conference on Knowledge Discovery and Data Mining, KDD 2001, pp. 57–66. ACM (2001)
4. Fox, J., Gilbert, M., Tang, W.Y.: Player experiences in a massively multiplayer online game: a diary study of performance, motivation, and social interaction. New Media Soc. (2018). https://doi.org/10.1177/1461444818767102
5. Han, M., Li, J., Cai, Z., Han, Q.: Privacy reserved influence maximization in GPS-enabled cyber-physical and online social networks. In: 2016 IEEE International Conferences on Big Data and Cloud Computing (BDCloud), pp. 284–292. IEEE (2016)
6. Iyer, R., Bilmes, J.: Algorithms for approximate minimization of the difference between submodular functions, with applications. In: Proceedings of the Twenty-Eighth Conference on Uncertainty in Artificial Intelligence, UAI 2012, pp. 407–417. AUAI Press, Arlington (2012)
7. Jung, K., Heo, W., Chen, W.: IRIE: scalable and robust influence maximization in social networks. In: 2012 IEEE 12th International Conference on Data Mining (ICDM), pp. 918–923. IEEE (2012)
8. Kempe, D., Kleinberg, J., Tardos, É.: Maximizing the spread of influence through a social network. In: International Conference on Knowledge Discovery and Data Mining, KDD 2003, pp. 137–146. ACM (2003)
9. Leskovec, J., Krause, A., Guestrin, C., Faloutsos, C., VanBriesen, J., Glance, N.: Cost-effective outbreak detection in networks. In: Proceedings of the 13th ACM SIGKDD International Conference on Knowledge Discovery and Data Mining, pp. 420–429. ACM (2007)

10. Li, Y., Zhang, D., Tan, K.-L.: Real-time targeted influence maximization for online advertisements. Proc. VLDB Endow. **8**(10), 1070–1081 (2015)
11. Narasimhan, M., Bilmes, J.: A submodular-supermodular procedure with applications to discriminative structure learning. In: Proceedings of the Twenty-First Conference on Uncertainty in Artificial Intelligence, UAI 2005, pp. 404–412. AUAI Press, Arlington (2005)
12. Richardson, M., Domingos, P.: Mining knowledge-sharing sites for viral marketing. In: Proceedings of the Eighth ACM SIGKDD International Conference on Knowledge Discovery and Data Mining, KDD 2002, pp. 61–70. ACM (2002)
13. Rodriguez, M.G., Schölkopf, B.: Influence maximization in continuous time diffusion networks. arXiv preprint arXiv:1205.1682 (2012)
14. Wang, Z., Yang, Y., Pei, J., Chu, L., Chen, E.: Activity maximization by effective information diffusion in social networks. IEEE Trans. Knowl. Data Eng. **29**(11), 2374–2387 (2017)
15. Wu, W.-L., Zhang, Z., Du, D.-Z.: Set function optimization. J. Oper. Res. Soc. China **7**, 1–11 (2018)

On Approximation Algorithm for the Edge Metric Dimension Problem

Yufei Huang[1], Bo Hou[1], Wen Liu[1], Lidong Wu[2], Stephen Rainwater[2], and Suogang Gao[1(✉)]

[1] College of Mathematics and Information Science, Hebei Normal University, Shijiazhuang 050024, People's Republic of China
sggao@mail.hebtu.edu.cn
[2] Department of Computer Science, Unibersity of Texas at Tyler, Tyler, USA
{lwu,srainwater}@uttyler.edu

Abstract. In this paper, we study the edge metric dimension problem (EMDP). We establish a potential function and give a corresponding greedy algorithm with approximation ratio $1 + \ln n + \ln(\log_2 n)$, where n is the number of vertices in the graph G.

Keywords: Edge metric generator · Edge metric dimension · Approximation algorithms · Submodular function

1 Introduction

The concepts of metric generators (originally called locating sets) and the concepts of metric dimension (originally called the location number) were introduced by Slater in [17] in connection with uniquely determining the position of an intruder in a network. Harary and Melter [11] discovered the same concepts independently.

We now recall the definition of the metric dimension. Let $G = (V, E)$ be a simple connected undirected graph. A vertex $v \in V$ is called to *resolve* or *distinguish* a pair of vertices $u, w \in V$ if $d(v, u) \neq d(v, w)$, where $d(\cdot, \cdot)$ denotes the distance between two vertices in G. A *metric generator* of G is a subset $V' \subseteq V$ such that for each pair $u, w \in V$ there exists some vertex $v \in V'$ that distinguishes u and w. The minimum cardinality of a metric generator is called the *metric dimension* of G, denoted by $\dim(G)$.

The metric dimension problem (MDP) has been widely investigated from the graph theoretical point of view. Cáceres et al. [3] studied the metric dimension of cartesian products $G \square H$, and proved that the metric dimension of $G \square G$ was tied in a strong sense to the minimum order of a so-called doubly resolving set in G. They established bounds on $G \square H$ for many examples of G and H. Chartrand et al. [7] studied resolvability in graphs and the metric dimension of a graph. It was shown that $\dim(H) \leq \dim(H \square K_2) \leq \dim(H) + 1$ for every connected graph H. Moreover, it was shown that for every positive real number ε, there exists a

D.-Z. Du et al. (Eds.): AAIM 2019, LNCS 11640, pp. 142–148, 2019.
https://doi.org/10.1007/978-3-030-27195-4_13

connected graph G and a connected induced subgraph H of G such that $\frac{\dim(G)}{\dim(H)} \leq \varepsilon$. Saputro et al. [16] studied the metric dimension of regular bipartite graphs, and determined the metric dimension of k-regular bipartite graphs $G(n, n)$ where $k = n - 1$ or $k = n - 2$. Chappell et al. [6] studied relationships between metric dimension, partition dimension, diameter, and other graph parameters. They constructed "universal examples" of graphs with given partition dimension, and they used these to provide bounds on various graph parameters based on metric and partition dimensions. They formed a construction showing that for all integers α and β with $3 \leq \alpha \leq \beta + 1$ there exists a graph G with partition dimension α and β. Cáceres et al. [5] studied the metric dimension of infinite locally finite graphs, i.e. those infinite graphs such that all its vertices have finite degree. They gave some necessary conditions for an infinite graph to have finite metric dimension and characterized infinite trees with finite metric dimension.

So far only a few papers have discussed the computational complexity issues of the MDP. The NP-hardness of the MDP was mentioned by Garey and Johnson [10]. An explicit reduction from the 3-SAT problem was given by Khuller et al. [14]. They also obtained for the Metric Dimension problem a $(2\ln(n) + \Theta(1))$-approximation algorithm based on the well-known greedy algorithm for the Set Cover problem and showed that the MDP is polynomial-time solvable for trees. Beerliova et al. [1] showed that the MDP (which they call the Network Verification problem) cannot be approximated within a factor of $O(\log(n))$ unless $P = NP$. Hauptmann et al. [12] gave a $(1+\ln(|V|)+\ln(\log_2(|V|)))$-approximation algorithm for the MDP in graphs.

The concept of a doubly resolving set of a graph G was introduced by Caceres et al. [4]. We say vertices u, v of the graph G *doubly resolve* vertices x, y of G, if $d(u, x) - d(u, y) \neq d(v, x) - d(v, y)$. A vertex set S is called a *doubly resolving set* of G if every two distinct vertices of G are doubly resolved by some two vertices of S.

Kratica et al. [15] proved that the minimal doubly resolving sets problem is NP-hard. Chen et al. [8] designed an $(1 + o(1)) \ln n$-approximation algorithm for the weighted minimum doubly resolving set problem.

The edge metric dimension is a variant of the metric dimension. We now recall the definition of the edge metric dimension. For any $v \in V$ and $e = uw \in E$, we use $d(e, v) = \min\{d(u, v), d(w, v)\}$ to denote the distance between the vertex v and the edge e. We say that two distinct edges $e_1, e_2 \in E$ are distinguished by the vertex $v \in V$ if $d(v, e_1) \neq d(v, e_2)$. A subset $S \subseteq V$ is said to be an *edge metric generator* of G if every two distinct edges of G can be distinguished by some vertex in S. An *edge metric basis* of G is an edge metric generator of G of the minimum cardinality and its cardinality is called the *edge metric dimension*, denoted by $\dim_e(G)$.

Kelenc et al. [13] proved that computing the edge metric dimension of connected graphs is NP-hard. As a response to an open problem presented in [13], Zhu et al. [18] considered the maximum edge metric dimension problem on graphs. Zubrilina [19] classified the graphs on n vertices for which $\dim_e(G) = n - 1$ and showed that $\frac{\dim_e(G)}{\dim(G)}$ is not bounded from above (here

$\dim(G)$ is the standard metric dimension of G). They computed $\dim_e(G \square P_m)$ and $\dim_e(G + K_1)$. Zubrilina [20] discussed the edge metric dimension of the random graph $G(n, p)$ and obtained $\dim_e(G(n, p)) = (1 + o(1)) \frac{4 \log(n)}{\log(\frac{1}{q})}$, where $q = 1 - 2p(1-p)^2(2-p)$. In this paper, we discuss the edge metric dimension problem.

The paper is organized as follows: In Sect. 2, we construct a normalized, monotone increasing and submodular potential function and give a greedy algorithm for the edge metric dimension problem. In Sect. 3, we show that the algorithm presented in this paper has approximation ratio $1 + \ln n + \ln(\log_2 n)$, where n is the number of vertices in the graph G.

2   Approximation Algorithm

Throughout this paper we assume that the graph $G = (V, E)$ is simple connected and undirected. In this section, we first construct a potential function and study the properties of the potential function. Then we give a greedy algorithm for the edge metric dimension of G.

Definition 2.1. Let Γ be a subset of V. We define the *equivalence relation* $\equiv_\Gamma$ for E as follows: for edges $e_1, e_2 \in E$,

$$e_1 \equiv_\Gamma e_2 \iff d(e_1, w) = d(e_2, w) \quad \forall w \in \Gamma.$$

Definition 2.2. Let Γ be a subset of V and $\{E_1, E_2, \ldots, E_k\}$ be the set of equivalence classes of $\equiv_\Gamma$ for E. We call the value $H(\Gamma) = \log_2(\prod_{i=1}^{k} |E_i|!)$ the *entropy* of Γ.

For any $v \in V$, let

$$\Delta_v H(\Gamma) := H(\Gamma) - H(\Gamma \cup \{v\}).$$

It is direct to see that any equivalence class of $\equiv_\Gamma$ is either an equivalent class of $\equiv_{\Gamma \cup \{v\}}$ or a union of several equivalence classes of $\equiv_{\Gamma \cup \{v\}}$.

Lemma 2.3. *Let Γ be a subset of V and $v \in V$. Then $\Delta_v H(\Gamma) = 0$ if each equivalence class of $\equiv_\Gamma$ is one of $\equiv_{(\Gamma \cup \{v\})}$; and $\Delta_v H(\Gamma) > 0$ otherwise.*

Lemma 2.4. *Let Γ be a subset of V. Then Γ is an edge metric generator of G if and only if $H(\Gamma) = 0$.*

Proof. Observe that each of the two assertions is equivalent with the assertion that every equivalent class of $\equiv_\Gamma$ is a singleton. The result follows. $\square$

Lemma 2.5. *For any two sets $\Gamma_0 \subseteq \Gamma_1 \subseteq V$ and any vertex $v \in V \backslash \Gamma_1$, we have*

$$\Delta_v H(\Gamma_0) \geq \Delta_v H(\Gamma_1). \tag{1}$$

Proof. If $\Gamma_0 = \Gamma_1$, then the lemma holds. If $\Gamma_0 \subset \Gamma_1$, we divide the proof into two cases: case 1, the vertex v partitions each equivalence class of $\equiv_{\Gamma_0}$ into at most two equivalence classes; case 2, the vertex v partitions some equivalence class of $\equiv_{\Gamma_0}$ into at least three equivalence classes.

Case 1. Since $\Delta_v H(\Gamma_0) = H(\Gamma_0) - H(\Gamma_0 \cup \{v\}) = \log_2 \frac{|\Pi_{\Gamma_0}|}{|\Pi_{\Gamma_0 \cup \{v\}}|}$, it suffices to show

$$\frac{|\Pi_{\Gamma_0}|}{|\Pi_{\Gamma_0 \cup \{v\}}|} \geq \frac{|\Pi_{\Gamma_1}|}{|\Pi_{\Gamma_1 \cup \{v\}}|}. \tag{2}$$

Write $S = \Gamma_1 \setminus \Gamma_0$. Let $\{E_1, E_2, \cdots, E_k\}$ be the equivalence classes of $\equiv_{\Gamma_0}$, $\{A_1, A_2, \cdots, A_n\}$ the equivalence classes of $\equiv_{\Gamma_0 \cup \{v\}}$ and $\{B_1, B_2, \cdots, B_t\}$ the equivalence classes of $\equiv_{\Gamma_1}$. By the comments above Lemma 2.3 and the assumption, for each i, $E_i = A_{i_1} \cup A_{i_2}$ and E_i is a union of some $B_{i_1}, \cdots, B_{i_t}$. Without loss of generality, assume $t = 2$. Let $F_i = A_{i_1} \cap B_{i_1}$, $H_i = A_{i_2} \cap B_{i_1}$, $C_i = (E_i \cap A_{i_1}) \setminus F_i$, $D_i = (E_i \cap A_{i_2}) \setminus H_i$. Let $|F_i| = f_i$, $|H_i| = h_i$, $|C_i| = c_i$, $|D_i| = d_i$. Then $|E_i| = f_i + h_i + c_i + d_i$. Since $\binom{f_i + c_i}{f_i + h_i + c_i + d_i} \geq \binom{f_i}{h_i + f_i}\binom{c_i}{c_i + d_i}$, we have

$$\prod_{i=0}^{k} \left(\frac{f_i + c_i}{f_i + h_i + c_i + d_i} \right) \geq \prod_{i=0}^{k} \left(\frac{f_i}{f_i + h_i} \right)\left(\frac{c_i}{c_i + d_i} \right),$$

i.e.

$$\prod_{i=0}^{k} \left(\frac{(f_i + h_i + c_i + d_i)!}{(f_i + c_i)!(h_i + d_i)!} \right) \geq \prod_{i=0}^{k} \left(\frac{(f_i + h_i)!(c_i + d_i)!}{(f_i)!(h_i)!(c_i)!(d_i)!} \right).$$

Thus

$$\frac{|\Pi_{\Gamma_0}|}{|\Pi_{\Gamma_0 \cup \{v\}}|} \geq \frac{|\Pi_{\Gamma_1}|}{|\Pi_{\Gamma_1 \cup \{v\}}|}.$$

Case 2. Assume that the vertex v partitions each E_j into k_j equivalence classes, where $j = 1, 2, \ldots, m$. Let $k = \max_j\{k_j\}$. Then by assumption, $k \geq 3$. For $1 \leq j \leq m$, there exist the vertices $x_1, x_2, \ldots, x_k$ such that x_1 divides E_j into E_{j_1} and $E_j \setminus E_{j_1}$, vertex x_2 divides $E_j \setminus E_{j_1}$ into E_{j_2} and $E_j \setminus (E_{j_1} \cup E_{j_2})$, $\ldots$, vertex x_k divides $E_j \setminus (E_{j_1} \cup E_{j_2} \cup \ldots \cup E_{j_{k_j-1}})$ into $E_{j_{k_j}}$ and $\emptyset$. Then by the argument in Case 1, we have

$$
\begin{aligned}
\Delta_v H(\Gamma_0) &= H(\Gamma_0) - H(\Gamma_0 \cup \{v\}) \\
&= (H(\Gamma_0) - H(\Gamma_0 \cup \{x_1\})) + (H(\Gamma_0 \cup \{x_1\}) - H(\Gamma_0 \cup \{x_1\} \cup \{x_2\})) \\
&\quad + \ldots + (H(\Gamma_0 \cup \{x_1\} \cup \{x_2\} + \ldots \cup \{x_{k-1}\}) - H(\Gamma_0 \cup \{x_1\} \cup \{x_2\} \cup \ldots \cup \{x_k\})) \\
&\geq (H(\Gamma_1) - H(\Gamma_1 \cup \{x_1\})) + (H(\Gamma_1 \cup \{x_1\}) - H(\Gamma_1 \cup \{x_1\} \cup \{x_2\})) \\
&\quad + \ldots + (H(\Gamma_1 \cup \{x_1\} \cup \{x_2\} \cup \ldots \cup \{x_{k-1}\}) - H(\Gamma_1 \cup \{x_1\} \cup \{x_2\} \cup \ldots \cup \{x_k\})) \\
&= H(\Gamma_1) - H(\Gamma_1 \cup \{v\}) \\
&= \Delta_v H(\Gamma_1).
\end{aligned}
$$

$\square$

Let $\mathbb{R}$ be the real number field. We define a function $f : 2^V \to \mathbb{R}$ by

$$f(\Gamma) = -H(\Gamma) + H(\emptyset) \quad \text{for } \Gamma \in 2^V.$$

Lemma 2.6. *The function f defined above is normalized, monotone increasing and submodular.*

Proof. It is easy to know that $f(\emptyset) = 0$, that is to say, the function f is normalized. By Lemma 2.3, f is monotone increasing. By Lemma 2.5 f is submodular. □

Based on the above lemmas, we give a greedy approximation algorithm for the EMDP.

Algorithm 1

Input: a simple connected undirected graph $G = (V, E)$.
Output: an edge metric generator of G.
1: Set $\Gamma \leftarrow \emptyset$.
2: **while** there exists a vertex $v \in V \setminus \Gamma$ such that $\Delta_v f(\Gamma) > 0$ **do**
3:     select a vertex $v \in V \setminus \Gamma$, that maximizes $\Delta_v f(\Gamma)$.
4:     $\Gamma \leftarrow \Gamma \cup \{v\}$.
5: **return** $\Gamma_g \leftarrow \Gamma$

---

3   Theoretical Analysis

To obtain the ratio of Algorithm 1. We first prove the following lemma.

Lemma 3.1. *Let $v_1, v_2, \cdots, v_k$ be the elements in Γ_g in the order of their selection into the set Γ_g. Denote $\Gamma_0 = \emptyset$ and $\Gamma_i = \{v_1, v_2, \cdots, v_i\}$, for $i = 1, \cdots, k$. Then for $i = 2, \cdots, k$, we have*

$$\Delta_{v_i} f(\Gamma_{i-1}) \geq 1.$$

Proof. By [2, Lemma 6], it is sufficient to prove $\Delta_{v_i} f(\Gamma_{i-1}) > 0$. Assume $\Delta_{v_i} f(\Gamma_{i-1}) = 0$ for some i ($2 \leq i \leq k$), for a contradiction. Then $H(\Gamma_{i-1} \cup \{v_i\}) = H(\Gamma_{i-1})$. By the greedy strategy, the vertex v_i can not be chosen in Γ_g. A contradiction. □

Theorem 3.2. *Algorithm 1 produces an approximate solution within a ratio $1 + \ln n + \ln(\log_2 n)$.*

Proof. Let Γ^* denote an optimal solution to the edge metric dimension problem. By Lemmas 2.6 and 3.1 and [9, Theorem 3.7], and since $f(\Gamma^*) = f(V) = \log_2(n!)$, the approximation ratio of Algorithm 1 is

$$1 + \ln(\frac{f(\Gamma^*)}{|\Gamma^*|})$$

$$= 1 + \ln(\frac{\log_2(n!)}{|\Gamma^*|})$$

$$\leq 1 + \ln(n \log_2 n) - \ln(|\Gamma^*|)$$

$$\leq 1 + \ln(\log_2 n) + \ln n.$$

$\square$

Acknowledgement. The authors would like to thank Professor Ding-Zhu Du for his many valuable advices during their study of approximation algorithm. This work was supported by the NSF of China (No. 11471097), Hebei Province Foundation for Returnees (CL201714) and Overseas Expertise Introduction Program of Hebei Auspices (25305008).

References

1. Beerliova, Z., et al.: Network discovery and verification. IEEE J. Sel. Area Commun. **24**, 2168–2181 (2006)
2. Berman, P., DasGupta, B., Kao, M.: Tight approximability results for test set problems in bioinformatics. J. Comput. Syst. Sci. **71**(2), 145–162 (2005)
3. Cáceres, J., et al.: On the metric dimension of Cartesian products of graphs. SIAM J. Discret. Math. **21**, 423–441 (2007)
4. Cáceres, J., Hernando, C., Mora, M., Pelayo, I.M., Puertas, M.l., Seara, C., et al.: On the metric dimension of Cartesian products of graphs. SIAM J. Discret. Math. **21**(2), 423–441 (2007)
5. Cáceres, J., Hernando, C., Mora, M., Pelayo, I., Puertas, M.: On the metric dimension of infinite graphs. Electron. Notes Discret. Math. **35**, 15–20 (2009)
6. Chappell, G., Gimbel, J., Hartman, C.: Bounds on the metric and partition dimensions of a graph. Ars Combin. **88**, 349–366 (2008)
7. Chartrand, G., Eroh, L., Johnson, M., Oellermann, O.: Resolvability in graphs and the metric dimension of a graph. Discret. Appl. Math. **105**, 99–133 (2000)
8. Chen, X.J., Hu, X.D., Wang, C.J.: Approximation for the minimum cost doubly resolving set problem. Theor. Comput. Sci. **609**, 526–543 (2016)
9. Du, D.Z., Ko, K.I., Hu, X.D.: Design and Analysis of Approximation Algorithms, vol. 62. Springer, New York (2011). https://doi.org/10.1007/978-1-4614-1701-9
10. Garey, M.R., Johnson, D.S.: Computers and Intractability: A Guide to the Theory of NP-Completeness. Freeman, New York (1979)
11. Harary, F., Melter, R.A.: On the metric dimension of a graph. Ars Combin. **2**, 191–195 (1976)
12. Hauptmann, M., Schmied, R., Viehmann, C.: Approximation complexity of metric dimension problem. J. Discret. Algorithms **14**, 214–222 (2012)
13. Kelenc, A., Tratnik, N., Yero, I.G.: Uniquely identifying the edges of a graph: the edge metric dimension. Discret. Appl. Math. **251**, 204–220 (2018)

14. Khuller, S., Raghavachari, B., Rosenfeld, A.: Landmarks in graphs. Discret. Appl. Math. **70**, 217–229 (1996)
15. Kratica, J., Čangalović, M., Kovačević-Vujčić, V.: Computing minimal doubly resolving sets of graphs. Comput. Oper. Res. **36**, 2149–2159 (2009)
16. Baca, M., Baskoro, E.T., Salman, A.N.M., Saputro, S.W., Suprijanto, D.: The metric dimension of regular bipartite graphs. Bull. Math. Soc. Sci. Math. Roumanie **54**(1), 15–28 (2011)
17. Slater, P.J.: Leaves of trees. In: Southeastern Conference on Combinatorics, Graph Theory, and Computing, Congressus Numerantium, vol. 14, pp. 549–559 (1975)
18. Zhu, E.Q., Taranenko, A., Shao, Z.H., Xu, J.: On graphs with the maximum edge metric dimension. Discret. Appl. Math. **257**, 317–324 (2019)
19. Zubrilina, N.: On edge dimension of a graph. Discret. Math. **341**, 2083–2088 (2018)
20. N. Zubrilina. On the edge metric dimension for the random graph (2016). arXiv:1612.06936 [math.CO]

The Seeding Algorithm for Spherical k-Means Clustering with Penalties

Sai Ji[1], Dachuan Xu[1], Longkun Guo[2(✉)], Min Li[3], and Dongmei Zhang[4]

[1] Department of Operations Research and Scientific Computing,
Beijing University of Technology, Beijing 100124, People's Republic of China
[2] College of Mathematics and Computer Science, Fuzhou University,
Fuzhou 350116, Fujian, People's Republic of China
longkun.guo@gmail.com
[3] School of Mathematics and Statistics, Shandong Normal University,
Jinan 250014, People's Republic of China
[4] School of Computer Science and Technology, Shandong Jianzhu University,
Jinan 250101, People's Republic of China

Abstract. Spherical k-means clustering is a generalization of k-means problem which is NP-hard and has widely applications in data mining. It aims to partition a collection of given data with unit length into k sets so as to minimize the within-cluster sum of cosine dissimilarity. In this paper, we introduce the spherical k-means clustering with penalties and give a $2 \max\{2, M\}(1 + M)(\ln k + 2)$-approximate algorithm, where M is the ratio of the maximal and the minimal penalty values of the given data set.

Keywords: Approximation algorithm · Spherical k-means clustering · Penalty

1 Introduction

Clustering problems arise in many applications such as data mining, data compression, machine learning and computer vision. These problems have been widely studied in the literature. For partial surveys, see e.g. [5,7,11,18,19]. The k-means problem is one of the most fundamental clustering tasks in combinatorial optimization and data mining. In this problem, we are given an n-point data set $X \in R^d$ and a positive integer k, the goal is to partition X into k disjoint subsets so as to minimize the total squared distances between each point and its closest center. This problem is NP-hard [2,8], one can not obtain the optimal solution in polynomial time unless $P = NP$. There are many research about k-means problem [1,6,13,20]. Lloyd [14] provides a local search heuristic for this problem which performs very well in practical and is still widely used today. Arthur and Vassilvitskii [4] further provide an $O(\log k)$-approximation algorithm called k-means++, which improve Lloyd's algorithm by choosing random initially centers with specific probabilities. The first constant approximation

D.-Z. Du et al. (Eds.): AAIM 2019, LNCS 11640, pp. 149–158, 2019.
https://doi.org/10.1007/978-3-030-27195-4_14

for k-means problem in general dimension is a $(9+\epsilon)$-approximation provided by Kanungo et al. [13] based on local search technique. Ahmadian et al. [3] improve the ratio to 6.357 by presenting a new primal-dual approach.

Spherical k-means clustering is a generalization of k-means problem which is NP-hard and widely used in data mining [17,18]. It aims to partition the given data with unit length into k sets so as to minimize the within-cluster sum of cosine dissimilarity. Hornik et al. [12] present a spherical k-means algorithm (SKM) based on primitive spherical k-means [9]. Endo et al. [10] provide a spherical k-means++ algorithm (SKM++) with theoretically guaranteed. Endo and Miyamoto [10] study the α-spherical k-means clustering, where the cosine dissimilarity is a little different from the one in spherical k-means clustering, and obtain an $O(\log_2 k)$-approximate algorithm. Li et al. [16] present an approximate algorithm with a constant factor for the spherical k-means clustering with separable sets. Moreover, they prove that their algorithm can be generalized to solve α-spherical k-means clustering with separable sets.

In this paper, we consider the spherical k-means clustering with penalties. In this problem, we are given a data set and a penalty cost for each data. All the data are on a unit sphere, that is the norm of each data is normalized to one. Each data is clustered to a cluster or be un-clustered by paying the corresponding penalty cost. The object of this problem is to partition the clustered data into k sets so as to minimize the total cost including the within-cluster sum of cosine dissimilarity and the sum of penalty cost. The main contribution is that we give a $2 \max\{2, M\}(1+M)(\ln k + 2)$-approximation seeding algorithm for the spherical k-means clustering with penalties based on [10,15], where M is the ratio of the maximal penalty value and the minimal one of the given data set.

The rest of this paper is organized as follows. In Sect. 2, the problems we studied and some notations are introduced. In Sect. 3, we give a detailed description of our seeding algorithm as well as our main results. The corresponding theoretical analysis of the algorithm is provided in Sect. 4. Some discussions are stated in Sect. 5.

2   Preliminaries

In this section, we will give a detailed description about spherical k-means clustering, spherical k-means clustering with penalties, as well as some symbols and notations used in this paper. For any two data $x, v \in R^d$, we use

$$d(x, v) = 1 - \langle x, v \rangle$$

to denote the cosine dissimilarity of x and v, where $\langle \cdot, \cdot \rangle$ means the inner product. Furthermore, for any data x and a set C, $d(x, C) = \min_{c \in C} d(x, c)$. For each integer n, denote $[n] = \{1, 2, \ldots, n\}$. For any cluster C, denote $\text{mean}(C) := \frac{\sum_{x \in C} x}{\|\sum_{x \in C} x\|}$ as the mean or center of mass of C.

Definition 1. Spherical k-means clustering: *Given a data set $\mathcal{X}$ in R^d with n points and an integer k. All the data are on a unit sphere, that is the norm* *of each data is normalized to one. The object is to find a k points center set C with unit norm so as to minimize the following objective function:*

$$\sum_{x \in \mathcal{X}} d(x, C).$$

Definition 2. Spherical k-means clustering with penalties: *Given an integer k and a data set $\mathcal{X}$ in R^d with n points, each data $x \in X$ is associated with a penalty cost $p(x)$. All the data are on a unit sphere, that is the norm of each data is normalized to one. The object is to find a k points center set C with unit form so as to minimize the following objective function:*

$$\phi_k(\mathcal{X}, C) := \sum_{x \in \mathcal{X}} \min\{d(x, C), p(x)\}.$$

We use $C_{OPT} := \{c_i^*, c_2^*, \ldots, c_k^*\}$ to denote the optimal solution of spherical k-means clustering, then we can simplify $\phi_k(\mathcal{X}, C_{OPT})$ as $\phi_k^*(\mathcal{X})$.

3   The Seeding Algorithm and Our Results

In this section, we will give our seeding algorithm for the spherical k-means clustering with penalties and our main result.

This algorithm is based on SKM++ algorithm [10] for spherical k-means clustering. This algorithm contains two main phase. In the first phase, we first choose a center uniformly at random from the given data set, then we choose $k-1$ centers one by one by specific probabilities. In the second phase, we partition the given data set into $k+1$ sets. One of the $k+1$ set is a penalty set that is all the data in this set are penalized. Next we update the centers of other k clusters and re-cluster all the data in the given set. We repeat the second phase until the center set no longer change.

Theorem 1. *For any instance $\mathcal{X}$ of spherical k-means clustering with penalties, let C' be the center set returned by Algorithm 1. We have*

$$\phi_k(\mathcal{X}, C') \leq 2 \max\{2, M\}(1 + M)(\ln k + 2)\phi_k^*(\mathcal{X}),$$

where M is the ratio of the maximal penalty value and the minimal one of the given data set $\mathcal{X}$.

4   Proof of Theorem 1

In this section, we will give a detailed proof of Theorem 1. From Lemma 3, we have that $\phi_k(\mathcal{X}, C') \leq \phi_k(\mathcal{X}, C)$. Therefore, if we have

$$\phi_k(\mathcal{X}, C) \leq 2 \max\{2, M\}(1 + M)(\ln k + 2)\phi_k^*(\mathcal{X}),$$

Theorem 1 is proved.

Here, we introduce two lemmas about the relaxed triangle inequalities in both spherical k-means clustering and k-means problem with penalties, which have been presented in [15, 16].

Algorithm 1

Input: A data set $\mathcal{X} \in R^d$ with n points, a penalty cost $p(x)$ for each data, a integer k.
Output: Data sets $C, C' \in R^d$ with k points and a penalty subset P.
1: Initialize $C := \emptyset, C' := \emptyset,\ i = 1$.
2: Choose the first center v_i uniformly at random from $\mathcal{X}$.
3: Update $C := C \cup \{c_i\}, i := i + 1$.
4: **while** $i \leq k$ **do**
5:    Choose the center c_i from $\mathcal{X}$ with probability $\frac{\min\{d(c_i, C),\ p(c_i)\}}{\sum_{x \in \mathcal{X}} \min\{d(x, C),\ p(x)\}}$.
6:    Update $C := C \cup \{c_i\}, i := i + 1$.
7: **end while**
8: Update $C' := C$.
9: Set $P := \{x \in \mathcal{X} : p(x) < d(x, C)\}$.
10: **for** i from 1 to k **do**
11:    Set $\mathcal{X}_i := \{x \in \mathcal{X} \backslash P : d(x, c_i) \leq d(x, c_j), j \in [k], i \neq j\}$. If there exists a set $C_1 \subseteq C$ which satisfies that $d(x, c_i) = d(x, c_j) < d(x, c_w), \forall i, j \in C_1, w \in C \backslash C_1$. Then, we randomly cluster the data x to one of $\mathcal{X}_i, i \in C_1$.
12: **end for**
13: **for** i from 1 to k **do**
14:    Update set C' by setting $c_i := \frac{\sum_{x \in \mathcal{X}_i} x}{||\sum_{x \in \mathcal{X}_i} x||}$,
15: **end for**
16: Repeat Step 10 to Step 15 until C' no longer changes.
17: **return** Data sets C, C', P.

Lemma 1. [16] *For any data $x, y, z \in \mathcal{X}$, we have*

$$d(x, y) \leq 2(d(x, z) + d(z, y)).$$

Lemma 2. [15] *For any data $x, y \in \mathcal{X}$, and any set Z, we have*

$$\min\{d(x, Z), p(x)\} \leq \max\{2, M\} \left[\min\{d(x, y), p(y)\} + \min\{d(y, Z), p(y)\}\right],$$

where

$$M := \frac{\max_{x \in \mathcal{X}} p(x)}{\min_{x \in \mathcal{X}} p(x)}.$$

Lemma 3. *For any cluster C and data $z \in C$, we have*

$$\sum_{x \in C} d(x, z) - \sum_{x \in C} d(x, mean(C)) \leq ||\sum_{x \in C} x|| d(z, mean(C)).$$

Proof.

$$\sum_{x \in C} d(x, z) - \sum_{x \in C} d(x, \text{mean}(C))$$

$$= \; < \sum_{x \in C} x, \text{mean}(C) > - < \sum_{x \in C} x, z >$$

$$= || \sum_{x \in C} || < \text{mean}(C), \text{mean}(C) > - || \sum_{x \in C} || < \text{mean}(C), z >$$

$$= || \sum_{x \in C} x || d(z, \text{mean}(C)).$$

$\square$

From step 10 of Algorithm 1, we can obtain a partition $\mathcal{X}_1, \mathcal{X}_2, \ldots, \mathcal{X}_k, P$ of set $\mathcal{X}$. Next, we construct a new partition of $\mathcal{X}$. First, we partition the penalty set P into k disjoint sets $P_1, P_2, \ldots, P_k$. Then the new partition can be described as follows:

$$\mathcal{X}_i^p := \mathcal{X}_i \cup P_i, \; i \in [k].$$

Let $\mathcal{X}_1^*, \mathcal{X}_2^*, \ldots, \mathcal{X}_k^*, P^*$ be the corresponding cluster in the optimal solution C_{OPT}. From Lemma 3, we have that $c_i^* = \text{mean}(\mathcal{X}_i^*), i \in [k]$. Similar to solution C, we can obtain a new partition $\mathcal{X}_1^{p*}, \mathcal{X}_2^{p*}, \ldots, \mathcal{X}_k^{p*}$ of $\mathcal{X}$ based on $\mathcal{X}_1^*, \mathcal{X}_2^*, \ldots, \mathcal{X}_k^*, P^*$.

Lemma 4. *Let $\mathcal{X}_i^{p*}$ be an arbitrary cluster in optimal solution, and z be an data selected uniformly at random from $\mathcal{X}_i^{p*}$, then the following inequality hold:*

$$E(\phi_k(\mathcal{X}_i^{p*}, z)) \leq 2\phi_k(\mathcal{X}_i^*, c_i^*) + (1 + M)\phi(P_i^*, c_i^*).$$

Proof. Recall the probability that z is selected. We have

$$E(\phi_k(\mathcal{X}_i^{p*}, z)) = \frac{1}{|\mathcal{X}_i^{p*}|} \sum_{z \in \mathcal{X}_i^{p*}} \phi_k(\mathcal{X}_i^{p*}, z)$$

$$= \frac{1}{|\mathcal{X}_i^{p*}|} \sum_{z \in \mathcal{X}_i^*} \phi_k(\mathcal{X}_i^{p*}, z) + \frac{1}{|\mathcal{X}_i^{p*}|} \sum_{z \in P_i^*} \phi_k(\mathcal{X}_i^{p*}, z)$$

$$= \frac{1}{|\mathcal{X}_i^{p*}|} \sum_{z \in \mathcal{X}_i^*} \sum_{x \in \mathcal{X}_i^{p*}} \min\{d(x, z), p(x)\} \tag{1}$$

$$+ \frac{1}{|\mathcal{X}_i^{p*}|} \sum_{z \in P_i^*} \sum_{x \in \mathcal{X}_i^{p*}} \min\{d(x, z), p(x)\} \tag{2}$$

From Lemma 3 and $||\sum_{x \in \mathcal{X}_i^*} x|| \leq |\mathcal{X}_i^*|$, one can know that

$$(1) \leq \frac{1}{|\mathcal{X}_i^{p*}|} \sum_{z \in \mathcal{X}_i^*} \sum_{x \in \mathcal{X}_i^*} d(x,z) + \frac{1}{|\mathcal{X}_i^{p*}|} \sum_{z \in \mathcal{X}_i^*} \sum_{x \in P_i^*} p(x)$$

$$\leq \frac{1}{|\mathcal{X}_i^{p*}|} \sum_{z \in \mathcal{X}_i^*} \left(\sum_{x \in \mathcal{X}_i^*} d(x, \mathrm{mean}(\mathcal{X}_i^*)) + || \sum_{x \in \mathcal{X}_i^*} x || d(z, \mathrm{mean}(\mathcal{X}_i^*)) \right)$$

$$+ \frac{|\mathcal{X}_i^*|}{|\mathcal{X}_i^{p*}|} \sum_{x \in P_i^*} p(x)$$

$$\leq (1 + \frac{|| \sum_{x \in \mathcal{X}_i^*} x ||}{|\mathcal{X}_i^{p*}|}) \phi_k(\mathcal{X}_i^*, c_i^*) + \frac{|\mathcal{X}_i^*|}{|\mathcal{X}_i^{p*}|} \sum_{x \in P_i^*} p(x)$$

$$\leq 2\phi_k(\mathcal{X}_i^*, c_i^*) + \phi_k(P_i^*, c_i^*).$$

By the definition of M, we have

$$(2) \leq \frac{1}{|\mathcal{X}_i^{p*}|} \sum_{z \in P_i^*} \sum_{x \in \mathcal{X}_i^{p*}} p(x)$$

$$\leq \frac{1}{|\mathcal{X}_i^{p*}|} \sum_{z \in P_i^*} \sum_{x \in \mathcal{X}_i^{p*}} M p(z)$$

$$= M \sum_{z \in P_i^*} p(z)$$

$$= M \phi_k(P_i^*, c_i^*).$$

Therefore

$$E(\phi_k(\mathcal{X}_i^{p*}, z)) \leq 2\phi_k(\mathcal{X}_i^*, c_i^*) + (1 + M)\phi_k(P_i^*, c_i^*).$$

$\square$

Lemma 5. *Assume that the t-th $(1 \leq t < k)$ iteration has finished in the first phase and let C^t be the center set selected by Algorithm 1, and $c_{t+1} \in \mathcal{X}_i^{p*}$ be the next selected center, $\mathcal{X}_i^{p*}$ be an arbitrary cluster in optimal solution. That is $C^{t+1} := C^t \cup \{c_{t+1}\}$, then we have*

$$E(\phi_k(\mathcal{X}_i^{p*}, C^{t+1})|C^t, c_{t+1} \in \mathcal{X}_i^{p*}) \leq 2\max\{2, M\}[2\phi_k(\mathcal{X}_i^*, c_i^*) + (1 + M)\phi_k(P_i^*, c_i^*)].$$

Proof. Recall the probability that c_{t+1} is selected and $z \in \mathcal{X}_i^{p*}$, we have

$$E(\phi_k(\mathcal{X}_i^{p*}, C^{t+1})|C^t, c_{t+1} \in \mathcal{X}_i^{p*})$$

$$= \sum_{c_{t+1} \in \mathcal{X}_i^{p*}} \frac{\min\{d(c_{t+1}, C^t), p(c_{t+1})\}}{\sum_{x \in \mathcal{X}_i^{p*}} \min\{d(x, C^t), p(x)\}} \sum_{x \in \mathcal{X}_i^{p*}} \min\{d(x, C^t), d(x, c_{t+1}), p(x)\}$$

From Lemma 2, we have

$$\min\{d(c_{t+1}, C^t), p(c_{t+1})\} \leq \frac{\max\{2, M\}}{|\mathcal{X}_i^{p*}|} \phi(\mathcal{X}_i^{p*}, c_{t+1})$$
$$+ \frac{\max\{2, M\}}{|\mathcal{X}_i^{p*}|} \phi(\mathcal{X}_i^{p*}, C^t). \tag{3}$$

Combined (3) and Lemma 4, we have

$$E(\phi_k(\mathcal{X}_i^{p*}, C^{t+1})|C^t, c_{t+1} \in \mathcal{X}_i^{p*})$$

$$\leq \sum_{c_{t+1} \in \mathcal{X}_i^{p*}} \frac{\frac{\max\{2, M\}}{|\mathcal{X}_i^{p*}|} \phi_k(\mathcal{X}_i^{p*}, c_{t+1})}{\sum_{x \in \mathcal{X}_i^{p*}} \min\{d(x, C^t), p(x)\}} \sum_{x \in \mathcal{X}_i^{p*}} \min\{d(x, C^t), d(x, c_{t+1}), p(x)\}$$

$$+ \sum_{c_{t+1} \in \mathcal{X}_i^{p*}} \frac{\frac{\max\{2, M\}}{|\mathcal{X}_i^{p*}|} \phi_k(\mathcal{X}_i^{p*}, C^t)}{\sum_{x \in \mathcal{X}_i^{p*}} \min\{d(x, C^t), p(x)\}} \sum_{x \in \mathcal{X}_i^{p*}} \min\{d(x, C^t), d(x, c_{t+1}), p(x)\}$$

$$\leq \frac{2\max\{2, M\}}{|\mathcal{X}_i^{p*}|} \sum_{c_{t+1} \in \mathcal{X}_i^{p*}} \phi_k(\mathcal{X}_i^{p*}, c_{t+1})$$

$$= 4\max\{2, M\}\phi_k(\mathcal{X}_i^*, c_i^*) + 2\max\{2, M\}(1 + M)\phi_k(P_i^*, c_i^*).$$

$\square$

For $1 \leq i \leq k$, denote H^t, U^t and W^t as the set of covered clusters, uncovered set and wasted set at the end of t-th iteration respectively [10].

$$H^t := \{i|1 \leq i \leq k, \mathcal{X}_i^{p*} \cap C^t \neq \emptyset\},$$
$$U^t := \{i|1 \leq i \leq k, \mathcal{X}_i^{p*} \cap C^t = \emptyset\},$$
$$W^t := t - |H^t|.$$

First, we can analyze the cost of covered sets by the following lemma.

Lemma 6. *For any $1 \leq t \leq k$, we have*

$$E\left(\sum_{i \in H^t} \phi_k(\mathcal{X}_i^{p*}, C^t)\right) \leq 2\max\{2, M\}(1 + M)\phi_k^*(\mathcal{X}).$$

Proof. Let $\mathcal{X}_t^{p*}$ be the cluster in optimal solution selected at t-th iteration and c_t be the center selected by Algorithm 1. Suppose $W^t = 0$, then by Lemma 5 the following inequality holds:

$$E\left(\sum_{i\in H^t} \phi_k(\mathcal{X}_i^{p*}, C^t)\right) = E(\phi_k(\mathcal{X}_1^{p*}, C^1)|C^0, \{c_1 \in \mathcal{X}_1^{p*}\}) + \dots$$

$$+ E(\phi_k(\mathcal{X}_t^{p*}, C^t)|C^{t-1}, \{c_t \in \mathcal{X}_t^{p*}\})$$

$$\leq 2\max\{2, M\}(1 + M)\sum_{i=1}^{t} \phi_k(\mathcal{X}_i^{p*}, c_i^*)$$

$$\leq 2\max\{2, M\}(1 + M)\phi_k^*(\mathcal{X}).$$

If $W^t \neq 0$, the proof is similar, we omit the proof. □

Next, we consider the cost of the uncovered sets by Lemmas 7–9.

Assume that the t-th iteration has been finished and the center set selected by the algorithm is C^t. For each $1 \leq t \leq k$, denote

$$\alpha_t := \frac{W^t \sum_{i\in U^t} \phi_k(\mathcal{X}_i^{p*}, C^t)}{|U^t|}.$$

Then we consider the following two cases for the $(t+1)$-th iteration:

- The center is selected from U^t.
- The center is selected from H^t.

Lemma 7. [10] *Let C^t be the center set selected by the algorithm at the end of the t-th iteration and assume that the next center c_j is selected from an uncovered cluster $\mathcal{X}_j^{p*}$. Then we have*

$$E(\alpha_{t+1} - \alpha_t|C^t, \ j \in U^t) \leq 0.$$

Lemma 8. [10] *Let C^t be the center set selected by the algorithm at the end of the t-th iteration. Assume that the next center c_j is selected from a covered cluster $\mathcal{X}_j^{p*}$. Then we have*

$$\alpha_{t+1} - \alpha_t \leq \frac{\sum_{i\in U^t} \phi_k(\mathcal{X}_i^{p*}, C^t)}{|U^t|}.$$

Combined Lemmas 7 and 8, we have the following lemma.

Lemma 9. [10] *For any $0 \leq t \leq k - 1$, we have*

$$E(\alpha_{t+1} - \alpha_t|C^t) \leq \frac{\sum_{i\in H^t} \phi_k(\mathcal{X}_i^{p*}, C^t)}{k - t}.$$

Now, we can finish the proof of Theorem 1 by the following two cases.

- $U^k = \emptyset$: From Lemma 6 we have

$$E(\phi_k(\mathcal{X}, C^k)) = E\left(\sum_{i\in H^k} \phi_k(\mathcal{X}_i^{p*}, C^k)\right) \leq 2\max\{2, M\}(1 + M)\phi_k^*(\mathcal{X}).$$

– $U^k \neq \emptyset$: Combining the definition of α_k, Lemmas 6 and 9, we have

$$
\begin{aligned}
E(\phi_k(\mathcal{X}, C^k)) &= E\left(\sum_{i \in H^k} \phi_k(\mathcal{X}_i^{p*}, C^k)\right) + E\left(\sum_{i \in U^k} \phi_k(\mathcal{X}_i^{p*}, C^k)\right) \\
&\leq 2\max\{2, M\}(1 + M)\phi_k^*(\mathcal{X}) + E(\alpha_k | C^{k-1}) \\
&= 2\max\{2, M\}(1 + M)\phi_k^*(\mathcal{X}) \\
&\quad + E(\alpha_k - \alpha_{k-1} | C^{k-1} + \ldots + \alpha_2 - \alpha_1 | C^1 + \alpha_1 | C^0) \\
&\leq 2\max\{2, M\}(1 + M)\phi_k^*(\mathcal{X}) + \frac{\sum_{i \in H^{k-1}} \phi_k(\mathcal{X}_i^{p*}, C^{k-1})}{1} \\
&\quad + \ldots + \frac{\sum_{i \in H^1} \phi_k(\mathcal{X}_i^{p*}, C^1)}{k - 1} \\
&\leq 2\max\{2, M\}(1 + M)(\ln k + 2)\phi_k^*(\mathcal{X}).
\end{aligned}
$$

Above all, we have

$$
E(\phi_k(\mathcal{X}, C^k)) = E(\phi_k(\mathcal{X}, C)) \leq 2\max\{2, M\}(1 + M)(\ln k + 2)\phi_k^*(\mathcal{X}).
$$

The Theorem 1 is proved.

5   Discussion

In this paper, we have studied the seeding algorithm for spherical k-means clustering with penalties and give a $2\max\{2, M\}(1 + M)(\ln k + 2)$-approximate algorithm for M being the ratio of the maximal and the minimal penalty values of the given object set. We are currently focusing on two tasks alongside the problem: One is to improve the ratio $2\max\{2, M\}(1 + M)(\ln k + 2)$ for spherical k-means clustering with penalties; The other is to study other variants of spherical k-means clustering.

Acknowledgements. The first and second authors are supported by National Natural Science Foundation of China (No. 11531014). The third author is supported by National Natural Science Foundation of China (No. 61772005) and Natural Science Foundation of Fujian Province (No. 2017J01753). The forth author is supported by Higher Educational Science and Technology Program of Shandong Province (No. J17KA171). The fifth author is supported by National Natural Science Foundation of China (No. 11871081).

References

1. Awasthi, P., Charikar, M., Krishnaswamy, R., Sinop, A.: The hardness of approximation of Euclidean k-means, arXiv preprint arXiv:1502.03316 (2015)
2. Aloise, D., Deshpande, A., Hansen, P.: NP-hardness of Euclidean sum-of-squares clustering. Mach. Learn. **75**(2), 245–248 (2009)

3. Ahmadian, S., Norouzi-Fard, A., Svensson, O., Ward, J.: Better guarantees for k-means and Euclidean k-median by primal-dual algorithms. In: Proceedings of the 58th Annual IEEE Symposium on Foundations of Computer Science (FOCS), pp. 61–72 (2017)

4. Arthur, D., Vassilvitskii, S.: k-means++: the advantages of careful seeding. In: Proceedings of the 18th Annual ACM-SIAM Symposium on Discrete Algorithms (SODA), pp. 1027–1035 (2007)

5. Blömer, J., Brauer, S., Bujna, K.: A theoretical analysis of the fuzzy k-means problem. In: Proceedings of the 16th IEEE International Conference on Data Mining (ICDM), pp. 805–810 (2017)

6. Blömer, J., Lammersen, C., Schmidt, M., Sohler, C.: Theoretical analysis of the k-means algorithm – a survey. In: Kliemann, L., Sanders, P. (eds.) Algorithm Engineering. LNCS, vol. 9220, pp. 81–116. Springer, Cham (2016). https://doi.org/10.1007/978-3-319-49487-6_3

7. Cohen-Addad, V., Klein, P.N., Mathieu, C.: Local search yields approximation schemes for k-means and k-median in Euclidean and minor-free metrics. SIAM J. Comput. **48**(2), 644–667 (2019)

8. Drineas, P., Frieze, A., Kannan, R., Vempala, V.: Clustering large graphs via the singular value decomposition. Mach. Learn. **56**(1–3), 9–33 (2004)

9. Dhillon, I., Modha, D.: Concept decompositions for large sparse text data using clustering. Mach. Learn. **42**(1–2), 143–175 (2001)

10. Endo, Y., Miyamoto, S.: Spherical k-means++ clustering. In: Torra, V., Narukawa, Y. (eds.) MDAI 2015. LNCS (LNAI), vol. 9321, pp. 103–114. Springer, Cham (2015). https://doi.org/10.1007/978-3-319-23240-9_9

11. Gupta, S., Kumar, R., Lu, K., Moseley, B., Vassilvitskii, S.: Local search methods for k-means with outliers. Proc. VLDB Endow. **10**(7), 757–768 (2017)

12. Hornik, K., Feinerer, I., Kober, M., Buchata, M.: Spherical k-means clustering. J. Stat. Softw. **50**(10), 1–22 (2015)

13. Kanungo, T., Mount, D., Netanyahu, N., Piatko, C., Silverma, R.: A local search approximation algorithm for k-means clustering. Comput. Geom. **28**(2–3), 89–112 (2004)

14. Lloyd, S.: Least squares quantization in PCM. IEEE Trans. **28**(2), 129–137 (1982)

15. Li, M., Xu, D., Yue, J., Zhang, D., Zhang, P.: The seeding algorithm for k-means with penalties. J. Comb. Optim. (under review)

16. Li, M., Xu, D., Zhang, D., Zou, J.: The seeding algorithms for spherical k-means clustering. J. Global Optim. 1–14 (2019)

17. Moriya, T., Roth, H., Nakamura, S., Oda, H., Kai, N., Oda, M.: Unsupervised pathology image segmentation using representation learning with spherical k-means. In: Digital Pathology, p. 36 (2018)

18. Tunali, V., Bilgin, T., Camurcu, A.: An improved clustering algorithm for text mining: multi-cluster spherical k-means. Int. Arab J. Inf. Technol. **13**(1), 12–19 (2016)

19. Xu, J., Han, J., Xiong, K., Nie F.: Robust and sparse fuzzy k-means clustering. In: Proceedings 25th International Joint Conference on Artificial Intelligence (IJCAI), pp. 2224–2230 (2016)

20. Xu, D., Xu, Y., Zhang, D.: A survey on algorithm for k-means and its variants. Oper. Res. Trans. **21**, 101–109 (2017)

Approximation Algorithm for the Correlation Clustering Problem with Non-uniform Hard Constrained Cluster Sizes

Sai Ji[1], Dachuan Xu[1(✉)], Min Li[2], and Yishui Wang[3]

[1] Department of Operations Research and Scientific Computing,
Beijing University of Technology, Beijing 100124, People's Republic of China
xudc@bjut.edu.cn

[2] School of Mathematics and Statistics, Shandong Normal University,
Jinan, People's Republic of China
liminemily@sdnu.edu.cn

[3] Shenzhen Institutes of Advanced Technology, Chinese Academy of Sciences,
Shenzhen 518055, People's Republic of China

Abstract. This paper considers the correlation clustering problem with non-uniform hard constrained cluster sizes, which is a generalization of correlation clustering problem. In this problem, we are given a positive integer U_v for each vertex v, and require $|C| \leq \min_{v \in C} U_v$ for any cluster C. We provide a $(2, 4)$-bicriteria approximation algorithm for this problem. Namely, the solution returned by the algorithm has the cost that is at most 4 times the optimum, and for each cluster C in the solution, we have $|C| \leq 2 \min_{v \in C} U_v$.

Keywords: Correlation clustering ·
Non-uniform hard constrained cluster sizes · Approximation algorithm

1 Introduction

Clustering problems arise in many applications such as data mining, data compression, machine learning and computer vision. These problems have been widely studied in the literature. For partial surveys, see e.g. [6,7,10,11,18,19].

In this paper, we study the correlation clustering problem which can be used in protein interaction networks, cross-lingual link detection, communication networks, and so on. The correlation clustering problem is introduced by Bansal et al. [9], motivated by both document clustering and agnostic learning. This problem can solve problems where we have conflicting measures among objects and the aim is to provide a consistent clustering. In this problem, we are given a complete graph $G = (V, E)$, each edge $(u, v) \in E$ is labeled by $+$ or $-$ depending on whether vertex u and vertex v have been deemed to be similar or different. There are two versions of this problem: minimizing disagreement and maximizing agreement. In the problem minimizing disagreement, we need to partition

© Springer Nature Switzerland AG 2019
D.-Z. Du et al. (Eds.): AAIM 2019, LNCS 11640, pp. 159–168, 2019.
https://doi.org/10.1007/978-3-030-27195-4_15

the vertices into some clusters so as to minimize the number of positive edges whose endpoints lie in different clusters plus the number of negative edges whose endpoints lie in the same cluster. In the problem of maximizing agreement, we need to partition the vertices into some clusters so as to maximize the total number of negative edges whose endpoints places in different clusters and positive edges whose endpoints places within the same cluster. The number of clusters in both versions is not limited. Bansal et al. [9] prove that the problem is NP-hard, one cannot obtain the optimal solution in polynomial time unless $P = NP$. There are many existing works on approximation algorithms for this problem [3,8,14,17,20]. Bansal et al. [9] give the first constant-factor approximation algorithm for the minimizing disagreement. For maximizing agreements, they give a polynomial time approximation scheme. Demaine et al. [14] provide an $O(\log n)$-approximation algorithm for general graphs. Charikar et al. [12] prove that the minimizing disagreement is APX-hard. They provide an LP-rounding 4-approximation algorithm for minimization disagreement on complete graphs as well as an $O(\log n)$-approximation algorithm for general graphs. Furthermore, they give a 0.766-approximation algorithm for the maximizing agreement by rounding a semidefinite programming relaxation. In the same year, Swamy [24] gives a 0.75-approximation algorithm for the weighted maximizing agreement based on the technique of [16], the number of clusters returned by this algorithm is at most 4. Then, he improves 0.75 to 0.766 by the technique of [15], the maximum number of clusters returned by the latter algorithm is 6. Ailon et al. [5] give a randomized 3-approximation algorithm and further improve the ratio to 2. Chawla et al. [13] provide the best deterministic LP-rounding 2.06-approximation algorithm. In particular, the graph is a complete bipartite graph, there is an LP-rounding 11-approximation algorithm provided by Amit [1]. Ailon et al. [2] give a deterministic LP-rounding 4-approximation algorithm as well as a randomized LP-rounding 4-approximation algorithm based on [5].

There are many variants of the correlation clustering problem [4,17,20,21,23]. In this paper, we are particularly interested in the correlation clustering problem with constrained cluster sizes which is introduced by Puleo et al. [22]. Puleo et al. [22] study the correlation clustering problem with soft constrained cluster sizes. In this problem, each vertex v has a penalty parameter μ_v. They provide a $\max\{\mu^*, 2/\alpha\}$-approximation algorithm for this problem, where μ^* is the maximum penalty cost of all vertices and $a \in (0, 1/2]$ is a given parameter. In this paper, we consider the minimizing version of correlation clustering problem with non-uniform hard constrained cluster sizes. In this problem, we are given a positive integer U_v for each vertex v, we need to partition the vertices into some clusters so as to minimize the number of positive edges whose endpoints lie in different clusters plus the number of negative edges whose endpoints lie in the same cluster. Furthermore, each cluster C should satisfies $|C| \leq \min_{v \in C} U_v$. Given a parameter $\alpha \in (1/2, 1)$, we provide an LP-rounding $(2, 4)$-bicriteria approximation algorithm for the problem based on [12]. Namely, the solution returned by the algorithm has the cost that is at most 4 times the optimum solution, and for each cluster C in the solution, we have $|C| \leq 2\min_{v \in C} U_v$.

The rest of this paper is organized as follows. In Sect. 2, we describe the formulation for minimizing version of correlation clustering problem with non-uniform hard constrained cluster sizes and the corresponding LP relaxation. In Sect. 3, we give our algorithm as well as the analysis of approximation ratio. Some discussions are given in Sect. 4.

2   The Correlation Clustering Problem with Non-uniform Hard Constrained Cluster Sizes

In the correlation clustering problem with non-uniform hard constrained cluster sizes, we are given a labeled complete graph $G = (V, E)$ together with a positive integer upper bound U_v for each vertex $v \in V$. We use E^+ and E^- to denote the set of positive edges and the set of negative edges, respectively. For each edge $(u, v) \in E$, we introduce a binary decision variable x_{uv} to denote whether the vertices u and v are in the same cluster. Then the minimizing version of correlation clustering problem with non-uniform hard constrained cluster sizes can be formulated as follows:

$$
\begin{aligned}
\min \quad & \sum_{(u,v)\in E^+} x_{uv} + \sum_{(u,v)\in E^-} (1 - x_{uv}) \\
\text{s. t.} \quad & x_{uv} + x_{vw} \geq x_{uw}, & \forall u, v, w \in V, \\
& \sum_{v\in V} (1 - x_{uv}) \leq U_u, & \forall u \in V, \\
& x_{uu} = 0, & \forall u \in V, \\
& x_{uv} \in \{0, 1\}, & \forall u, v \in V. \quad (1)
\end{aligned}
$$

The first constraint ensures that if the vertices $u, v \in V$ are in the same cluster and $v, w \in V$ are also in the same cluster, then the vertices u and w must be in the same cluster. The second constraint ensures that the number of vertices in each cluster is no more than the corresponding upper bound. The third constraint is natural since any vertex in a cluster should be counted. The LP relaxation of (1) is given as follows:

$$
\begin{aligned}
\min \quad & \sum_{(u,v)\in E^+} x_{uv} + \sum_{(u,v)\in E^-} (1 - x_{uv}) \\
\text{s. t.} \quad & x_{uv} + x_{vw} \geq x_{uw}, & \forall u, v, w \in V, \\
& \sum_{v\in V} (1 - x_{uv}) \leq U_u, & \forall u \in V, \\
& x_{uu} = 0, & \forall u \in V, \\
& 0 \leq x_{uv} \leq 1, & \forall u, v \in V. \quad (2)
\end{aligned}
$$

3   Algorithm

In this section, we present our $(2, 4)$-bicriteria approximation algorithm for the minimizing version of correlation clustering problem with non-uniform hard constrained cluster sizes. The detailed algorithm is shown in Subsect. 3.1 and theo-retical analysis of the algorithm are provided in Subsect. 3.2.

3.1 Algorithm

In this subsection, we provide our algorithm for the correlation clustering problem with non-uniform hard constrained cluster sizes. First, we give a high level of our algorithm. Let $\alpha \in (1/2, 1)$ be a parameter which will be specified later, the algorithm consists of two main steps. In the first step, we solve (2) to obtain the fractional optimal solution x^*. We can see the value of x_{uv} as the distance between the vertices u and v. The second step is an iterative clustering process. We initialize the un-clustered set $S := V$. In each iteration, we choose a vertex u with the minimum upper bound from the un-clustered set S as a center vertex. Let T_u be the set of vertices from the un-clustered set whose distance from u is no greater than the parameter α. We let $C_u = \{u\}$ or $C_u = T_u$ according to the distances between the vertices in T_u and u. We repeat the iterative clustering process until all the vertices have been clustered. The detailed algorithm is shown as follows:

Algorithm 1

Input: A labeled complete graph $G = (V, E)$, positive integer U_v for each vertex v, and parameter $\alpha \in \left(\frac{1}{2}, 1\right)$.

Output: A partition of vertices.

1: Solve the LP relaxation of (1) to obtain the optimal solution x^*.
2: Let $S = V$.
3: **while** $S \neq \emptyset$ **do**
4:    Select a vertex u with the minimum upper bound from S, i. e. $u = \mathrm{argmin}_{v \in S} U_v$.
5:    Let $T_u := \{v \in S : x_{uv}^* \leq \frac{1}{2}\}$.
6:    **if** $\dfrac{\sum_{v \in T_u} x_{uv}^*}{|T_u|} > 1 - \alpha$, **then**
7:       Let $C_u = \{u\}$. (Type 1 of cluster)
8:       Let $S := S - \{u\}$.
9:    **else**
10:       Let $C_u = T_u$. (Type 2 of cluster)
11:       Let $S := S - C_u$.
12:    **end if**
13: **end while**
14: **return** the partition $\{C_u\}$

3.2 Analysis

In this subsection, we give the theoretical analysis of Algorithm 1. For the conveniens of statement, we call a positive edge whose endpoints lie in different clusters or a negative edge whose endpoints lie in the same cluster as an "error".

Algorithm 1 is an iterative processes. In each iteration, we choose a center vertex u and construct a cluster C_u from the set of un-clustered vertices. Thus the errors are contained in the set of the negative edges whose endpoints lie in C_u and the positive cut edges between cluster C_u and set $S - C_u$. Each error only needs to be counted once, that is, we don't need to consider the errors later. The total number of errors is the total number of new errors produced by each iteration. The analysis of our algorithm contains three parts. In the first part, we analyze the upper bound of the size of each cluster (cf. Lemma 1). In the second part, we give the upper bound of the number of errors in type 1 of cluster (cf. Lemmas 2 and 3). In the last part, we give the upper bound of the number of errors in type 2 of cluster (cf. Lemmas 4 and 7).

Lemma 1. *For each cluster C associated with center vertex u, we have*

$$|C| \le 2 \min_{v \in C} U_v.$$

Proof. From the second constraint of (2), for each center vertex u we have

$$\sum_{v \in V} (1 - x_{uv}^*) \le U_v.$$

Recall the definition of T_u and step 4 of Algorithm 1. We have

$$|C| \le |T_u| \le 2 \sum_{v \in T_u} (1 - x_{uv}^*) \le 2 \sum_{v \in V} (1 - x_{uv}^*) \le 2 U_u \le 2 \min_{v \in C} U_v.$$

The lemma is concluded.

Type 1 of Cluster. In this case, we have $\sum_{v \in T_u} x_{uv}^* / |T_u| > 1 - \alpha$ and we make $C_u = \{u\}$ a singleton cluster. The errors are shown in Fig. 1.

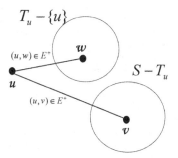

Fig. 1. Errors in type 1 of cluster

There are two kinds of errors in this type of cluster. One is the positive edge (u, w) with $w \in T_u - \{u\}$. Another is the positive edge (u, v) with $v \in S - T_u$. Lemmas 2 and 3 show the upper bounds of the numbers of these two kinds of errors, respectively.

Lemma 2. *For each positive edge $(u, v) \in E^+, v \in S - T_u$, the number of errors can be bounded by $2x^*_{uv}$.*

Lemma 3. *The total number of errors associated with $(u, w) \in E^+$ and $w \in T_u$ can be bounded by*

$$\frac{1}{1-\alpha} \left[\sum_{(u,v) \in E^+, v \in T_u} x^*_{uv} + \sum_{(u,v) \in E^-, v \in T_u} (1 - x^*_{uv}) \right].$$

Proof. From Algorithm 1, we have

$$\frac{\sum_{w \in T_u} x^*_{uw}}{|T_u|} > 1 - \alpha.$$

Furthermore, for each negative edge (u, v) with $v \in T_u$, we have $1 - x^*_{uv} \geq x^*_{uv}$. Thus,

$$\text{the number of errors} \leq |T_u| < \frac{1}{1-\alpha} \sum_{v \in T_u} x^*_{uv}$$

$$\leq \frac{1}{1-\alpha} \left[\sum_{(u,v) \in E^+, v \in T_u} x^*_{uv} + \sum_{(u,v) \in E^-, v \in T_u} (1 - x^*_{uv}) \right].$$

The lemma is concluded.

Type 2 of Cluster. Recall Algorithm 1, there are two kinds of errors. One is the negative edge whose endpoints both lie in T_u. Another is the positive edge whose endpoints lie in T_u and $S - T_u$, respectively. The specific errors refer to Fig. 2.

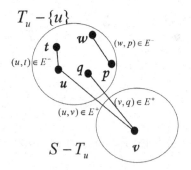

Fig. 2. Errors in type 2 of cluster

Lemma 4. *For each negative edge (w, p) with $w, p \in T_u$, if $x_{uw}^*, x_{up}^* \leq 1 - \alpha$, then the number of errors can be bounded by*

$$\frac{1}{2\alpha - 1} \left(1 - x_{wp}^*\right).$$

For other negative edges, we relabel the vertices (other than u) so that $p < w$ if $x_{up}^* < x_{uw}^*$, breaking ties arbitrarily. Then, we have

Lemma 5. *For each vertex $p \in T_u$, if $x_{up}^* > 1 - \alpha$, then the total number of errors produced by all the negative edges (w, p) with $w < p$ can be bounded by the*

$$\frac{2}{2\alpha - 1} \left[\sum_{(w,p) \in E^+, w<p} x_{wp}^* + \sum_{(w,p) \in E^-, w<p} (1 - x_{wp}^*) \right].$$

Proof. The proof is similar to that of [12]. Denote by P_p the number of positive edges $(w, p) \in E^+$ with $w < p$, and N_p the number of negative edges $(w, p) \in E^-, w < p$. Furthermore, the average distance of the vertices w with $w < p$ from u is less than the average distance of the vertices in T_u from u. Then, we have

$$\sum_{(w,p) \in E^+, w<p} x_{wp}^* + \sum_{(w,p) \in E^-, w<p} (1 - x_{wp}^*)$$

$$\geq \sum_{(w,p) \in E^+, w<p} (x_{up}^* - x_{uw}^*) + \sum_{(w,p) \in E^-, w<p} (1 - x_{uw}^* - x_{up}^*)$$

$$\geq P_p x_{up}^* + N_p (1 - x_{up}^*) - \sum_{w<p} x_{uw}^*$$

$$\geq P_p x_{up}^* + N_p (1 - x_{up}^*) - (1 - \alpha)(P_p + N_p)$$

$$> (1 - \alpha) P_p + \frac{1}{2} N_p - (1 - \alpha)(P_p + N_p)$$

$$\geq \left[\frac{1}{2} - (1 - \alpha) \right] N_p$$

$$= \frac{2\alpha - 1}{2} N_p.$$

Therefore,

the number of errors $= |N_p|$

$$< \frac{2}{2\alpha - 1} \left[\sum_{(w,p) \in E^+, w<p} x_{wp}^* + \sum_{(w,p) \in E^-, w<p} (1 - x_{wp}^*) \right].$$

The lemma is concluded.

Next, we analyze the number of errors produced by positive edges in this type of cluster.

Lemma 6. *For a vertex* $v \in S - T_u$, *if* $(q, v) \in E^+, q \in T_u$ *and* $x^*_{uv} \geq \alpha$, *then the number of errors can be bounded by*

$$\frac{2}{2\alpha - 1} x^*_{qv}.$$

Lemma 7. *For a vertex* $v \in S - T_u$, *if* $\frac{1}{2} < x^*_{uv} < \alpha$, *then number of errors associated with* v *can be bounded by*

$$\frac{2}{2\alpha - 1} \left[\sum_{(q,v) \in E^+, q \in T_u} x^*_{qv} + \sum_{(q,v) \in E^-, q \in T_u} (1 - x^*_{qv}) \right].$$

Proof. Similar to that of Lemma 5. Denote by P_v the number of positive edges $(q, v) \in E^+$ with $q \in T_u$, and N_v the number of negative edges $(q, v) \in E^-$ with $q \in T_u$. For each vertex $v \in S - T_u$, we have

$$\sum_{(q,v) \in E^+, q \in T_u} x^*_{qv} + \sum_{(q,v) \in E^-, q \in T_u} (1 - x^*_{qv})$$

$$\geq \sum_{(q,v) \in E^+, q \in T_u} (x^*_{uv} - x^*_{uq}) + \sum_{(q,v) \in E^-, q \in T_u} (1 - x^*_{uv} - x^*_{uq})$$

$$\geq P_v x^*_{uv} + N_v (1 - x^*_{uv}) - \sum_{q \in T_u} x^*_{uq}$$

$$\geq P_v x^*_{uv} + N_v (1 - x^*_{uv}) - (1 - \alpha)(P_v + N_v)$$

$$> \frac{1}{2} P_v + (1 - \alpha) N_v - (1 - \alpha)(P_v + N_v)$$

$$= \frac{2\alpha - 1}{2} P_v.$$

Therefore,

the number of errors $= |P_v|$

$$< \frac{2}{2\alpha - 1} \left[\sum_{(q,v) \in E^+, q \in T_u} x^*_{qv} + \sum_{(q,v) \in E^-, q \in T_u} (1 - x^*_{qv}) \right].$$

The lemma is concluded.

Combining Lemmas 2–7, we can obtain the following lemma which is an upper bound of the total number of errors.

Lemma 8. *Given a labeled complete graph* $G = (V, E)$ *together with a positive integer* U_v *for each vertex* v. *The total number of errors produced by Algorithm 1 can be bounded by*

$$\max \left\{ 2, \frac{1}{1 - \alpha}, \frac{2}{2\alpha - 1} \right\} \left[\sum_{(u,v) \in E^+} x^*_{uv} + \sum_{(u,v) \in E^-} (1 - x^*_{uv}) \right],$$

where E^+, E^- *and* x^* *denote the set of positive edges, the set of negative edges and the optimal fractional solution of (2), respectively.*

Combining Lemmas 1, 8, we can easily obtain the following theorem.

Theorem 1. *Setting $\alpha = 3/4$, Algorithm 1 is a $(2,4)$-bicriteria approximation algorithm for the minimizing version of correlation clustering problem with non-uniform hard constrained cluster sizes.*

4 Discussions

In this paper, we present a bicriteria approximation algorithm for the minimizing version of correlation clustering problem with non-uniform hard constrained cluster sizes. There are two possible future research questions to pursue. One is to study the maximizing version of correlation clustering problem with constrained cluster sizes. Another is to design an algorithm whose solution satisfies the constraint of cluster sizes.

Acknowledgements. The second author is supported by National Natural Science Foundation of China (No. 11531014). The third author is supported by Higher Educational Science and Technology Program of Shandong Province (No. J17KA171). The forth author is supported by Natural Science Foundation of China (No. 61433012), Shenzhen Research Grant (KQJSCX2018033017 0311901, JCYJ20180305180840138 and GGFW201707311403 1767), and Shenzhen Discipline Construction Project for Urban Computing and Data Intelligence.

References

1. Amit, N.: The bicluster graph editing problem. Diss, Tel Aviv University (2004)
2. Ailon, N., Avigdor-Elgrabli, N., Liberty, E., Zuylen, A.V.: Improved approximation algorithms for bipartite correlation clustering. SIAM J. Comput. **41**(5), 1110–1121 (2012)
3. Achtert, E., Böhm, C., David, J., Kröger, P., Zimek, A.: Global correlation clustering based on the hough transform. Stat. Anal. Data Min. **1**(3), 111–127 (2010)
4. Ahn, K.J., Cormode, G., Guha, S., Mcgregor, A., Wirth, A.: Correlation clustering in data streams. In: Proceedings of the 32th International Conference on International Conference on Machine Learning (ICML), pp. 2237–2246 (2015)
5. Ailon, N., Charikar, M., Newman, A.: Aggregating inconsistent information: ranking and clustering. J. ACM, **55**(5), Article No. 23 (2008)
6. Arthur, D., Vassilvitskii, S.: k-Means++: the advantages of careful seeding. In: Proceedings of the 18th Annual ACM-SIAM Symposium on Discrete Algorithms (SODA), pp. 1027–1035 (2007)
7. Ahmadian, S., Norouzi-Fard, A., Svensson, O., Ward, J.: Better guarantees for k-means and Euclidean k-median by primal-dual algorithms. In: Proceedings of the 58th Annual Symposium on Foundations of Computer Science (FOCS), pp. 61–72 (2017)
8. Bonchi, F.: Overlapping correlation clustering. Knowl. Inf. Syst. **35**(1), 1–32 (2013)
9. Bansal, N., Blum, A., Chawla, S.: Correlation clustering. Mach. Learn. **56**(1–3), 89–113 (2004)

10. Byrka, J., Fleszar, K., Rybicki, B., Spoerhase, J.: Bi-factor approximation algorithms for hard capacitated k-median problems. In: Proceedings of the 26th Annual ACM-SIAM Symposium on Discrete Algorithms (SODA), pp. 722–736 (2015)
11. Braverman, V., Lang, H., Levin, K., Monemizadeh, M.: Clustering problems on sliding windows. In: Proceedings of the 27th Annual ACM-SIAM Symposium on Discrete Algorithms (SODA), pp. 1374–1390 (2016)
12. Charikar, M., Guruswami, V., Wirth, A.: Clustering with qualitative information. J. Comput. Syst. Sci. **71**(3), 360–383 (2005)
13. Chawla, S., Makarychev, K., Schramm, T., Yaroslavtsev, G.: Near optimal LP rounding algorithm for correlationclustering on complete and complete k-partite graphs. In: Proceedings of the 47th Annual ACM Symposium on Theory of Computing (STOC), pp. 219–228 (2015)
14. Demaine, E., Emanuel, D., Fiat, A., Immorlica, N.: Correlation clustering in general weighted graphs. Theoret. Comput. Sci. **361**(2), 172–187 (2006)
15. Frieze, A., Jerrum, M.: Improved approximation algorithms for maxk-cut and max bisection. Algorithmica **18**(1), 67–81 (1997)
16. Goemans, M.X., Williamson, D.P.: Improved approximation algorithms for maximum cut and satisfiability problems using semidefinite programming. J. ACM **42**(6), 1115–1145 (1995)
17. Giotis, I., Guruswami, V.: Correlation clustering with a fixed number of clusters. In: Proceedings of the 17th Annual ACM-SIAM Symposium on Discrete Algorithms (SODA), pp. 1167–1176 (2006)
18. Li, S.: On uniform capacitated k-median beyond the natural LP relaxation. ACM Trans. Algorithms, **13**(2), Article No. 22 (2017)
19. Li, M., Xu, D., Zhang, D., Zhang, T.: A streaming algorithm for k-means with approximate coreset. Asia Pac. J. Oper. Res. **36**, 1–18 (2019)
20. Mathieu, C., Schudy, W.: Correlation clustering with noisy input. In: Proceedings of the 21th Annual ACM-SIAM Symposium on Discrete Algorithms (SODA), pp. 712–728 (2010)
21. Mathieu, C., Sankur, O., Schudy, W.: Online correlation clustering. Comput. Stat. **21**(2), 211–229 (2010)
22. Puleo, G.J., Milenkovic, O.: Correlation clustering with constrained cluster sizes and extended weights bounds. SIAM J. Optim. **25**(3), 1857–1872 (2015)
23. Puleo, G.J., Milenkovic, O.: Correlation clustering and biclustering with locally bounded errors. IEEE Trans. Inf. Theory **64**(6), 4105–4119 (2018)
24. Swamy, C.: Correlation clustering: maximizing agreements via semidefinite programming. In: Proceedings of the 15th Annual ACM-SIAM Symposium on Discrete Algorithms (SODA), pp. 526–527 (2004)

Two-Way Currency Trading Algorithms in the Discrete Setting

Fei Li[(✉)]

Department of Computer Science, George Mason University, Fairfax, VA 22030, USA
lifei@cs.gmu.edu

Abstract. In an one-way currency trading model, a player trades all his dollars to yen with the objective of maximizing the total amount of yen got at the end of this game. Exchange rates from dollar to yen vary over time. An optimal offline algorithm trades all dollars using the highest trading rate. In the online setting, any traded yen cannot be converted back to dollars before this game terminates. Under the assumption of knowing the lower bound and the upper bound of exchange rates, as well as the trading duration beforehand, a threat-based online algorithm proposed in [5] has been proved optimal. In an two-way currency trading model, the player is allowed to trade dollars and yen back and forth over time before this game ends. In this paper, we focus on the two-way currency trading model. We re-study an optimal offline algorithm and present a one-step look-ahead optimal algorithm for this model. We further study a setting in which the number of trades cannot exceed a given number k. We provide a few optimal algorithms, analyze their running time, and show output-sensitive time bounds under this restriction for the two-way currency trading problem.

Keywords: Currency trading problem · Exact algorithm · Output-sensitive time bound

1 Introduction

Currency trading problems (also called, *conversion problems*) have been studied extensively in the past decade. Such problems are mainly on how to trade one asset into another, for example, dollar for yen. The objective is to optimize the amount of the converted wealth under various constraints. Paper [6] surveys the recent results on online conversion algorithms. In this paper, we consider the currency trading problems in the *discrete setting*, that is, time is discrete (for example, in terms of days) so that an exchange rate is given at the beginning of each day and this rate does not change until the next day starts. The *duration* of trading currencies is defined as the total number of days.

One-Way Currency Trading Problem. The one-way currency trading problem was initially introduced in [5]. In this problem, there are two currencies, say

© Springer Nature Switzerland AG 2019
D.-Z. Du et al. (Eds.): AAIM 2019, LNCS 11640, pp. 169–178, 2019.
https://doi.org/10.1007/978-3-030-27195-4_16

dollar and yen. A player has an initial asset in dollars. On one day t, a currency exchange rate (from dollar to yen) r_t is known to the player. An online player is allowed to make currency trade at day t, to determine how many, from the remaining dollars if any, are to be traded into yen using the exchange rate r_t. At the end of this trading duration, any left dollars will be traded using the last exchange rate before the game ends. During this process, any traded yen cannot be converted back to dollars. The upper bound M and the lower bound m of currency exchange rates, as well as the trading duration n, are a priori information known to online players. The player's objective is to maximize the total amount of yen traded at the end of this game.

Both the offline and the online versions of this problem have been studied in [5]. An optimal offline algorithm is to trade all initial asset at the highest exchange rate, that is, at the rate $\max_{1 \le t \le n} r_t$. A threat-based online algorithm proposed in [5] has been proved optimal with a competitive ratio c_n, where c_n is the root of

$$n \left(1 - \left(\frac{c_n - 1}{\frac{M}{m} - 1} \right)^{\frac{1}{n}} \right) - c_n = 0 \tag{1}$$

Two-Way Currency Trading Problem. A related problem to the one-way currency trading problem is the two-way currency trading problem. In this problem, there are two currencies, say dollar and yen. A player has an initial asset in dollars. On one day t, two currency exchange rates are known, r_t and r_t^{-1}—the rate r_t (respectively, r_t^{-1}) denotes the number of yen (respectively, dollars) that can be purchased using one dollar (respectively, yen)[1]. The online player is allowed to make currency trading at time t, to determine how much amount of dollar or yen is to be traded into the other currency. At the end of this trading duration, any left dollars will be traded into yen using the last dollar-to-yen exchange rate before this game ends. The upper bound M ($1/m$, respectively) and the lower bounds m ($1/M$, respectively) of currency exchange rates r_t (r_t^{-1}, respectively), as well as the trading duration n, are priori information known to online players. The player's objective is to maximize the total amount of yen got at the end of this game.

For the offline version of this problem, the authors of [5] claimed briefly that "The optimal algorithm for the offline player is to convert all his dollars to yen at the end of each upward run, and all his yen to dollars at the end of each downward run." However, no rigorous proof has been provided in [5]. The

---

[1] A remark on the exchange rates r_t and r_t^{-1} is as below. On one hand, if at some time t, the exchange rate r from dollars to yen and the exchange rate r' from yen to dollars satisfy $r \times r' > 1$, then we can iteratively run many rounds of exchanging all dollars to yen and all yen to dollars back and forth to increase the amount of dollars that we can have. In each round, we increase the total amount of dollars by $r \times r'$ times. On the other hand, if $r \times r' < 1$, then it is unnecessary to exchange any dollars to yen at time t unless t is the last step of the game. So, in the two-way currency trading problem, assuming $r \times r' \neq 1$ does not lead to a reasonable and non-trivial different problem from the one with $r \times r' = 1$.

online version of this problem has been studied in [3] [5]. For the *continuous setting* (in which the exchange rates are allowed to vary at any time during a day) of this two-way currency trading problem, the best known lower bound of competitive ratio is $\bar{c}^g$ [3] and the best known upper bound of competitive ratio is $\tilde{c}^g$ [3], where $\bar{c}$ is the root of $\bar{c}(1 + (\bar{c} - 1)e^{\bar{c}})^2 - \frac{M}{m} = 0$ and $\tilde{c}$ is the root of $\ln \frac{\frac{M}{\tilde{c} \cdot m} - 1}{\tilde{c} - 1} - \tilde{c} = 0$, the number of local exchange rate maxima is $g/2$, and the number of local exchange rate minima is $g/2$ in the input instance. For the *discrete setting*, the lower bound of competitive ratio is $(c_n)^{\frac{g}{2}}$ [5] where g is the number of maxima and minima in the input instance, and c_n is the value as the one in Eq. (1). The authors [3] improved the lower bound of competitive ratios for the continuous setting but unfortunately this improved bound cannot be held in the discrete setting. The upper bound of competitive ratios for the discrete setting is the same as the one for the continuous setting [3].

Our Contributions. In this paper, we focus on the two-way currency trading problem in the discrete setting. We re-study an optimal offline algorithm for this problem and provide a detailed proof in Sect. 2. Based on the properties of the offline algorithm, we design a one-step look-ahead optimal algorithm. In Sect. 3, we consider the two-way currency trading problem in a setting in which the number of trades is restricted no more than a number k. We provide a few algorithms and provide running time analysis as well as instance optimality.

2   An One-Step Look-Ahead Algorithm for the Two-Way Currency Trading Problem

In this section, we provide a proof of an optimal offline algorithm for the two-way currency trading problem. From this offline algorithm we derive an intuition of designing online algorithms with look-ahead.

An Optimal Offline Algorithm. Let $i = 1, 2, \ldots, n$, denote the n time points at the beginning of the n trading days in an increasing time order at which the currency exchange rates may be updated. The currency exchange rates $r_1, r_2, \ldots, r_n$ at time $1, 2, \ldots, n$ respectively consist of an input instance. Without loss of generality, we assume that any two neighboring exchange rates are *distinct*. (If some consecutive days have the same currency exchange rate, then we merge these days together as one 'super-day' with only one exchange rate.) Without loss of generality, we assume that a player has an initial normalized asset of 1 dollar. Let r_i and r_i^{-1} denote the currency exchange rates (*dollar to yen* and *yen to dollar* respectively) at time i. We employ two variables, d_i (≥ 0) and y_i (≥ 0) respectively, to denote the total amount of dollars and the total amount of yen that a player has just before the time point i. Let x_i (≥ 0) and z_i (≥ 0) denote the amount of dollars and yen to be traded at time i respectively. We have the following key observations.

Theorem 1. *There exists an optimal offline algorithm having the following 6 properties.*

$$P_1: \qquad x_i \neq 0 \Rightarrow z_i = 0, \qquad \forall i = 1, \ldots, n$$
$$P_2: \qquad z_i \neq 0 \Rightarrow x_i = 0, \qquad \forall i = 1, \ldots, n$$
$$P_3: \qquad x_{i-1} \neq 0 \Rightarrow x_i = 0, \qquad \forall i = 2, \ldots, n$$
$$P_4: \qquad z_{i-1} \neq 0 \Rightarrow z_i = 0, \qquad \forall i = 2, \ldots, n$$
$$P_5: \qquad d_i \neq 0 \Rightarrow y_i = 0, \qquad \forall i = 1, \ldots, n$$
$$P_6: \qquad y_i \neq 0 \Rightarrow d_i = 0, \qquad \forall i = 1, \ldots, n$$

The above properties uniquely define an algorithm and therefore, an algorithm satisfying these properties is an optimal offline algorithm.

Theorem 1 indicates that in an optimal offline algorithm, the following properties hold.

1. Property P_1 and Property P_2 imply that the player trades at most one type of currency at any time point.
2. Property P_3 and Property P_4 imply that the player trades dollars and yen in an alternative manner in increasing time order. Note that trading currency may not happen at all the time points, that is, it is possible that the algorithm does nothing at a time.
3. Property P_5 and Property P_6 imply that the player only needs to keep one type of currency on hand at any time.

Proof (Proof of Theorem 1). We use a contradiction method and a mathematical induction method to prove Theorem 1.

On P_1 and P_2. Assume that time i is the first time point such that in an optimal offline algorithm we have $x_i \neq 0$ and $z_i \neq 0$. At time i, z_i/r_i units of dollars are generated by trading z_i units of yen.

- Assume $x_i \geq \frac{z_i}{r_i}$.
  We reset $x_i' \leftarrow x_i - \frac{z_i}{r_i}$ and $z_i' \leftarrow 0$. These new x-values and z-values at time i satisfy the property P_1 of Theorem 1 but do not hurt the optimality.
- Assume $x_i < \frac{z_i}{r_i}$.
  We reset $x_i' \leftarrow 0$ and $z_i' \leftarrow z_i - x_i \cdot r_i$. Again, these new x-values and z-values satisfy the property P_2 of Theorem 1 but do not hurt the optimality.

We can repeat doing the above procedures in increasing time order to make sure that at all the time i, Property P_1 and Property P_2 hold.

On P_3 and P_4. Now, assume that in an optimal offline algorithm, two neighboring time points i and $i+1$ are the first two time points such that $x_i \neq 0$ and $x_{i+1} \neq 0$.

- Assume $r_i \leq r_{i+1}$.
  We reset $x_i' \leftarrow 0$ and $x_{i+1}' \leftarrow x_i + x_{i+1}$. The above change does not decrease the amount of yen traded using $x_i + x_{i+1}$ units of dollars at time $i+1$.

– Assume $r_i > r_{i+1}$.

Property P_1 implies that only dollars are traded into yen at these two time points. It is safe to reset $x_i' \leftarrow x_i + x_{i+1}$ and $x_{i+1}' \leftarrow 0$. So, Property P_3 holds.

Similarly, we conclude that Property P_4 holds by replacing the x-values using z-values in the above expressions.

On P_5 and P_6. Now, we prove Property P_5 and Property P_6. Recall that we already proved Properties P_1, P_2, P_3, and P_4 such that at any time, the player trades at most one type of currency and he trades dollars into/from yen in an alternative manner in increasing time order.

Without loss of generality, we assume that trades happen at each time point $1', 2', \ldots, m'$, where $1 \leq 1' < 2' < \ldots < m' \leq n$. Based on Properties P_3 and P_4, we know that m' is an odd number. Trades from dollars to yen should be made at time points $1', 3', 5', \ldots, m'$. Trades from yen to dollars should be made at time points $2', 4', \ldots, m' - 1$. Let f_i denote the fraction that the amount of currency is traded at time i. We need to show

$$f_i = 1, \ \forall i = 1', 2', \ldots, m'. \tag{2}$$

We are using a mathematical induction method. Instead of showing that Properties P_5 and P_6 hold in an increasing time order, we prove Eq. (2) in a *backward manner* (in decreasing time order). This way of induction will ease the presentation of our proof significantly.

Let the last trade time be m' and all the dollars are traded into yen. So, Property P_6 holds at time m'. We now trace back to the prior time slot $(m-1)'$ in which a trade of yen to dollars is made (based on Property P_4). At time $(m-1)'$, only yen is traded into dollars (based on Property P_2). Now, we examine the objective, the total number of yen Y got at the end of time point m. We have

$$Y = d_{m'} \times r_{m'} + y_{m'} \tag{3}$$

$$= \left(\frac{f_{(m-1)'} \times y_{(m-1)'}}{r_{(m-1)'}} \right) r_{m'} + y_{m'} \tag{4}$$

$$= \left(\frac{f_{(m-1)'} \times y_{(m-1)'}}{r_{(m-1)'}} \right) r_{m'} + \left(1 - f_{(m-1)'}\right) y_{(m-1)'}$$

$$= y_{(m-1)'} \left(1 + f_{(m-1)'} \left(\frac{r_{m'}}{r_{(m-1)'}} - 1 \right) \right) \tag{5}$$

Note that Eq. (4) holds due to Property P_2 and Property P_4. We also remark here that $r_{m'} > r_{(m-1)'}$. (If $r_{m'} < r_{(m-1)'}$, then the offline algorithm does not need to make any trades at time $(m-1)'$ and time m' since any amount of yen traded into dollars at time $(m-1)'$ and then traded into yen at time m' will become $r_{m'}/r_{(m-1)'}$ times less.) We trade dollars to yen at higher exchange rates and yen to dollars at lower exchange rates. In order to maximize Y in Eq. (5), we need to set $f_{(m-1)'} = 1$ and to maximize $y_{(m-1)'}$. Thus, Property P_5 holds at time m' and Property P_6 holds at time $(m-1)'$. Now, we shall maximize

$y_{(m-1)'}$. Maximizing $y_{(m-1)'}$ (without considering optimizing $d_{(m-1)'}$) is similar to maximizing Y in Eq. (3). We can repeat the above procedure in a backward manner and thus inductively prove that Property P_5 holds. Similarly, Property P_6 can be proved.

In the following, we describe an optimal offline algorithm. We have the initial setting, $d_1 = 1$ and $y_1 = 0$. Based on Theorem 1, the following recurrence allows us to use a dynamic programming approach to solve the offline version of the two-way trading problem:

$$d_{i+1} = \max \left\{ d_i, \frac{y_i}{r_i} \right\} \tag{6}$$

$$y_{i+1} = \max \left\{ y_i, d_i \times r_i \right\} \tag{7}$$

The optimal solution should be y_{n+1}, the y-value at the end of the last time point n. This dynamic programming based optimal offline algorithm has its running time of $O(n)$ and space requirement of $O(n)$. For this offline algorithm, it does not need to know the currency exchange duration or the upper/lower bounds of exchange rates. Recall that $\{d_i | i = 1, \ldots, n\}$ and $\{y_i | i = 1, \ldots, n\}$ are two sets of variables that are used in our algorithm. The actual amounts of dollars and yen kept in hand by an offline optimal algorithm can be traced back from the d-values and y-values generated by the above recursive Equalities (6) and (7). We trace back from the last time point $n + 1$. If $d_{i+1} = \frac{y_i}{r_i}$, then an optimal offline algorithm trades y_i yen into d_{i+1} dollars at time i: If $y_{i+1} = d_i \times r_i$, then an optimal offline algorithm trades d_i dollars to y_{i+1} yen at time i. Otherwise, the algorithm does nothing.

An Optimal Algorithm with the Power of Look-Ahead. How to trade currencies is essentially an online decision-making problem. In this following, we present an online algorithm which is equipped with the look-ahead power for the two-way currency trading problem. Employing the power of s-step look-ahead for online algorithms has been used to offer online algorithms a little power of "seeing" the future s input. The larger value s is, the more powerful an algorithm is. This approach has been used to analyze traditional online problems such as the paging problem [2] and the packet scheduling problem [1]. It is clearly that in the setting $s = n$, such algorithms are offline ones with a complete set of n exchange rates. More examples and applications can be found in [4].

Note that for the one-way currency trading problem, an optimal offline algorithm trades all the dollars using the global maximum exchange rate. Looking-ahead does not improve online algorithms' competitiveness for the one-way trading problem. (For example, an instance with setting $n = s + 1$ shows that with s-step look-ahead, an online algorithm cannot be benefited at all in terms of competitive ratios.)

We study 1-step look-ahead online algorithms. At time i, in additional to knowing the exchange rates (r_i from dollar to yen and $1/r_i$ from yen to dollar), an one-step look-ahead online player knows the next time step's exchange rate.

Theorem 2. *With the power of 1-step look-ahead, an algorithm achieves the optimal amount of traded currency for the two-way currency trading problem.*

Proof. We study the following 1-step look-ahead online algorithm. At time i, look at the next exchange rate.

- Assume the next exchange rate is the same as the current exchange rate that we just experienced.
  Then the algorithm does nothing and we postpone the decision of trading currency till the next time point.
- Assume the next exchange rate is larger than the current one.
  Then we trade all the current currency (if in yen) from yen to dollars or (if in dollars) we keep the total amount of dollars and plan to trade them into yen at a future time point.
- Assume the next exchange rate is smaller than the current one.
  Then we trade all the current currency (if in dollars) from dollars to yen or (if in yen) we keep the total amount of yen and plan to trade them into dollars in a future time point.

We can assume $r_{i+1} = 1$ if there is no exchange rate update from time i till the end of this game so that we can always get yen at the end of this game. It is clear that this algorithm satisfies all the properties described in Theorem 1. Note that Properties P_1 to P_6 uniquely define an algorithm. Based on the optimal offline algorithm described this section, Theorem 2 holds.

We remark here that this 1-competitive algorithm with one-step look-ahead is the same as the one described in [5]. It takes the advantages of postponing decisions to make. Therefore, the type of currency to be traded as well as the amount of currency to be traded can be determined later. These decisions are made at local maximum and local minimum exchange rates; such exchange rates are known to an algorithm with one-step look-ahead. We hope our observations can help improve the competitiveness of online algorithms without look-ahead.

3 Optimal Algorithms for a Setting in Which the Number of Currency Trades Is Bounded

In this section, we consider the setting in which an algorithm has its number of currency trades bounded by a number k. Initially, an user has dollars only and let this amount be a. Let the number of currency trading days be n and let r_i denote the exchange rate from dollar to yen (respectively, $1/r_i$ from yen to dollar) on the i-th day, $\forall i = 1, 2, \ldots, n$. Note that the algorithm outputs yen so that the total number of trades must be an odd number. The objective is the maximize the total amount of yen, with the number of currency trades bounded by k. If $k = 1$, then we trade all dollars using the highest exchange rate only once. The highest trading rate $\max_{i=1}^{n} r_i$ can be searched in a linear time. If $k = 1$, then the algorithm's running time is $O(n)$. In the following, we consider the general case in which $k > 1$.

3.1   k Is a Constant

We provide a dynamic programming approach to solve the problem with a constant k bound of the number of currency trades. Here we assume $k = o(\log n)$.

Algorithm. This algorithm generalizes the optimal offline algorithm (where $k = n$) described in Sect. 2. Using the same analysis as in Sect. 2, we conclude that even if the player is allowed to trade at most k times, for the optimal algorithm which maximizes the amount of traded yen, we still have

1. the player trades at most one type of currency at any time point;
2. that the player trades dollars and yen in an alternative manner in increasing time order; and
3. the player only needs to keep one type of currency on hand at any time.

For each currency exchange rate r_i given at time i, we associate it with two values to indicate the best amount of dollars and yen that we can get at the end of time i respectively, if there are no more currency trades after this rate r_i and if the number of exchange rates up to time i is bounded by k_i, where $k_i \le k$. Let $OPT^d(i, k_i)$ denote the highest amount of dollars that one can get and $OPT^y(i, k_i)$ the highest amount of yen that one can get. Note that $OPT^y(n, k)$ is the optimal amount of yen traded. Let r_i^* denote the highest exchange rate among $\{r_1, \dots, r_i\}$. Scanning the array of all the values of r_i, we can get all the values r_i^* ($\forall i = 1, 2, \dots, n$) in linear time $O(n)$. Also, the values r_i^* are non-decreasing with the value i increasing. We have the following base cases:

$$OPT^d(1, k_i) = a, \qquad\qquad \forall k_i = 1, 2, \dots, k \qquad (8)$$
$$OPT^y(1, k_i) = r_1^*, \qquad\qquad \forall k_i = 1, 2, \dots, k \qquad (9)$$
$$OPT^d(i, 1) = a, \qquad\qquad \forall i = 1, \dots, n \qquad (10)$$
$$OPT^y(i, 1) = r_i^*, \qquad\qquad \forall i = 1, \dots, n \qquad (11)$$

Similar to the algorithms shown in Sect. 2, we consider the recursion to calculate $OPT^d(i, k_i)$ and $OPT^y(i, k_i)$ and have

$$OPT^d(i, k_i) = \max\left(\frac{OPT^y(i-1, k_i - 1)}{r_i}, OPT^d(i-1, k_i)\right) \qquad (12)$$

$$OPT^y(i, k_i) = \max\left(OPT^d(i-1, k_i - 1) \times r_i, OPT^y(i-1, k_i)\right) \qquad (13)$$

The dynamic programming based algorithm has the initial step covering Eq. (8) to Eq. (11). Here this algorithm has two **for** loops. For the first **for** loop, i goes from 2 to n. For the second **for** loop, k_i goes from 1 to k. Within the second **for** loop, we have Eqs. 12 and 13.

Analysis. The analysis of the running time of this algorithm is straightforward. We have $n \times k$ values $OPT^d(i, k_i)$ and $n \times k$ values $OPT^y(i, k_i)$. In total, we have $2n \times k$ values. For each value, it takes a constant time to calculate and thus, the total running time is $O(n \times k)$. Note that k is a constant, so, the running time of this algorithm is $O(n)$, which is asymptotically optimal.

3.2 k Is Not a Constant

If k is not a constant, then the dynamic programming algorithm described in Sect. 3.1 has its running-time bounded by $O(n \cdot k)$, which is $O(n^2)$. In this section, we design a new algorithm using a different idea from the one used when k is a constant and analyze the output-sensitive time bound.

Algorithm. Note that we have described an optimal offline algorithm without the constraints of having at most k trades in Sect. 2. We run this algorithm once and get an optimal trading strategy. If the total number of trades is no more than k in this strategy, then we get an optimal algorithm under the constraint of k trades and its running time is $O(n)$. In the following, we focus on the case in which the optimal offline algorithm got from the algorithm in Sect. 2 has more than k trades. We use a greedy algorithm (to be described below) to reduce the number of trades till we get at most k currency exchanges.

Based on the properties that we have for the optimal algorithm without the constraint of number of trades, we have a series of exchange rates and name them in increasing time order as $r'_1, r'_2, \ldots, r'_l$. In this trading strategy, we trade all dollars to yen using rates r'_{2i-1}, and all yen to dollars using rates r'_{2i}, where $i = 1, 2, \ldots, \frac{l+1}{2}$. Let a denote the initial asset that the player has. Thus, this player's amount of yen is maximized as the following value

$$a \left(\frac{r'_1}{r'_2} \frac{r'_3}{r'_4} \cdots \frac{r'_{l-2}}{r'_{l-1}} \frac{r'_l}{1} \right) = a \prod_{i=1}^{\frac{l-1}{2}} \frac{r'_{2i-1}}{r'_{2i}} r'_l \tag{14}$$

To reduce the number of trades from l to k, we remove two neighboring currency exchange rates, either r'_i and r'_{i+1} or r'_i and r'_{i-1}. This pair of trading rates are removed from the trading strategy in increasing order the ratios in the set $\bigcup_i \left\{ \frac{r_i}{r_{i-1}}, \frac{r_i}{r_{i+1}} \right\}$. After removing each pair of rates, the remaining currency trades consist of a new trading strategy and we update the union $\bigcup_i \left\{ \frac{r_i}{r_{i-1}}, \frac{r_i}{r_{i+1}} \right\}$ correspondingly. We repeat removing exchange rates until we have only k currency trades in the schedule.

Correctness Analysis. Let OPT denote an optimal algorithm with the number of currency exchanges bounded by k. We first argue that (c_1) this greedy algorithm does not drop any time points that OPT has traded at. We then argue that (c_2) the more number of exchange rates are allowed, the more (no less-than) amount of yen that OPT can get.

First, we assume OPT and the greedy algorithm start with the same strategy with the number of trades l. OPT follows a strategy to remove the trading times and using an exchange argument, OPT can remove the same pair of trading times as what our algorithm does. Thus, we repeatedly apply this operation and c_1 is proved. Second, we know that the strategy alternatives trades dollars and yen and either $r_i/r_{i-1} > 1$ or $r_i/r_{i+1} > 1$ in the optimal strategy. So, removing a pair $r_i/r_{i-1} > 1$ or $r_i/r_{i+1} > 1$ from the trading strategy results in a total amount yen reduced by $r_i/r_{i-1} > 1$ or $r_i/r_{i+1} > 1$ times. Thus, c_2 is proved.

$O(n \log n)$ *running-time analysis.* Recall that $k < n$. We build a min-heap of n exchange rates in the trading strategy. We remove the min-value in each round and update the neighboring values. We remove at most $n - k$ pairs of trading time points and the total running time of this algorithm is bounded by $O((n - k) \log n) = O(n \log n)$.

A *tighter* $O(n \log k)$ *running-time analysis.* We further improve the algorithm's running time as $O(n \log k)$ using a different implementation. In this modified implementation version, we do not remove the min-ratio exchange rates after we get a complete trading strategy. Instead, we construct a schedule with a bounded exchange number k and implement the step of removing the smallest ratio at the time when the number of trading currencies in the current trading strategy is over k. We maintain a min-heap with the size of k. Following the increasing order of the currency trade times in the optimal offline algorithm described in Sect. 2, at any time when the number of trades is over than k, we remove the min-value in this heap and update the heap correspondingly. With the same argument, this modified greedy algorithm is also optimal under the constraint of k currency exchanges. However, this algorithm maintains a min-heap with size $\leq k$ at any time and therefore, its total *output-sensitive time bound*[2] is $O(n \log k)$.

References

1. Bohm, M., Chrobak, M., Jez, L., Li, F., Sgall, J., Vesely, P.: Online packet scheduling with bounded delay and lookahead. Theor. Comput. Sci. (TCS), (2019, to appear)
2. Breslauer, D.: On competitive on-line paging with lookahead. Theor. Comput. Sci. (TCS) **209**(1–2), 365–375 (1998)
3. Dannoura, E., Sakurai, K.: An improvement on El-Yaniv-Fiat-Karp-Turpin's money-making bi-directional trading strategy. Inf. Process. Lett. (IPL) **66**(1), 27–33 (1998)
4. Dunke, F., Nickel, S.: A general modeling approach to online optimization with lookahead. Omega **63**, 134–153 (2015)
5. El-Yaniv, R., Fiat, A., Karp, R.M., Turpin, G.: Optimal search and one-way trading online algorithms. Algorithmica **30**(1), 101–139 (2001)
6. Mohr, E., Ahmad, I., Schmidt, G.: Online algorithms for conversion problems: a survey. Surv. Oper. Res. Manag. Sci. **19**, 87–104 (2014)
7. Nielsen, F.: Grouping and querying: a paradigm to get output-sensitive algorithms. In: Akiyama, J., Kano, M., Urabe, M. (eds.) JCDCG 1998. LNCS, vol. 1763, pp. 250–257. Springer, Heidelberg (2000). https://doi.org/10.1007/978-3-540-46515-7_21
8. Sharir, M., Overmars, M.H.: A simple output-sensitive algorithm for hidden surface removal. ACM Trans. Graph. (TOG) **11**(1), 1–11 (1992)

---

[2] An *output-sensitive algorithm* is an algorithm whose running time depends on the size of the output in addition to the size of the input, for example [7,8].

Approximation Algorithms for the Minimum Power Partial Cover Problem

Menghong Li, Yingli Ran, and Zhao Zhang$^{(\boxtimes)}$

College of Mathematics and Computer Science, Zhejiang Normal University,
Jinhua 321004, Zhejiang, China
hxhzz@sina.com

Abstract. In this paper, we study the minimum power partial cover problem (MinPowerPartCov). Suppose X is a set of points and $\mathcal{S}$ is a set of sensors on the plane, each sensor can adjust its power, the covering range of a sensor s with power $p(s)$ is a disk centered at s which has radius $r(s)$ satisfying $p(s) = c \cdot r(s)^{\alpha}$. Given an integer $k \leq |X|$, the MinPowerPartCov problem is to determine the power assignment on each sensor such that at least k points are covered and the total power consumption is the minimum. We present an approximation algorithm with approximation ratio 3^{α}, using a local ratio method, which coincides with the best known ratio for the minimum power (full) cover problem. Compared with the paper [9] which studies the MinPowerPartCov problem for $\alpha = 2$, our ratio improves their ratio from $12 + \varepsilon$ to 9.

Keywords: Power · Partial cover · Approximation algorithm · Local ratio

1 Introduction

With the rapid development of wireless sensor networks (WSNs), intensive studies on WSNs have emerged, especially on the coverage problem. In a coverage problem, the most basic requirement is to keep all points under monitoring. In a typical WSN, the service area of a sensor is a disk centered at the sensor whose radius is determined by the power of the sensor. A typical relation between the power $p(s)$ of sensor s and the radius $r(s)$ of its service area is

$$p(s) = c \cdot r(s)^{\alpha}, \tag{1}$$

where c and $\alpha \geq 1$ are some constants (α is usually called the *attenuation facor*). So, the greater power a sensor possesses, the larger service it can provide. In other words, the consumption of energy and the quality of service are two conflicting factors. The question is how to balance these two conflicting factors by adjusting power at the sensors so that the desired service can be accomplished using the minimum total power. This question is motivated by the intention to extend the lifetime of WSN under limited energy supply, and we call it the *minimum power coverage* problem (MinPowerCov).

© Springer Nature Switzerland AG 2019
D.-Z. Du et al. (Eds.): AAIM 2019, LNCS 11640, pp. 179–191, 2019.
https://doi.org/10.1007/978-3-030-27195-4_17

In the real world, it is often too costly to satisfy the covering requirement of every point. So, it is beneficial to study the *minimum power partial coverage* problem (MinPowerPartCov), in which it is sufficient to cover at least k (instead of all) points. The problem is motivated by the purpose of further saving energy while keeping an acceptable quality of service.

The MinPowerPartCov problem can be viewed as a special case of the *minimum weight partial set cover* problem (MinWPSC). Given a set E of elements, a collection of sets $\mathcal{S}$, a weight function $w : \mathcal{S} \mapsto \mathbb{R}^+$, and an integer $k \leq |E|$, the MinWPSC problem is to find the minimum weight sub-collection of sets $\mathcal{F} \subseteq \mathcal{S}$ such that at least k elements are covered by $\mathcal{F}$, i.e., $|\bigcup_{S \in \mathcal{F}} S| \geq k$ and $w(\mathcal{F}) = \sum_{S \in \mathcal{F}} w(S)$ is minimum. Notice that in a MinPowerParCov problem, the power at a sensor can be discretized by assuming that there is a point on the boundary of the disk supported by the assigned power. We call such a disk as a *canonical disk*. So, if we associate with each sensor $|X|$ canonical disks, where X is the set of points, each disk corresponds to the set of points contained in it, and the weight of the disk equals the power supporting the disk which is determined by Eq. (1), then the MinPowerParCov problem is reduced to the MinWPSC problem.

It is known that the MinWPSC problem has an f-approximation [2], where f is the maximum frequency of an element, that is, the maximum number of sets containing a common element. For the MinWPSC problem obtained by the above reduction from a MinPowerParCov problem, f equals the number of sensors, which is too large to be a good approximation factor. So, the main purpose of this paper is to explore geometric properties of the MinPowerParCov problem to obtain a better approximation.

1.1 Related Works

The *minimum weight set cover* problem (MinWSC) is a classic combinatorial problem. It is well-known that MinSC admits approximation ratio $H(\Delta)$ [7,14], where $H(\Delta) = 1 + \frac{1}{2} + \ldots + \frac{1}{\Delta}$ is the Harmonic number and Δ denotes the size of the largest set. It is also known that a simple LP-rounding algorithm can achieve an approximation ratio of f, where f is the maximum number of sets containing a common element (see for example Chapter 12 of the book [23]).

For the *minimum weight partial set cover* problem (MinWPSC), Slavík [21] obtained an $H(\min\{\lceil k \rceil, \Delta\})$-approximation using a greedy strategy, Bar-Yehuda [2] obtained an f-approximation using local ratio method, Gandhi [10] also obtained f approximation using a primal-dual method. Very recently, Inamdar *et al.* [13] designed an LP-rounding algorithm, obtaining approximation ratio $2\beta + 2$, where β is the integrality gap for the natural linear program of the minimum weight (full) set cover problem.

For the *geometric minimum weight set cover* problem, much better approximation factors can be achieved. Using a partition and shifting method, Hochbaum *et al.* [12] obtained a PTAS for the minimum unit disk cover problem in which the disks are uniform and there are no prefixed locations for the disks. For the minimum disk cover problem in which disks may have different sizes,

Mustafa *et al.* [15] designed a PTAS using a local search method. This PTAS was generalized by Roy *et al.* [20] to non-piercing regions including pseudo-disks. These are results for the cardinality version of the geometric set cover problem. Considering weight, Varadarajan [22] presented a clever quasi-uniform sampling technique, which was improved by Chan *et al.* [8], yielding a constant approximation for the minimum weight disk cover problem. This constant approximation was generalized by Bansal *et al.* [4] for the minimum weight disk multi-cover problem in which every point has to be covered multiple times. Using a separator framework, Mustafa *et al.* [16] obtained a quasi-PTAS for the minimum weight disk cover problem.

To our knowledge, there are two papers studying the *geometric minimum partial set cover* problem. The first paper is [10], in which Gandhi *et al.* presented a PTAS for the minimum (cardinality) partial unit disk cover problem using a partition and shifting method. Notice that this result only works for the case when the centers of the disks are not prefixed. Another paper is due to Inamdar *et al.* [13], in which a $(2\beta + 2)$-approximation was obtained for the *general* minimum weight partial set cover problem, where β is the integrality gap of the natural linear program for the minimum weight (full) set cover problem. As a consequence, for those geometric set cover problems (including the disk cover problem) in which β is a constant, the approximation ratio for the partial version is also a constant (but the constant is large).

Recently, there are a lot of works studying the *minimum power multi-cover* problem (MinPowerMC), in which every point p is associated with a covering requirement cr_p, and the goal is to find a power assignment with the minimum total power such that every point p is covered by at least cr_p disks. Let cr_{max} be the maximum number of times that a point requires to be covered. Using a local ratio method, Bar-Yehuda *et al.* [3] presented a $3^\alpha \cdot cr_{max}$-approximation algorithm. The dependence on cr_{max} was removed by Bhowmick *et al.* [5], achieving an approximation ratio of $4 \cdot (27\sqrt{2})^\alpha$. This result was further generalized to any metric space in [6], the approximation ratio is at most $2 \cdot (16 \cdot 9)^\alpha$. For the minimum power (*single*) cover problem, the best known ratio is 3^α (as a consequence of [3] and the fact $cr_{max} = 1$ in this case).

There is only one paper [9] studying the minimum power *partial* (single) cover problem (MinPowerPartCov), and the study is on the special case when $\alpha = 2$. The approximation ratio obtained in [9] is $(12 + \varepsilon)$, where ε is an arbitrary constant greater than zero, by a reduction to a prize-collecting coverage problem.

1.2   Contribution

In this paper, we show that the MinPowerPartCov problem can be approximated within factor 3^α, which coincides with the best known ratio for the MinPower-Cov problem (the full version of the minimum power coverage problem). When applied to the case when $\alpha = 2$, our ratio is 9, which is better than $12 + \varepsilon$ obtained in [9].

Our algorithm is inspired by the local ratio method used in [3] to study the MinPowerCov problem. New ideas have to be explored to surmount the difficulty.

2   The Problem and a Preprocessing

The problem studied in this paper is formally defined as follows.

Definition 1 (Minimum Power Partial Cover (MinPowerPartCov)).
Suppose X is a set of n points and S is a set of m sensors on the plane, k is an integer satisfying $0 \leq k \leq n$. A point $x \in X$ is covered by a sensor $s \in S$ with power $p(s)$ if x belongs to the disk supported by $p(s)$, that is $x \in Disk(s, r(s))$, where $Disk(s, r(s))$ is the disk centered at s whose radius $r(s)$ is determined by $p(s)$ through equation $p(s) = c \cdot r(s)^{\alpha}$. A point is covered by a power assignment $p : S \mapsto \mathbb{R}^{+}$ if it is covered by some disk supported by p. The goal of MinPower-PartCov is to find a power assignment p covering at least k points such that the total power $\sum_{s \in S} p(s)$ is as small as possible.

In an optimal solution, we may assume that for any sensor s, there is at least one point that is on the boundary of the disk $Disk(s, p(s))$, since otherwise we may reduce $p(s)$ to cover the same set of points, resulting in a lower power. Therefore, at most mn disks need to be considered. We denote the set of such disks by $\mathcal{D}$. In the following, denote by $(X, \mathcal{D}, k)$ an instance of the MinPowerPartCov problem, and use $opt(X, \mathcal{D}, k)$ to denote the optimal power for the instance $(X, \mathcal{D}, k)$. To simplify the notation, we use D to represent both a disk in $\mathcal{D}$ and the set of points covered by D, and use $r(D)$ and $p(D)$ to denote the radius and the power of disk D, where $p(D) = c \cdot r(D)^{\alpha}$. For a set of disks $\mathcal{D}$, we shall use $\mathcal{C}(\mathcal{D}) = \bigcup_{D \in \mathcal{D}} D$ to denote the set of points covered by $\mathcal{D}$.

In order to control the approximation factor of our algorithm, we need a preprocessing step: guessing the maximum power of a sensor (or equivalently, the radius of a maximum disk) in an optimal solution. Suppose $D_{\max} \in \mathcal{D}$ is the guessed disk. Denote by $\mathcal{D}_{\leq r(D_{\max})}$ the subset of disks of $\mathcal{D}$ whose radii are no greater than the radius of $D_{\max}$ (excluding $D_{\max}$), and denote by $(X \setminus D_{\max}, \mathcal{D}_{\leq r(D_{\max})}, k - |D_{\max}|)$ the *residual instance* after guessing $D_{\max}$. The following lemma is obvious.

Lemma 1. *Suppose $D_{\max}$ is the correctly guessed disk with the maximum power in an optimal solution of instance $(X, \mathcal{D}, k)$. Then*

$$opt(X, \mathcal{D}, k) = opt(X \setminus D_{\max}, \mathcal{D}_{\leq r(D_{\max})}, k - |D_{\max}|) + p(D_{\max}).$$

3   A Local Ratio Algorithm

In this section, we first present an algorithm for the MinPowerPartCov problem on the instance $(X \setminus D_{\max}, \mathcal{D}_{\leq r(D_{\max})}, k - |D_{\max}|)$. And then show how to make use of it to find a power assignment for the original MinPowerPartCov problem.

3.1 Algorithm After the Preprocessing

For simplicity of notation in this section, we still use $(X, \mathcal{D}, k)$ to denote the residual instance, assuming that every disk in $\mathcal{D}$ has radius at most $r(D_{\max})$.

The algorithm consists of three steps.

(i) In the first step, a local ratio method is employed to find a *minimal partial cover* $\bar{\mathcal{D}}$, that is, $\bar{\mathcal{D}}$ covers at least k points, while for any disk $D \in \bar{\mathcal{D}}$, the number of points covered by $\bar{\mathcal{D}} - \{D\}$ is strictly less than k.

(ii) Before going into the second step, remove a disk D_{rmv} from $\bar{\mathcal{D}}$ which is chosen in the last call of the local ratio method in the first step. Then, in the second step, a *maximal independent set of disks* $\mathcal{I} \subseteq \bar{\mathcal{D}} \setminus \{D_{rmv}\}$ is computed in a greedy manner, that is, disks in $\mathcal{I}$ are mutually disjoint, while any disk $D \in \bar{\mathcal{D}} \setminus \{D_{rmv}\}$ which is not picked into $\mathcal{I}$ intersects some disk in $\mathcal{I}$.

(iii) In the third step, every disk in $\mathcal{I}$ has its radius enlarged three times. Such set of disks together with $\{D_{\max}, D_{rmv}\}$ are the output of the algorithm.

The first step is accomplished by Algorithm 1, in which the MinPowerPartCov instance $(X, \mathcal{D}, k)$ is viewed as an instance of the minimum weight partial set cover problem, where X serves as the set of elements to be covered, $\mathcal{D}$ serves as the collection of sets, and the weight of each $D \in \mathcal{D}$ is $p(D)$. The local ratio method was first proposed by Bar-Yehuda and Even in [1]. The idea is to recursively peel off a special weight from the original weight. If the problem with the special weight admits an α-approximation, then one can assemble an α-approximate solution for the problem with respect to the original weight. In this paper, the special weight peeled off in each iteration (denoted by $\bar{p}$) is proportional to the number of uncovered points of a disk, and then the disks of residual weight zero are put into $\bar{\mathcal{D}}$.

Algorithm 1. $LR(X, \mathcal{D}, p, k)$.

Input: A set of points X, a set of disks $\mathcal{D}$, a weight function $p : \mathcal{D} \mapsto \mathbb{R}^+$, a covering requirement k.

Output: A *minimal* subset of disks $\bar{\mathcal{D}}$ covering at least k points.

1: If $k = 0$, then return $\bar{\mathcal{D}} \leftarrow \emptyset$
2: $\gamma \leftarrow \min_{D \in \mathcal{D}} p(D)/|X \cap D|$
3: $\bar{p}(D) \leftarrow \gamma \cdot |X \cap D|$ for each $D \in \mathcal{D}$
4: $p(D) \leftarrow p(D) - \bar{p}(D)$ for each $D \in \mathcal{D}$
5: $\mathcal{D}_{=0} \leftarrow \{D \in \mathcal{D} : p(D) = 0\}$
6: $X \leftarrow X \setminus \mathcal{C}(\mathcal{D}_{=0})$, $\mathcal{D} \leftarrow \mathcal{D} \setminus \mathcal{D}_{=0}$, $k \leftarrow \max\{0, k - |\mathcal{C}(\mathcal{D}_{=0})|\}$
7: $\bar{\mathcal{D}}' \leftarrow LR(X, \mathcal{D}, p, k)$
8: Let $\bar{\mathcal{D}}_{=0}$ be a minimal subset of $\mathcal{D}_{=0}$ such that $\bar{\mathcal{D}}' \cup \bar{\mathcal{D}}_{=0}$ covers at least k points.
9: Return $\bar{\mathcal{D}} \leftarrow \bar{\mathcal{D}}' \cup \bar{\mathcal{D}}_{=0}$

Algorithm 1 is in fact a function which will be recursively called. In the algorithm, after peeling off a special weight $\bar{p}$, we use $\mathcal{D}_{=0}$ to denote the set of disks with residual weight $p - \bar{p}$ being zero. Since taking disks of zero cost seems to be a free meal, we take all of them *temporarily* and consider the residual instance, the goal of which is to satisfy the residual covering requirement using the residual disks. Line 6 of the algorithm is to construct the residual instance. Having found a *minimal* solution $\bar{\mathcal{D}}'$ to the residual instance, the algorithm adds a *minimal* subset of disks of $\mathcal{D}_{=0}$, denoted as $\bar{\mathcal{D}}_{=0}$, into $\bar{\mathcal{D}}'$ to cover at least k points. This step is to guarantee that the resulting set of disks $\bar{\mathcal{D}}$ is *minimal*, which is very crucial to the control of the approximation factor.

Suppose the function LR is called $t + 1$ times. Denote by $\bar{\mathcal{D}}^{(i)}$, $p^{(i)}$, $\bar{p}^{(i)}$ etc. those objects at the end of the i-th calling of function LR. Then we have the following relations.

(i) $X^{(0)} = X$, $\mathcal{D}^{(0)} = \mathcal{D}$, $p^{(0)} = p$, and $k^{(0)} = k$.

(ii) For $i = 1, \ldots, t$,

$$\gamma^{(i)} = \min\{p^{(i-1)}(D)/|X^{(i-1)} \cap D|\} \text{ for each } D \in \mathcal{D}^{(i-1)}$$

$$\bar{p}^{(i)}(D) = \gamma^{(i)} \cdot |X^{(i-1)} \cap D| \text{ for each } D \in \mathcal{D}^{(i-1)} \tag{2}$$

$$p^{(i)}(D) = p^{(i-1)}(D) - \bar{p}^{(i)}(D) \text{ for each } D \in \mathcal{D}^{(i-1)} \tag{3}$$

$$\mathcal{D}_{=0}^{(i)} = \{D \in \mathcal{D}^{(i-1)} : p^{(i)}(D) = 0\}$$

$$X^{(i)} = X^{(i-1)} \setminus \mathcal{C}(\mathcal{D}_{=0}^{(i)})$$

$$\mathcal{D}^{(i)} = \mathcal{D}^{(i-1)} \setminus \mathcal{D}_{=0}^{(i)}$$

$$k^{(i)} = \max\{0, k^{(i-1)} - |\mathcal{C}^{(i-1)}(\mathcal{D}_{=0}^{(i)})|\} \tag{4}$$

Here $\mathcal{C}^{(i-1)}(\mathcal{D}_{=0}^{(i)}) = \mathcal{C}(\mathcal{D}_{=0}^{(i)}) \cap X^{(i-1)}$. As a consequence of the above relations,

$$k^{(i)} = \max\{0, k - |\mathcal{C}(\bigcup_{j=1}^{i} \mathcal{D}_{=0}^{(j)})|\}. \tag{5}$$

It should be noticed that in expressions (4) and (5), except for $i = t$, the value of $k^{(i)}$ equals the second term.

(iii) $k^{(t)} = 0$, $\bar{\mathcal{D}}^{(t+1)} = \emptyset$. And for $i = t, t-1, \ldots, 1$,

$$\bar{\mathcal{D}}^{(i)} = \bar{\mathcal{D}}^{(i+1)} \cup \bar{\mathcal{D}}_{=0}^{(i)}.$$

As a consequence

$$\bar{\mathcal{D}}^{(i)} = \bigcup_{j=i}^{t} \bar{\mathcal{D}}_{=0}^{(j)} \subseteq \bigcup_{j=i}^{t} \mathcal{D}_{=0}^{(j)}. \tag{6}$$

The above relation can be illustrated by the following figure.

Remark 1. If a disk D has its weight reduced to zero in the i-th call of LR, that is, if $p^{(i-1)}(D) > 0$ and $p^{(i)}(D) = 0$, then D does not play roles in the deeper calls of LR. In this case, we may view $p^{(j)}(D) = \bar{p}^{(j)}(D) = 0$ for any j with $i + 1 \le j \le t$. By such a point of view, for any $0 \le i \le t$, we may extend the definition of functions $p^{(i)}$ and $\bar{p}^{(i)}$ on any disk $D \in \mathcal{D}$.

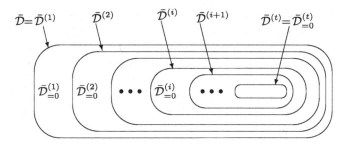

Fig. 1. Illustration for the structure of $\bar{\mathcal{D}}^{(i)}$.

Lemma 2. *For any* $i = 1, \ldots, t+1$, *the set* $\bar{\mathcal{D}}^{(i)}$ *is a* minimal *set of disks covering* $k^{(i-1)}$ *points of* $X^{(i-1)}$.

Proof. We prove the lemma by a backward induction on i. The base step when $i = t+1$ is obvious, since $k^{(t)} = 0$ and $\bar{\mathcal{D}}^{(t+1)} = \emptyset$.

For the induction step, suppose $i \leq t$ and $\bar{\mathcal{D}}^{(i+1)}$ is a minimal set of disks covering $k^{(i)}$ points of $X^{(i)}$. By expression (4) and the remark below it, we have

$$k^{(i-1)} = k^{(i)} + |\mathcal{C}^{(i-1)}(\mathcal{D}^{(i)}_{=0})|. \tag{7}$$

So $\bar{\mathcal{D}}^{(i+1)} \cup \mathcal{D}^{(i)}_{=0}$ can cover $k^{(i-1)}$ elements of $X^{(i-1)}$, which implies that a minimal subset $\bar{\mathcal{D}}^{(i)}_{=0} \subseteq \mathcal{D}^{(i)}_{=0}$ exists such that $\bar{\mathcal{D}}^{(i+1)} \cup \bar{\mathcal{D}}^{(i)}_{=0}$ can cover $k^{(i-1)}$ elements of $X^{(i-1)}$ (Fig. 1).

What remains to show is that $\bar{\mathcal{D}}^{(i+1)} \cup \bar{\mathcal{D}}^{(i)}_{=0}$ is minimal. By line 8 of Algorithm 1, no disk in $\bar{\mathcal{D}}^{(i)}_{=0}$ can be removed without violating the covering requirement $k^{(i-1)}$. For any disk $D \in \bar{\mathcal{D}}^{(i+1)}$, by the minimality of $\bar{\mathcal{D}}^{(i+1)}$, we have $|\mathcal{C}^{(i)}(\bar{\mathcal{D}}^{(i+1)} \setminus \{D\})| < k^{(i)}$. Then by (7), we have $|\mathcal{C}^{(i-1)}((\bar{\mathcal{D}}^{(i+1)} \setminus \{D\}) \cup \bar{\mathcal{D}}^{(i)}_{=0})| < k^{(i-1)}$. The minimality of $\bar{\mathcal{D}}^{(i)}$ is proved. $\qquad\square$

The second step is realized by Algorithm 2. Given a set of disks $\mathcal{D}$, Algorithm 2 finds a maximal independent set of disks by recursively choosing the disk with the maximum radius and deleting those disks intersecting it.

Algorithm 2. $IS(\mathcal{D})$.

Input: A set of disks $\mathcal{D}$.
Output: A maximal independent set of disks $\mathcal{I}$.

1: $\mathcal{I} \leftarrow \emptyset$
2: **while** $\mathcal{D} \neq \emptyset$ **do**
3:     $D' \leftarrow \arg\max_{D \in \mathcal{D}} r(D)$
4:     $\mathcal{I} \leftarrow \mathcal{I} \cup \{D'\}$
5:     $\mathcal{N} \leftarrow$ the set of disks of $\mathcal{D}$ that intersect D'
6:     $\mathcal{D} \leftarrow \mathcal{D} \setminus \mathcal{N}$
7: **end while**
8: Return $\mathcal{I}$

Algorithm 3 combines the above two algorithms to compute a feasible solution $\mathcal{M}$ to the residual instance. We use $c(D)$ and $r(D)$ to denote the center and the radius of disk D, respectively. So, $Disk(c(D), 3r(D))$ represents the disk with center $c(D)$ and radius $3r(D)$ (which a disk obtained from D by enlarging its radius by three times). Notice that $\mathcal{M}$ is not a subset of $\mathcal{D}$. Before calling Algorithm 2, a disk D_{rmv} is deleted from $\bar{\mathcal{D}}$, where D_{rmv} belongs to the set of disks added in the deepest call of LR. This is to control the approximation ratio which will be clear in the latter proofs.

Algorithm 3. $Cov(X, \mathcal{D}, k)$

Input: A residual instance $(X, \mathcal{D}, k)$.
Output: a set of disks $\mathcal{M}$ covering at least k points.
  1: $\bar{\mathcal{D}} \leftarrow LR(X, \mathcal{D}, k)$
  2: $D_{rmv} \leftarrow$ an arbitrary disk in $\bar{\mathcal{D}}^{(t)}_{=0}$ where t is the last call of LR
  3: $\mathcal{I} \leftarrow IS(\bar{\mathcal{D}} \setminus \{D_{rmv}\})$
  4: $\mathcal{M} \leftarrow \{Disk(c(D), 3r(D)): D \in \mathcal{I}\} \cup \{D_{rmv}\}$
  5: Return $\mathcal{M}$

The next theorem shows that Algorithm 3 computes a feasible solution to the residual instance.

Theorem 1. *The set of disks $\mathcal{M}$ computed by Algorithm 3 covers at least k points.*

Proof. The set of disks in $\bar{\mathcal{D}}$ computed in line 1 of the algorithm cover at least k points. For any point x which is covered by $\bar{\mathcal{D}}$, if x is covered by D_{rmv} or any disk in $\mathcal{I}$, then it is also covered by $\mathcal{M}$. Otherwise, x is covered by a disk D which is removed in line 6 of Algorithm 2. This disk D is removed because it intersects a disk $D' \in \mathcal{I}$. Because of the greedy choice of disk D' in line 3 of Algorithm 2, we have $r(D) \leq r(D')$. Hence $d(x, c(D')) \leq d(x, c(D)) + d(c(D), c(D')) \leq r(D) + (r(D) + r(D')) \leq 3r(D')$, where $d(\cdot, \cdot)$ denotes the Euclidean distance. So, x is covered by $disk(c(D'), 3r(D')) \in \mathcal{M}$. □

The following lemma is a key lemma towards the analysis of the approximation ratio.

Lemma 3. *Suppose $\mathcal{D}^*$ is an optimal solution for $(X, \mathcal{D}, k)$. Then the independent set of disks $\mathcal{I}$ output by Algorithm 2 satisfies $p(\mathcal{I}) \leq p(\mathcal{D}^*)$.*

Proof. We prove
$$p^{(i)}(\mathcal{I}) \leq p^{(i)}(\mathcal{D}^*) \tag{8}$$
by a backward induction on $i = t, t-1, \ldots, 0$. Since $p^{(0)} = p$, what is required by the lemma is exactly $p^{(0)}(\mathcal{I}) \leq p^{(0)}(\mathcal{D}^*)$.

For the base step, we have $p^{(t)}(\mathcal{I}) = 0$ because every disk $D \in \mathcal{I} \subseteq \bar{\mathcal{D}}^{(1)}$ belongs to some $\bar{\mathcal{D}}^{(j)}_{=0}$ (by (6)) and thus $p^{(t)}(D) = 0$. So (8) holds for $i = t$.

For the induction step, suppose (8) is true for some $i \leq t$. We are going to prove

$$p^{(i-1)}(\mathcal{I}) \leq p^{(i-1)}(\mathcal{D}^*). \tag{9}$$

By (3), inequality (9) is equivalent with

$$p^{(i)}(\mathcal{I}) + \bar{p}^{(i)}(\mathcal{I}) \leq p^{(i)}(\mathcal{D}^*) + \bar{p}^{(i)}(\mathcal{D}^*).$$

Combining this with the induction hypothesis, it suffices to prove

$$\bar{p}^{(i)}(\mathcal{I}) \leq \bar{p}^{(i)}(\mathcal{D}^*). \tag{10}$$

By (6) and Remark 1,

$$\text{for any disk } D \in \bar{\mathcal{D}}^{(1)} \setminus \bar{\mathcal{D}}^{(i)}, \text{ we have } \bar{p}^{(i)}(D) = 0. \tag{11}$$

Combining this with (2) and the fact $\mathcal{I} \subseteq \bar{\mathcal{D}}^{(1)}$, we have

$$\bar{p}^{(i)}(\mathcal{I}) = \sum_{D \in \mathcal{I}} \bar{p}^{(i)}(D) = \sum_{D \in \mathcal{I} \cap \bar{\mathcal{D}}^{(i)}} \gamma^{(i)} \cdot |X^{(i-1)} \cap D|$$

$$= \sum_{x \in X^{(i-1)}} \gamma^{(i)} \cdot |\{D \in \mathcal{I} \cap \bar{\mathcal{D}}^{(i)} : x \in D\}|.$$

Since no disks in $\mathcal{I}$ can intersect, we have

$$\sum_{x \in X^{(i-1)}} |\{D \in \mathcal{I} \cap \bar{\mathcal{D}}^{(i)} : x \in D\}| = |X^{(i-1)} \cap \mathcal{C}(\mathcal{I} \cap \bar{\mathcal{D}}^{(i)})|.$$

Since $D_{rmv} \notin \mathcal{I}$, we have $\mathcal{I} \cap \bar{\mathcal{D}}^{(i)} \subseteq \bar{\mathcal{D}}^{(i)} \setminus \{D_{rmv}\}$. Combining this with Lemma 2 and the observation that $D_{rmv} \in \bar{\mathcal{D}}_{=0}^{(t)} \subseteq \bar{\mathcal{D}}^{(i)}$, we have

$$|X^{(i-1)} \cap \mathcal{C}(\mathcal{I} \cap \bar{\mathcal{D}}^{(i)})| < k^{(i-1)}.$$

Hence,

$$\bar{p}^{(i)}(\mathcal{I}) \leq \gamma^{(i)} k^{(i-1)} \tag{12}$$

On the other hand, because of (11),

$$\bar{p}^{(i)}(\mathcal{D}^*) = \sum_{D \in \mathcal{D}^*} \bar{p}^{(i)}(D) = \sum_{D \in \mathcal{D}^* \setminus \left(\bar{\mathcal{D}}^{(1)} \setminus \bar{\mathcal{D}}^{(i)} \right)} \gamma^{(i)} \cdot |X^{(i-1)} \cap D|.$$

Combining the facts

$$|\mathcal{C}(\mathcal{D}^*)| \geq k$$

$$\bar{\mathcal{D}}^{(1)} \setminus \bar{\mathcal{D}}^{(i)} \subseteq \bigcup_{j=1}^{i-1} \mathcal{D}_{=0}^{(j)} \text{ by (6), and}$$

$$k^{(i-1)} = \max\{0, k - |\mathcal{C}(\bigcup_{j=1}^{i-1} \mathcal{D}_{=0}^{(j)})|\} \text{ by (5),}$$

we have

$$\sum_{D \in \mathcal{D}^* \setminus \left(\bar{\mathcal{D}}^{(1)} \setminus \bar{\mathcal{D}}^{(i)} \right)} |X^{(i-1)} \cap D| \geq |X^{(i-1)} \cap \mathcal{C}(\mathcal{D}^* \setminus (\bar{\mathcal{D}}^{(1)} \setminus \bar{\mathcal{D}}^{(i)}))|$$

$$\geq \left| X^{(i-1)} \cap \mathcal{C} \left(\mathcal{D}^* \setminus \bigcup_{j=1}^{i-1} \mathcal{D}_{=0}^{(j)} \right) \right| \geq k^{(i-1)}.$$

Hence,

$$\bar{p}^{(i)}(\mathcal{D}^*) \geq \gamma^{(i)} k^{(i-1)}. \tag{13}$$

Then inequality (10) follows from (12) and (13), and the lemma is proved. $\square$

The next theorem estimates the approximation effect of Algorithm 3.

Theorem 2. *Suppose $\mathcal{C}^*$ is an optimal solution on instance $(X, \mathcal{D}, p, k)$, and $\mathcal{M}$ is the output of Algorithm 3. Then*

$$p(\mathcal{M}) \leq 3^\alpha p(\mathcal{C}^*) + p(D_{rmv}).$$

Proof. For each disk $D \in \mathcal{M} \setminus \{D_{rmv}\}$, it comes from a disk $D' \in \mathcal{I}$ by expanding the radius by three times. Hence by (1), $p(D) = 3^\alpha p(D')$. So $p(\mathcal{M}) \leq 3^\alpha p(\mathcal{I}) + p(D_{rmv})$, and the theorem follows from Lemma 3. $\square$

By Theorem 2, the approximate effect is related with $p(D_{rmv})$. The reason why we should guess a disk $D_{\max}$ with the largest radius in an optimal solution is now clear: to control the term $p(D_{rmv})$ to be not too large. The algorithm combining the guessing technique is presented as follows.

3.2   The Whole Algorithm

Algorithm 4 is the whole algorithm for the MinPowerPartCov problem. It first guesses a disk $D_{\max}$ with the maximum radius in an optimal solution, takes it, and then uses Algorithm 3 on the residual instance. For a guessed disk D, the residual instance consists of all those disks $\mathcal{D}_{\leq r(D)}$ whose radii are no larger than $r(D)$ (excluding D itself), and the goal is to cover the remaining elements $X \setminus D$ beyond the remaining covering requirement $\max\{0, k - |D|\}$. The weight function, denoted as p_D, is determined by (1). If for a guessed disk D, Algorithm 3 does not return a feasible solution, then we regard the solution to have cost ∞. Algorithm 4 returns the best solution among all the guesses.

Algorithm 4. $MinPowerPartCov(X, \mathcal{D}, k, p)$

Input: A set of points X, a set of sensors $\mathcal{S}$, a covering requirement k.
Output: A power assignment p to cover at least k points.
1: Construct the set $\mathcal{D}$ of canonical disks, determine the weight of each disk by (1).
2: **for** $D \in \mathcal{D}$ **do**
3:     $\mathcal{M}_D \leftarrow Cov(X \setminus D, \mathcal{D}_{\leq r(D)}, p_D, \max\{0, k - |D|\})$
4:     $\mathcal{F}_D \leftarrow \mathcal{M}_D \cup \{D\}$
5: **end for**
6: $\widetilde{D} \leftarrow \arg\min_{D \in \mathcal{D}}\{p(\mathcal{F}_D)\}$
7: Return the power assignment corresponding to $\mathcal{F}_{\widetilde{D}}$

Theorem 3. *Algorithm 4 is an 3^α-approximation algorithm for the MinPower-PartCov problem.*

Proof. Suppose $D_{\max}$ is the disk with the maximum radius in an optimal solution. By Theorem 2 and the fact $p(D_{max,rmv}) \leq p(D_{max})$, we have

$$p(\mathcal{F}_{D_{\max}}) = p(\mathcal{M}_{D_{\max}}) + p(D_{\max}) \leq 3^\alpha p(\mathcal{C}^*_{D_{\max}}) + 2p(D_{\max})$$
$$\leq 3^\alpha \left(p(\mathcal{C}^*_{D_{\max}}) + p(D_{\max}) \right) = 3^\alpha opt,$$

where *opt* is the optimal power. Since the set $\mathcal{F}_{\widetilde{D}}$ computed by Algorithm 4 satisfies $p(\mathcal{F}_{\widetilde{D}}) \leq p(\mathcal{F}_{D_{\max}})$, the theorem is proved. □

4 Conclusion

In this paper, we presented an approximation algorithm for the minimum power partial cover problem achieving approximation ratio 3^α, using a local ratio method. This ratio improves the ratio of $(12 + \varepsilon)$ in [9], and matches the best known ratio for the minimum power (full) cover problem in [3].

Recently, there are a lot of studies on the minimum power multi-cover problem [5,6]. A problem which deserves to be explored is the minimum power partial multi-cover problem (adding partial covering requirement). According to current studies on the minimum partial set multi-cover problem [17–19], it seems that studying the combination of multi-cover and partial cover in a general setting is very difficult. An interesting question is whether geometry can make the situation better?

Acknowledgment. This research is supported in part by NSFC (11771013, 61751303, 11531011) and the Zhejiang Provincial Natural Science Foundation of China (LD19A010001, LY19A010018).

References

1. Bar-Yehuda, R., Even, S.: A local-ratio theorem for approximating the weighted vertex cover problem. Ann. Discret. Math. **25**, 27–46 (1985)
2. Bar-Yehuda, R.: Using homogeneous weights for approximating the partial cover problem. J. Algorithms **39**(2), 137–144 (2001)
3. Bar-Yehuda, R., Rawitz, D.: A note on multicovering with disk. Comput. Geom. **46**(3), 394–399 (2013)
4. Bansal, N., Pruhs, K.: Weighted geometric set multi-cover via quasi-uniform sampling. In: Epstein, L., Ferragina, P. (eds.) ESA 2012. LNCS, vol. 7501, pp. 145–156. Springer, Heidelberg (2012). https://doi.org/10.1007/978-3-642-33090-2_14
5. Bhowmick, S., Varadarajan, K., Xue, S.-K.: A constant-factor approximation for multi-covering with disks. Comput. Geom. **6**(1), 220–24 (2015)
6. Bhowmick, S., Inamdar, T., Varadarajan, K.: On metric multi-covering problems. Computational Geometry, arxiv:1602.04152 (2017)
7. Chvatal, V.: A greedy heuristic for the set-covering problem. Math. Oper. Res. **4**(3), 233–235 (1979)
8. Chan, T.M., Granty, E., Konemanny, J., Sharpe, M.: Weighted capacitated, priority, and geometric set cover via improved quasi-uniform sampling. In: SODA, pp. 1576–1585 (2012)
9. Freund, A., Rawitz, D.: Combinatorial interpretations of dual fitting and primal fitting. CiteSeer (2011). http://citeseerx.ist.psu.edu/viewdoc/summary?doi=10.1.1.585.9484
10. Gandhi, R., Khuller, S., Srinivasan, A.: Approximation algorithms for partial covering problems. J. Algorithms **53**(1), 55–84 (2004)
11. Gibson, M., Pirwani, I.A.: Algorithms for dominating set in disk graphs: breaking the $\log n$ barrier. In: de Berg, M., Meyer, U. (eds.) ESA 2010. LNCS, vol. 6346, pp. 243–254. Springer, Heidelberg (2010). https://doi.org/10.1007/978-3-642-15775-2_21
12. Hochbaum, D.S., Maas, W.: Approximation schemes for covering and packing problems in image processing and VLSI. J. ACM **32**, 130–136 (1985)
13. Inamdar, T., Varadarajan, K.: On partial covering for geometric set system. Comput. Geom. **47**, 1–14 (2018)
14. Johnson, D.S.: Approximation algorithms for combinatorial problems. J. Comput. Syst. Sci. **9**, 256–278 (1974)
15. Mustafa, N.H., Ray, S.: Improved results on geometric hitting set problems. Discret. Comput. Geom. **44**, 883–895 (2010)
16. Mustafa, N.H., Raman, R., Ray, S.: Quasi-polynomial time approximation scheme for weighted geometric set cover on pseudodisks. SIAM J. Comput. **44**(6), 1650–1669 (2015)
17. Ran, Y., Zhang, Z., Du, H., Zhu, Y.: Approximation algorithm for partial positive influence problem in social network. J. Comb. Optim. **33**, 791–802 (2017)
18. Ran, Y., Shi, Y., Zhang, Z.: Local ratio method on partial set multi-cover. J. Comb. Optim. **34**(1), 1–12 (2017)
19. Ran, Y., Shi, Y., Zhang, Z.: Primal dual algorithm for partial set multi-cover. In: Kim, D., Uma, R.N., Zelikovsky, A. (eds.) COCOA 2018. LNCS, vol. 11346, pp. 372–385. Springer, Cham (2018). https://doi.org/10.1007/978-3-030-04651-4_25
20. Roy, A.B., Govindarajan, S., Raman, R., Ray, S.: Packing and covering with non-piercing regions. Discret. Comput. Geom. **60**, 471–492 (2018)

21. Slavík, P.: Improved performance of the greedy algorithm for partial cover. Inf. Process. Lett. **64**(5), 251–254 (1997)
22. Varadarajan, K.: Weighted geometric set cover via quasi-uniform sampling. In: STOC 2010, pp. 641–648 (2010)
23. Vazirani, V.V.: Approximation Algorithms. Springer, Heidelberg (2001). https:// doi.org/10.1007/978-3-662-04565-7

On Approximations for Constructing Required Subgraphs Using Stock Pieces of Fixed Length

Junran Lichen[1], Jianping Li[1(✉)], Ko-Wei Lih[2], and Xingxing Yu[3]

[1] Department of Mathematics, Yunnan University,
Kunming 650504, People's Republic of China
junranlichen@hotmail.com, jianping@ynu.edu.cn
[2] Institute of Mathematics, Academia Sinica, Taipei 10617, Taiwan
makwlih@sinica.edu.tw
[3] School of Mathematics, Georgia Institute of Technology, Atlanta, USA
yu@math.gatech.edu

Abstract. In this paper, we consider the problem of constructing required subgraphs using stock pieces of fixed length (CRS-SPFL, for short), which is a variant of the problem of minimum-cost edge-weighted subgraph constructions (MCEWSC, for short). This new problem has many important applications in our reality life, and it is defined as follows. In the MCEWSC problem Q, the objective is to choose a minimum-cost subset of edges from a graph such that these edges form a required subgraph (e.g., a spanning tree). In the CRS-SPFL problem Q', these edges are further required to be constructed by some stock pieces of fixed length L, the new objective is to minimize the total cost to construct such a required subgraph G', where the total cost is sum of the cost to buy necessary these stock pieces and the cost to construct all edges in such a subgraph G'.

We obtain the following three main results. (1) Whenever the MCEWSC problem Q can be approximated by an α-approximation algorithm (for the case $\alpha = 1$, the MCEWSC problem Q is solved optimally by a polynomial-time exact algorithm), we can design a 2α-approximation algorithm to solve the CRS-SPFL problem Q'; (2) In addition, when the MCEWSC problem Q is to find a minimum spanning tree, we provide a $\frac{3}{2}$-approximation algorithm and an AFPTAS to solve the CRS-SPFL problem Q', respectively; (3) Finally, when the MCEWSC problem Q is to find a single-source shortest paths tree, we present a $\frac{3}{2}$-approximation algorithm and an AFPTAS to solve the CRS-SPFL problem Q', respectively.

Keywords: Subgraph constructions · Stock pieces of fixed length L ·
Bin packing · Approximation algorithms · AFPTAS

Supported by the National Natural Science Foundation of China [No. 11861075], IRTSTYN, and Key Joint Project of Yunnan Provincial Science and Technology Department and Yunnan University [No. 2018FY001(-014)], Innovation Team Fostering Project of Yunnan Province.

D.-Z. Du et al. (Eds.): AAIM 2019, LNCS 11640, pp. 192–202, 2019.
https://doi.org/10.1007/978-3-030-27195-4_18

1   Introduction

Many graph optimization problems, especially the minimum-cost edge-weighted subgraph construction problems, are motivated from our reality life applications, e.g., the minimum spanning tree problem, the shortest path problem, and the minimum Steiner tree problem [6]. The objective in each of such problems is to choose a minimum-cost subset of edges from a weighted graph so that these edges form a required subgraph or structure (for example, a spanning tree structure or a Steiner tree structure). In addition, there exist some optimization problems that can be regarded as some combinations of other well-known graph optimization problems. For example, the single-source shortest paths tree problem can be regarded as a combination of the shortest path problem and the minimum arborescence problem. These optimization problems have a number of interesting applications in electronic power transportation networks, fiber optic networks, Quality of Service (QoS) routing in high-speed networks, and there exist many polynomial-time exact or approximation algorithms to solve these problems [6].

The Euclidean Steiner tree problem (EST, for short) is to find a tree with minimum Euclidean length that spans a set of points fixed in the Euclidean plane while allowing the addition of some so-called Steiner points. The EST problem [3] is one of the well-known NP-hard problems, and it is widely used to design real-world structures like highways and oil pipelines. There are many approximation algorithms to solve the EST problem [6].

Lin and Xue [5] addressed a variant of the EST problem that is called the Steiner tree problem with minimum number of Steiner points and bounded edge-length (STP-MSPBEL, for short). The STP-MSPBEL problem is to find a Steiner tree interconnecting all pre-assigned terminals with the minimum number of Steiner points such that each edge in this Steiner tree has Euclidean length no more than a given constant d. They showed that the STP-MSPBEL problem is NP-hard, and they presented a 4-approximation algorithm. Furthermore, Chen et al. [1] designed a 3-approximation algorithms to resolve the STP-MSPBEL problem.

In our reality life, stock materials to be used to construct all edges in such a Steiner tree are sold with length constraints, i.e., each stock piece has a bounded length L ($d \leq L$). When we assume that the selling price of each stock piece of length L is a constant, we should consider the minimum cost of necessary stock pieces to construct all edges in a Steiner tree, rather than the minimum sum of lengths of all edges in the same Steiner tree. In general, the sum of lengths of necessary stock pieces to be used to construct all edges in a Steiner tree is at least the sum of lengths of all edges in the same Steiner tree, by the reason that some unused parts of stock pieces are "wasted".

Furthermore, a more general problem arises naturally when some edges in a Steiner tree may have lengths greater than the length L of stock piece. When we plan to construct an edge e of length $w(e) > L$ in a Steiner tree, we may use $i(e) = \lceil \frac{w(e)}{L} \rceil - 1$ whole stock pieces of length L and a part of length $w'(e) = w(e) - i(e) \cdot L$ from a stock piece of length L to construct that edge e. In addition, we may consider the cost to assemble these $i(e) + 1$ parts together to construct that edge e. Finally, we should also consider the construction cost of all edges in the whole Steiner tree desired. To sum up, we should consider the total cost of such a Steiner tree that includes three components: (1) the cost of purchasing necessary stock pieces of length L, (2) for each edge e in such a Steiner tree, the cost of assembling all parts to construct the edge e (if any), and (3) the cost of constructing all edges in such a Steiner tree.

Motivated by the previous problems, especially the STP-MSPBEL problem and the length constraint on stock pieces, we are interested in the following useful variant of the problem of minimum-cost edge-weighted subgraph constructions, where this new problem has many important applications in our reality life. This new variant problem is posed as follows. For a minimum-cost edge-weighted subgraph construction problem Q (MCEWSC, for short), the objective is to choose a minimum-cost subset of edges such that these edges form a required subgraph or structure. We now consider the new variant problem Q' of Q such that these edges are further required to be constructed by some stock pieces of length L, where such each stock piece of length L is sold at the price c_0. The new objective, however, is to minimize the total cost of such a subgraph or structure G', i.e., $\min_{G'}\{\sum_{e \in G'} c(e) + k(G') \cdot c_0\}$, where $c(e)$ is the construction cost of edge e and $k(G') \cdot c_0$ is the cost to buy necessary stock pieces of length L to construct all edges in such same subgraph or structure G'.

For convenience, we refer to the new variant problem Q' of Q as the problem of constructing required subgraphs using stock pieces of fixed length (CRS-SPFL, for short), and simply denote it as the CRS-SPFL problem Q'. We also emphasize the following process in this paper how to construct an edge in a specific subgraph or structure. When we plan to construct an edge e of length $w(e) > L$ in such a required subgraph or structure, we may always use $i(e) = \lceil \frac{w(e)}{L} \rceil - 1$ whole stock pieces of length L and then a part of length $w'(e) = w(e) - i(e) \cdot L$ from a stock piece of length L together to construct that edge e. Using this process, we should permit at most one piece of length less than L in any edge in our edge-construction process, i.e., we should not permit at least two pieces of length less than L in some edge in our edge-construction process.

Our new problem can be roughly treated as a combination of the MCEWSC problem and a generalization of the bin-packing problem. In the description of our problem, it seems that we have ignored the cost of assembling all parts to construct each edge in a specific subgraph or structure (if needed). We shall show in the next section that the assembling cost of each edge e may become a part of construction cost of that edge e by using a polynomial reduction. (Proposition 1 will provide the details.) So it is sufficient for us to consider the total cost of a specific subgraph or structure so that the cost can be split into two components: (1) the cost to buy necessary stock pieces of length L, and (2) the cost to construct all edges in such same subgraph or structure.

Now, there are three special cases to be noted. (1) If the sale price of each stock piece is not taken into consideration, i.e., $c_0 = 0$, our new problem becomes the MCEWSC problem Q in [6]. (2) If the length L of each stock piece is large, for example, $L \geq \sum_{e \in G} w(e)$, i.e., $k(G') = 1$, our new problem also becomes the

MCEWSC problem Q. (3) If the construction cost of each edge is not taken into consideration, i.e., $c(e) = 0$ for each edge e in G, we treat this new problem as the problem of constructing specific subgraph with minimum number of stock pieces of fixed length (CSS-MSP, for short), which is similarly defined as the STP-MSPBEL problem addressed by Lin and Xue [5].

As far as what we have known, these CRS-SPFL problems have not appeared in our reality life, and there are no known polynomial-time exact or approximation algorithms to solve these new optimization problems.

Our paper is organized as follows. In Sect. 2, we present terminology for describing our algorithms and provide key lemmas to ensure the correctness of approximation algorithms. In Sect. 3, whenever the MCEWSC problem Q can be approximated by an α-approximation algorithm, we design a 2α-approximation algorithm to solve the CRS-SPFL problem Q'; In Sect. 4, we provide a $\frac{3}{2}$-approximation algorithm and an fully asymptotic polynomial-time approximation scheme (AFPTAS) to solve the problem Q' of constructing a minimum-cost spanning tree using stock pieces of fixed length, respectively; In Sect. 5, we further present a $\frac{3}{2}$-approximation algorithm and an AFPTAS to solve the problem Q' of constructing a minimum-cost single-source shortest paths tree using stock pieces of fixed length, respectively. In Sect. 6, we state our conclusion and further research.

2   Terminology, Reduction and Key Lemmas

The notations and terminology to be used to describe our algorithms are similar to solve a generalization of the bin-packing problem [2,7]. We treat each stock piece of length L as a "bin" with capacity L. Let e be an edge of a graph $G = (V, E; w, c)$ equipped with length function $w : E \rightarrow R^+$ and cost function $c : E \rightarrow R_0^+$. We treat this edge e as an "item" of size $w(e)$. For each edge e of length $w(e)$, let $i(e) = \lceil \frac{w(e)}{L} \rceil - 1$ and $w'(e) = w(e) - i(e) \cdot L$. To construct edge e, we may use $i(e)$ whole stock pieces of length L and a part of length $w'(e)$ from such a stock piece. The construction cost of an edge e is set to be $c(e)$. Note that the value $c(e)$ does not include the cost to buy some stock piece(s) to construct that edge e. Using a similar terminology to describe the process to pack items into bins for the bin-packing problem, we describe such a process as "packing an edge e of length $w(e)$ or an item of size $w(e)$ into some bin(s) with capacity L".

For convenience, denote $G = (V, E; w; c; L)$ as an instance of the CRS-SPFL problem Q', i.e., a graph $G = (V, E; w; c)$ equipped with length function $w : E \rightarrow R^+$ and cost function $c : E \rightarrow R_0^+$ together with some stock pieces of length L. Similarly, denote $H = (V, E; f)$ as an instance of the MCEWSC problem Q equipped with cost function $f : E \rightarrow R^+$.

For a specific subgraph $G' = (V', E'; w; c)$ of a weighted graph $G = (V, E; w; c)$ for the CRS-SPFL problem Q', let $w(E') = \sum_{e \in E'} w(e)$ and $c(E') = \sum_{e \in E'} c(e)$. Similarly, for a specific subgraph $H' = (V', E'; f)$ of an instance $H = (V, E; f)$ for the MCEWSC problem Q, let $f(E') = \sum_{e \in E'} f(e)$. Other notations and terminology can be found in the reference [6].

When we consider the CRS-SPFL problem Q', if an edge e chosen in the specific subgraph G' has $w(e) > L$, using the process mentioned-above, we have to use $i(e) = \lceil \frac{w(e)}{L} \rceil - 1$ stock pieces of length L and then a part of length $w'(e) = w(e) - i(e) \cdot L$ together to construct an edge e. At the same time, we should pay a cost $i(e)p(e)$ to patch these $i(e)+1$ parts together to construct that edge e, where $p(e)$ is an unit cost of assembling together two any parts in such an edge e (if any). The value $i(e)p(e)$ is called as patching cost of that edge e. We shall show by a polynomial reduction that this patching cost of each edge e may become a part of construction cost of the same edge e in a weighted graph $G = (V, E; w; c; L)$ equipped with patching cost function $p : E \to R_0^+$.

Let Q'' denote the general version of the CRS-SPFL problem Q', where the construction of each edge e needs a patching cost (if any). For convenience, given the MCEWSC problem Q, we refer to new general problem as the general CRS-SPFL problem Q'' of Q. More precisely, if $G = (V, E; w; c; L)$ is an instance of the CRS-SPFL problem Q', let $H = (V, E; w; c; p; L)$ be an instance of the general CRS-SPFL problem Q'' with the same functions $w(\cdot)$ and $c(\cdot)$ and patching cost function $p : E \to R_0^+$. Then, we can obtain the following important proposition.

Proposition 1. *The general CRS-SPFL problem Q'' is polynomially equivalent to the CRS-SPFL problem Q'.*

In the view of Proposition 1, we will consider the CRS-SPFL problem Q' rather than the general CRS-SPFL problem Q'' in the remainder of this paper. For any instance $G = (V, E; w; c; L; c_0)$ of the CRS-SPFL problem Q', where c_0 is the price of each stock piece of length L, we always assume patching cost $p(e) = 0$ for each edge e in the instance G. In the construction of edges of specific subgraph G' of G, "the necessary number of stock pieces of bounded length L" is rephrased as "the necessary number of bins with capacity L", or simply "the number of bins used".

We need the following two key lemmas when we design better approximation algorithms to solve two special versions of the CRS-SPFL problem.

Lemma 1. [6] *Let $G = (V, E; l)$ be a weighted graph equipped with weight function $l : E \to R^+$ and $T = (V, E_T)$ a minimum spanning tree of G with edge-set $E_T = \{e_{i_1}, e_{i_2}, \ldots, e_{i_{n-1}}\}$ satisfying $l(e_{i_1}) \leq l(e_{i_2}) \leq \cdots \leq l(e_{i_{n-1}})$. If T^* is any spanning tree of G with edge-set $E_{T^*} = \{e_{j_1}, e_{j_2}, \ldots, e_{j_{n-1}}\}$ satisfying $l(e_{j_1}) \leq l(e_{j_2}) \leq \cdots \leq l(e_{j_{n-1}})$, then the inequality $l(e_{i_k}) \leq l(e_{j_k})$ holds for each $k = 1, 2, \ldots, n-1$.*

Lemma 2. [4] *For any real number $\varepsilon > 0$ and an instance of the bin-packing problem involving m items $a_1, a_2, \ldots, a_m$ having sizes $s(a_1), s(a_2), \ldots, s(a_m)$, where $0 < s(a_i) \leq L$ for each $i = 1, 2, \ldots, m$, there is a polynomial-time algorithm $\mathcal{A}_\varepsilon$ in time $\mathcal{O}(T(\varepsilon^{-2}, m))$ to find a packing using at most $(1 + \varepsilon)OPT + \mathcal{O}(\varepsilon^{-2})$ bins with capacity L, where OPT is the minimum number of bins used for the same instance and $T(x, y) = \mathcal{O}(x^8 \log x \log^2 y + x^4 y \log x \log y)$, i.e., the algorithm $\mathcal{A}_\varepsilon$ is an AFPTAS to solve the bin-packing problem.*

3   Approximation Algorithm for the CRS-SPFL Problem

In this section, we consider the problem Q' of constructing required sub-graph using stock pieces of fixed length (CRS-SPFL, for short), meanwhile the minimum-cost edge-weighted subgraph construction problem Q (MCEWSC, for short) can be approximated by an α-approximation algorithm $\mathcal{A}_\alpha$ (for the case $\alpha = 1$, the MCEWSC problem Q is solved optimally by a polynomial-time exact algorithm), where the objective in Q is to choose a minimum-cost subset of edges such that the edges form a specific subgraph.

We use the following strategy to design an approximation algorithm to solve the CRS-SPFL problem Q'. (1) For any instance $G = (V, E; w; c; L)$ of the CRS-SPFL problem Q', define a new cost $c'(e)$ on each edge e of the instance $H = (V, E; c')$ of the MCEWSC problem Q, then use an α-approximation algorithm $\mathcal{A}_\alpha$ on the instance $H = (V, E; c')$ to find a specific subgraph $G' = (V', E')$ of H having minimum total cost $c'(E')$; (2) Use the First-Fit algorithm (FF) [2] to pack the items $w(e)$ of all edges e in E' into bins with capacity L.

Our first approximation algorithm to solve the CRS-SPFL problem Q' is described as follows.

Algorithm:3.1

Input:	a weighted graph $G = (V, E; w; c; L)$ equipped with $w : E \to R^+$ and $c : E \to R_0^+$.
Output:	a subgraph $G' = (V', E'; w; c; L)$ of G such that the total construction cost of G' is as small as possible.

Begin

Step 1 From an instance $G = (V, E; w; c; L)$ of the CRS-SPFL problem Q', construct the instance $H = (V, E; c')$ of the MCEWSC problem Q, where $c'(e) = c(e) + c_0 \cdot \frac{w(e)}{L}$ for each edge e in E.

Step 2 For the instance $H = (V, E; c')$ of the MCEWSC problem Q, use the α-approximation algorithm $\mathcal{A}_\alpha$ to find a specific subgraph $G' = (V', E'; c')$ of $H = (V, E; c')$, having cost $c'(E')$ as small as possible. Denote $E' = \{e_{i_1}, e_{i_2}, \ldots, e_{i_m}\}$.

Step 3 For each edge e in E', first use $i(e)$ stock pieces of length L to construct the part of edge e having length $i(e) \cdot L$, where $i(e) = \lceil \frac{w(e)}{L} \rceil - 1$ and $w'(e) = w(e) - i(e) \cdot L$ as before. And let $k_1 = \sum_{e \in E'} i(e)$ denote the number of stock pieces used at this step.

Step 4 Use the FF algorithm [2] to pack the items $w'(e_{i_1})$, $w'(e_{i_2})$, $\ldots$, $w'(e_{i_m})$ into k_2 bins with capacity L, i.e., k_2 stock pieces of bounded length L.

Step 5 Output the subgraph $G' = (V', E'; w; c; L)$ produced in Step 2 and the total construction cost $OUT = \sum_{e \in E'} c(e) + (k_1 + k_2) \cdot c_0$.

End

Theorem 1. *Algorithm 3.1 is an 2α-approximation algorithm to solve the CRS-SPFL problem Q', and the complexity of Algorithm 3.1 is $\mathcal{O}(\max\{f(n), n^2\})$, where $f(n)$ is the complexity of an α-approximation algorithm $\mathcal{A}_\alpha$ to solve the MCEWSC problem Q.*

Proof. For the instance $G = (V, E; w; c; L)$ of the CRS-SPFL problem Q', suppose that the specific subgraph $G' = (V', E'; w; c; L)$ is a feasible solution produced by Algorithm 3.1 with the output value $OUT = \sum_{e \in E'} c(e) + (k_1 + k_2) \cdot c_0$ and a specific subgraph $G^* = (V^*, E^*; w; c; L)$ is an optimal solution with the optimal value $OPT = \sum_{e \in E^*} c(e) + k^* \cdot c_0$. Let $k = k_1 + k_2$ be the number of bins used in Algorithm 3.1 and k^* the minimum number of stock pieces used for the instance $G = (V, E; w; c; L)$. We may also assume that a specific subgraph $G^{**} = (V^{**}, E^{**}; c')$ is an optimal solution to the instance $H = (V, E; c')$ of the MCEWSC problem Q with the optimal value $c'(E^{**})$. We are going to show $OUT \leq 2\alpha \cdot OPT$. Since Algorithm 3.1 executes the α-approximation algorithm $\mathcal{A}_\alpha$ to find a specific subgraph $G' = (V', E'; c')$ to the instance $H = (V, E; c')$ of the MCEWSC problem Q, we obtain $c'(E') \leq \alpha \cdot c'(E^{**})$. Since G^* is an optimal solution to the instance $G = (V, E; c; L)$ of the CRS-SPFL problem Q', the specific subgraph G^* is also a feasible solution to the instance $H = (V, E; c')$ of the MCEWSC problem Q. We thus obtain $c'(E^{**}) \leq c'(E^*)$, implying $c'(E') \leq \alpha \cdot c'(E^*)$.

We have $k_1 = \sum_{e \in E'} i(e)$ at Step 3. When we pack the items $\{w'(e) \mid e \in E'\}$ into k_2 bins with capacity L at Step 4, there are $k_2 - 1$ bins with total size at least $\frac{L}{2}$. (Otherwise the number k_2 may be reduced.) In addition, Step 4 in Algorithm 3.1 implies $\sum_{e \in E'} w'(e) > L + \frac{L}{2} \cdot (k_2 - 2) = \frac{L \cdot k_2}{2}$. Hence, we obtain the following

$$\alpha \cdot OPT = \alpha \cdot \left(\sum_{e \in E^*} c(e) + k^* \cdot c_0 \right)$$

$$\geq \alpha \cdot \left(\sum_{e \in E^*} c(e) + \frac{\sum_{e \in E^*} w(e)}{L} \cdot c_0 \right)$$

$$= \alpha \cdot \sum_{e \in E^*} \left(c(e) + c_0 \cdot \frac{w(e)}{L} \right)$$

$$\geq \sum_{e \in E'} \left(c(e) + c_0 \cdot \frac{w(e)}{L} \right)$$

$$= \sum_{e \in E'} c(e) + \frac{c_0}{L} \cdot \sum_{e \in E'} (i(e) \cdot L + w'(e))$$

$$= \sum_{e \in E'} c(e) + c_0 \cdot \sum_{e \in E'} i(e) + \frac{c_0}{L} \cdot \sum_{e \in E'} w'(e)$$

$$\geq \sum_{e \in E'} c(e) + c_0 \cdot k_1 + \frac{c_0}{L} \cdot \frac{L \cdot k_2}{2}$$

$$\geq \frac{1}{2} \cdot \left(\sum_{e \in E'} c(e) + (k_1 + k_2) \cdot c_0 \right)$$

$$= \frac{1}{2} \cdot OUT$$

implying $OUT \leq 2\alpha \cdot OPT$.

The complexity of Algorithm 3.1 can be determined as follows. (1) Step 1 needs at most time $\mathcal{O}(n^2)$ to define the new function $c'(\cdot)$; (2) Step 2 executes the algorithm $\mathcal{A}_\alpha$ to find a specific subgraph $G' = (V', E'; c')$ in time $f(n)$; (3) For each edge $e \in E'$, treating one step to use $i(e)$ stock pieces of length L to construct the part of edge e having length $i(e) \cdot L$, Step 3 needs at most time $\mathcal{O}(|E'|)$ to obtain k_1 stock pieces of bounded length L; (4) The FF algorithm [2] in Step 4 needs at most time $\mathcal{O}(|E'|)$ to pack the items $w'(e_{i_t})$ into k_2 bins. Hence, the complexity of Algorithm 3.1 is $\mathcal{O}(\max\{f(n), n^2\})$.

This establishes the theorem. ∎

4   Approximation Algorithms for the MCST-LBSP Problem

In this section, we consider the problem Q' of constructing a spanning tree using stock pieces of fixed length (MCST-LBSP, for short). It is easy to show that the MCST-LBSP problem cannot be approximated within performance ratio $\frac{3}{2} - \varepsilon$ for any smaller number $\varepsilon > 0$, unless $\mathcal{P} = \mathcal{NP}$.

In addition, we may assume that $c(e) \le c(e')$ always holds whenever $w(e) \le w(e')$ for any two edges e and e' in an instance $G = (V, E; w, c; L)$ of the MCST-LBSP problem. Using this property, we shall provide a better approximation algorithm and an AFPTAS to solve the MCST-LBSP problem.

The strategy to solve the MCST-LBSP problem is described as follows. (1) With respect to the weight function $w : E \to R^+$, we use a polynomial-time exact algorithm [6] (referred as the MST algorithm) to find a minimum spanning tree $T = (V, E_T; w)$; (2) Use the First-Fit-Decreasing algorithm (FFD) [7] to pack the sizes of all edges in T into bins with capacity L.

Our approximation algorithm to solve the MCST-LBSP problem is described as follows.

Algorithm:4.1

Input:	a weighted graph $G = (V, E; w, c; L)$.
Output:	a spanning tree $T = (V, E_T)$ and the total cost $OUT = \sum_{e \in E_T} c(e) + k \cdot c_0$.

Begin	
Step 1	With respect to the weight function $w : E \to R^+$, use the MST algorithm [6] to find a minimum spanning tree $T = (V, E_T; c')$. Let $E_T = \{e_{i_1}, e_{i_2}, \ldots, e_{i_{n-1}}\}$.
Step 2	For each edge e_{i_r} in E_T, first use $i(e_{i_r})$ stock pieces of bounded length L to construct the part of edge e_{i_r} having length $i(e_{i_r}) \cdot L$. Recall $i(e_{i_r}) = \lceil \frac{w(e_{i_r})}{L} \rceil - 1$ and $w'(e_{i_r}) = w(e_{i_r}) - i(e_{i_r}) \cdot L$. And let $k_1 = \sum_{e_{i_r} \in E_T} i(e_{i_r})$ denote the number of stock pieces used at this step.
Step 3	Use the FFD algorithm [7] to pack the items $w'(e_{i_1})$, $w'(e_{i_2})$, ..., $w'(e_{i_{n-1}})$ into k_2 bins with capacity L.
Step 4	Output the spanning tree $T = (V, E_T)$ and the total cost $OUT = \sum_{e \in E_T} c(e) + k \cdot c_0$, where $k = k_1 + k_2$.
End	

Theorem 2. *Algorithm 4.1 is a $\frac{3}{2}$-approximation algorithm to solve the MCST-LBSP problem, and its complexity is $\mathcal{O}(n^2)$.*

Proof. Suppose that $T = (V, E_T)$ is a spanning tree produced by Algorithm 4.1 with the output value $OUT = \sum_{e \in T} c(e) + (k_1 + k_2) \cdot c_0$. Let T^* be an optimal spanning tree with the value $OPT = \sum_{e \in T^*} c(e) + k^* \cdot c_0$ to the instance $G = (V, E; w, c; L)$ for the MCST-LBSP problem. Note that k_1 and k_2 are determined at Steps 2–3 of Algorithm 4.1 and k^* is the minimum number of stock pieces used. We are going to show that $OUT \leq \frac{3}{2} \cdot OPT$.

Since Algorithm 4.1 uses the MST algorithm [6] to find a minimum spanning tree $T = (V, E_T)$ of G, we may assume that the edge set $E_T = \{e_{i_1}, e_{i_2}, \dots, e_{i_{n-1}}\}$ of $T = (V, E_T)$ satisfies $w(e_{i_1}) \leq w(e_{i_2}) \leq \cdots \leq w(e_{i_{n-1}})$.

For any optimal spanning tree T^* with the value OPT for the MCST-LBSP problem, we may assume that the edge set $E_{T^*} = \{e^*_{j_1}, e^*_{j_2}, \dots, e^*_{j_{n-1}}\}$ of $T^* = (V, E_{T^*})$ satisfies $w(e^*_{j_1}) \leq w(e^*_{j_2}) \leq \cdots \leq w(e^*_{j_{n-1}})$. Then Lemma 1 implies that the inequality $w(e_{i_r}) \leq w(e^*_{j_r})$ holds for each $r = 1, 2, \dots, n-1$. Using our general assumption that $c(e) \leq c(e')$ always holds whenever $w(e) \leq w(e')$ for any two edges e and e' in $G = (V, E; w, c; L)$, we have the fact that $c(e_{i_r}) \leq c(e^*_{j_r})$ holds for each $r = 1, 2, \dots, n-1$.

Let $k(E_T)$ be the minimum number of bins with length L for the items $w(e_{i_1})$, $w(e_{i_2})$, $\dots$, $w(e_{i_{n-1}})$. We are going to show $k(E_T) \leq k^*$.

For each edge e_{i_r} in E_T, we know that $i(e_{i_r}) = \lceil \frac{w(e_{i_r})}{L} \rceil - 1$, $w'(e_{i_r}) = w(e_{i_r}) - i(e_{i_r}) \cdot L$ and $k_1 = \sum_{e \in E_T} i(e)$. For each edge $e^*_{j_r}$ in $E_{T^*} = \{e^*_{j_1}, e^*_{j_2}, \dots, e^*_{j_{n-1}}\}$, let $w^*(e^*_{j_r}) = w(e^*_{j_r}) - i(e_{i_r}) \cdot L$. Then we still have $w'(e_{i_r}) \leq w^*(e^*_{j_r})$ for each $r = 1, 2, \dots, n-1$.

Since $k_1 = \sum_{e_{i_r} \in E_T} i(e_{i_r})$, $w'(e_{i_r}) \leq w^*(e^*_{j_r})$ $(r = 1, 2, \dots, n-1)$ and the items $w(e^*_{j_1})$, $w(e^*_{j_2})$, $\dots$, $w(e^*_{j_{n-1}})$ are packed into k^* bins B_1, B_2, $\dots$, B_{k^*} with capacity L, we may assume that these $n-1$ items $w^*(e^*_{j_1})$, $w^*(e^*_{j_2})$, $\dots$, $w^*(e^*_{j_{n-1}})$ can be packed into the first $k^* - k_1$ bins B_1, B_2, $\dots$, $B_{k^* - k_1}$ with capacity L, and the others $i(e_{i_r}) \cdot L$ $(r = 1, 2, \dots, n-1)$ can be packed into the last k_1 bins $B_{k^* - k_1 + 1}$, $B_{k^* - k_1 + 2}$, $\dots$, B_{k^*} with capacity L. Then we can pack the items $w(e_{i_1})$, $w(e_{i_2})$, $\dots$, $w(e_{i_{n-1}})$ into these k^* bins in the following way. For each item $w'(e_{i_r})$, $1 \leq r \leq n-1$, if the item $w^*(e^*_{j_r})$ is packed into B_t, where $t \in \{1, 2, \dots, k^* - k_1\}$, then we pack the item $w'(e_{i_r})$ into B_t. We finally pack the other items of sizes $i(e_{i_r}) \cdot L$ $(r = 1, 2, \dots, n-1)$ into the last k_1 bins $B_{k^* - k_1 + 1}$, $B_{k^* - k_1 + 2}$, $\dots$, B_{k^*}. This shows that the items $w(e_{i_1})$, $w(e_{i_2})$, $\dots$, $w(e_{i_{n-1}})$ can be packed into these k^* bins, in which the items $w(e^*_{j_1})$, $w(e^*_{j_2})$, $\dots$, $w(e^*_{j_{n-1}})$ are packed. Hence, $k(E_T) \leq k^*$.

For the items $w'(e_{i_1}), w'(e_{i_2}), \dots, w'(e_{i_{n-1}})$ of all edges in E_T produced at Step 2, since the FFD algorithm [7] is a $\frac{3}{2}$-approximation algorithm for the bin-packing problem, the FFD algorithm can produce a feasible packing with the output value k_2 to satisfy $k_2 \leq \frac{3}{2}(k(E_T) - k_1)$. By the fact $k(E_T) \leq k^*$, we obtain $k_1 + k_2 \leq k_1 + \frac{3}{2}(k(E_T) - k_1) \leq \frac{3}{2}k(E_T) \leq \frac{3}{2}k^*$. Hence, we obtain the following

$$OUT = \sum_{r=1}^{n-1} c(e_{i_r}) + (k_1 + k_2) \cdot c_0 \leq \sum_{r=1}^{n-1} c(e_{j_r}^*) + \frac{3}{2}k^* \cdot c_0 \leq \frac{3}{2} \cdot OPT$$

The complexity of Algorithm 4.1 can be determined as follows. (1) Step 1 executes the MST algorithm [6] in time $\mathcal{O}(n^2)$ to find a minimum spanning tree $T = (V, E_T; w)$; (2) For each edge $e_{i_r} \in E_T$, treating one step to use $i(e_{i_r})$ stock pieces of length L to construct the part of edge e_{i_r} having length $i(e_{i_r}) \cdot L$, Step 2 needs at most running time $\mathcal{O}(|E_T|)$ to obtain k_1 stock pieces of length L; (3) Step 3 uses the FFD algorithm [7] in at most $\mathcal{O}(n \log_2 n)$ time to pack the items $w'(e_{i_r})$ ($r = 1, 2, \ldots, n-1$) into k_2 bins with capacity L. Hence, the total running time of Algorithm 4.1 is $\mathcal{O}(n^2)$.

This establishes the theorem. ∎

Using an AFPTAS, saying $\mathcal{A}_\varepsilon$, due to Karmarkar and Karp [4] to solve the bin-packing problem, we are going to design an AFPTAS, denoted as Algorithm 4.2, to solve the MCST-LBSP problem.

The pseudo-code description of Algorithm 4.2 is modified from Algorithm 4.1 by replacing "Step 3. Use the FFD algorithm [7] to pack the items $w'(e_{i_1})$, $w'(e_{i_2})$, ..., $w'(e_{i_{n-1}})$ into k_2 bins with capacity L" with "Step 3. Use the Karmarkar-Karp's AFPTAS [4], saying $\mathcal{A}_\varepsilon$, to pack the items $w'(e_{i_1})$, $w'(e_{i_2})$, ..., $w'(e_{i_{n-1}})$ into k_2 bins with capacity L". We omit the explicit description of Algorithms 4.2.

Theorem 3. *For any real number $\varepsilon > 0$ and any instance $G = (V, E; w; c; L)$ of the MCST-LBSP problem, Algorithm 4.2 can produce a feasible solution with value at most $(1 + \varepsilon)OPT + \mathcal{O}(\varepsilon^{-2})$. The running time is $\mathcal{O}(T(\varepsilon^{-2}, m))$, where $T(x, y) = \mathcal{O}(x^8 \log x \log^2 y + x^4 y \log x \log y)$, i.e., Algorithm 4.2 is an AFPTAS to solve the MCST-LBSP problem.*

5   Approximation Algorithms for the MCSSSPT-LBSP Problem

In this section, we consider the problem Q' of minimum-cost single-source shortest paths tree using length-bounded stock pieces (MCSSSPT-LBSP, for short).

The strategy to solve the MCSSSPT-LBSP problem is described as follows. (1) For the fixed vertex v_1 in D, use the Dijkstra algorithm [6] to construct an auxiliary acyclic digraph $D_{v_1} = (V, A_{v_1}; w, c; v_1)$ that consists of the union of all shortest (v_1, v_r)-paths in D for all vertices v_r in $V - \{v_1\}$; (2) For each vertex $v_r \in V$, choose an arc entering v_r in D_{v_1} with minimum length, such that we can construct a minimum arborescence T at v_1 in D_{v_1}; (3) Use FFD algorithm [7] to pack the items of lengths of all arcs in T into bins with capacity L.

Our two approximation algorithms to solve the MCSSSPT-LBSP problem are similar to Algorithms 4.1 and 4.2, denoted as Algorithms 5.1 and 5.2, and we omit their details here. We can only state the main results to solve the MCSSSPT-LBSP problem as follows.

Theorem 4. *Algorithm 5.1 is a $\frac{3}{2}$-approximation algorithm to solve the MCSSSPT-LBSP problem and its complexity is $\mathcal{O}(n^2)$.*

Theorem 5. *For any real number $\varepsilon > 0$ and any instance $D = (V, A; w; c; L)$ of the MCSSSPT-LBSP problem, Algorithm 5.2 can produce a feasible solution with value at most $(1 + \varepsilon)OPT + \mathcal{O}(\varepsilon^{-2})$. The running time of Algorithm 5.2 is $\mathcal{O}(T(\varepsilon^{-2}, m))$, where $T(x, y) = \mathcal{O}(x^8 \log x \log^2 y + x^4 y \log x \log y)$. Hence, Algorithm 5.2 is an AFPTAS to solve the MCSSSPT-LBSP problem.*

6   Conclusion and Further Research

In this paper, we consider the problem of constructing required subgraph constructions using stock pieces of fixed length (CRS-SPFL) and then obtain three results.

1. Whenever there exists an α-approximation algorithm to solve the MCEWSC problem Q, we can design a 2α-approximation algorithm to solve the CRS-SPFL problem Q';
2. When the problem Q is to find a minimum spanning tree, we provide a $\frac{3}{2}$-approximation algorithm and an AFPTAS to solve the problem Q' of constructing a spanning tree using stock pieces of fixed length;
3. When the problem Q is to find a single-source shortest paths tree, we also give a $\frac{3}{2}$-approximation algorithm and an AFPTAS to solve the problem Q' of constructing a single-source shortest paths tree using stock pieces of fixed length.

A challenging task for further research is to design a $\frac{3\alpha}{2}$-approximation algorithm to solve the CRS-SPFL problem Q', for which the MCEWSC problem Q is solved by an α-approximation algorithm.

References

1. Chen, D.H., Du, D.Z., Hu, X.D., Lin, G.H., Wang, L.S., Xue, G.: Approximation for Steiner trees with minimum number of Steiner points. J. Global Optim. **18**, 17–33 (2000)
2. Coffman, E.G., Garey, M.R., Johnson, D.S.: Approximation algorithms for bin packing: a survey. In: Hochbaum, D. (ed.) Approximation Algorithms for NP-Hard Problems, pp. 46–93. PWS Publishing, Boston (1996)
3. Garey, M.R., Johnson, D.S.: Computers and Intractability: A Guide to the Theory of NP-Completeness. W.H Freeman, San Francisco (1979)
4. Karmarkar, N., Karp, R.M.: An efficient approximation scheme for the one-dimensional bin-packing problem. In: Proceedings of 23rd Annual IEEE Symposium on Foundations of Computer Science, 3–5 November 1982, pp. 312–320 (1982)
5. Lin, G.H., Xue, G.L.: Steiner tree problem with minimum number of Steiner points and bounded edge-length. Inform. Process. Lett. **69**, 53–57 (1999)
6. Schrijver, A.: Combinatorial Optimization: Polyhedra and Efficiency. Springer, The Netherlands (2003)
7. Simchi-Levi, D.: New worst-case results for the bin-packing problem. Naval Res. Logist. **41**, 579–585 (1994)

A Primal Dual Approximation Algorithm for the Multicut Problem in Trees with Submodular Penalties

Xiaofei Liu[1] and Weidong Li[2,3(⌂)]

[1] School of Electronic Engineering and Computer Science, Peking University,
Beijing 100871, People's Republic of China
[2] School of Mathematics and Statistics, Yunnan University,
Kunming 650504, People's Republic of China
weidongmath@126.com
[3] Dianchi College of Yunnan University,
Kunming 650000, People's Republic of China

Abstract. In this paper, we introduce the multicut problem in trees with submodular penalties, which generalizes the prize-collecting multicut problem in trees and vertex cover with submodular penalties. We present a combinatorial 3-approximation algorithm, based on the primal-dual scheme for the multicut problem in trees.

Keywords: Multicut · Submodular functions ·
Approximation algorithm

1 Introduction

The multicut problem proposed by Hu [7] is one of the most classical and active topics in combinatorial optimization and approximation algorithms for a long time. It has been found a lot of applications in many areas such as telecommunication, routing, transportation and VLSI design [17]. Given an undirected graph with weight edges and a list of vertex pairs, the multicut problem is to find a minimum weight set of edges separating each pair of vertices in the list. Obviously, the multicut problem is a generalization of the classical s-t minimum cut problem. Also, the multicut problem is a generalization of the multiway cut problem [1], which is to find a minimum weight set of edges whose removal separates every pair of a given set of vertices. For the general graphs, the multicut problem admits a polynomial-time $O(\log k)$-approximation algorithm based on the region growing scheme [4], where k is the number of given pairs. For the trees, the multicut problem admits a polynomial-time 2-approximation algorithm which is based on the primal-dual method [5]. In the same paper, the authors proved that the multicut problem in trees is at least as hard to approximate as vertex cover which can not be approximated within $2 - \epsilon$ for any $\epsilon > 0$ under the unique game conjecture [12], implying that the algorithm in [5] maybe the best possible

© Springer Nature Switzerland AG 2019
D.-Z. Du et al. (Eds.): AAIM 2019, LNCS 11640, pp. 203–211, 2019.
https://doi.org/10.1007/978-3-030-27195-4_19

approximation algorithm. Zhang et al. [17] introduced the generalized multicut problem in trees and presented an approximation algorithm. Liu and Zhang [15] introduced the generalized multiway cut problem in trees and showed that it is fixed parameter tractable according to the optimal value, which is improved by Kanj et al. [11].

Levin and Segev [13] proposed the prize-collecting multicut problem in trees, which is a generalization of the multicut problem in trees. In this variant we are not required to separate all pairs. However, if the set of edges we pick does not separate a pair (s_i, t_i), we incur a penalty of π_i. The objective is to find a subset M of edges that minimizes the cost of M plus the penalties of unseparated pairs. If $\pi_i = \infty$ for every pair i, the prize-collecting multicut problem in trees is exactly the multicut problem in trees [5]. Levin and Segev [13] proved that the prize-collecting multicut problem in trees is equivalent to the multicut problem in trees by modifying the original tree and collection of pairs. Thus, the prize-collecting multicut problem admits a 2-approximation algorithm.

In combinatorial optimization, submodular functions play a key role somewhat similar to that played by convex/concave functions in continuous optimization. Submodular function has the property of decreasing marginal return and occur in many mathematical models in operations research, economics, engineering, computer science, and management science [3]. Hayrapetyan et al. [6] introduced the facility location problem with submodular penalties and presented a non-combinatorial 2.50-approximation algorithm. Du et al. [2] designed a combinatorial 3-approximation algorithm for the same problem. More related results can be found in [14].

Recently, Xu et al. [16] introduced the submodular vertex cover problem with submodular penalties, which is to find a vertex subset to cover some edges and penalize the uncovered edges such that the total cost including covering and penalty is minimized. They presented a non-combinatorial 2.54-approximation algorithm by using the ellipsoid method and the techniques proposed in [14], and a combinatorial 4-approximation algorithm by relaxing the dual programs of the linear program relaxations for the primal problem to a slightly weaker version. Subsequently, Kamiyama [10] presented a combinatorial 3-approximation algorithm based on the approximation algorithm of Iwata and Nagano [9] for the submodular cost set cover problem.

Since the multicut in trees is a generalization of the vertex cover problem, it is natural to consider the multicut in trees with submodular penalties (MCTSP, for short), which is to find a subset M of edges that minimizes the cost of M plus the penalty of the set of unseparated pairs which is defined by a submodular function. Clearly, MCTSP is a generalization of the prize-collecting multicut problem in trees proposed in [13]. However, it is not clear whether MCTSP is equivalent to the multicut problem in trees. In this paper, we present a combinatorial 3-approximation algorithm for MCTSP based on the primal-dual scheme in [5].

The remainder of this paper is organized as follows. In Sect. 2, we describe the definition of MCTSP and some preliminaries. In Sect. 3, we give a primal-dual 3-approximation algorithm for MCTSP. We present some conclusions and possible directions for future research in the last section.

2   Preliminaries

We are given a tree $T = (V, E)$, a cost function $c : E \rightarrow Z^+$, a set $P = \{(s_1, t_1),$ $(s_2, t_2), \ldots, (s_k, t_k)\}$ of k source-sink pairs of vertices where $s_i, t_i \in V$, and a monotone increasing submodular function $\pi(\cdot) : 2^P \rightarrow R_+$. Here, the set function $\pi(\cdot)$ is *submodular* implying that

$$\pi(X \cup Y) + \pi(X \cap Y) \leq \pi(X) + \pi(Y), \forall X, Y \subseteq P.$$

The set function $\pi(\cdot)$ is *monotone increasing* implying that

$$\pi(X) \leq \pi(Y), \forall X \subseteq Y \subseteq P.$$

Without loss of generality, assume that $\pi(\emptyset) = 0$. The multicut in trees with submodular penalties (MCTSP, for short) is to find a partial multicut $M \subseteq E$ whose removal disconnects some source-sink pairs. Let R be the set of *rejected* source-sink pairs, which are still connected after removing the edges in M. The objective is to minimize the cost $c(M) = \sum_{e \in M} c(e)$ of the partial multicut M plus the submodular penalty cost $\pi(R)$. Clearly, if $\pi(R) > c(E)$ for arbitrary $R \neq \emptyset$ which implies that any pair cannot be rejected, the MCTSP problem is exactly the multicut problem in trees considered in [5]. If $\pi(R) = \sum_{i:(s_i,t_i) \in R} \pi_i$ where π_i is the penalty cost of rejecting the pair $(s_i, t_i) \in P$, the MCTSP problem is exactly the prize-collecting multicut problem in trees studied in [13].

Let P_i denote the unique path from s_i to t_i in the tree. We introduce a binary variable x_e for each $e \in E$, where

$$x_e = \begin{cases} 1, & \text{if } e \in M, \\ 0, & \text{otherwise.} \end{cases}$$

It implies that

$$c(M) = \sum_{e:e \in M} c(e) = \sum_{e:e \in E} c(e)x_e.$$

The MCTSP problem can be formulated as the following program:

$$\min \sum_{e:e \in E} c(e)x_e + \sum_{R:R \subset P} \pi(R)z_R$$

$$s.t. \sum_{e \in E:e \in P_i} x_e + \sum_{R \subseteq P:(s_i,t_i) \in R} z_R \geq 1, \forall(s_i, t_i) \in P, \qquad (1)$$

$$x_e, z_R \in \{0, 1\}, \forall e \in E, R \subseteq P.$$

The first set of constraints of (1) guarantees that each pair $(s_i, t_i) \in P$ is either disconnected after removing the edges in $M = \{e \in E \mid x_e = 1\}$ or rejected. Due to the submodularity of $\pi(\cdot)$, there must exist exactly one subset $R \subseteq P$ such that $z_R = 1$ in the optimal solution of (1). Relaxing the integer constraints, we obtain

$$\min \sum_{e:e \in E} c(e)x_e + \sum_{R:R \subset P} \pi(R)z_R$$

$$s.t. \quad \sum_{e \in E:e \in P_i} x_e + \sum_{R \subseteq P:(s_i,t_i) \in R} z_R \geq 1, \text{ for } i = 1, 2, \ldots, k, \tag{2}$$

$$x_e, z_R \geq 0, \forall e \in E, R \subseteq P.$$

The corresponding dual program is

$$\max \sum_{i=1}^{k} y_i$$

$$s.t. \quad \sum_{i:e \in P_i} y_i \leq c(e), \quad \forall e \in E.$$

$$\sum_{i:(s_i,t_i) \in R} y_i \leq \pi(R), \quad \forall R \subseteq P. \tag{3}$$

$$y_i \geq 0, \text{ for } i = 1, 2, \ldots, k.$$

For an instance of the MCTSP, let OPT be the optimal value of (1). Similarly, let OPT_R be the optimal value of (2) and OPT_D be the optimal value of (3), respectively. By the famous strong duality theorem, we have

Lemma 1. $OPT_D = OPT_R \leq OPT.$

3   The Prima-Dual Approximation Algorithm

Following the primal-dual scheme designed in [5], we begin by rooting the tree at an arbitrary vertex, say r. The *level* of a vertex is its distance from the root r. The level of root r is 0. The level of an edge $e = (u, v)$ with $u, v \in V$ is the minimum level of u and v. A source-sink pair $(s_i, t_i) \in P$ is *contained in the subtree T_v rooted at v* if the corresponding path P_i lies completely within this subtree. A pair (s_i, t_i) is *contained in level l* if it is contained in a subtree rooted at some vertex in level l. Let l_{max} be the maximum level which contains at least a pair in P.

An edge e_1 is an *ancestor* of an edge e_2 if e_1 lies on the path from e_2 to the root. For each $v \in V$, let P_v be the set of pairs which are contained in the subtree T_v rooted at v. For an edge $e \in E$, it is called *tight* if it satisfies $\sum_{i:e \in P_i} y_i = c(e)$. Similarly, a set $R \subset P$ is called *tight* if it satisfies $\sum_{i:(s_i,t_i) \in R} y_i = \pi(R)$.

Initially, set $M = R = \emptyset$, $y_i = 0$ for $i = 1, 2, \ldots, k$, $P^{disc} = \emptyset$ and $R^{temp} = \emptyset$, where P^{disc} denotes the set of pairs which are disconnected after removing the edges in M, and R^{temp} denotes the set of pairs which are rejected temporarily. The algorithm makes two phases over the tree.

Phase 1. We move up the tree from level l_{max} to level 0, one level at a time, picking some edges (a subset of these edges will be retained as the partial multicut). At each level $l = l_{max}, l_{max} - 1, \ldots, 0$, for every vertex v in level l such that T_v contains at least one pair in $P \setminus (P^{disc} \cup R^{temp})$, let P_v contain all the pair of $P \setminus (P^{disc} \cup R^{temp})$ in the subtree T_v. For each pair $(s_i, t_i) \in P_v$, we increase the dual variable y_i as much as possible. According the constraints of (3), we distinguish the following two cases in this process.

CASE 1. There is an edge $e \in P_i$ becomes tight. Add (s_i, t_i) to the set P^{disc}. If there are two "new" tight edges such that one is an ancestor of the other, then one of these edges is redundant for disconnecting (s_i, t_i). We add the edge that is the ancestor to the set of **frontier**(v), which the *frontier* of vertex v. Else, add all the "new" tight edges (at most two edges) to **frontier**(v). If there are two edges in **frontier**(v) such that one is an ancestor of the other, then we only need the ancestor in the set of **frontier**(v).

CASE 2. There is a subset $R \subseteq P$ becomes tight. Add all the pairs in these "new" tight subsets to the temporary set R^{temp}.

Phase 2. We move down the tree one level from level 0 to level l_{max}, one level at a time, and build the partial multicut M. At each level $l = 0, 1, \ldots, l_{max}$, for every vertex v in level l such that **frontier**$(v) \neq \emptyset$, consider the edges in **frontier**(v). For each edge $e \in$ **frontier**(v), if no edge along the path from e to v is already included in the partial multicut M, add e to M. Finally, for each pair $(s_i, t_i) \in R^{temp}$, if there is no edge $e \in M \cap P_i$, then add (s_i, t_i) to R.

Lemma 2. *The above algorithm can be implemented in polynomial time.*

Proof. Consider the process in **Phase 1**. When we compute the value of y_i corresponding to the pair $(s_i, t_i) \in P_v$, to ensure the first set of constraints of (3), the value of y_i is at most

$$\Delta_{i,e} = \min_{e:e \in P_i} \left(c(e) - \sum_{j:e \in P_j, j \neq i} y_j \right).$$

Clearly, the value of $\Delta_{i,e}$ can be found in polynomial time by computing $c(e) - \sum_{j:e \in P_j, j \neq i} y_j$ for every edge $e \in P_i$.

For any subset $R \subseteq P$, we define $y(R) = \sum_{i:(s_i,t_i) \in R}(-y_i)$. Clearly, $y(\cdot)$ is a modular function. To ensure the second set of constraints of (3), the value of y_i is at most

$$\Delta_{i,R} = \min_{R \subseteq P:(s_i,t_i) \in R} \left(\pi(R) - \sum_{j:(s_j,t_j) \in R, j \neq i} y_j \right) = \min_{R \subseteq P:(s_i,t_i) \in R} (\pi(R) + y(R)),$$

where the last equality follows the fact $y_i = 0$ before increasing the value of y_i. Since $\pi(\cdot)$ is a submodular function and $y(\cdot)$ is a modular function, $\pi(\cdot) + y(\cdot)$ is a submodular function, which is easy to verify. Therefore, the value of $\Delta_{i,R}$ can be found in polynomial time by using the combinatorial algorithm for the submodular minimization problem [8].

This implies that

$$y_i = \min\{\Delta_{i,e}, \Delta_{i,R}\}$$

can be computed in polynomial time. It is easy to verify other processes can be implemented in polynomial time. Thus, the lemma holds.   ∎

Lemma 3. The temporary set R^{temp} of rejected pairs satisfies

$$\pi(R^{temp}) = \sum_{i:(s_i,t_i)\in R^{temp}} y_i. \tag{4}$$

Proof. Let $\mathscr{S}$ be the set of tight subsets of R^{temp} such that for any $S \in \mathscr{S}$, we have $S \subseteq R^{temp}$ and $\pi(S) = \sum_{i:(s_i,t_i)\in S} y_i$. Consider any two different tight subsets S_1 and S_2 in $\mathscr{S}$. By the definition, we have

$$\pi(S_1) = \sum_{i:(s_i,t_i)\in S_1} y_i, \text{ and } \pi(S_2) = \sum_{i:(s_i,t_i)\in S_2} y_i.$$

Therefore,

$$\sum_{i:(s_i,t_i)\in S_1\cup S_2} y_i + \sum_{i:(s_i,t_i)\in S_1\cap S_2} y_i = \sum_{i:(s_i,t_i)\in S_1} y_i + \sum_{i:(s_i,t_i)\in S_2} y_i$$
$$= \pi(S_1) + \pi(S_2)$$
$$\geq \pi(S_1 \cup S_2) + \pi(S_1 \cap S_2)$$
$$\geq \pi(S_1 \cup S_2) + \sum_{i:(s_i,t_i)\in S_1\cap S_2} y_i,$$

where the first inequality follows from the submodularity of $\pi(\cdot)$, and the second inequality follows from the second set of constraints of (3). It implies that

$$\sum_{i:(s_i,t_i)\in S_1\cup S_2} y_i \geq \pi(S_1 \cup S_2).$$

Moreover, $\sum_{i:(s_i,t_i)\in S_1\cup S_2} y_i = \pi(S_1 \cup S_2)$, following from the second set of constraints of (3). It means that $S_1 \cup S_2$ is a tight subset. Repeating merging the tight subsets, we obtain that $\cup_{S\in\mathscr{S}} S$ is a tight subset. As every pair in must be in a tight subset $S \in \mathscr{S}$, $R^{temp} = \cup_{S\in\mathscr{S}} S$ is a tight subset, i.e., $\pi(R^{temp}) = \sum_{i:(s_i,t_i)\in R^{temp}} y_i$. ∎

Lemma 4. For any pair $(s_i, t_i) \in P$, either $|M \cap P_i| \geq 1$ or $(s_i, t_i) \in R$.

Proof. If not, we need to increase the dual variable y_i corresponding to (s_i, t_i) until Case 1 or Case 2 in **Phase 1** occurs. Thus, the lemma holds according to the process in two phases. ∎

Lemma 4 implies that (M, R) is a feasible solution for the MCTSP problem.

Lemma 5. For any pair $(s_i, t_i) \in P$, the partial multicut M satisfies $|M \cap P_i| \leq 2$.

Proof. Similar to the proof of the Lemma 5.1 in [5]. ∎

Theorem 6. (M, R) is a 3-approximation solution for the MCTSP problem.

Proof. For an instance of the MCTSP, the objective value of (M, R) is

$$OUT = \sum_{e:e \in M} c(e) + \pi(R) = \sum_{e:e \in M} \sum_{i:e \in P_i} y_i + \pi(R)$$

$$= \sum_{i=1}^{k} \sum_{e:e \in M \cap P_i} y_i + \pi(R) = \sum_{i=1}^{k} y_i |M \cap P_i| + \pi(R)$$

$$\leq 2 \sum_{i=1}^{k} y_i + \pi(R) \leq 2 \sum_{i=1}^{k} y_i + \pi(R^{temp})$$

$$= 2 \sum_{i=1}^{k} y_i + \sum_{i:(s_i,t_i) \in R^{temp}} y_i \leq 3 \sum_{i=1}^{k} y_i$$

$$\leq 3OPT_D \leq 3OPT,$$

where the first inequality follows from Lemma 5, the second inequality follows from the fact $R \subseteq R^{temp}$, the last equality follows from Lemma 3, and the last inequality follows from Lemma 1.   ∎

4   Conclusion

We introduce the multicut in trees with submodular penalties, which generalizes the vertex cover problem with submodular penalties, the multicut in trees, and the prize-collecting multicut problem in trees. By extending the primal-dual scheme in [5], we obtain a combinatorial 3-approximation algorithm.

Noting that the recent papers [10, 16] studied the submodular vertex cover problem with submodular penalties, it is interesting to consider the submodular multicut in trees with submodular penalties, which can be formulated as

$$\min \quad \sum_{M:M \subseteq E} c(M)x_M + \sum_{R:R \subseteq P} \pi(R)z_R$$

$$s.t. \quad \sum_{M \subseteq E: P_i \cap M \neq \emptyset} x_M + \sum_{R \subseteq P:(s_i,t_i) \in R} z_R \geq 1, \text{ for } i = 1, 2, \ldots, k,$$

$$x_M, z_R \in \{0, 1\}, \forall M \subseteq E, R \subseteq P.$$

Here, both $c(\cdot)$ and $\pi(\cdot)$ are submodular functions. Relaxing the integer constraints, we obtain

$$\min \quad \sum_{M:M \subseteq E} c(M)x_M + \sum_{R:R \subseteq P} \pi(R)z_R$$

$$s.t. \quad \sum_{M \subseteq E: P_i \cap M \neq \emptyset} x_M + \sum_{R \subseteq P:(s_i,t_i) \in R} z_R \geq 1, \text{ for } i = 1, 2, \ldots, k,$$

$$x_M, z_R \geq 0, \forall M \subseteq E, R \subseteq P.$$

The corresponding dual program is

$$\max \ \sum_{i=1}^{k} y_i$$

$$s.t. \ \sum_{i:P_i \cap M \neq \emptyset} y_i \leq c(M), \quad \forall M \subseteq E.$$

$$\sum_{i:(s_i,t_i)\in R} y_i \leq \pi(R), \quad \forall R \subseteq P.$$

$$y_i \geq 0, \ \text{for} \ i = 1, 2, \ldots, k.$$

The combinatorial 3-approximation algorithm for the submodular vertex cover problem with submodular penalties [10] can not be extended to the submodular multicut in trees with submodular penalties directly. It is challenging to design a novel combinatorial 3-approximation algorithm.

Acknowledgements. The work is supported in part by the National Natural Science Foundation of China [No. 61662088], Program for Excellent Young Talents of Yunnan University, Training Program of National Science Fund for Distinguished Young Scholars, IRTSTYN, and Key Joint Project of the Science and Technology Department of Yunnan Province and Yunnan University [No. 2018FY001(-014)].

References

1. Dahlhaus, E., Johnson, D.S., Papadimitriou, C.H., Seymour, P.D., Yannakakis, M.: The complexity of multiterminal cuts. SIAM J. Comput. **23**(4), 864–894 (1994)
2. Du, D., Lu, R., Xu, D.: A primal-dual approximation algorithm for the facility location problem with submodular penalties. Algorithmica **63**, 191–200 (2012)
3. Fujishige, S.: Submodular Functions and Optimization, 2nd edn. Elsevier, Amsterdam (2005)
4. Garg, N., Vazirani, V.V., Yannakakis, M.: Approximate max-flow min-(multi) cut theorems and their applications. SIAM J. Comput. **25**(2), 235–251 (2006)
5. Garg, N., Vazirani, V.V., Yannakakis, M.: Primal-dual approximation algorithms for integral flow and multicut in trees. Algorithmica **18**(1), 3–20 (1997)
6. Hayrapetyan, A., Swamy, C., Tardos, E.: Network design for information networks. In: Proceedings of the Sixteenth Annual ACM-SIAM Symposium on Discrete Algorithms, pp. 933–942 (2005)
7. Hu, T.C.: Integer Programming and Network Flows. Addison-Wesley, Reading (1969)
8. Iwata, S., Fleischer, L., Fujishige, S.: A combinatorial strongly polynomial algorithm for minimizing submodular functions. J. ACM **48**(4), 761–777 (2001)
9. Iwata, S., Nagano, K.: Submodular function minimization under covering constraints. In: The 50th Annual Symposium on Foundations of Computer Science, FOCS, pp. 671–680 (2009)
10. Kamiyama, N.: A note on the submodular vertex cover problem withsubmodular penalties. Theor. Comput. Sci. **659**, 95–97 (2017)
11. Kanj, I., et al.: Improved parameterized and exact algorithms for cut problems on trees. Theor. Comput. Sci. **607**, 455–470 (2015)

12. Khot, S., Regev, O.: Vertex cover might be hard to approximateto with $2 - \epsilon$. J. Comput. Syst. Sci. **74**(3), 335–349 (2008)
13. Levin, A., Segev, D.: Partial multicuts in trees. Theor. Comput. Sci. **369**(1–3), 384–395 (2006)
14. Li, Y., Du, D., Xiu, N., Xu, D.: Improved approximation algorithms for the facility location problems with linear/submodular penalties. Algorithmica **73**(2), 460–482 (2015)
15. Liu, H., Zhang, P.: On the generalized multiway cut in trees problem. J. Comb. Optim. **27**(1), 65–77 (2014)
16. Xu, D., Wang, F., Du, D., Wu, C.: Approximation algorithms for submodular vertex cover problems with linear/submodular penalties using primal-dual technique. Theor. Comput. Sci. **630**, 117–125 (2016)
17. Zhang, P., Zhu, D., Luan, J.: An approximation algorithm for the generalized k-multicut problem. Discrete Appl. Math. **160**(7–8), 1240–1247 (2012)

Algorithmic Aspect on the Minimum (Weighted) Doubly Resolving Set Problem of Graphs

Changhong Lu$^{(\boxtimes)}$, Qingjie Ye, and Chengru Zhu

School of Mathematical Sciences, Shanghai Key Laboratory of PMMP, East China Normal University, Shanghai 200241, People's Republic of China
chlu@math.ecnu.edu.cn, mathqjye@qq.com

Abstract. Let G be a simple graph, where each vertex has a nonnegative weight. A vertex subset S of G is a doubly resolving set (DRS) of G if for every pair of vertices u, v in G, there exist $x, y \in S$ such that $d(x, u) - d(x, v) \neq d(y, u) - d(y, v)$. The minimum weighted doubly resolving set (MWDRS) problem is finding a doubly resolving set with minimum total weight. We establish a linear time algorithm for the MWDRS problem of all graphs in which each block is complete graph or cycle. Hence, the MWDRS problems for block graphs and cactus graphs can be solved in linear time. We also prove that k-edge-augmented tree (a tree with additional k edges) with minimum degree $\delta(G) \geq 2$ admits a doubly resolving set of size at most $2k + 1$. This implies that the DRS problem on k-edge-augmented tree can be solved in $O(n^{2k+3})$ time.

Keywords: Doubly resolving set · Block graph · Cactus graph · k-edge-augmented trees

1 Introduction

Let G be a finite, connected, simple and undirected graph with vertex set $V = V(G)$ and edge set $E = E(G)$. The distance between vertices u and v is denoted by $d(u, v)$. The minimum degree of G is denoted by $\delta(G)$.

A set of vertices S resolves a graph G if every vertex is uniquely determined by its vector of distances to the vertices in S. Resolving sets were independently defined in the 1970s by Slater [16], Harary and Melter [9]. Resolving sets have since been widely investigated, see [1,3,6,7,10].

Locating the source of a diffusion in complex networks is an intriguing challenge [8,14,15]. Placing an observer at vertex v incurs a cost, and the observer with a clock can record the time at which the state of v is changed. Typically, the time when the single source originates an information is unknown. The observers

Supported in part by National Natural Science Foundation of China (Nos. 11371008, 11871222) and Science and Technology Commission of Shanghai Municipality (No. 18dz2271000).

D.-Z. Du et al. (Eds.): AAIM 2019, LNCS 11640, pp. 212–222, 2019.
https://doi.org/10.1007/978-3-030-27195-4_20

can only report the times they receive the information. The information is diffused from the source to any vertex through shortest paths in the network. Our goal is to select a subset S of vertices with minimum total cost such that the source can be uniquely located by the "infected" times of vertices in S. This problem is equivalent to finding a minimum weighted doubly resolving set in networks [4].

Doubly resolving sets were introduced by Cáceres et al. [2] as a tool for researching resolving sets. Let G be a graph of order $n \geq 2$, and each vertex $v \in V$ has a nonnegative *weight* $w(v)$. We say that $\{x, y\}$ doubly resolves $\{u, v\}$, if

$$d(x, u) - d(x, v) \neq d(y, u) - d(y, v).$$

For any $S \subseteq V$, we define $w(S) = \sum_{s \in S} w(s)$. A vertex subset S of G is a *doubly resolving set* (DRS) of G if every pair of vertices in G is doubly resolved by some pair of vertices in S. The *minimum weighted doubly resolving set* (MWDRS) problem is finding a doubly resolving set with minimum total weight. In the special case where all vertex weights are equal to 1, the problem is referred to as the minimum doubly resolving set (MDRS) problem.

As far as general graphs are concerned, the MDRS problem has been proved to be NP-hard [11]. A polynomial time algorithm was given by Chen et al. [5] to find the MWDRS of given trees, cycles, wheels and k-edge-augmented tree (a tree with additional k edges).

Other researchers have contributed to the other special graphs, such as Hamming graphs [12], prism graphs [17] and convex polytopes [13].

A *cut-vertex* is a vertex whose removal increases the number of connected components. A graph G is *2-connected* if $|V| > 2$ and G is a connected graph that has no cut-vertex. A *block* of a graph G is a maximal connected subgraph of G that has no cut-vertex. If each block of G is a clique, then G is a *block graph*. If each block of G is either an edge or a cycle, then G is a *cactus graph*. Note that a tree is a block graph and also is a cactus graph since each block of a tree is K_2.

In Sect. 2, we give a main theorem (Theorem 2) concerning on the MDRS problem for graphs with cut-vertices. From Theorem 2, we can establish a linear time algorithm for the MWDRS problem of G if each block of G is complete graph or cycle. This implies that the MWDRS problem for block graphs and cactus graphs can be solved in linear time. In Sect. 3, We prove that k-edge-augmented tree with minimum degree $\delta(G) \geq 2$ admits a doubly resolving set of size at most $2k + 1$ by using an algorithmic proof. This result implies that the DRS problem on k-edge-augmented tree can be solved in $O(n^{2k+3})$ time.

2   Algorithm for Graphs with Cut-Vertices

In this section, we research the graph with cut-vertices. If G has a cut-vertex r, then G can divide into two graphs G_1 and G_2.

Lemma 1. *Let* $G_1 = (V_1, E_1), G_2 = (V_2, E_2), G = (V_1 \cup V_2, E_1 \cup E_2)$, *where* $V_1 \cap V_2 = \{r\}$.

(a) If S_i is a DRS of $G_i(i = 1, 2)$, then $S = S_1 \cup S_2 \backslash \{r\}$ is a DRS of G.
(b) If S is a DRS of G, then $(S \cap V_i) \cup \{r\}$ is a DRS of $G_i(i = 1, 2)$.
(c) If S is a DRS of G, then $S \backslash \{r\}$ is also a DRS of G.

Proof.(a) First observe that if $x \in V_1, y \in V_2$, then $d(x, y) = d(x, r) + d(r, y)$. If x, y belong to the same V_i, say $x, y \in V_1$. Since S_1 is a DRS of G_1, there exist $u, v \in S_1$, such that $d(u, x) - d(u, y) \neq d(v, x) - d(v, y)$. If $u, v \neq r$, then $u, v \in S$ and we are done. Otherwise, without loss of generality, we assume that $u = r$. Choose $z \in S_2 \backslash \{r\}$, then

$$d(z, x) - d(z, y) = d(z, u) + d(u, x) - (d(z, u) + d(u, y)) = d(u, x) - d(u, y),$$

saying $\{x, y\}$ is doubly resolved by $\{z, v\}$.
If x, y belong to the different V_i, say $x \in V_1, y \in V_2$. Since S_1 is a DRS of G_1, there exist $u, v \in S_1$, such that $d(u, x) - d(v, x) \neq d(u, r) - d(v, r)$. If $u, v \neq r$, then $u, v \in S$ and

$$d(u, y) - d(v, y) = d(u, r) + d(r, y) - (d(v, r) + d(r, y)) = d(u, r) - d(v, r),$$

saying $\{x, y\}$ is doubly resolved by $\{u, v\}$. Otherwise, without loss of generality, we assume that $u = r$. We have $d(v, x) - d(r, x) \neq d(v, r)$. Choose $z \in S_2 \backslash \{r\}$, then

$$d(v, x) - d(z, x) < d(v, r) + d(r, x) - (d(z, r) + d(r, x)) = d(v, r) - d(z, r),$$

$$d(v, y) - d(z, y) \geq d(v, r) + d(r, y) - (d(z, r) - d(r, y)) = d(v, r) - d(z, r),$$

saying $\{x, y\}$ is doubly resolved by $\{z, v\}$.
(b) For each $x, y \in V_1$, there exist $u, v \in S$, such that $d(u, x) - d(u, y) \neq d(v, x) - d(v, y)$. If $u, v \in S_1$, we have finished. If $u, v \in S_2$, then

$$d(u, x) - d(u, y) = d(u, r) + d(r, x) - (d(u, r) + d(r, y)) = d(r, x) - d(r, y),$$

$$d(v, x) - d(v, y) = d(v, r) + d(r, x) - (d(v, r) + d(r, y)) = d(r, x) - d(r, y),$$

a contradiction. If $u \in S_1, v \in S_2$, then

$$d(v, x) - d(v, y) = d(v, r) + d(r, x) - (d(v, r) + d(r, y)) = d(r, x) - d(r, y),$$

saying $\{x, y\}$ is doubly resolved by $\{u, r\}$.
(c) If $r \notin S$, we have done. If $r \in S$, then $S_i = S \cap V_i$ is a DRS of G_i by (b). Thus, $S \backslash \{r\} = S_1 \cup S_2 \backslash \{r\}$ is a DRS of G by (a).
$\square$

According to Lemma 1, when we consider the doubly resolving set problem in each block of G, we can assume that cut-vertices are belong to S. Thus, we have to introduce the following definition.

Given $D \subseteq V$, we define S to be a *D-doubly resolving set* (D-DRS) of G if $D \subseteq S$ and S is a DRS of G. If S is a D-DRS of G with minimum total weight, then S is a *D-minimum weighted doubly resolving set* (D-MWDRS). In particular, if $D = \emptyset$, then D-DRS is equivalent to DRS.

Theorem 1. *Let* $G_1 = (V_1, E_1), G_2 = (V_2, E_2), G = (V_1 \cup V_2, E_1 \cup E_2)$, *where* $V_1 \cap V_2 = \{r\}$. *Let* D *be a vertex subset of* G *and* $D_i = D \cap V_i (i = 1, 2)$.

(a) If $r \notin D$ *and* S_i *is a* $D_i \cup \{r\}$-*MWDRS of* G_i *(i = 1, 2), then* $S_1 \cup S_2 \backslash \{r\}$ *is a* D-*MWDRS of* G.

(b) If $r \in D$ *and* S_i *is a* D_i-*MWDRS of* G_i *(i = 1, 2), then* $S_1 \cup S_2$ *is a* D-*MWDRS of* G.

Proof. If T is a D-DRS of G, then $(T \cap V_i) \cup \{r\}$ is a $D_i \cup \{r\}$-DRS of G_i by Lemma 1(b). Since S_i is a $D_i \cup \{r\}$-MWDRS of G_i, we have

$$w(T \backslash \{r\}) = \sum_{i=1}^{2} [w((T \cap V_i) \cup \{r\}) - w(r)] \geq \sum_{i=1}^{2} [w(S_i) - w(r)] = w(S_1 \cup S_2 \backslash \{r\}).$$

Lemma 1(a) shows that $S_1 \cup S_2 \backslash \{r\}$ is a $D \backslash \{r\}$-DRS of G. Therefore, when $r \notin D$, $S_1 \cup S_2 \backslash \{r\}$ is a D-MWDRS of G. If $r \in D$, then $r \in T$ and $T \backslash \{r\}$ is a $D \backslash \{r\}$-DRS of G. In this case, we have

$$w(T) = w(r) + w(T \backslash \{r\}) \geq w(r) + w(S_1 \cup S_2 \backslash \{r\}) = w(S_1 \cup S_2).$$

Thus, $S_1 \cup S_2$ is a D-MWDRS of G when $r \in D$. □

If G is not 2-connected, then there are at least two end-blocks that each contains exactly one cut-vertex of G. Finding all blocks of G can be finished by Breadth First Search algorithm in linear time.

Theorem 2. *Let* $G = (V, E)$ *be a graph which is not 2-connected. Let* $V = \bigcup_{i=1}^{p} V_i$ *such that* $G_i = G[V_i]$ *is a block of* G *and* $E(G_i) \cap E(G_j) = \emptyset (i \neq j)$. *Let* R *be the set of all cut-vertices of* G *and* $R_i = R \cap V_i$ *(i = 1, ..., p). Let* D *be a vertex subset of* G *and* $D_i = D \cap V_i (i = 1, ..., p)$. *Let* $U = R \cap D$ *and* S_i *be a* $(D_i \cup R_i)$-*MWDRS of* $G_i (i = 1, ..., p)$. *Then* $S = \bigcup_{i=1}^{p} (S_i \backslash R_i) \cup U$ *is a* D-*MWDRS of* G.

Proof. We use induction on p. For $p = 1$, it is trivial. For $p > 1$, without loss of generality, we assume that G_p is an end-block which means $|R_p| = 1$. Let $R_p = \{r\}, H = G\left[\bigcup_{i=1}^{p-1} V_i\right]$. Let $D_H = (D \cap V(H)) \cup \{r\}$ and $U_H = D_H \cap U$. By induction, $S' = \bigcup_{i=1}^{p-1} (S_i \backslash R_i) \cup U_H$ is a D_H-MWDRS of H. Applying Theorem 1 for H and G_p, we have finished the proof. □

From Theorem 2, we can establish a linear time algorithm for the MWDRS problem of G if the D-MWDRS problem for each block of G can be computed in linear time. Now we focus on the graph in which each block is a complete graph or a cycle.

Lemma 2. *Let* K_n *be a complete graph of order* $n \geq 2$. *Let* D *be a vertex subset of* K_n *and* u *be the vertex with minimum weight in* $V \backslash D$. *Then* S *is a* D-*MWDRS of* K_n, *where*

$$S = \begin{cases} V & \text{if } n = 2 \text{ or } D = V, \\ V \backslash \{u\} & \text{if } n > 2 \text{ and } D \neq V. \end{cases}$$

Proof. It is trivial if $n = 2$ or $D = V$. Now we assume that $n \geq 3$ and $D \neq V$. Let x, y be the two distinct vertices of S. Then $\{x, y\}$ can doubly resolves $\{x, y\}$, since $d(x, x) - d(x, y) = -1$ and $d(y, x) - d(y, y) = 1$. Besides, $\{x, y\}$ doubly resolves $\{u, x\}$ since $d(u, x) - d(u, y) = 0$ and $d(x, x) - d(x, y) = -1$. Thus S is a D-DRS of K_n. Let T be a D-DRS of G. If $|T| \leq n - 2$, then for any $u, v \notin T$ and $x, y \in T$, we have $d(u, x) - d(u, y) = d(v, x) - d(v, y) = 0$, a contradiction. By the definition of u, we have $w(T) \geq w(S)$. Thus S is a D-MWDRS of K_n. □

Recall that the block is a graph in which each block is a complete graph. By Theorem 2 and Lemma 2, we get the following theorem.

Theorem 3. *The MWDRS problem on block graphs can be solved in $O(|V|+|E|)$ time.*   □

Now we consider the D-MWDRS problem for cycles. Chen et al. [5] proved the following lemma and corollary.

Lemma 3 ([5]). *Let C_n be a cycle of order n and S be a subset of V with at least two vertices. Let $\mathcal{P}_S$ be a set of edge-disjoint paths such that they are internally disjoint from S and their union is C_n. Then S is a DRS of C_n if and only if no path $\mathcal{P}_S$ has length longer than $\lceil n/2 \rceil$ and at least one path in $\mathcal{P}_S$ has length shorter than $n/2$.*

Corollary 1 ([5]). *If an MWDRS of C_n has cardinality 3, then there exists an MWDRS that contains the minimum weight vertex in C_n.*

We design the Algorithm 1 to solve the D-MWDRS problem of cycle C_n. Now we prove the correctness of our algorithm.

Lemma 4. *Algorithm 1 finds a D-MWDRS of C_n in $O(n)$ time.*

Proof. It is easy to check the algorithm runs in $O(n)$ time by Lemma 3. We need to prove the correctness of the algorithm.

If $|D| \leq 1$, then $|S| \in \{2, 3\}$ by Lemma 3. If $|S| = 2$, the only situation is that n is odd and $|D| \leq 1$. This situation is dealt in lines 4–5 and 8–9. If n is even, then for each $u \in V - \{v_1, v_{1+n/2}\}$, $S = \{u, v_1, v_{1+n/2}\}$ is a DRS of C_n. This situation is dealt in lines 10–11. If $|S| = 3$ and $|D| \leq 1$, then $v_1 \in S$ by Corollary 1 and the definition. Assume $S = \{v_1, v_j, v_i\}$ with $j < i$. Then $2 \leq j \leq \lfloor n/2 \rfloor$ and $2 + \lceil n/2 \rceil \leq i \leq j + \lceil n/2 \rceil$ since $j = 1 + \lfloor n/2 \rfloor$ or $i = 1 + \lceil n/2 \rceil$ has been consider before. This situation is dealt in lines 12–17. Note that our algorithm confirms that v_i is the minimum weight vertex in $\{v_{2+\lceil n/2 \rceil}, \ldots, v_{j+\lceil n/2 \rceil}\}$.

If $|D| \geq 2$, then $|S| \in \{|D|, |D| + 1, |D| + 2\}$ by Lemma 3. If $|S| = |D|$, then D is a D-MWDRS. Algorithm would stop in line 21. We deal the situation if $|S| = |D| + 1$ in lines 23–26 and deal the situation if $|S| = |D| + 2$ in lines 27–32. Note that if $|S| = |D| + 2$ then $l \geq 3 + \lceil n/2 \rceil$.   □

Recall that if each block of G is either an edge or a cycle, then G is a *cactus graph*. By Theorem 2 and Lemma 4, we get the following theorem.

Algorithm 1. Finding a D-MWDRS in cycle C_n.

Input: A cycle $C_n = (V, E, w)$ and a vertex set $D \subseteq V$.

Output: The D-MWDRS S.

1 **if** $|D| \leq 1$ **then**
2   **if** $|D| = 0$ **then**
3   let $G := v_1 v_2 \ldots v_n v_1$ with $w(v_1) := \min\{w(v)|v \in V\}$;
4   **if** n *is odd* **then**
5           $S := \arg\min w(\{v_i, v_{i+(n-1)/2}\}), W := w(S)$;
6   **else if** $|D| = 1$ **then**
7   let $G := v_1 v_2 \ldots v_n v_1$, where v_1 is the vertex in D;
8   **if** n *is odd* **then**
9              $S := \arg\min\{w(\{v_1, v_{(n+1)/2}\}), w(\{v_1, v_{(n+3)/2}\})\}, W := w(S)$;
10   **if** n *is even* **then**
11          $u := \arg\min\{w(v)|v \in G - \{v_1, v_{1+n/2}\}\}, S := \{u, v_1, v_{1+n/2}\}, W := w(S)$;
12       $i := 2 + \lceil n/2 \rceil$;
13   **for** $j := 2$ **to** $\lfloor n/2 \rfloor$ **do**
14   **if** $w(v_{j+\lceil n/2 \rceil}) < w(v_i)$ **then**
15             $i := j + \lceil n/2 \rceil$;
16   **if** $w(v_j) + w(v_i) + w(v_1) < W$ **then**
17             $S := \{v_1, v_j, v_i\}, W := w(S)$;
18 **else**
19   **if** D *is a DRS of* C_n **then**
20       $S := D$;
21   **return**;
22   let $G := v_1 v_2 \ldots v_n v_1$ with v_1, v_l are endpoints of the longest path in $\mathcal{P}_D$;
23   **if** n *is even* **and** $D = \{v_1, v_{1+n/2}\}$ **then**
24       $u := \arg\min\{w(v)|v \in V - D\}, S := \{u, v_1, v_{1+n/2}\}, W := w(S)$;
25   **else**
26       $u := \arg\min\{w(v)|v \in \{v_{l-\lceil n/2 \rceil}, \ldots, v_{1+\lceil n/2 \rceil}\}\}, S := \{u\} \cup D, W := w(S)$;
27       $i := 2 + \lceil n/2 \rceil$;
28   **for** $j := 2$ **to** $l - 1 - \lceil n/2 \rceil$ **do**
29   **if** $w(v_{j+\lceil n/2 \rceil}) < w(v_i)$ **then**
30             $i := j + \lceil n/2 \rceil$;
31   **if** $w(v_j) + w(v_i) + w(D) < W$ **then**
32             $S := \{v_j, v_i\} \cup D, W := w(S)$;

Theorem 4. *The MWDRS problem on cactus graphs can be solved in $O(|V| + |E|)$ time.* □

In fact, by Theorem 2 and Lemmas 2, 4, the MWDRS problem can be solved in linear time when each block of G is cycle or complete graph. Here is an example.

Example 1. Consider the graph of Fig. 1 which all vertex weights is 1. A solid black circle means cut-vertex. It is easy to see that the set of all hollow circle is an MWDRS of G.

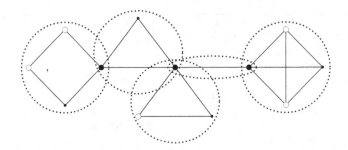

Fig. 1. A graph G with five blocks.

3 k-Edge-Augmented Trees

Recall that a connected graph is called a k-edge-augmented tree if the removal of at most k edges from the graph leaves a spanning tree. In [5], the following theorem is proved.

Theorem 5 ([5]). *For $k \geq 2$, let G be a k-edge-augmented tree with $\delta(G) \geq 2$. If S is an MWDRS of G, then $|S| \leq 12k - 12$.*

In order to get a better upper bound when G is an unweighted graph, we design the Algorithm 2.

Theorem 6. *For $k \geq 1$, let G be a k-edge-augmented tree with $\delta(G) \geq 2$. Then we can get a doubly resolving set S of G such that $|S| \leq 2k + 1$ by Algorithm 2 in $O(|V| + |E|)$ time.*

Proof. Our algorithm use the technique of breadth-first search(BFS) except lines 18–26. Because each vertex change the key value at most once and lines 18–26 run k times, the total time spent in lines 18–26 is $O(k + |V|) = O(|V| + |E|)$. Since the breadth-first search takes $O(|V| + |E|)$ time, the total running time is $O(|V| + |E|)$.

Since line 18 runs k times and line 8 runs once, the cardinality of S is at most $1 + 2k$.

Claim. For each vertex v, $d(u, v.key) = d(u, v) + d(v, v.key)$.

Proof. Since each edge belongs to a cycle by $\delta(G) \geq 2$, $v.key \neq NIL$. Let $v_0 = v$, $v_{i+1} = v_i.\pi$ and $P_{v,u} = v_0 v_1 \ldots v_{d(u,v)}$. According to the property of BFS, $P_{v,u}$ is a shortest path from v to u. Because v belongs to $P_{v.key,u}$, we have $d(u, v.key) = d(u, v) + d(v, v.key)$.   ($\square$)

Let v, w be the two distinct vertices of V. Suppose without loss of generality that $d(u, v) \leq d(u, w)$. If $\{u, w.key\}$ doubly resolves $\{v, w\}$, we have done. Otherwise, we have

$$d(v, w.key) - d(v, u) = d(w, w.key) - d(w, u).$$

Algorithm 2. Finding a DRS in the k-edge-augmented tree G with $\delta(G) \geq 2$.

Input: A k-edge-augmented tree $G = (V, E)$ with $\delta(G) \geq 2$ and a vertex u.

Output: A doubly resolving set S.

```
1  for each vertex v ∈ V do
2      v.visit := false;
3      v.π := NIL;
4      v.key := NIL;
5  u.visit := true;
6  u.key := u;
7  put u in a queue;
8  S := {u};
9  while the queue is not empty do
10     remove the first vertex v from queue;
11     for all unmarked edge (v, w) do
12         mark (v, w);
13         if w.visit = false then
14             w.visit := true;
15             w.π := v;
16             put w in a queue;
17         else
18             S := S ∪ {v, w};
19             x := v;
20             while x.key = NIL do
21                 x.key := v;
22                 x := x.π;
23             x := w;
24             while x.key = NIL do
25                 x.key := w;
26                 x := x.π;
```

Since

$$d(u, v) + d(v, w.key)$$
$$\geq d(u, w.key)$$
$$= d(u, w) + d(w, w.key)$$
$$= d(u, v) + d(v, w.key) + (d(u, w) - d(u, v)) + (d(w, w.key) - d(v, w.key))$$
$$= d(u, v) + d(v, w.key) + 2(d(u, w) - d(u, v))$$
$$\geq d(u, v) + d(v, w.key),$$

we have $d(u, w) = d(u, v)$ and $d(v, w.key) = d(w, w.key)$. Let x be the vertex such that $d(v, w.key) = d(v, x) + d(x, w.key)$ and $x.key = w.key$ with $d(v, x)$ minimal. Let y be the vertex such that $d(v, x) = d(v, y) + d(y, x)$ and $(x, y) \in E$. By definition, $y \neq x.\pi$. Thus, when (x, y) visits, it leads to $x, y \in S$. If $x \neq w.key$, then $d(u, x) < d(u, w.key)$, a contradiction. If $x = w.key$ and $d(w, y) = d(v, y)$, then since $d(u, y) < d(u, w.key)$, it leads to a contradiction. Thus, $y \in S$ and $d(w, y) \neq d(v, y)$. We have

$$d(v, w.key) - d(v, y) = d(w, w.key) - d(v, y) \neq d(w, w.key) - d(w, y).$$

It implies that $\{w.key, y\}$ doubly resolves $\{v, w\}$.   □

Theorem 6 implies that the DRS problem on k-edge-augmented tree can be solved in $O(n^{2k+3})$ time. Firstly, we can compute S by Algorithm 2 in $O(n + k)$ time. Secondly, if S isn't an MDRS, then the MDRS must belongs to $U = \{T \subseteq V | 2 \leq |T| < |S|\}$. Finally, since we can verify whether $T \in U$ is a DRS in $O(n^3)$ time and $|U| \leq n^{2k}$, we can solve it in $O(n^{2k+3})$ time.

Corollary 2. *Let G be a k-edge-augmented tree with $\delta(G) \geq 2$ and $k \geq 2$. If G has a cut-vertex u, then there exists a doubly resolving set S with $|S| \leq 2k$.*

Proof. By Theorem 6, we can get a doubly resolving set S with $u \in S$ and $|S| \leq 1 + 2l$. By Lemma 1(c), $T = S \backslash \{u\}$ is also a doubly resolving set of G and $|T| \leq 2k$.   □

The Corollary 2 is tight. To see this, let T_k be the graph arises from k C_4 by pasting these graphs together along a vertex. The graph of T_4 is shown in Fig. 2. It is easy to check that $dr(T_k) = 2k$ by Theorem 4.

In fact, we can prove that Theorem 6 is not tight when $k = 2$.

Theorem 7. *If G be a 2-edge-augmented tree with $\delta(G) \geq 2$, then there exists a doubly resolving set S with $|S| \leq 4$.*

Proof. If G is a cut-vertex, we have done by Corollary 2. Thus we assume that G is a 2-connected graph. By ear decomposition theorem, G can be decomposed three paths P_1, P_2, P_3 with same end vertices s, t. Suppose without loss of generality that $|P_1| \leq |P_2| \leq |P_3|$. Let $P_1 = su_1 \ldots u_a t, P_2 = sv_1 \ldots v_b t, P_3 = sw_1 \ldots w_c t$. Figure 3 is shown the graph when $a = 1, b = 3, c = 5$.

Let $S = \{s, t, v_{\lceil b/2 \rceil}, w_{\lceil c/2 \rceil}\}$, then we prove that S is a DRS of G. Firstly, since $C_1 = sv_1 \ldots v_b t u_a \ldots u_1 s, C_2 = sw_1 \ldots w_c t u_a \ldots u_1 s$ satisfying $d_G(x, y) =$

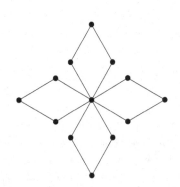

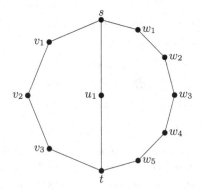

Fig. 2. The graph T_4

Fig. 3. A 2-connected 2-edge-augmented tree

$d_{C_i}(x, y)$ when $x, y \in C_i$. According to Lemma 3, S can doubly resolving $\{x, y\}$. Therefore, we just need to prove that $\{v_i, w_j\}$ can be doubly resolved.

Since $d(v_i, w_{\lceil c/2 \rceil}) > \lceil c/2 \rceil \geq d(w_j, w_{\lceil c/2 \rceil})$ and $d(v_i, v_{\lceil b/2 \rceil}) \leq \lceil b/2 \rceil < d(w_j, v_{\lceil b/2 \rceil})$, we have

$$d(v_i, v_{\lceil b/2 \rceil}) - d(v_i, w_{\lceil b/2 \rceil}) < \lceil b/2 \rceil - \lceil c/2 \rceil < d(w_j, v_{\lceil c/2 \rceil}) - d(w_j, w_{\lceil c/2 \rceil}).$$

That implies that $\{v_i, w_j\}$ can be doubly resolved by $\{v_{\lceil b/2 \rceil}, w_{\lceil c/2 \rceil}\}$.   □

4   Open Problem

We pose the following conjecture that we have yet to settle.

Conjecture 1. For $k \geq 2$, if G be a k-edge-augmented tree with $\delta(G) \geq 2$, then there exists a doubly resolving set S with $|S| \leq 2k$.

We remark that if $k = 2$ in the statement of Conjecture 1, then the conjecture is true by Theorem 7. When $k \geq 3$, we have yet to settle the conjecture when G is a 2-connected graph by Corollary 2.

References

1. Bailey, R.F., Cameron, P.J.: Base size, metric dimension and other invariants of groups and graphs. Bull. Lond. Math. Soc. **43**(2), 209–242 (2011)
2. Cáceres, J., et al.: On the metric dimension of Cartesian products of graphs. SIAM J. Discrete Math. **21**(2), 423–441 (2007)
3. Chartrand, G., Eroh, L., Johnson, M.A., Oellermann, O.R.: Resolvability in graphs and the metric dimension of a graph. Discrete Appl. Math. **105**(1–3), 99–113 (2000)
4. Chen, L., Lu, C., Zeng, Z.: Labelling algorithms for paired-domination problems in block and interval graphs. J. Comb. Optim. **19**(4), 457–470 (2010)
5. Chen, X., Wang, C.: Approximability of the minimum weighted doubly resolving set problem. In: Cai, Z., Zelikovsky, A., Bourgeois, A. (eds.) COCOON 2014. LNCS, vol. 8591, pp. 357–368. Springer, Cham (2014). https://doi.org/10.1007/978-3-319-08783-2_31
6. Díaz, J., Pottonen, O., Serna, M., van Leeuwen, E.J.: On the complexity of metric dimension. In: Epstein, L., Ferragina, P. (eds.) ESA 2012. LNCS, vol. 7501, pp. 419–430. Springer, Heidelberg (2012). https://doi.org/10.1007/978-3-642-33090-2_37
7. Epstein, L., Levin, A., Woeginger, G.J.: The (weighted) metric dimension of graphs: hard and easy cases. Algorithmica **72**(4), 1130–1171 (2015)
8. Gomez-Rodriguez, M., Leskovec, J., Krause, A.: Inferring networks of diffusion and influence. ACM Trans. Knowl. Discov. Data (TKDD) **5**(4), 21 (2012)
9. Harary, F., Melter, R.: On the metric dimension of a graph. Ars Comb. **2**(191–195), 1 (1976)
10. Khuller, S., Raghavachari, B., Rosenfeld, A.: Landmarks in graphs. Discrete Appl. Math. **70**(3), 217–229 (1996)
11. Kratica, J., Čangalović, M., Kovačević-Vujčić, V.: Computing minimal doubly resolving sets of graphs. Comput. Oper. Res. **36**(7), 2149–2159 (2009)

12. Kratica, J., Kovačević-Vujčić, V., Čangalović, M., Stojanović, M.: Minimal doubly resolving sets and the strong metric dimension of Hamming graphs. Appl. Anal. Discrete Math. **6**(1), 63–71 (2012)
13. Kratica, J., Kovačević-Vujčić, V., Čangalović, M., Stojanović, M.: Minimal doubly resolving sets and the strong metric dimension of some convex polytopes. Appl. Math. Comput. **218**(19), 9790–9801 (2012)
14. Pinto, P.C., Thiran, P., Vetterli, M.: Locating the source of diffusion in large-scale networks. Phys. Rev. Lett. **109**(6), 068702 (2012)
15. Shah, D., Zaman, T.: Rumors in a network: who's the culprit? IEEE Trans. Inform. Theory **57**(8), 5163–5181 (2011)
16. Slater, P.J.: Leaves of trees. Congr. Numer. **14**(549–559), 37 (1975)
17. Čangalović, M., Kratica, J., Kovačević-Vujčić, V., Stojanović, M.: Minimal doubly resolving sets of prism graphs. Optimization **62**(8), 1037–1043 (2013)

Trajectory Optimization of UAV for Efficient Data Collection from Wireless Sensor Networks

Chuanwen Luo[1], Lidong Wu[2], Wenping Chen[1], Yongcai Wang[1], Deying Li[1(✉)], and Weili Wu[3]

[1] School of Information, Renmin University of China,
Beijing 100872, People's Republic of China
chuanwen_luo@163.com, {chenwenping,ycw,deyingli}@ruc.edu.cn
[2] Department of Computer Science, University of Texas at Tyler,
Tyler, TX 75799, USA
lwu@uttyler.edu
[3] Department of Computer Science, University of Texas at Dallas,
Richardson, TX 75080, USA
weiliwu@utdallas.edu

Abstract. Unmanned Aerial Vehicles (UAVs) are expected to be an important component in the upcoming wireless communication field, which are increasingly used as data collectors to gather sensing data from Wireless Sensor Networks (WSNs) due to their high mobility. Since the storage capacity and lifetime of sensors are increasing with the development of science and technology, sensors can store more and more sensing data about the monitoring area. However, due to the energy limitation of UAVs, we can not collect all data from WSN in limited time. Therefore, in this paper, we investigate the Maximizing Data Collection Proportion (MDCP) problem: given the limited budget of UAV, the objective is to find the trajectory of UAV such that the minimum data collection proportion of collected data to the stored data among all sensors is maximized. We first prove that the MDCP problem is NP-hard. Then we propose two approximation algorithms to design the trajectory of UAV, and give the theoretical analysis for the algorithms. Finally, we present numerical results in different scenarios to evaluate the effectiveness of the proposed algorithms.

Keywords: Wireless Sensor Network · Data collection · Unmanned Aerial Vehicle · Mobile collector · Trajectory optimization

1 Introduction

In Wireless Sensor Networks (WSNs), sensors are mainly powered by batteries and each sensor has a limited energy source, which restricts the lifetime of the

This work was supported in part by the National Natural Science Foundation of China under Grants (11671400, 61672524).

© Springer Nature Switzerland AG 2019
D.-Z. Du et al. (Eds.): AAIM 2019, LNCS 11640, pp. 223–235, 2019.
https://doi.org/10.1007/978-3-030-27195-4_21

network, such as [1,2]. The WSN adopts multi-hop communication for forwarding packets, which can make sensors around base station deplete much faster than other sensors, and shorten the lifetime of the network, such as [3]. Moreover, some sensors that are deployed in the detection area may be disconnected to the network and cannot forward data to the base station. To solve these problems, many data gathering strategies with ground mobile collectors are proposed to extend the network lifetime of WSNs, such as [4,5]. However, in many cases, sensors are deployed in complex ground environments, especially in rugged and hilly terrain where obstacles could inhibit ground mobile collectors to complete a mission. Moreover, due to the slow speed of movement, ground mobile collectors need more time to collect sensing data, which increases the communication latency and restricts the time-real applications of WSNs.

To overcome these problems, Unmanned aerial vehicle (UAV) has gained a lot of attention as a mobile collector in WSNs, such as [6,7]. Compared with the ground data collectors, UAV is not restricted by roadway and can be used in specific monitoring areas, which has a higher speed of movement and enables users to interact with the networking environment more quickly and accurately. In this paper, we focus on the Maximizing Data Collection Proportion (MDCP) problem. The MDCP problem is to find the optimal trajectory of UAV while maximizing the minimum proportion of collected data to the carried data among all sensors with limited energy of UAV. The contributions of this paper can be summarized as follows:

(1) We identify the problem as Maximizing Data Collection Proportion (MDCP) and prove that it is a new NP-hard problem.
(2) For the MDCP problem, we propose two approximation algorithms to optimize the trajectory of UAV. And we give the theoretical analysis for the proposed algorithms.
(3) Numerical simulation results under different scenarios are presented to verify the validity of the proposed algorithms.

The remainder of this paper is organized as follows. Section 2 introduces related works. Section 3 introduces network models and the definition of the problem. Section 4 proposes two approximation algorithms for the MDCP problem. Simulations are shown in Sect. 5. Section 6 concludes this paper.

2   Related Works

Recently, there has been a growing interest in employing UAVs as mobile collectors for data collection in WSNs, such as [8–10].

In [8], Gong et al. investigated the time minimization problem of UAV by considering both traveling and communication of UAV, in which the UAV collects data from a set of energy constrained sensors. But they only considered the scenario where a UAV collects data from a set of sensors on a straight line. In [9], Zhan et al. studied the joint optimization of sensors' wake-up schedule and UAV's trajectory to achieve reliable and energy efficient data collection in WSNs such that the maximum energy consumption of all sensors is minimized. In [10], Yang et al. investigated the Ground-to-UAV (G2U) communication system, where a UAV is to collect a given amount of data from a ground terminal at a fixed location. They proposed two practical UAV trajectories to balance the energy consumption of ground terminals and UAV.

3   Models and Definition

In this section, we introduce the models and the definition of the MDCP problem.

3.1   Network Model

Assume that all sensors in WSN are randomly deployed in the two-dimensional plane and that they have the same three-dimensional transmission range R. The network is denoted by the set $V = \{v_1, v_2, \cdots, v_n\}$ in which each sensor v_i has S_i units of data. Let a UAV with fixed flight speed s and the height H $(H \leq R)$ be the mobile collector to collect data from sensors. For any sensor $v_i \in V$, let $SN(v_i)$ represent the hemispheric region above the ground which centers at v_i and whose radius is normalized to R. The UAV can collect data from sensor v_i only when it hovers in $SN(v_i)$.

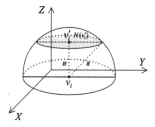

Fig. 1. The grey shaded area $N(v_i')$ is the *data collection area* for the sensor v_i, which centers at v_i' and whose radius is r.

We build a three-dimensional coordinate system, XYZ, to represent the three-dimensional space. Without loss of generality, we assume that the sensors are deployed in the first quadrant of the coordinate system and the Z coordinates of them are zero. Therefore, for each sensor $v_i \in V$, the cross section of $SN(v_i)$ cut by the plane $Z = H$ is a circular area, denoted $N(v_i')$ which centers at v_i' and whose radius is $r = \sqrt{R^2 - H^2}$, where v_i' is the projection of v_i, and the X and Y coordinates of v_i' are the same as v_i and its Z coordinate is H, as shown in Fig. 1. Since the flight height of UAV is H and the UAV can collect data from sensor v_i only when it hovers in $SN(v_i)$, the circular area $N(v_i')$ is called the *data collection area* of v_i, i.e., the data of v_i can only be collected by UAV within the region $N(v_i')$. Let $D = \{N(v_1'), N(v_2'), \cdots, N(v_n')\}$.

Since clustering algorithm could enable basic data gathering work in WSNs, and the UAV could focus on gathering sensing data from cluster-head sensors [7], in this paper, we assume that the sensors are sparsely distributed in the detection area, and that any two circular areas $N(v_i') \in D$ and $N(v_j') \in D$ are disjoint with each other. If U is the flight trajectory of UAV, then let $Len(U)$ be the length of U.

3.2   Data Transmission Model

The UAV collects data from v_i when it hovers in $N(v_i')$. As the transmission distance changes during flying, the transmission power and the data transmission rate of v_i should also adapt to the varying path loss. The LOS ground-to-air channel model between the UAV and sensors with path loss exponent is adopted [8,11]. Since the sensors in the network have the same transmission power, the data transmission rate from sensor v_i to UAV can be expressed as

$$C_i = \frac{1}{2} B \log_2(1 + \frac{\beta}{d_i^{\alpha}}),$$

where d_i is the Euclidean distance between v_i and UAV, B denotes the wireless channel bandwidth, β represents the reference signal-to-noise ratio (SNR) at the reference distance $d_i = 1\,\text{m}$ and $\alpha \geq 2$ is the path loss exponent.

3.3   Definition for the Problem

In this section, we give the detailed definition of the Maximizing Data Collection Proportion (MDCP) problem.

In the MDCP problem, given the limited active time T for the UAV, we aim to find a closed continuous curve U and a hovering set P of UAV on U such that each hovering point $P_{v_i} \in P$ is located in $N(v_i')$ and all or part of the data carried by each sensor $v_i \in V$ is collected by UAV and transported to the base station. The objective of the MDCP problem is to maximize the proportion of data $\mu = \min\{\frac{S_i'}{S_i} | v_i \in V\}$ for all sensors in V, where S_i' is the amount of data collected by UAV from sensor v_i and S_i is the amount of data carried by v_i. We formulate the problem as the Definition 1 shown.

Definition 1 (MDCP). *Given a set $V = \{v_1, v_2, ..., v_n\}$ of n sensors in which each sensor v_i has S_i units of sensing data, a set of data collection areas $D = \{N(v_1'), N(v_2'), \cdots, N(v_n')\}$ and a UAV with the corresponding speed s, the flight height H, the initial active time T and the initial location L, the Maximizing Data Collection Proportion (MDCP) problem is to find a tour U and a hovering set P of UAV such that*

(1) the tour U starts from and ends at L and for each $v_i \in V$, U passes through $N(v_i')$,
(2) UAV collects data during hovering in $N(v_i')$,
(3) the sum of the traveling time and hovering time of UAV is less than or equal to T,
(4) for each $v_i \in V$, there exists a hovering point $P_{v_i} \in P$ on U and the UAV collects S_i' units of data from v_i on P_{v_i} with hovering time $t_h^{v_i}$, and
(5) $\mu = \min\{\frac{S_i'}{S_i} | v_i \in V\}$ is maximum, where S_i' is the amount of data collected by UAV from sensor v_i.

Theorem 1. *The MDCP problem is NP-hard.*

Proof. Let $S_i = 0$ for each sensor $v_i \in V$ and $T = \infty$ for the UAV. Then the MDCP problem can be reduced to the Travelling Salesman Problem with Neighbors (TSPN), which is proved to be NP-hard by [12]. Since a special case is NP-hard, the MDCP problem is NP-hard.

4   Algorithms for the MDCP Problem

In this section, we put forward two approximation algorithms, called MDCP-A1 and MDCP-A2 to solve the MDCP problem.

Algorithm 1. MDCP-A1

Data: $D = \{N(v_1'), N(v_2'), ..., N(v_n')\}$, $V = \{v_1, v_2, \cdots, v_n\}$ and S_i for each $v_i \in V$, the speed s, flight height H and the initial active T of UAV;

Result: U, P and μ;

1 Using the $(1 + \epsilon)$-approximation algorithm for the TSPN problem to compute the flight trajectory U of UAV for D [12];

2 For each $N(v_i') \in D$, select the intersection point of $N(v_i')$ and U as the hovering point P_{v_i} of UAV;

3 $T_r = T - \frac{Len(U)}{s}$;

4 **if** $T_r < \lambda$ **then**

5 | There is no solution for the problem;

6 **else**

7 | $\mu = 1$, $\mu_1 = \mu_2 = 0$;

8 | **while** $T_r \geq \lambda$ **do**

9 | | for each $v_i \in V$, compute $t_h^{v_i} = \frac{2\mu S_i}{B \log_2(1+\frac{\beta}{R^\alpha})}$, and

                 $T_{total}^h = \sum_{v_i \in V} t_h^{v_i}$;

10 | | **if** $T_{total}^h > T_r$ **then**

11 | | | $\mu_1 = \mu$, $\mu = \frac{\mu_1 + \mu_2}{2}$;

12 | | **else**

13 | | | **if** $T_r - T_{total}^h \geq \lambda$ **then**

14 | | | | $\mu_2 = \mu$, $\mu = \frac{\mu_1 + \mu_2}{2}$;

15 | | | **else**

16 | | | | $T_r = T_r - T_{total}^h$, **return** μ;

17 | | | **end**

18 | | **end**

19 | **end**

20 **end**

21 **for** *each* $v_i \in V$ **do**

22 | $t_h^{v_i} = \frac{2\mu S_i}{B \log_2(1+\frac{\beta}{R^\alpha})}$, $P = P \cup \{(P_{v_i}, t_h^{v_i})\}$;

23 **end**

4.1   The MDCP-A1 Algorithm

The MDCP-A1 algorithm consists of three steps. In the first step, we apply the $(1 + \epsilon)$-approximation algorithm for the TSPN problem proposed by [12] to compute the flight trajectory U of UAV. Afterwards, for each $N(v_i') \in D$, we compute an intersection point of $N(v_i')$ and U as the hovering point P_{v_i} of UAV. Then we obtain the remaining active time of UAV is $T_r = T - Len(U)/s$. In the second step, we use the bisection method to compute the proportion of data $\mu = \min\{\frac{S_i'}{S_i}|v_i \in V\}$ based on the remaining active time of UAV. We establish two variables μ_1 and μ_2 representing the upper and lower bounds of μ, respectively. Initially, we set $\mu = 1$ and $\mu_1 = \mu_2 = 0$. We use the very small constant λ to control the end of the while loop. If $T_r < \lambda$, then there is no solution for the MDCP problem. Otherwise, the algorithm repeats the following two steps until $T_r < \lambda$: (1) Compute the total hovering time of UAV for all sensors in V, $T_{total}^h = \sum_{v_i \in V} t_h^{v_i}$, where $t_h^{v_i} = \frac{2\mu S_i}{B \log_2(1+\frac{\beta}{H^\alpha})}$; (2) If $T_r < T_{total}^h$, then we set $\mu_1 = \mu$ and $\mu = \frac{\mu_1 + \mu_2}{2}$. Otherwise, if $T_r - T_{total}^h \geq \lambda$, then we set $\mu_2 = \mu$ and $\mu = \frac{\mu_1 + \mu_2}{2}$, or else we set $T_r = T_r - T_{total}^h$ and return μ. Finally, for each $v_i \in V$, we compute the hovering time $t_h^{v_i}$ of UAV on the point P_{v_i} based on the value of returned μ, and $P = P \cup \{(P_{v_i}, t_h^{v_i})\}$. The detailed description is given by the Algorithm 1.

Now we analyze the performance ratio of the MDCP-A1 algorithm. Suppose U^* and $P^* = \{P_{v_1}^*, P_{v_2}^*, \cdots, P_{v_n}^*\}$ are the optimal trajectory and hovering set of UAV, respectively and $\mu^* = \min\{\frac{S_i^*}{S_i}|v_i \in V\}$ is the optimal solution for the MDCP problem, where S_i^* is the optimal amount of data collected from v_i by UAV and C_i^* is the data transmission rate from v_i to UAV when it hovers on $P_{v_i}^*$. For each $v_i \in V$, let $\mu_i^* = \frac{S_i^*}{S_i}$.

Lemma 1. $\mu^* = \mu_1^* = \mu_2^*, \cdots, = \mu_n^*$.

Proof. We prove the lemma by contradiction. For simplicity, we assume that $\mu_i^* = \mu^*$ and there exist a sensor $v_j \in V$ such that $\mu_i^* < \mu_j^*$. Since $\mu_i^* = \frac{S_i^*}{S_i} = \frac{t_h^{v_i} \cdot C_i^*}{S_i}$ and $\mu_j^* = \frac{S_j^*}{S_j} = \frac{t_h^{v_j} \cdot C_j^*}{S_j}$, we can obtain $\frac{t_h^{v_i} \cdot C_i^*}{S_i} < \frac{t_h^{v_j} \cdot C_j^*}{S_j}$. When U^* and P^* are fixed, the data transmission rates C_i^* and C_j^* are fixed. Therefore, we can increase μ^* by increasing the hovering time $t_h^{v_i}$ of v_i while decreasing the hovering time $t_h^{v_j}$ of v_j, which is contradict to the assumption. Therefore, we can obtain $\mu^* = \mu_1^* = \mu_2^*, \cdots, = \mu_n^*$.

Lemma 2. $\mu^* \leq \frac{B \log_2(1+\frac{\beta}{H^\alpha})(T - \frac{Len(U^*)}{s})}{2\sum_{i=1}^n S_i}$.

Proof. Based on Lemma 1, we have $\sum_{i=1}^n \frac{\mu^* S_i}{C_i^*} + \frac{Len(U^*)}{s} \leq T$. For each sensor $v_i \in V$, since the distance d_i between v_i and UAV is greater than or equal to H, we have $C_i^* = \frac{1}{2}B \log(1 + \frac{\beta}{d_i^\alpha}) \leq \frac{1}{2}B \log(1 + \frac{\beta}{H^\alpha})$. Therefore, we can derive

$$\mu^* \leq \frac{T - \frac{Len(U^*)}{s}}{\sum_{i=1}^n \frac{S_i}{C_i^*}} \leq \frac{T - \frac{Len(U^*)}{s}}{\sum_{i=1}^n \frac{2S_i}{B \log_2(1+\frac{\beta}{H^\alpha})}} = \frac{B \log_2(1 + \frac{\beta}{H^\alpha})(T - \frac{Len(U^*)}{s})}{2\sum_{i=1}^n S_i}.$$

Lemma 3. $Len(U) \le (1 + \epsilon) \cdot Len(U^*)$, where U is got by MDCP-A1.

Proof. Suppose U_T^* is the optimal trajectory for the TSPN problem on D. According to the MDCP-A1 algorithm, we can get $Len(U) \le (1 + \epsilon) \cdot Len(U_T^*)$. Since U^* is a feasible solution for the TSPN problem, we have $Len(U_T^*) \le Len(U^*)$. Therefore, we can obtain $Len(U) \le (1 + \epsilon) \cdot Len(U^*)$.

Theorem 2. $\mu \ge \dfrac{\log_2(1 + \frac{\beta}{R^\alpha})}{\log_2(1 + \frac{\beta}{H^\alpha})} \cdot \mu^*$, where μ is got by the MDCP-A1 algorithm.

Proof. According to the MDCP-A1 algorithm, we can obtain

$$\sum_{i=1}^{n} \frac{2\mu S_i}{B \log_2(1 + \frac{\beta}{R^\alpha})} + \frac{Len(U)}{s} \ge T - \lambda.$$

Therefore, based on Lemmas 1, 2 and 3, we have

$$
\begin{aligned}
\mu &\ge \frac{B \log_2(1 + \frac{\beta}{R^\alpha})(T - \frac{Len(U)}{s} - \lambda)}{2\sum_{i=1}^{n} S_i} \ge \frac{B \log_2(1 + \frac{\beta}{R^\alpha})(T - (1 + \epsilon)\frac{Len(U^*)}{s} - \lambda)}{2\sum_{i=1}^{n} S_i} \\
&= \frac{\log_2(1 + \frac{\beta}{R^\alpha})}{\log_2(1 + \frac{\beta}{H^\alpha})} \cdot \frac{B \log_2(1 + \frac{\beta}{H^\alpha})(T - (1 + \epsilon)\frac{Len(U^*)}{s} - \lambda)}{2\sum_{i=1}^{n} S_i} \\
&\ge \frac{\log_2(1 + \frac{\beta}{R^\alpha})}{\log_2(1 + \frac{\beta}{H^\alpha})} \cdot \mu^* - \frac{B \log_2(1 + \frac{\beta}{R^\alpha})(\frac{\epsilon \cdot Len(U^*)}{s} + \lambda)}{2\sum_{i=1}^{n} S_i} \\
&\ge \frac{\log_2(1 + \frac{\beta}{R^\alpha})}{\log_2(1 + \frac{\beta}{H^\alpha})} \cdot \mu^* - \frac{B \log_2(1 + \frac{\beta}{R^\alpha})(\epsilon \cdot T + \lambda)}{2\sum_{i=1}^{n} S_i}.
\end{aligned}
$$

Since λ and ϵ are very small constants, the approximation ratio is

$$\frac{\log_2(1 + \frac{\beta}{R^\alpha})}{\log_2(1 + \frac{\beta}{H^\alpha})} \cdot \mu^* - \frac{B \log_2(1 + \frac{\beta}{R^\alpha})(\epsilon \cdot T + \lambda)}{2\sum_{i=1}^{n} S_i} \approx \frac{\log_2(1 + \frac{\beta}{R^\alpha})}{\log_2(1 + \frac{\beta}{H^\alpha})} \cdot \mu^*.$$

The theorem has been proved.

4.2   The MDCP-A2 Algorithm

We next present an approximation algorithm MDCP-A2 for the MDCP problem.

The algorithm is composed of five steps. In the first step, we employ the $(1 + \epsilon)$-approximation algorithm for the TSPN problem proposed by [12] to compute the flight trajectory U' of UAV. Afterwards, for each $N(v_i') \in D$, we compute the first intersection point $P_{v_i}^c$ between $N(v_i')$ and U'. Then, we connect two points v_i' and $P_{v_i}^c$ and obtain the line segment $(P_{v_i}^c, v_i')$. In the second step, we change the line segment $(P_{v_i}^c, v_i')$ to two (parallel) edges between the same pair of vertices. And let U^{v_i} denote the trajectory of UAV that is composed of the two parallel edges. In the third step, we compute the trajectory U of UAV by combining U' with U^{v_i} for each $v_i \in V$ and let v_i' be the hovering point P_{v_i} of

v_i. Then we can obtain the remaining active time of UAV is $T_r = T - Len(U)/s$. In the fourth step, we employ the bisection method to compute the proportion of data μ based on the remaining active time of UAV. We establish two variables μ_1 and μ_2 representing the upper and lower bounds of μ and let the very small constant λ dominate the end of the while loop. Initially, we set $\mu = 1$ and $\mu_1 = \mu_2 = 0$. If $T_r < \lambda$, then there is no solution for the MDCP problem. Otherwise, the algorithm repeats the following two steps until $T_r < \lambda$: (1) Compute the total hovering time of UAV for all sensors in V, $T^h_{total} = \sum_{v_i \in V} t^{v_i}_h$, where $t^{v_i}_h = \frac{2\mu S_i}{B \log_2(1 + \frac{\beta}{H^\alpha})}$; (2) If $T_r < T^h_{total}$, then we set $\mu_1 = \mu$ and $\mu = \frac{\mu_1 + \mu_2}{2}$. Otherwise, if $T_r - T^h_{total} \geq \lambda$, then we set $\mu_2 = \mu$ and $\mu = \frac{\mu_1 + \mu_2}{2}$, or else we set $T_r = T_r - T^h_{total}$ and return μ. Finally, for each sensor $v_i \in V$, we compute the hovering time $t^{v_i}_h$ of UAV on the point P_{v_i} based on the value of returned μ, and $P = P \cup \{P_{v_i}, t^{v_i}_h\}$.

The detailed description of the algorithm is given by the Algorithm 2.

Lemma 4. $Len(U) \leq (1 + \epsilon + \frac{8}{\pi}) \cdot Len(U^*) + 8r$, where U is got by MDCP-A2.

Proof. According to the MDCP-A2 algorithm, we can obtain U is composed of U' and U^{v_i} for each $v_i \in V$. Similar to Lemma 3, we can obtain $Len(U') \leq (1 + \epsilon) \cdot Len(U^*)$. Let A_{U^*} be the area scanned by a disk of radius $2r$ whose center moves along U^*. Since U^* visits all disks, A_{U^*} can covers all disks in D. Since when $n = 1$, $Len(U^*) = 0$ and $A_{U^*} = 4\pi r^2$. This area is bounded as

$$n\pi r^2 \leq A_{U^*} \leq 4r \cdot Len(U^*) + 4\pi r^2.$$

Thus, $n \leq \frac{4}{\pi r} \cdot Len(U^*) + 4$. The length of tour U at most

$$Len(U) = Len(U') + \sum_{v_i \in V} Len(U^{v_i}) \leq (1 + \epsilon) \cdot Len(U^*) + 2nr$$

$$\leq (1 + \epsilon + \frac{8}{\pi}) \cdot Len(U^*) + 8r.$$

The Lemma has been proved.

Theorem 3. $\mu \geq (1 - (1 + \frac{8}{\pi}) \cdot \eta^* - \frac{s\lambda + 8r}{sT - Len(U^*)})\mu^*$, where μ is got by MDCP-A2 and η^* is the ratio of the optimal flight time to the remaining time of UAV.

Proof. According to the MDCP-A2 algorithm, we can obtain

$$\sum_{i=1}^n \frac{2\mu S_i}{B \log_2(1 + \frac{\beta}{H^\alpha})} + \frac{Len(U)}{s} \geq T - \lambda.$$

Therefore, based on Lemmas 1 and 4, we have

$$\sum_{i=1}^n \frac{2\mu S_i}{B \log(1 + \frac{\beta}{H^\alpha})} + (1 + \epsilon + \frac{8}{\pi})\frac{Len(U^*)}{s} + \frac{8r}{s} \geq T - \lambda.$$

Algorithm 2. MDCP-A2

Data: $D = \{N(v'_1), N(v'_2), ..., N(v'_n)\}$, $V = \{v_1, v_2, \cdots, v_n\}$ and S_i for each
$\quad\quad v_i \in V$, the speed s, flight height H and the initial active T of UAV;
Result: U, P and μ;

1 Using the $(1 + \varepsilon)$-approximation algorithm for the TSPN problem to compute
   the traveling trajectory U' of UAV for D [12];
2 For each $N(v'_i) \in D$, compute the first intersection point $P^c_{v_i}$ between $N(v'_i)$ and
   U';
3 Change the line segment $(P^c_{v_i}, v'_i)$ to two line segments between the same pair of
   points and $U^{v_i} = (P^c_{v_i}, v'_i) + (v'_i, P^c_{v_i})$;
4 Compute the traveling trajectory U of UAV by combining U' and U^{v_i} for each
   $N(v'_i) \in D$, and let v'_i be the hovering point P_{v_i} of v_i;
5 $T_r = T - \frac{Len(U)}{s}$;
6 **if** $T_r < \lambda$ **then**
7 $\quad$ | $\quad$ There is no solution for the problem;
8 **else**
9 $\quad$ | $\quad$ $\mu = 1$, $\mu_1 = \mu_2 = 0$;
10 $\quad$ | $\quad$ **while** $T_r \geq \lambda$ **do**
11 $\quad$ | $\quad$ | $\quad$ **for** each $v_i \in V$, compute $t^{v_i}_h = \frac{2\mu S_i}{B \log_2(1+\frac{\beta}{H^\alpha})}$, and $T^h_{total} = \sum_{v_i \in V} t^{v_i}_h$;
12 $\quad$ | $\quad$ | $\quad$ **if** $T^h_{total} > T_r$ **then**
13 $\quad$ | $\quad$ | $\quad$ | $\quad$ $\mu_1 = \mu$, $\mu = \frac{\mu_1 + \mu_2}{2}$;
14 $\quad$ | $\quad$ | $\quad$ **else**
15 $\quad$ | $\quad$ | $\quad$ | $\quad$ **if** $T - T^h_{total} \geq \lambda$ **then**
16 $\quad$ | $\quad$ | $\quad$ | $\quad$ | $\quad$ $\mu_2 = \mu$, $\mu = \frac{\mu_1 + \mu_2}{2}$;
17 $\quad$ | $\quad$ | $\quad$ | $\quad$ **else**
18 $\quad$ | $\quad$ | $\quad$ | $\quad$ | $\quad$ $T_r = T_r - T^h_{total}$, **return** μ;
19 $\quad$ | $\quad$ | $\quad$ | $\quad$ **end**
20 $\quad$ | $\quad$ | $\quad$ **end**
21 $\quad$ | $\quad$ **end**
22 **end**
23 **for** each $v_i \in V$ **do**
24 $\quad$ | $\quad$ $t^{v_i}_h = \frac{2\mu S_i}{B \log_2(1+\frac{\beta}{H^\alpha})}$, $P = P \cup \{(P_{v_i}, t^{v_i}_h)\}$;
25 **end**

Consequently, according to Lemma 2, we can get

$$\mu \geq (T - \lambda - \frac{(1 + \epsilon + \frac{8}{\pi}) \cdot Len(U^*)}{s} - \frac{8r}{s}) \cdot \frac{B \log_2(1 + \frac{\beta}{H^\alpha})}{2\sum_{i=1}^n S_i}$$

$$\geq (T - \lambda - \frac{(1 + \epsilon + \frac{8}{\pi}) \cdot Len(U^*)}{s} - \frac{8r}{s}) \cdot \frac{1}{T - \frac{Len(U^*)}{s}} \cdot \mu^*$$

$$= (1 - \frac{(\epsilon + \frac{8}{\pi})\frac{Len(U^*)}{s}}{T - \frac{Len(U^*)}{s}} - \frac{\lambda + \frac{8r}{s}}{T - \frac{Len(U^*)}{s}})\mu^* \geq (1 - (1 + \frac{8}{\pi})\eta^* - \frac{s\lambda + 8r}{sT - Len(U^*)})\mu^*.$$

The theorem has been proved.

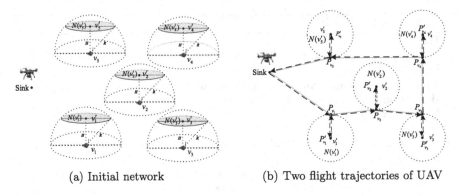

(a) Initial network   (b) Two flight trajectories of UAV

Fig. 2. Two flight trajectories of UAV got by MDCP-A1 and MDCP-A2, respectively.

4.3 Discussion

Note that the MDCP-A2 algorithm is different from the MDCP-A1 algorithm. Although the trajectory of UAV got by MDCP-A1 is less than MDCP-A2, the data transmission rate from sensors to UAV got by MDCP-A1 is less than MDCP-A2. Therefore, the proposed two algorithms can deal with certain cases well. To illustrate the differences between the two algorithms, we give an example as follows. Assume that the network consists of five sensors and a UAV and let $V = \{v_1, v_2, v_3, v_4, v_5\}$ in which each sensor has S_i units of data and $D = \{N(v_1'), N(v_2'), N(v_3'), N(v_4'), N(v_5')\}$, as shown in Fig. 2(a).

We assume the hovering set and trajectory of UAV got by MDCP-A1 are $P = \{P_{v_1}, P_{v_2}, P_{v_3}, P_{v_4}, P_{v_5}\}$ and U respectively, as the blue solid line shown in Fig. 2(b), and the hovering set and trajectory of UAV got by MDCP-A2 are $P' = \{P_{v_1}', P_{v_2}', P_{v_3}', P_{v_4}', P_{v_5}'\}$ and U' respectively, as the black dotted line shown in Fig. 2(b), where the position of P_{v_i}' is the same as v_i'. Therefore, we have $Len(U') = Len(U) + 10 \cdot r$. The data transmission rates on P_{v_i} and P_{v_i}' from v_i to UAV are $C_i = \frac{1}{2}B\log(1 + \frac{\beta}{R^\alpha})$ and $C_i' = \frac{1}{2}B\log(1 + \frac{\beta}{H^\alpha})$, respectively. Since $H \le R$, we have $C_i \le C_i'$. Therefore, based on different network configurations, we can obtain that if $\frac{T - \frac{Len(U)}{s}}{2\sum_{i=1}^{5} S_i} \cdot B\log_2(1 + \frac{\beta}{R^\alpha}) \ge \frac{T - \frac{Len(U')}{s}}{2\sum_{i=1}^{5} S_i} \cdot B\log_2(1 + \frac{\beta}{H^\alpha})$, the MDCP-A1 algorithm performs better than MDCP-A2, otherwise, the performance of MDCP-A1 algorithm is worse than MDCP-A2.

5   Simulation Setup and Results

In this section, we construct extensive simulations to validate the proposed algorithms. In the simulations, we assume that sensors are randomly deployed in the 2000 m × 2000 m square area. For simplicity, we assume that all sensors have the same amount of data S. We verify the values of the $\mu = \min\{\frac{S_i'}{S_i} | v_i \in V\}$ got by algorithms MDCP-A1 and MDCP-A2 under different network configurations. Each subgraph in Fig. 3 illustrates the simulation results of the algorithms

MDCP-A1 and MDCP-A2. Each experimental result is the average of 100 runs. The version of MATLAB R2013a for all simulations is used.

In Fig. 3(a), when we set $N = 100$, $S = 2000$ KB, $R = 100$ m, $\beta = 150$ dB, $B = 2$ kHz, $H = 80$ m, $s = 5$ m/s and change T from 7000 to 12000 s, the results show that if $T \leq 9000$ s, then the MDCP-A1 algorithm outperforms MDCP-A2, otherwise, the MDCP-A2 algorithm outperforms MDCP-A1, which show that each of them can deal with certain cases.

In Fig. 3(b) when we set $T = 8000$ s, $S = 2000$ KB, $R = 100$ m, $\beta = 150$ dB, $B = 2$ kHz, $H = 80$ m, $s = 5$ m/s and change N from 60 to 110, the results show that when $N \leq 83$, the MDCP-A2 outperforms MDCP-A1 and when $N > 83$, the MDCP-A1 algorithm is better than MDCP-A2.

In Fig. 3(c), when we set $T = 10000$ s, $N = 100$, $R = 100$ m, $\beta = 150$ dB, $B = 2$ kHz, $H = 80$ m, $s = 5$ m/s and change S from 1000 to 6000 KB, the results show that the algorithm MDCP-A2 outperforms MDCP-A1. This is because when other configurations are fixed, the flight trajectory and the amount of the collected data by either of the two algorithms are determined. Therefore, if the performance of one algorithm is better than another initially, then it's going to be fine all the time as S grows.

In Fig. 3(d), when we set $T = 5500$ s, $N = 60$, $S = 2000$ KB, $\beta = 150$ dB, $B = 2$ kHz, $H = 30$ m, $s = 5$ m/s and change R from 40 to 80 m, the results show that the algorithm MDCP-A1 outperforms MDCP-A2. And we can observe that both the proportions of data decrease as R increases. This is because the data transmission rate from sensors to UAV decreases with R increasing.

In Fig. 3(e), when we set $T = 10000$ s, $N = 100$, $S = 2000$ KB, $R = 100$ m, $B = 2$ kHz, $H = 80$ m, $s = 5$ m/s and change β from 100 to 600 dB, the results show that the performance of MDCP-A1 is slightly better than MDCP-A2. We also find that both proportions of data increase logarithmic with β increasing since the data transmission rate from sensors to UAV grows logarithmic with β increasing.

In Fig. 3(f), when we set $T = 12000$ s, $N = 100$, $S = 6000$ KB, $\beta = 150$ dB, $R = 100$ m, $H = 80$ m, $s = 5$ m/s and change B from 1 to 6 kHz, we can find that the performance of MDCP-A2 is better than MDCP-A1. And we also find that μ increases linearly as B grows. This is because the data transmission rate from sensors to UAV grows linearly with B increasing.

In Fig. 3(g), when we set $T = 6000$ s, $N = 60$, $S = 2000$ KB, $R = 100$ m, $\beta = 150$ dB, $B = 2$ kHz, $s = 5$ m/s and change H from 40 to 90 m, the results show that if $H \leq 67$ m, the performance of MDCP-A1 is better than MDCP-A2, otherwise, the MDCP-A1 outperforms worse than MDCP-A2, which show that each of the proposed algorithms can deal with certain cases.

In Fig. 3(h), when we set $T = 8000$ s, $N = 100$, $S = 2000$ KB, $R = 100$ m, $\beta = 150$ dB, $B = 2$ kHz, $H = 80$ m and change s from 5 to 10 m/s, we can observe that if $s \leq 5.8$ m/s, the MDCP-A1 algorithm outperforms MDCP-A2, otherwise, the MDCP-A1 is worse than MDCP-A2, which show that each of the proposed algorithms can deal with certain cases.

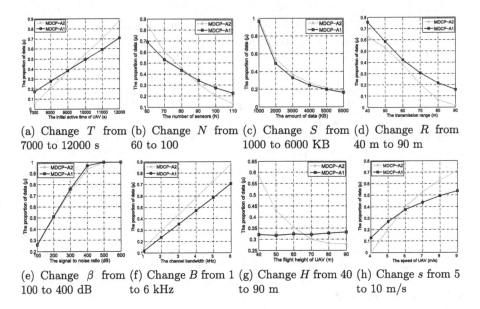

(a) Change T from 7000 to 12000 s

(b) Change N from 60 to 100

(c) Change S from 1000 to 6000 KB

(d) Change R from 40 m to 90 m

(e) Change β from 100 to 400 dB

(f) Change B from 1 to 6 kHz

(g) Change H from 40 to 90 m

(h) Change s from 5 to 10 m/s

Fig. 3. Comparison the performance of MDCP-A1 with MDCP-A2.

6   Conclusion

In this paper, we identify the problem Maximizing Data Collection Proportion (MDCP). The MDCP problem is to find the trajectory of UAV such that the minimum data collection proportion of collected data to the stored data among all sensors is maximized with the limited energy of UAV. We first prove that the MDCP problem is NP-hard. Then we propose two approximation algorithms to solve the problem, and each of which can deal with certain cases. Finally, we give the theoretical analysis and simulations for the proposed algorithms to verify the effectiveness.

References

1. Li, J., Cheng, S., Cai, Z., Yu, J., Wang, C., Li, Y.: Approximate holistic aggregation in wireless sensor networks. ACM Trans. Sens. Netw. (TOSN) **13**(2), 11 (2017)
2. Cheng, S., Cai, Z., Li, J., Gao, H.: Extracting kernel dataset from big sensory data in wireless sensor networks. IEEE Trans. Knowl. Data Eng. **29**(4), 813–827 (2016)
3. Luo, C., Chen, W., Yu, J., Wang, Y., Li, D.: A novel centralized algorithm for constructing virtual backbones in wireless sensor networks. EURASIP J. Wirel. Commun. Netw. **2018**(1), 55 (2018)
4. Mehrabi, A., Kim, K.: Maximizing data collection throughput on a path in energy harvesting sensor networks using a mobile sink. IEEE Trans. Mob. Comput. **15**(3), 690–704 (2015)
5. Xia, N., Wang, C., Yu, Y., Du, H., Xu, C., Zheng, J.: A path forming method for water surface mobile sink using Voronoi diagram and dominating set. IEEE Trans. Veh. Technol. **67**(8), 7608–7619 (2018)

6. Kim, D., Xue, L., Li, D., Zhu, Y., Wang, W., Tokuta, A.O.: On theoretical trajectory planning of multiple drones to minimize latency in search-and-reconnaissance operations. IEEE Trans. Mob. Comput. **16**(11), 3156–3166 (2017)

7. Wu, T., Yang, P., Yan, Y., Rao, X., Li, P., Xu, W.: ORSCA: optimal route selection and communication association for drones in WSNs. In: 2017 Fifth International Conference on Advanced Cloud and Big Data (CBD), pp. 420–424. IEEE (2017)

8. Gong, J., Chang, T.-H., Shen, C., Chen, X.: Flight time minimization of UAV for data collection over wireless sensor networks. IEEE J. Sel. Areas Commun. **36**(9), 1942–1954 (2018)

9. Zhan, C., Zeng, Y., Zhang, R.: Energy-efficient data collection in UAV enabled wireless sensor network. IEEE Wirel. Commun. Lett. **7**(3), 328–331 (2017)

10. Yang, D., Wu, Q., Zeng, Y., Zhang, R.: Energy tradeoff in ground-to-UAV communication via trajectory design. IEEE Trans. Veh. Technol. **67**(7), 6721–6726 (2018)

11. Ahmed, N., Kanhere, S.S., Jha, S.: On the importance of link characterization for aerial wireless sensor networks. IEEE Commun. Mag. **54**(5), 52–57 (2016)

12. Dumitrescu, A., Mitchell, J.S.B.: Approximation algorithms for TSP with neighborhoods in the plane. J. Algorithms **48**(1), 135–159 (2003)

Locality Sensitive Algotrithms for Data Mule Routing Problem

Pablo L. A. Munhoz[1], Felipe P. do Carmo[2], Uéverton S. Souza[2(⊠)],
Lúcia M. A. Drummond[2], Pedro Henrique González[3], Luiz S. Ochi[2],
and Philippe Michelon[4]

[1] Instituto de Ciências Exatas e Tecnológicas,
Universidade Federal de Viçosa, Viçosa, Brazil
`pablo.munhoz@ufv.br`
[2] Instituto de Computação,
Universidade Federal Fluminense, Niterói, Brazil
`fpcarmo@id.uff.br`, `{ueverton,lucia,satoru}@ic.uff.br`
[3] Departamento de Informática,
Centro Federal de Educação Tecnológica Celso Suckow da Fonseca,
Rio de Janeiro, Brazil
`pegonzalez@eic.cefet-rj.br`
[4] Laboratoire de Mathématiques d'Avignon,
Université d'Avignon et des Pays de Vaucluse, Avignon, France
`philippe.michelon@univ-avignon.fr`

Abstract. A usual way to collect data in a Wireless Sensor Network
(WSN) is by the support of a special agent, called data mule, that moves
between sensor nodes and performs all communication between them. In
this work, the focus is on the construction of the route that the data
mule must follow to serve all nodes in the WSN. This paper deals with
the case when the data mule does not have a global view of the network,
i.e., a prior knowledge of the network as a whole. Thus, at each node, the
data mule makes a decision about the next node to be visited based only
on a limited local knowledge of the WSN. Considering this realist scenario, two locality sensitive algorithms are proposed. These algorithms
differ by the criterion of choice of the next visited node, while the first
one uses a simpler greedy choice, the second one uses the geometric concept of convex hull. They were executed in instances of the literature and
their results were compared both in terms of route length and in number of sent messages as well. Some theoretical results, a mathematical
formulation, and some lower bounds for the global view scenario are also
proposed, in order to provide some parameters to evaluate the quality
of the solutions given by the proposed algorithms. The obtained results
show that the proposed algorithms give good solutions in a reasonable
time when compared with the optimal solutions and lower bounds.

Keywords: Data mule · Routing Problem · Locality sensitive ·
Convex hull

The authors acknowledges the support from CAPES, CNPq and FAPERJ.

© Springer Nature Switzerland AG 2019
D.-Z. Du et al. (Eds.): AAIM 2019, LNCS 11640, pp. 236–248, 2019.
https://doi.org/10.1007/978-3-030-27195-4_22

1   Introduction

Wireless Sensor Networks, WSN, have received much attention in last decades
due to its flexibility and its easy and fast deployment capability. In addition,
there are many practical applications where WSN can be used, as, for example,
environmental monitoring and military applications [4,7]. In this kind of network, communication is exclusively wireless and information exchange is accomplished when there is intersection between sensors spatial coverages. The information routing is one of the main problems of WSN [6,8]. In some cases, a
mobile agent, called date mule, is responsible for performing the network communication. Data mule is a mobile agent that has greater processing, memory
capacities and energy availability than regular sensors of the WSN. It is responsible for collecting data from all sensors and take them to a base station, reducing
the number of exchanged messages in the network and, consequently, the spent
energy for data transmission.

This work considers that sensors are distributed in a bi-dimensional space
and have communication range equal to r. In that scenario, the data mule has to
serve each node of the WSN, by sending or receiving data to/from each of them.
In the first service, the data mule has no knowledge about the global network and
should visit a minimum number of nodes necessary to serve all nodes demands.
After that, the mule could storage the position of the nodes and execute a global
view algorithm. Thus, with no knowledge, at each node, it has to decide the next
one to be visited, aiming at the minimization of the total route. The goal of using
a data mule and minimizing its route is to minimize the energy consumption of
the WSN. Note that the visited nodes form a connected dominant set. Remark
also that at each node, the data mule covers only a limited set of other sensor
nodes, and having that limited knowledge about the nodes, it has to choose the
best one to be visited next.

Thus, our problem follows the next basic assumptions. Let $G = (V, E)$ be a
graph representing a network and the geographical position of a node is given
by euclidean coordinates, i.e., $V(G)$ is a set of points placed in an Euclidean
plan, and each edge $(i, j) \in E(G)$ between two vertexes i and j exists if the corresponding sensors, that they represent, are within their communication range.
The set $N(i)$ contains the neighbour nodes of vertex i, and the corresponding
sensor only knows the other sensors in the neighbourhood and their corresponding euclidean coordinates. Edges have no weights. Let $s \in V$ be a vertex, from
where the data mule initiates and finishes the route, called here, base station.
The data mule moves between nodes, and only serves a node i when located in
some node $j \in N(i)$. The data mule can serve every node that is in the neighborhood, and not only the node where it is. When a mule moves to a node from
another one, the corresponding edge is included in the route. An edge can be
used by the data mule more than once. Each time the edge is used, it is included
in the route.

The objective of the problem treated in this work is minimizing the route
length traversed by the data mule, that visits a subset of nodes $D \subseteq V$ forming a connected dominant set of G. The remaining of this work is organized as

follows. Related works are presented in Sect. 2. Section 3 presents some theoretical remarks on Data Mule Routing problem (DMRP) and a mathematical formulation for the global view case is also presented in this section. In Sect. 4, we present two algorithms to work on the realistic scenario where the data mule has only a local view of the network. Results and analyses are shown in Sect. 5. Due to space constraints, some proofs are omitted.

2   Related Work

The communication in WSNs can be performed basically either by using virtual backbones and its own network infra-structure, or by using a mobile agent, called data mule. The problem of constructing virtual backbones for communication can be modeled as a Minimum Connected Dominating Set Problem, MCDS, as can be seen in [3,9].

Papers from related literature of WSN that aim at solving the MCDS, usually present distributed algorithms and consider simultaneous communication among sensor nodes, and local processing at each node. When the communication is performed through a data mule, most works focus on the Data Mule Routing Problem, an $\mathcal{NP}$-hard problem [11]. Usually, those related papers consider that the data mule has the complete knowledge of the WSN. Considering that approximate heuristics were proposed to solve the problem [1,10,12].

3   Theoretical Remarks

In order to demonstrate how difficult it is to find a route for a data mule, we will first perform an analysis of the problem when the data mule has a global view of the network, that is, the data mule knows the underlying graph G representing the network. Some proofs are omitted and presented in the appendices.

In such a scenario, the data mule's aim is to solve the following problem.

DATA MULE WITH GLOBAL VIEW

Input: A graph G, and a base station node $v \in V(G)$.
Goal: Determine a minimum closed walk W of G such that $v \in V(W)$, and for all node $x \in V(G)$, $N[x] \cap V(W) \neq \emptyset$. That is, either $x \in V(W)$ or some neighbor y of x belongs to $V(W)$.

Next result shows that if it is assumed that G is a general graph then it is very unlikely to exist an algorithm to find in polynomial time some routes for a data mule with cost near to the optimal solution.

Theorem 1. DATA MULE WITH GLOBAL VIEW *on general graphs cannot be approximated in polynomial time to within a factor of* $(1 - o(1)) \ln n$, *where* $n = |V(G)|$, *unless* $NP \subseteq DTIME(n^{O(\log \log n)})$.

The above result illustrates the hardness of approximate data mule problems on general graphs. Next result shows that the realistic instances of our problem belongs to a very special graph class.

Definition 1. *A* unit disk *graph is the intersection graph of a family of disks of unit size in the Euclidean plane.*

Lemma 1. *Any realistic instance G of* DATA MULE WITH GLOBAL VIEW *is a unit disk graph.*

The next result illustrates an interesting property of our problem in practical instances.

Lemma 2. DATA MULE WITH GLOBAL VIEW *on Unit Disk Graphs admits a $(2 + \epsilon)$-approximation algorithm, $\epsilon > 0$.*

3.1   Lower Bounds

Now, we are interested to recognize some lower bounds for a solution of DATA MULE WITH GLOBAL VIEW that can be found by efficient algorithms.

Given G be a simple graph, and a base station $v \in V(G)$. We denote $d(v, w)$ as the distance between v and w in G, and $d_v = \max_{w \in V(G)} d(v, w)$.

Lemma 3. *Let $OPT(G, v)$ be an optimal solution value for* DATA MULE WITH GLOBAL VIEW *on G with base station v. It holds that $OPT(G, v) \geq 2(d_v - 1)$. And such lower bound can be found in $O(m)$ time.*

Proof. Let w be a vertex of $V(G)$ such that $d_v = d(v, w)$ (w and d_v can be found by a breadth-first search). Clearly the data mule needs to travel through at least $d_v - 1$ edges in order to attend w. After that, the data mule will travel through some edges, and will return to base station which spend at least more $d_v - 1$ steps.

Now we present a hierarchy of lower bounds for the problem, whose values and performance to be found depends of an input integer k.

Lemma 4. *Given G, v and an integer $k \geq 1$. Let S be a set composed by the k most distant vertices of v, and T_k be a steiner tree to connect $\{v\} \cup S$. Let $LB_k = |T_k| - k + \min_{w \in S} d(v, w) - 1$. For all $k \geq 1$, it holds that*

$$OPT(G, v) \geq LB_k.$$

Proof. Note that $LB_1 = 2(d_v - 1)$. As previously, the data mule needs to travel through some edges in order to attend all vertices in S, which walks at least $|T_K| - k$ edges. As the data mule must return to base station then at least more $\min_{w \in S} d(v, w) - 1$ steps must be done.

To compute LB_k, $k > 1$, we have to solve a Steiner instance as subroutine. As Steiner Tree is NP-hard then such strategy is viable only for very small values of k. However even $k = 2$ is already able to improve our previous lower bound. In special, LB_2 can also be quickly found as described below.

Lemma 5. $LB_{k=2}$ *can be computed in* $O(m.n)$ *time.*

Proof. As $k = 2$ then the Steiner subroutine must connect only three terminals. Thus, at most one vertex, say u, of such Steiner tree has degree greater than two (equal to 3). Hence finding all shortest path between all pair of vertices, which can be done in $O(n.m)$ time, such Steiner tree can be constructed. Note that, given u, if any, the path from a terminal to u is a shortest path, and there exists only $O(n)$ possibilities for u.

3.2 Mathematical Formulation

A mathematical formulation, that considers all the characteristics of the problem is proposed in this subsection. Although the use of mathematical formulation is not the most appropriate approach for the Data Mule Routing Problem, it is a key technique for obtaining bounds. It is important to notice that in the problem solved using the formulation all information about the network is available, allowing to obtain a better path and consequently a reduction in the number of mule movements used to serve all sensor nodes.

Let $G'(V, A)$ be a bidirected graph formed by bidirectiong the edges of original graph $G(V, E)$. The set of vertices $V(G)$ are maintained as points in an Euclidean plane as the original graph. The set of arcs $(i, j) \in A$, represents the possible paths that the mule can choose. Two sets of variables are defined, x_{ij} and y_i. The variables x_{ij} are 1 if the data mule uses the arc (i, j) in his path, and 0 otherwise. The other binary variables y_i are set to 1 if the vertex $i \in V$ are visited by the mule, and 0 if it is attended by another node.

$$\min \quad \sum_{(i,j)\in A} x_{ij} \tag{1}$$

$$\text{s.t.} \quad \sum_{j\in N(i)\cup\{i\}} y_j \geq 1 \qquad \forall i \in V \tag{2}$$

$$\sum_{j\in\delta^+(i)} x_{ij} \geq y_i \qquad \forall i \in V \tag{3}$$

$$\sum_{j\in\delta^-(i)} x_{ji} \geq y_i \qquad \forall i \in V \tag{4}$$

$$\sum_{j\in\delta^+(i)} x_{ij} = \sum_{j\in\delta^-(i)} x_{ji} \qquad \forall i \in V \tag{5}$$

$$\sum_{j\in\delta^+(i)} x_{ij} \leq |N(i)|y_i \qquad \forall i \in V \tag{6}$$

$$\sum_{j\in\delta^-(i)} x_{ji} \leq |N(i)|y_i \qquad \forall i \in V \tag{7}$$

$$y_0 = 1 \tag{8}$$

$$\sum_{i\in\bar{S}}\sum_{j\in S} x_{ij} \geq y_s \qquad \forall S \subseteq V \setminus \{0\}, s \in S \tag{9}$$

$$x_{ij} \in \{0,1\} \qquad\qquad\qquad \forall (i,j) \in A \qquad (10)$$
$$y_i \in \{0,1\} \qquad\qquad\qquad \forall i \in V \qquad (11)$$

The objective function (1) aims to minimize the number of moves used by the mule to serve all nodes. Constraints (2) ensures that all nodes will be attended either by the mule's or by a neighbor in the mule's path. Constraints (3) and (4) guarantee that if one node is in the mule's path, at least one edge must enter and at least one edge must leave this node. The set of constraints (5) ensures that the number of edges entering and leaving one node must be the same. Constraints (6) and (7) imposes the limits of edges entering and leaving a node i by the number of their neighbours ($N(i)$). The constraint (8) ensures that the Base Station belongs in the mule's path. The set of constraints (9) eliminate sub-cycles. Finally, constraints (10) and (11) define the domain of the variables.

4   Algorithms for Data Mule with Local View

Now, we present two strategies to deal with the Data Mule Routing Problem, when the data mule does not have a global view, i.e, a prior knowledge of the network as a whole. In the first one, the mule decides his path based on the number of uncovered nodes nearby the current sensor node. In the second, the mule decision is based on the computation of convex-hulls of the current sensor [3].

4.1   Algorithms Based on Number of Uncovered Neighbours

The data mule begins its path in the sensor node that represents the base station s and decides which will be the next sensor to be visited by using a greedy method that will be next explained. The edges traversed by the data mule forms a tree, where the nodes of the tree represent the sensors visited by the data mule.

We consider here that a sensor node can be in one of the following states: (i) *dominator*, when the data mule is or was located in the same position of this sensor node, (ii) *covered*, when the sensor is or was within the communication range of a *dominator*, indicating that it has already been served by the data mule, and (iii) *uncovered*, when it is not in any of the previous described states. Initially, except for the Base Station, all nodes are in the *uncovered* state. In the end of the algorithm, every node will be in either a *dominator* or a *covered* state.

The proposed algorithms, one for the data mule and the other for the regular sensor nodes are described in Algorithms 1 and 2. They use the following types of messages to decide about the data mule moving: *msg_request*, contains a request for number of uncovered neighbours, sent from data mule to a regular sensor node; *msg_numernodes*, contains the number of uncovered neighbours, sent from regular sensor node to data mule; *msg_serve*, informs that the node can already be served, sent from data mule to regular sensor node; and *msg_served*, informs that the node has been served, sent among regular sensor nodes.

Algorithm 1. ALGNUM – DATA MULE ALGORITHM

---

1 **Variables:**
2      $parent \leftarrow \emptyset$
3      $states \leftarrow \emptyset$
4 **Upon reaching** node u coming from node v
5 **if** u *is not in dominator state* **then**
6      $states \leftarrow states \setminus \{< u, covered >\}$
7      $states \leftarrow states \cup \{< u, dominator >\}$
8      $parent \leftarrow parent \cup \{< u, v >\}$
9 **foreach** $v \in N(u)$ **do**
10      $states \leftarrow states \cup \{< v, covered >\}$
11   **Send** msg_serve to v
12   **Send** $msg\_request$ to v
13
14 **Upon** receiving all $msg\_numbernodes$ of $N(u)$
15 **if** $\exists\, v \in N(u) \mid numbernodes(v) \neq 0$ **then**
16   **Data Mule** moves to sensor node $t \in N(u)$ with the greatest $numbernodes$
17 **else**
18   **if** u *is the Base Station* **then**
19   **Data Mule** stop moving
20   **else**
21          $v \leftarrow get\_parent(parent, u)$
22   **Data Mule** moves to v

The data mule starts and finishes in a Base Station. Initially and upon reaching a sensor node u, the data mule sends the message msg_serve to all neighbours $N(u)$, indicating the node can be already served, updating their states to *covered* (lines 10–11 Algorithm 1). After that, the data mule also sends $msg\_request$ to them (line 12 Algorithm 1). Upon receiving that message (line 6 of Algorithm 2), a sensor node replies with the message $msg\_numernodes$ containing the number of neighbours in *uncovered* state (line 7 of Algorithm 2). The data mule, then, moves to the sensor node with the greatest number of uncovered neighbours (line 16 of Algorithm 1). When the mule arrives at a node u from a node v, if u is not in *dominator* state, it updates its state with *dominator* and the variable $parent(u)$ with v (lines 5–8 Algorithm 1). In this way, the data mule movement tree is being formed. When the data mule receives messages $msg\_numernodes$ containing only zeros, indicating that there is no neighbours in *uncovered* state, it moves to the parent of the current node. When this occurs in the Base Station, the data mule stops moving.

Regarding the regular sensor node, when it receives the msg_serve from data mule, it sends the message msg_served to all neighbours (lines 4–5 Algorithm 2). A sensor node when receives msg_served, decreases its variable containing the number of uncovered neighbours (lines 8–9 Algorithm 2). These steps allow for each sensor to keep the number of uncovered neighbours updated.

Algorithm 2. REGULAR SENSOR NODE u ALGORITHM

1 **Variables:**
2      $uncoveredNodes \leftarrow |N(u)|$
3 **Upon** receiving a message M
4 **case** $M = msg\_serve$
5 | **Send** msg_served to all $N(u)$
6 **case** $M = msg\_request$
7 | **Send** $msg\_numbernodes(uncoveredNodes)$ to mule
8 **case** $M = msg\_served$
9 | $uncoveredNodes--$

4.2 Algorithms Based on Convex-Hull

The second proposed approach, AlgCH, follows the same steps of AlgNUM. However the criterion used to decide the data mule's movement is based on the notion of convex-hull.

The convex hull of a set Q of points, called $CH(Q)$, is the smallest convex polygon P for which each point in Q is either on the boundary of P or in its interior. It is assumed that all points in Q are unique and that Q contains at least three no co-linear points [2]. An example of Q and the corresponding $CH(Q)$ are shown in Fig. 1.

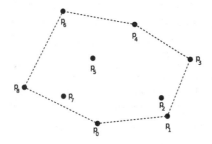

Fig. 1. $Q = \{p_0, p_1, \ldots, p_8\}$ and the corresponding $CH(Q)$ in dashed line

In this work, Graham's scan algorithm was used to calculate the convex-hull. Its running time is $O(n \; lg \; n)$, where n is the total number of points [2].

Data Mule Algorithm. In AlgCH approach, regular sensor nodes execute also Algorithm 2, described in Subsect. 4.1. Messages exchanged among the data mule and sensor nodes are the same previously used in AlgNUM. Only the data mule algorithm is modified, as presented in Algorithm 3. In both data mule algorithms, upon reaching a sensor node u, the date mule sends the message msg_serve to all nodes in $N(u)$, updating their status to *covered* (lines 10–12). However, in AlgCH before sending the next messages $msg\_request$, the mule calculates the convex-hull of $N(u) \cup \{u\}$ (line 13 Algorithm 3), by using Graham's scan algorithm [2]. Here, the mule only sends $msg\_request$ to the regular sensors of the convex-hull (lines 14–15 Algorithm 3), which, in turn, reply with the number of uncovered neighbors. The mule, as in the previous algorithm, will move to the sensor that replied with the larger number of uncovered neighbours.

That approach aims at reducing the number of exchanged message between the data mule and sensor nodes. That happens because nodes inside the convex-hull do not receive $msg\_request$. Another desired effect is the reduction in the number of data mule movements. The idea is that the data mule could cover more sensor nodes, with less movements, if it moved on the boundary of the convex-hull.

Algorithm 3. ALGCH – DATA MULE ALGORITHM

---

1 **Variables:**
2      $parent \leftarrow \emptyset$
3      $states \leftarrow \emptyset$
4      $CHnodes \leftarrow \emptyset$
5 **Upon *reaching* node u coming from node v**
6 | **if** u *is not in dominator state* **then**
7 | | $states \leftarrow states \setminus \{< u, covered >\}$
8 | | $states \leftarrow states \cup \{< u, dominator >\}$
9 | | $parent \leftarrow parent \cup \{< u, v >\}$
10 | **foreach** $v \in N(u)$ **do**
11 | | $states \leftarrow states \cup \{< v, covered >\}$
12 | | **Send** msg_serve to v
13 | $CHnodes \leftarrow Graham's\ Scan(N(u) \cup \{u\})$
14 | **foreach** $v \in CHnodes$ **do**
15 | | **Send** $msg\_request$ to v
16 **Upon *receiving* all $msg\_numbernodes$ of $CHnodes$**
17 | **if** $\exists v \in CHnodes \mid numbernodes(v) \neq 0$ **then**
18 | | **Data Mule** moves to sensor node $t \in CHnodes$ with the greatest $numbernodes$
19 | **else**
20 | | **if** u *is the Base Station* **then**
21 | | | **Data Mule** stop moving
22 | | **else**
23 | | | $v \leftarrow get\_parent(parent, u)$
24 | | | **Data Mule** moves to v

---

5   Computational Experiments and Analysis

In order to evaluate the proposed algorithms, we used instances generated for the Close-enough Traveling Salesman Problem [5]. We selected ten instances that respect our problem assumptions: (i) nodes are located in an Euclidean plan, (ii) nodes have the same acting range, in our case, communication range, and (iii) the instances are connected graphs. The number of nodes of the selected instances varies from 100 to 1000.

We implemented the heuristics in two scenarios. In the first one, the data mule waits for an acknowledgment, from each sensor node to which it sent a *msg_serve*, before sending *msg_request*. In its turn, the sensor node waits for acknowledgements, from all sensors to which it sent *msg_served*, to send the corresponding acknowledgments to the data mule. Note that, in this scenario, all sent messages *msg_serve* and *msg_served* are delivered and processed, and, consequently, every sensor knows the correct number of covered nodes, when it receives a *msg_request*. The second scenario is the same described in Sects. 4.1 and 4.2, and no acknowledgment message is employed.

We also implemented the mathematical formulation and the algorithm for calculating a lower bound, as presented in Sects. 3.2 and 3.1. Those results are used as baseline to evaluate the quality of results and execution times of the proposed algorithms.

The algorithms were implemented in the programming language C++, and used MPI for message-passing. The mathematical formulation was implemented in the programming language C++ and used IBM ILOG CPLEX Optimizer v12.5.1 as mixed integer programming solver. The algorithm to calculate the lower bound (LB_3) calculation was implemented in C++ and used a graph library, LEMON[1]. Our tests were executed in a Intel Core i7 3.6 Ghz computer, with 16 GB of RAM and Linux Mint 18 as its operating system.

Table 1 presents the results obtained by the methods that have the global view of the network. It shows instance names in column **Inst.**; the lower bounds, in column Sol_{LB_3}, found by using the Steiner Tree with $k = 3$, as presented in Lemma 4; the results obtained by the mathematical formulation are presented in columns, **LR**, linear relaxation on root node, $T_{LR}(s)$, time to obtain this linear relaxation, Sol_{Mat}, solution of entire formulation presented in Sect. 3.2, and the corresponding time, **T(s)**.

It was not possible to prove the optimality of solution found for instance bonus1000, so only the feasible solution found was presented in the table. Note that the results showed that very good lower bounds were generated, matching with the optimal solutions in 5 of 10 instances and presenting better results concerning the linear relaxation of the mathematical formulation. Those results indicate that the lower bound algorithm gives good solutions, that can be also used as a baseline to our analyse.

Table 2 summarizes the results obtained by the proposed algorithms. Column **Inst.** also presents instance names and the other columns present the obtained results by **AlgNum** and **AlgCH** and the corresponding number of exchanged messages, **msgs**. Note that in **Case 2 – Without ACK**, the results are averages of 10 executions, because, as the algorithms do not use messages of acknowledgments, different solutions can be found at each execution.

---

[1] LEMON – Library for Efficient Modeling and Optimization in Networks, available on https://lemon.cs.elte.hu.

Table 1. Computational results – global view

Inst.	LB_3		Mathematical Formulation			
	Sol_{LB_3}	T(s)	LR	$T_{LR}(s)$	Sol_{Math}	T(s)
kro100	4	0.01	3	0.01	4	1.31
rat195	4	0.07	3	0.02	4	4.27
team2_200	4	0.06	4	0.03	5	14.35
team3_300	32	0.09	19	0.07	74	17996.7
lin318	4	0.18	3.67	0.27	5	6012.51
rd400	6	0.28	5	0.97	6	7336.95
pcb442	6	0.38	4.14	0.88	6	37180.6
team6_500	3	0.66	3	2.20	3	225.29
dsj1000	6	3.00	4	2.50	8	24842.1
bonus1000	8	2.12	7.86	38.46	22*	86400

* an optimal solution was not found in a limit of 24 h

Remark that the mathematical formulation has complete knowledge of the network while the heuristics are locality sensitive and have incomplete knowledge. Then, it was already expected that the results given by the algorithms are worse than the ones given by the mathematical formulation.

Table 2. Computational results – locality sensitive heuristics

Inst.	Case 1 – With ACK				Case 2 – Without ACK			
	Sol_{NUM}	msgs	Sol_{CH}	msgs	Sol_{NUM}	msgs	Sol_{CH}	msgs
kro100	6	11108	6	10538	10.0	6664.7	15.3	5852.4
rat195	4	43006	4	42108	6.4	23040.8	10.4	21795.9
team2_200	10	30562	12	29016	14.6	17515.6	18.4	15212.2
team3_300	74	21742	74	19556	94.2	13586.5	88.6	10876.1
lin318	8	94250	10	91874	13.2	51019.1	13.4	46772.0
rd400	14	112694	14	108482	22.8	62300.5	17.2	54981.5
pcb442	12	169950	12	165778	17.4	90869.9	17.8	85027.3
team6_500	6	312252	6	307818	12.0	164828.5	9.8	154971.2
dsj1000	8	837514	10	828926	17.8	431716.1	19.9	418260.2
bonus1000	22	450712	26	441438	36.4	238016.4	35.4	222859.8

Results presented in Table 2 show that the scenario **With ACK** produces better results, concerning the number of edges of the data mule path, than its counterpart. This occurs because in this scenario the data mule makes decisions based on a consistent state of the network. On the other hand, the number of exchanged messages increases thanks to ACK messages. In the scenario **With ACK**, *AlgNum* outperforms *AlgCH* in four instances and gives the same results in the other six, using, however, a greater number of messages. In the scenario **Without ACK**, *AlgNum* outperforms *AlgCH* in seven instances. However, *AlgCH* finds better solutions in three instances. Here, *AlgCH* also employs less messages than *AlgNum*. The proposed algorithms gave the optimal solutions in six cases (4 in AlgNUM and 2 in AlgCH) and they presented an average worsening percentage of 61.10% and 164.87% in **With ACK** and **Without ACK** cases, respectively, when compared with the mathematical formulation results. The results obtained by the proposed locality sensitive heuristics are good, since the mathematical formulation results were obtained in an unrealistic scenario where the data mule had the complete knowledge of the network.

References

1. Bin Tariq, M.M., Ammar, M., Zegura, E.: Message ferry route design for sparse ad hoc networks with mobile nodes. In: Proceedings of the 7th ACM International Symposium on Mobile Ad Hoc Networking and Computing. MobiHoc 2006, pp. 37–48. ACM, New York (2006)
2. Cormen, T.H., Stein, C., Rivest, R.L., Leiserson, C.E.: Introduction to Algorithms, 2nd edn. McGraw-Hill Higher Education, New York (2001)
3. Islam, K., Akl, S.G., Meijer, H.: A constant factor localized algorithm for computing connected dominating sets in wireless sensor networks. In: 14th IEEE International Conference on Parallel and Distributed Systems. ICPADS 2008, pp. 559–566. IEEE (2008)
4. Jang, H.C., Lien, Y.N., Tsai, T.C.: Rescue information system for earthquake disasters based on MANET emergency communication platform. In: Proceedings of the 2009 International Conference on Wireless Communications and Mobile Computing: Connecting the World Wirelessly, IWCMC 2009, pp. 623–627. ACM, New York (2009)
5. Mennell, W.K.: Heuristics for solving three routing problems: close-enough traveling salesman problem, close-enough vehicle routing problem, sequence-dependent team orienteering problem. Ph.D. thesis, University of Maryland (College Park, Md.), College Park, Maryland, USA (2009)
6. Puccinelli, D., Haenggi, M.: Wireless sensor networks: applications and challenges of ubiquitous sensing. IEEE Circuits Syst. Mag. **5**, 2005 (2005)
7. Sahin, C.S., et al.: Uniform distribution of mobile agents using genetic algorithms for military applications in MANETs. In: 2008 Military Communications Conference. MILCOM 2008, pp. 1–7. IEEE (2008)
8. Sharma, S., Bansal, R.K., Bansal, S.: Issues and challenges in wireless sensor networks. In: 2013 International Conference on Machine Intelligence and Research Advancement (ICMIRA), pp. 58–62. IEEE (2013)
9. Stojmenovic, I., Seddigh, M., Zunic, J.: Dominating sets and neighbor elimination-based broadcasting algorithms in wireless networks. IEEE Trans. Parallel Distrib. Syst. **13**(1), 14–25 (2002)
10. Sugihara, R., Gupta, R.K.: Path planning of data mules in sensor networks. ACM Trans. Sens. Netw. **8**(1), 1:1–1:27 (2011)

11. Zhao, W., Ammar, M.: Message ferrying: proactive routing in highly-partitioned wireless ad hoc networks. In: 2003 Proceedings of the Ninth IEEE Workshop on Future Trends of Distributed Computing Systems. FTDCS 2003, pp. 308–314 (2003)
12. Zhao, W., Ammar, M., Zegura, E.: Controlling the mobility of multiple data transport ferries in a delay-tolerant network. In: Proceedings IEEE 24th Annual Joint Conference of the IEEE Computer and Communications Societies, vol. 2, pp. 1407–1418. IEEE (2005)

Maximize a Monotone Function
with a Generic Submodularity Ratio

Qingqin Nong[1]([envelope])[ORCID], Tao Sun[1], Suning Gong[1], Qizhi Fang[1], Dingzhu Du[2][ORCID],
and Xiaoyu Shao[1]

[1] School of Mathematical Science, Ocean University of China,
Qingdao 266100, Shandong, People's Republic of China
qqnong@ouc.edu.cn
[2] Department of Computer Science, University of Texas, Dallas 75083, USA

Abstract. Generic submodularity ratio γ is a general measurement to characterize how close a nonnegative monotone set function is to be submodular. In this paper, we make a systematic analysis of greedy algorithms for maximizing a monotone and normalized set function with a generic submodularity ratio γ under Cardinality constraints, Knapsack constraints, Matroid constraints and K-intersection constraints.

Keywords: Non-submodular · Greedy · Independent system

1 Introduction

Many combinatorial optimization problems may be formulated as the maximization of a set-function f defined on a ground set N. For any pair of $S, T \subseteq N$, denote by $f_S(T) = f(S \cup T) - f(S)$ the marginal gain of set T in S. Specially, denote by $f_S(j) = f(S + j) - f(S)$ the marginal gain of singleton set $\{j\}$ in S. The function $f(\cdot)$ is called submodular, if $f_S(j) \geq f_T(j)$ for any $S \subseteq T$ and any $j \in N \setminus T$. Many set functions that occur in practical problems turn out to be submodular functions or slight modifications thereof. Thus, submodular maximization applies in many areas of computer science and applied mathematics, such as machine learning [1], computer vision [10], operations research [9].

Extensive work has been conducted in the area of optimizing submodular maximization problems. The greedy approach is a basic method for the problems: start from an empty set; iteratively add to the current solution set one element that results in the largest marginal gain of the objective function while satisfying the constraints. Historically, one of the very first problems examined was maximizing a monotone ($f(S) \leq f(T)$ whenever $S \subseteq T$) and normalized

This research was supported in part by the National Natural Science Foundation of China under grant numbers 11201439 and 11871442, and was also supported in part by the Natural Science Foundation of Shandong Province under grant number ZR2019MA052 and the Fundamental Research Funds for the Central Universities.

D.-Z. Du et al. (Eds.): AAIM 2019, LNCS 11640, pp. 249–260, 2019.
https://doi.org/10.1007/978-3-030-27195-4_23

($f(\emptyset) = 0$) submodular function under various constraints. We mention here the most relevant results.

For Cardinality constraints, Fisher et al. [7] and Nemhauser et al. [16] proved that for submodular maximization, if f is nonnegative, monotone, greedy approach yields a $(1 - \frac{1}{e})$-approximation. Feige [6] proved that unless $P = NP$, no polynomial time algorithm can achieve an approximation ratio better than $(1 - \frac{1}{e})$ for the cardinality constrained maximization of set cover functions.

For Knapsack constraints, Wolsey [18] presented a modified greedy algorithm, with performance guarantee $1 - \frac{1}{e^\beta} \approx 0.35$, where β is a unique root of equation $e^x = 2 - x$. Khuller et al. [12] were the first to obtain an approximation ratio of $(1 - \frac{1}{e})$ using partial enumeration in the case of a set cover function. The result extended to any submodular function by Sviridenko [19].

For Matroid constraints, Fisher et al. [7] proved that greedy approach yields a 1/2-approximation. This result was improved to $(1 - \frac{1}{e})$ by Calinescu et al. [4] using continuous optimization. A combinatorial algorithm was later constructed by Filmus and Ward [8]. A matching lower bound was due to [16,17]. By introducing the total curvature $c = 1 - \min\limits_{S, j \notin S} \frac{f_S(e)}{f_\emptyset(j)}$, Conforti and Cornuéjols [3] proved that, when f additionally has total curvature c, greedy approches yield a $(1 - e^{-c})/c$-approximation for a uniform matroid and a $1/(1 + c)$-approximation for a general matroid.

For K-intersection constraints, Fisher et al. [7] showed that a greedy algorithm has an approximation factor of $\frac{1}{K+1}$. This result was improved by Lee et al. [15] to $\frac{1}{K+\delta}$, for any constant $\delta > 0$, using a local search approach that exploits exchange properties of the underlying combinatorial structure. Some more general classes like K-systems, K-extendible set systems and etc. were also considered [7,11].

Submodular maximization plays an important role in combinatorial optimization. However, for many applications, including experimental design and sparse Gaussian processes [14], $f(\cdot)$ is in general not submodular [13]. Based on one of equivalent definitions of submodular functions, Das and Kempe [5] proposed the submodularity ratio, $\widetilde{\gamma} = \min\limits_{\Omega, S \subseteq N} \frac{\sum_{j \in \Omega \setminus S} f_S(j)}{f_S(\Omega)}$. It is a quantity characterizing how close a set function is to being submodular. Bian et al. [2] combined and generalized the ideas of curvature and submodularity ratio and gave a $(1 - e^{-c\widetilde{\gamma}})/c$-approximation for maximization problem of nonsubmodular function f with curvature c and submodularity ratio $\widetilde{\gamma}$ under Cardinality constraints.

We say that a set function $f(\cdot)$ is strictly monotone if it is monotone and $f(S) < f(T)$ whenever $S \subset T$. Generic submodularity ratio γ is another quantity characterizing how close a nonnegative nondecreasing set function is to be submodular. It is derived from a different equivalent definition of submodular functions. In this paper we make a systematic analysis of greedy algorithms for maximizing a strictly monotone and normalized set function with a generic submodularity ratio γ under Cardinality constraints, Knapsack constraints, Matroid constraints and K-intersection constraints.

2 Preliminaries

Generic submodularity ratio measures how many times the marginal gains of joining an element to a set than that of a superset.

Definition 1 (Generic Submodularity Ratio). *Given a ground set N and a nondecreasing set function $f : 2^N \to \mathcal{R}_+$, the generic submodularity ratio of f is the largest scalar γ such that for any $S \subseteq T \subseteq N$ and any $j \in N \setminus T$,*

$$f_S(j) \geq \gamma \cdot f_T(j).$$

Remarks:

- (a) $\gamma \in [0,1]$. Due to nonnegative and nondecreasing properties of $f(\cdot)$, $f_A(B) = f(A \cup B) - f(A) \geq 0$ for any $A, B \subseteq N$. This implies $\gamma \geq 0$. Observe that the equality $f_S(j) = f_T(j)$ holds when $S = T$, we have γ can not be larger than 1.
- (b) $f(\cdot)$ is submodular iff $\gamma = 1$. f is a submodular set function if and only if the inequality $f_A(j) \geq f_B(j)$ holds for any pair of A, B with $A \subseteq B$ and $j \in N \setminus B$.

Thus $\gamma \geq 1$. Combining with $\gamma \leq 1$, we obtain the result.

For Convenience, henceforth we say that a function $f(\cdot)$ is γ-submodular if its generic submodularity ratio is γ. We say that a function $f(\cdot)$ is γ-subadditive if for any pair subsets of N, A and B, satisfying

$$f(A) + f(B) \geq \gamma f(A \cup B).$$

We have the following results.

Proposition 1. *If $f(\cdot)$ is γ-submodular, for all $S \subseteq N$, $f_S(\cdot)$ is γ-submodular.*

Proof. Consider A and B two subsets of N. Assume that $A \subseteq B$. The marginal contribution of an element j to A under function $f_S(\cdot)$ is

$$\begin{aligned}
f_{S,A}(j) &= f_S(A + j) - f_S(A) \\
&= [f(A + j \cup S) - f(S)] - [f(A \cup S) - f(S)] \\
&= f(A + j \cup S) - f(A \cup S) \\
&= f_{A \cup S}(j).
\end{aligned}$$

Similarly, the marginal contribution of an element j to B under function $f_S(\cdot)$ is $f_{S,B}(j) = f_{B \cup S}(j)$. Since $f(\cdot)$ is γ-submodular and $A \cup S$ is a subset of $B \cup S$, we have

$$f_{S,A}(j) = f_{A \cup S}(j) \geq \gamma f_{B \cup S}(j) = \gamma f_{S,B}(j).$$

Proposition 2. *If $f(\emptyset) = 0$ and $f(\cdot)$ is γ-submodular, for all $S \subseteq N$, $f_S(\cdot)$ is γ-subadditive.*

Proof. Consider A and B two subsets of N. Write $B = \{e_1, \ldots, e_b\}$, $B_i = \{e_1, \ldots, e_i\}$ $(1 \le i \le b)$ and $B_0 = \emptyset$.

$$f(A \cup B) = f(A) + \sum_{i=1}^{b} f_{A \cup B_{i-1}}(e_i)$$
$$\le f(A) + \sum_{i=1}^{b} \frac{1}{\gamma} f_{B_{i-1}}(e_i)$$
$$= f(A) + \frac{1}{\gamma} f(B)$$
$$\le \frac{1}{\gamma}(f(A) + f(B)),$$

where the first inequality holds from the assumption that $f(\cdot)$ is γ-submodular and the last one holds from the fact that $\gamma \le 1$.

Corollary 1. *Let $f(\cdot)$ be a γ-submodular function, then:*

$$\forall S \subseteq T \subseteq N, f(T) \le f(S) + \frac{1}{\gamma} \sum_{j \in T \setminus S} f_S(j)$$

Proof. Since $f_S(\emptyset) = 0$ and $f(\cdot)$ is γ-submodular implying that $f_S(\cdot)$ is γ-subadditive, we have

$$f(T) = f(S) + f_S(T \setminus S) \le f(S) + \frac{1}{\gamma} \sum_{j \in T \setminus S} f_S(j).$$

Clearly, the generic submodularity ratio of a strictly monotone function is greater than 0. In the sequel, we consider the problem of maximizing a non-negative strictly monotone and normalized set function $f(\cdot)$ under Cardinality constraints, Knapsack constraints, Matroid constraints and K-intersection constraints respectively.

3   Cardinality Constraint

Let $f(\cdot)$ be a strictly monotone γ-submodular function and assume that it is normalized, that is, $f(\emptyset) = 0$. Consider the maximization problem $\max_{|S| \le k} f(S)$ and a greedy algorithm for it. Let S^* be the optimal solution of the problem.

Theorem 1. *Let S_G be the set returned by Algorithm 1. Then*

$$f(S_G) \ge (1 - e^{-\gamma}) f(S^*).$$

Proof. Denote by $S_t = \{\tilde{e}_1, \tilde{e}_2, \cdots, \tilde{e}_t\}$ the value of S_G after the tth time line 4 of Algorithm 1 is executed, where $\tilde{e}_t$ is the tth element chosen in S_G. Let $S_0 = \emptyset$. Then

$$f(S^*) \le f(S_{t-1} \cup S^*)$$
$$\le f(S_{t-1}) + \frac{1}{\gamma} \cdot \sum_{e \in S^* \setminus S_{t-1}} f_{S_{t-1}}(e)$$
$$\le f(S_{t-1}) + \frac{1}{\gamma} \cdot \sum_{e \in S^* \setminus S_{t-1}} f_{S_{t-1}}(\tilde{e}_t)$$
$$= f(S_{t-1}) + \frac{1}{\gamma} \cdot \sum_{e \in S^* \setminus S_{t-1}} (f(S_t) - f(S_{t-1}))$$
$$\le f(S_{t-1}) + \frac{k}{\gamma} \cdot (f(S_t) - f(S_{t-1})),$$

Algorithm 1. Greedy Algorithm (Cardinality Constraint)

Input: A ground set N, value query oracle for a strictly monotone γ−submodular set
function $f : 2^N \to \mathcal{R}_+$, k

Output: A solution of Problem $\max\limits_{|S| \leq k} f(S)$

1: Initialize: $S_G \leftarrow \emptyset$
2: **while** $|S_G| < k$ **do**
3:     $\tilde{e} \leftarrow \arg\max_{e \in N} f_{S_G}(e)$
4:     $S_G \leftarrow S_G + \tilde{e}$
5:     $N \leftarrow N - \tilde{e}$
6: **end while**
7: **return** S_G

---

where the first inequality follows from the monotonicity of f, the second inequality follows from Corollary 1, the third inequality follows from the greediness of Algorithm 1 and the last inequality follows from $|S^*| \leq k$. Rearrange the terms of the above inequality, we have

$$f(S_t) - f(S^*) \geq (1 - \frac{\gamma}{k})[f(S_{t-1}) - f(S^*)],$$

implying

$$f(S_t) - f(S^*) \geq (1 - \frac{\gamma}{k})^t [f(S_0) - f(S^*)].$$

Together with the fact that $S_0 = \emptyset$ and $f(\emptyset) = 0$, the above inequality implies

$$f(S_t) \geq (1 - (1 - \frac{\gamma}{k})^t) f(S^*).$$

Let $t = k$ and using $(1 - \frac{\gamma}{k})^k \leq e^{-\gamma}$, we have $f(S_G) \geq (1 - e^{-\gamma}) f(S^*)$.

4   Knapsack Constraint

There is a cost function $c : N \to \mathcal{R}^+$ and a budget $B \in \mathcal{R}^+$. For any $S \subseteq N$, let $c(S) = \sum\limits_{e \in S} c(e)$. Consider the optimization problem $\max\limits_{c(S) \leq B} f(S)$. A natural greedy way is, without violating the budget, in each iteration picking the element with maximum marginal contribution divided by the cost.

If $S^* \setminus S_G = \emptyset$, we have $f(S_G) \geq f(S^*)$. Thus assume that $S^* \setminus S_G \neq \emptyset$. List the elements of N in the order that is deleted from N in Algorithm 2. Suppose that $\ddot{e}$ is the first element that is in $S^* \setminus S_G$. We have the following theorem.

Lemma 1. *Let $\dot{e}$ be an arbitrary element that is scheduled not later than $\ddot{e}$ and line 4 of Algorithm 2 evaluates false. Then*

$$f(S_G + \dot{e}) \geq (1 - e^{-\gamma}) f(S^*).$$

Algorithm 2. Greedy Algorithm (Knapsack Constraint)

Input: A ground set N, value query oracle for a monotone γ-submodular set function
   $f : 2^N \to \mathcal{R}^+$, cost function $c : N \to \mathcal{R}^+$ and a budget B
Output: A solution of Problem $\max\limits_{c(S) \leq B} f(S)$

1: Initialize: $S_G \leftarrow \emptyset$
2: **while** $N \neq \emptyset$ **do**
3:    $\tilde{e} \leftarrow \arg\max_{e \in N} \frac{f_{S_G}(e)}{c(e)}$
4:    **if** $c(S_G) + c(\tilde{e}) \leq B$ **then**
5:       $S_G \leftarrow S_G + \tilde{e}$
6:    **end if**
7:    $N \leftarrow N - \tilde{e}$
8: **end while**
9: **return** S_G

Proof. Suppose that the value of S_G is S_t (t elements have been selected) exactly before $\dot{e}$ is deleted from N in the algorithm. From the definitions of $\dot{e}$ and $\ddot{e}$ respectively, one can see that each element in either $S_G \setminus S_t$ or $S^* \setminus S_G$ is schedule after $\dot{e}$. This means that each element in $S^* \setminus S_t$ is scheduled after $\dot{e}$. Note that, for each element e scheduled after $\dot{e}$, $\frac{f_{S_t}(e)}{c(e)} \leq \frac{f_{S_t}(\dot{e})}{c(\dot{e})}$. We have

$$
\begin{aligned}
f(S^*) &\leq f(S_t \cup S^*) \\
&\leq f(S_t) + \frac{1}{\gamma} \cdot \sum_{e \in S^* \setminus S_t} f_{S_t}(e) \\
&= f(S_t) + \frac{1}{\gamma} \cdot \sum_{e \in S^* \setminus S_t} c(e) \frac{f_{S_t}(e)}{c(e)} \\
&\leq f(S_t) + \frac{1}{\gamma} \cdot \sum_{e \in S^* \setminus S_t} c(e) \frac{f_{S_t}(\dot{e})}{c(\dot{e})} \\
&\leq f(S_t) + \frac{B}{\gamma c(\dot{e})} \cdot [f(S_t + \dot{e}) - f(S_t)],
\end{aligned}
$$

where the first inequality holds from the monotonicity of f, the second inequality holds from Corollary 1, the third inequality holds from the greediness of Algorithm 2 and the fact the each element in $S^* \setminus S_t$ is deleted from N later than $\dot{e}$, the last inequality holds from the fact that $c(S^*) \leq B$. Rearrange the terms of the above inequality, we have

$$
f(S_t + \dot{e}) - f(S^*) \geq (1 - \frac{\gamma c(\dot{e})}{B})[f(S_t) - f(S^*)]. \tag{1}
$$

Similarly, consider the induction relation between S_t and S_{t-1}. As the proof of Theorem 1, let $\tilde{e}_t$ be the tth element chosen in S_G.

$$
\begin{aligned}
f(S^*) &\leq f(S_{t-1} \cup S^*) \\
&\leq f(S_{t-1}) + \frac{1}{\gamma} \cdot \sum_{e \in S^* \setminus S_{t-1}} f_{S_{t-1}}(e) \\
&= f(S_{t-1}) + \frac{1}{\gamma} \cdot \sum_{e \in S^* \setminus S_{t-1}} c(e) \frac{f_{S_{t-1}}(e)}{c(e)}
\end{aligned}
$$

$$\leq f(S_{t-1}) + \frac{1}{\gamma} \cdot \sum_{e \in S^* \setminus S_{t-1}} c(e) \frac{f_{S_{t-1}}(\tilde{e}_t)}{c(\tilde{e}_t)}$$
$$\leq f(S_{t-1}) + \frac{B}{\gamma c(\tilde{e}_t)} \cdot [f(S_t) - f(S_{t-1})]$$

where the third inequality follows from the fact that $S^* \setminus S_{t-1} \subseteq (S^* \setminus S_t) + \tilde{e}_t$ and thus each element in $S^* \setminus S_{t-1}$ is deleted from N not earlier than $\tilde{e}_t$. Thus

$$f(S_t) - f(S^*) \geq (1 - \frac{\gamma c(\tilde{e}_t)}{B})[f(S_{t-1}) - f(S^*)], \tag{2}$$

From inequalities (1) and (2), we have

$$f(S_t + \dot{e}) - f(S^*) \geq (1 - \frac{\gamma c(\dot{e})}{B}) \prod_{i=1}^{t} (1 - \frac{\gamma c(\tilde{e}_i)}{B})[f(S_0) - f(S^*)].$$

Using that $1 - x \leq e^{-x}$,

$$f(S_t + \dot{e}) - f(S^*) \geq \exp(-\frac{\gamma c(\dot{e})}{B}) \prod_{i=1}^{t} \exp(-\frac{\gamma c(\tilde{e}_i)}{B}))[f(S_0) - f(S^*)].$$
$$= \exp(-\frac{\gamma c(S_t + \dot{e})}{B})[f(S_0) - f(S^*)]$$

Since $f(S_0) = 0$ and $c(S_t + \dot{e}) > B$, we have $f(S_t + \dot{e}) \geq (1 - e^{-\gamma})f(S^*)$. Since f is monotone, $f(S_G + \dot{e}) \geq f(S_t + \dot{e})$ and concludes the proof.

Algorithm 3. Greedy Algorithm (Knapsack Constraint, simple fix)

Input: A ground set N, value query oracle for a monotone γ−submodular set function
   $f : 2^N \to \mathcal{R}^+$, cost function $c : N \to \mathcal{R}^+$ and a budget B
Output: A solution of Problem $\max_{c(S) \leq B} f(S)$
1: $e^* \leftarrow \arg\max_{e \in N, c(e) \leq B} f(\{e\})$
2: $S_G \leftarrow$ result of Algorithm 2
3: **return** $\arg\max\{f(S_G), f(e^*)\}$

Theorem 2. *Let S be the set returned by Algorithm 3. Then*

$$f(S) \geq \frac{\gamma(1 - e^{-\gamma})}{2} f(S^*).$$

Proof. As before, let $\dot{e}$ be an arbitrary element that is scheduled not later than $\ddot{e}$ and line 4 of Algorithm 2 evaluates false. Then by Lemma 1

$$(1 - e^{-\gamma})f(S^*) \leq f(S_G + \dot{e}) = f(S_G) + f_{S_G}(\dot{e}) \leq f(S_G) + \frac{1}{\gamma}f(\dot{e})$$
$$\leq \frac{1}{\gamma}(f(S_G) + f(\dot{e})) \leq \frac{1}{\gamma}(f(S_G) + f(e^*)) \leq \frac{2}{\gamma}f(S).$$

The result follows.

Algorithm 4. Greedy Algorithm 4 (Knapsack Constraint, Partial Enumeration)

Input: A ground set N with n elements, value query oracle for a monotone γ-submodular set function $f : 2^N \to \mathcal{R}^+$, cost function $c : N \to \mathcal{R}^+$, budget B

Output: A solution of Problem $\max\limits_{c(S) \leq B} f(S)$

1: $\tilde{S} \leftarrow \arg\max\limits_{S \subseteq N, c(S) \leq B, |S| < d} f(S)$
2: $\bar{S} \leftarrow \emptyset$
3: **for all** $S \subseteq N, |S| = d, c(S) \leq B$ **do**
4: $\quad N' \leftarrow N \setminus S$
5: $\quad S_G \leftarrow$ Algorithm 2 for N' with initialization $S_G \leftarrow S$, $f \leftarrow f_S$ and $B \leftarrow B - c(S)$
6: $\quad$ **if** $f(S_G) > f(\bar{S})$ **then**
7: $\quad\quad \bar{S} \leftarrow S_G$
8: $\quad$ **end if**
9: **end for**
10: **return** $\arg\max\{f(\tilde{S}), f(\bar{S})\}$

Theorem 3. *For* $d \geq \frac{e^\gamma}{\gamma^2}$, *Algorithm 4 returns a set* T *with* $f(T) \geq (1 - e^{-\gamma})f(S^*)$.

Proof. If $|S^*| \leq d$, Algorithm 4 finds the optimal solution. Consider the case that $|S^*| > d$. Order the elements in $S^* = \{e_1^*, \ldots, e_l^*\}$ such that

$$e_i^* \in \arg\max\limits_{e \in S^* \setminus S_{i-1}^*} f_{S_{i-1}^*}(e),$$

where $S_{i-1}^* = \{e_1^*, \ldots, e_{i-1}^*\}$ and $S_0^* = \emptyset$. Consider the iteration of Algorithm 4 where the initialization is $f \leftarrow f_{S_d^*}$, $S_G \leftarrow S_d^*$ and $B \leftarrow B - c(S_d^*)$. Then the optimal solution of problem $\max\limits_{c(S) \leq B - c(S_d^*)} f_{S_d^*}(S)$ is $S^* \setminus S_d^*$. Let us consider the moment that Algorithm 2 evaluated to false for the element $\ddot{e}$, the first element that is in $S^* \setminus S_G$. By Lemma 1

$$f_{S_d^*}((S_G \setminus S_d^*) + \ddot{e}) \geq (1 - e^{-\gamma})f_{S_d^*}(S^* \setminus S_d^*),$$

equivalently,

$$f(S_G + \ddot{e}) - f(S_d^*) \geq (1 - e^{-\gamma})[f(S^*) - f(S_d^*)].$$

Thus $f(S_G) + f_{S_G}(\ddot{e}) \geq (1 - e^{-\gamma})f(S^*) + \frac{1}{e^\gamma}f(S_d^*)$, implying

$$f(S_G) \geq (1 - e^{-\gamma})f(S^*) + \frac{1}{e^\gamma}f(S_d^*) - f_{S_G}(\ddot{e})$$
$$\geq (1 - e^{-\gamma})f(S^*) + \frac{1}{e^\gamma}f(S_d^*) - \frac{1}{\gamma}f_{S_d^*}(\ddot{e}), \tag{3}$$

where the last inequality holds from $S_d^* \subseteq S_G$ and thus $f_{S_d^*}(\ddot{e}) \geq \gamma f_{S_G}(\ddot{e})$. Since $f_{S_d^*}(\ddot{e}) \leq \frac{1}{\gamma}f_{S_{j-1}^*}(\ddot{e}) \leq \frac{1}{\gamma}f_{S_{j-1}^*}(e_j^*)$ for each $j \in \{1, \ldots, d\}$,

$$f_{S_d^*}(\ddot{e}) \leq \frac{1}{\gamma} \min\limits_{1 \leq j \leq d} f_{S_{j-1}^*}(e_j^*).$$

Together with $f(S_d^*) = \sum_{j=1}^{d} f_{S_{j-1}^*}(e_j^*)$, we have

$$f_{S_d^*}(\ddot{e}) \leq \frac{1}{\gamma d} f(S_d^*).$$

This means that if $d \geq \frac{e^\gamma}{\gamma^2}$, Algorithm 4 returns a set, denoted by T, with

$$\begin{aligned}
f(T) &\geq f(S_G) \\
&\geq (1 - e^{-\gamma}) f(S^*) + \frac{1}{e^\gamma} f(S_d^*) - \frac{1}{\gamma} f_{S_d^*}(\ddot{e}) && \text{[By Inequality (3)]} \\
&\geq (1 - e^{-\gamma}) f(S^*) + \frac{1}{e^\gamma} f(S_d^*) - \frac{1}{\gamma^2 d} f(S_d^*) \\
&\geq (1 - e^{-\gamma}) f(S^*).
\end{aligned}$$

The result follows.

5 K-Intersection Constraint

Given a ground set N, a pair $(N, \mathcal{I})$ is called an *independent system* if $\mathcal{I} \subseteq 2^N$ is hereditary (that is, for every set $S \in (N, \mathcal{I})$, every set $S' \subseteq S$ is also in $\mathcal{I}$.) Independent systems are further divided into a few known classes. The following classes of independent systems are probably most highly researched

Definition 2 *(Matroid). An independent system is a matroid if for every two sets $S, T \in \mathcal{I}$ such that $|T| > |S|$, there exists an element $x \in T \setminus S$, such that $S \cup \{x\} \in \mathcal{I}$. This property is called the augmentation property of matroid.*

Definition 3 *(K-intersection). An independent system $\mathcal{M} = (N, \mathcal{I})$ is $K-$ intersection if there exist K matroids $\mathcal{M}_i = (N, \mathcal{I}_i)(1 \leq i \leq K)$ such that a set $S \subseteq N$ is in $\mathcal{I}$ if and only if $S \in \bigcap_{i=1}^{K} \mathcal{I}_i$.*

Let $\mathcal{M} = (N, \mathcal{I})$ be an independent system. A base of set V $(V \subseteq N)$ is a maximal independent subset of V.

Definition 4 *(K-system). An independent system $\mathcal{M} = (N, \mathcal{I})$ is a $K-$system if for every set $V \subseteq N$ the ratio of the sizes of the largest base of V to the smallest base of V is at most K.*

Here is the known hierarchy of set systems:

$$\text{matroid} \subseteq K - \text{intersection} \subseteq K - \text{system}.$$

Clearly, matroid is a special case of K-intersection with $K = 1$. In this section we first consider the maximization of strictly monotone $\gamma-$submodular set function $f : 2^N \to \mathcal{R}^+$ under K-intersection constraints. We will obtain a byproduct of the optimization problem under matroid constraints. We assume that each matroid in the K-intersection is given by an independence oracle that answers whether or not $S \in \mathcal{I}_i$. We first describe the greedy heuristic and then analyze the approximate ratio of this approach.

Algorithm 5. Greedy Algorithm 5 (K-Intersection Constraint)

Input: A ground set N, value query oracle for strictly monotone γ-submodular set
  function $f : 2^N \to \mathcal{R}^+$, a K-intersection $\mathcal{M} = (N, \mathcal{I})$
Output: A solution of Problem $\max_{S \in \mathcal{I}} f(S)$
1: Initialization: $S_0 \leftarrow \emptyset$ and $t \leftarrow 1$
2: **while** $N \neq \emptyset$ **do**
3:     $\tilde{e} \leftarrow \max_{e \in N} f_{S_{t-1}}(e)$
4:     **if** $S_{t-1} + \tilde{e} \in \mathcal{I}$ **then**
5:        $\tilde{e}_t \leftarrow \tilde{e}$, $S_t \leftarrow S_{t-1} + \tilde{e}_t$ and $t \leftarrow t + 1$
6:     **end if**
7:     $N \leftarrow N - \tilde{e}$
8: **end while**
9: **return** S_t

Suppose that $t = g$ when the algorithm terminates. Let $S_g = \{\tilde{e}_1, \tilde{e}_2, \cdots, \tilde{e}_g\}$ be the set returned by the greedy algorithm, where $\tilde{e}_t$ is the element chosen in Iteration t. Denote $S_t = \{\tilde{e}_1, \tilde{e}_2, \cdots, \tilde{e}_t\}$. Let $U_{[t,t+1)}$ $(t = 1, 2, \cdots, g-1)$ be the set of the elements including $\tilde{e}_t$ and those considered in the $t+1$ iteration of the greedy approach before the addition of $\tilde{e}_{t+1}$. Let $U_{[g,g+1)} = V \setminus \bigcup_{k=1}^{g-1} U_{[k,k+1)}$.

Proposition 3. *(i) for each $e \in U_{[t-1,t)} \setminus \{\tilde{e}_{t-1}\}$, $f_{S_{t-1}}(e) \geq f_{S_{t-1}}(\tilde{e}_t)$ but $S_{t-1} \cup \{e\} \notin \mathcal{I}$; (ii) $f_{S_{t-1}}(e) \leq f_{S_{t-1}}(\tilde{e}_t)$ for each $e \in \bigcup_{k=t}^{g} U_{[k,k+1)}$.*

Note that S_t is a maximal independent set and thus a base of $\bigcup_{k=1}^{t} U_{[k,k+1)}$ and $S^* \cap (\bigcup_{k=1}^{t} U_{[k,k+1)})$ is an independent set of $\bigcup_{k=1}^{t} U_{[k,k+1)}$. Let $a_k = |S^* \cap U_{[k,k+1)}|$ $(k = 1, 2, \cdots, g)$. Since a $K-$ intersection system is a K-system, we have the following result.

Lemma 2. $\sum_{k=1}^{t} a_k \leq Kt$ *for $t = 1, 2, \cdots, g$.*

Theorem 4. *Let S_g be the set returned by Algorithm 5. Then*

$$\frac{f(S_g)}{f(S^*)} \geq \frac{\gamma}{K + \gamma}.$$

Proof. Index the elements of the optimal solution S^* in the order that is deleted in the greedy algorithm. Suppose $S^* = \{e_1^*, \ldots, e_l^*\}$. Let $b = \lceil \frac{l}{K} \rceil$. From Lemma 2, we have $b \leq g$. Define

$$\begin{aligned}
S_1^* &= \{e_1^*, \ldots, e_K^*\}; \\
S_2^* &= \{e_{K+1}^*, \ldots, e_{2K}^*\}; \\
&\vdots \\
S_{b-1}^* &= \{e_{(b-2)K+1}^*, \ldots, e_{(b-1)K}^*\}; \\
S_b^* &= S^* \setminus \bigcup_{i=1}^{b-1} S_i^*
\end{aligned}$$

From Lemma 2, one can see that $S_i^* \subseteq \bigcup_{k=i}^{g} U_{[k,k+1)}$. By Proposition 3, this means that for each $e \in S_i^*$, the inequality

$$f_{S_{i-1}}(e) \leq f_{S_{i-1}}(\tilde{e}_i) \tag{4}$$

holds. From the monotonicity of f, we have

$$
\begin{aligned}
f(S^*) &\leq f(S^* \cup S_g) \\
&= f(S_g \cup (\bigcup_{i=1}^{b} S_i^*)) \\
&= f(S_g) + \sum_{i=1}^{b} \sum_{e_j^* \in S_i^*} [f(S_g \cup \{e_1^*, \ldots, e_j^*\}) - f(S_g \cup \{e_1^*, \ldots, e_{j-1}^*\})] \\
&\leq f(S_g) + \frac{1}{\gamma} \cdot \sum_{i=1}^{b} \sum_{e_j^* \in S_i^*} f_{S_{i-1}}(e_j^*) \\
&\leq f(S_g) + \frac{K}{\gamma} \cdot \sum_{i=1}^{b} f_{S_{i-1}}(e_j^*) \\
&\leq f(S_g) + \frac{K}{\gamma} \cdot \sum_{i=1}^{g} f_{S_{i-1}}(\tilde{e}_j) \\
&= (1 + \frac{K}{\gamma})f(S_g),
\end{aligned}
$$

where the first inequality holds from the monotonicity of f, the second inequality holds from the fact that f is a γ−submodular function, the third and fourth inequalities follow by inequality (4) and the fact that $b \leq g$ respectively. This completes the proof of the theorem.

Corollary 2. *Apply Algorithm 5 to maximize a nonnegative strictly monotone set γ−submodular functions under matroid constraint. Let S_g be the set returned. Then*

$$
\frac{f(S_g)}{f(S^*)} \geq \frac{\gamma}{1 + \gamma}.
$$

6   Conclusion

The objective functions for many applications in practice are in general not submodular. To solve these optimization problems, an important research method is to introduce some parameters to describe the characteristics of the non-submodular functions, such as submodularity ratio, curvature, supermodular degree, etc., and then design algorithms for the problems and analyze the performances of the algorithms with these parameters.

References

1. Balcan, M.F., Harvey, N.J.A.: Learning submodular functions. In: Proceedings of the 43rd ACM Symposium on Theory of Computing, pp. 793–802. ACM, San Jose (2011)
2. Bian, A.A., Buhmann, J.M., Krause, A., Tschiatschek, S.: Guarantees for greedy maximization of non-submodular functions with applications. In: International Conference on Machine Learning, pp. 498–507 (2017)
3. Conforti, M., Cornuéjols, G.: Submodular set functions, matroids and the greedy algorithm: tight worst-case bounds and some generalizations of the Rado-Edmonds theorem. Discret. Appl. Math. **7**(3), 251–274 (1984)
4. Calinescu, G., Chekuri, C., Pál, M., Vondrák, J.: Maximizing a submodular set function subject to a matroid constraint (extended abstract). In: Fischetti, M., Williamson, D.P. (eds.) IPCO 2007. LNCS, vol. 4513, pp. 182–196. Springer, Heidelberg (2007). https://doi.org/10.1007/978-3-540-72792-7_15

5. Das, A., Kempe, D.: Submodular meets spectral: greedy algorithms for subset selection, sparse approximation and dictionary selection. In: Proceedings of the 28th International Conference on International Conference on Machine Learning, pp. 1057–1064, Omni press, Bellevue (2011)

6. Feige, U.: A threshold of $\ln n$ for approximating set cover. J. ACM **45**(4), 634–652 (1998)

7. Fisher, M.L., Nemhauser, G.L., Wolsey, L.A.: An analysis of approximations for maximizing submodular set functions II. Math. Program. Study **8**, 73–87 (1978)

8. Filmus, Y., Ward, J.: A tight combinatorial algorithm for submodular maximization subject to a matroid constraint. In: the 53rd Annual Symposium on Foundations of Computer Science Foundations of Computer Science (FOCS), pp. 659–668, IEEE, New Brunswick (2012)

9. Hochbaum, D.S., Hong, S.P.: About strongly polynomial time algorithms for quadratic optimization over submodular constraints. Math. Program. **69**(1), 269–309 (1995)

10. Hochbaum, D.S.: An efficient algorithm for image segmentation, Markov random fields and related problems. J. ACM **48**(4), 686–701 (2001)

11. Jenkyns, T.: The efficacy of the greedy algorithm. Cong. Num. **17**, 341–350 (1976)

12. Khuller, S., Moss, A., Naor, J.S.: The budgeted maximum coverage problem. Inf. Process. Lett. **70**(1), 39–45 (1999)

13. Krause, A., Singh, A., Guestrin, C.: Nearoptimal sensor placements in gaussian processes: theory, efficient algorithms and empirical studies. J. Mach. Learn. Res. **9**, 235–284 (2008)

14. Lawrence, N., Seeger, M., Herbrich, R.: Fast sparse Gaussian process methods: the informative vector machine. In: Advances in Neural Information Processing Systems, vol. 15, pp. 625–632 (2003)

15. Lee, J., Sviridenko, M., Vondrák, J.: Submodular maximization over multiple matroids via generalized exchange properties. Math. Oper. Res. **35**(4), 795–806 (2010)

16. Nemhauser, G.L., Wolsey, L.A., Fisher, M.L.: An analysis of approximations for maximizing submodular set functions I. Math. Program. **14**(1), 265–294 (1978)

17. Nemhauser, G.L., Wolsey, L.A.: Best algorithms for approximating the maximum of a submodular set function. Math. Oper. Res. **3**(3), 177–188 (1978)

18. Wolsey, L.A.: Maximising real-valued submodular functions: primal and dual heuristics for location problems. Math. Oper. Res. **7**, 410–425 (1982)

19. Sviridenko, M.: A note on maximizing a submodular set function subject to a knapsack constraint. Oper. Res. Lett. **32**(1), 41–43 (2004)

Approximation Algorithm for Stochastic Prize-Collecting Steiner Tree Problem

Jian Sun[1,2], Haiyun Sheng[1], Yuefang Sun[3], and Xiaoyan Zhang[1(✉)]

[1] School of Mathematical Science and Institute of Mathematics,
Nanjing Normal University, Nanjing 210023, Jiangsu, People's Republic of China
`zhangxiaoyannjnu@126.com`
[2] Department of Operations Research and Scientific Computing,
Beijing University of Technology, Beijing 100124, People's Republic of China
[3] School of Mathematical Information, Shaoxing University,
Shaoxing 312000, Zhejiang, People's Republic of China

Abstract. Steiner tree problem is a typical NP-hard problems in combinatorial optimization, which has comprehensive application background and is a hot topic in recent years. In this paper, we study the stochastic prize-collecting Steiner tree problem. Before the actual requirements materialize, we can choose (purchase) some edges in the first stage. When actual requirements are revealed, drawn from a prespecified probability distribution, then there are more edges may be chosen (purchased) for the actual requirements. The goal is to minimize the sum of the first stage cost, the expected second stage cost and the expected penalty cost. We propose a primal-dual 3-approximation algorithm for the stochastic prize-collecting Steiner tree problem.

Keywords: Combinatorial optimization ·
Stochastic prize-collecting Steiner tree · Approximation algoritorithm

1 Introduction

The Steiner tree problem (STP) is a classical network design problem. For an undirected graph $G = (V, E)$ with edge costs $c_e \in \mathbb{R}_{\geq 0}$, $\forall e \in E$, and a set of terminals $\emptyset \neq T \subseteq V$. It asks for a minimum cost edge set $E' \subseteq E$ such that $G[E']$ connects T. The decision problem of the STP is NP-complete [13], even in the case of edge weights 1 and 2 [1]. Moreover, the STP is approximable with a constant factor and the currently best ratio is $\ln 4 + \epsilon < 1.39$ [2]. In addition, integer linear programmings and their polytopes have been studied intensively in the 1990's, see [3–5].

In this paper, we consider the prize-collecting Steiner tree problem and the two stage stochastic Steiner tree problem. Firstly, the prize-collecting Steiner tree problem is an extension of the Steiner tree problem where each vertex left out of the tree pays a penalty. The goal is to find a tree that minimizes the sum of its edge costs and the penalties for the vertices left out of the tree. The problem

D.-Z. Du et al. (Eds.): AAIM 2019, LNCS 11640, pp. 261–271, 2019.
https://doi.org/10.1007/978-3-030-27195-4_24

has many applications in network design and has been used to approximate many other problems. Secondly, the two stage stochastic Steiner tree problem is also a natural extension of the STP to a two stage stochastic combinatorial optimization problem.

It is well known that the primal-dual scheme has been used to provide approximation algorithms for many problems. Goemans and Williamson gave a $(2 - \frac{1}{n-1})$-approximation for the prize-collecting Steiner tree [8]. The best approximation algorithms known for the prize-collecting Steiner tree problem are based on the primal-dual scheme [7]. It is well known that the different linear programming formulations of the problem may lead to different algorithms. Gupta and Pál [9] have given a 4-approximation algorithm by boosted sampling technique for the two stage stochastic Steiner tree problem.

Formally, we define a two stage stochastic optimization problem with recourse as follows [15]. When only the partial information is available, let vector x_0 be the set of decision variables in the first stage, which have to be fixed. Later, we choose some second stage (or recourse) variables x_1 to augment the first stage solution x_0 when the full information is available. ξ denotes the random vector and it defines the constraint matrix T, cost vector q and requirement vector h when the full information is available. Let A, c, and b denote the same for the first stage. Given a vector (or matrix) a, we use a' to denote the transpose of a. Let P represent additional constraints such as nonnegativity or integrality, the components of x_0 and x_1 need to satisfy the additional constraints. The stochastic program can be written according to [15] as follows:

$$\min \quad c'x_0 + E_\xi Q(x_0, \xi)$$
$$\text{s.t.} \quad Ax_0 = b,$$
$$x_0 \in P,$$

where

$$Q(x_0, \xi) = \min q'x_1$$
$$\text{s.t.} \quad T(x_0, x_1) = h,$$
$$x_1 \in P.$$

Here $Q(x_0, \xi)$ denotes the optimal cost of the second stage, which is conditioned on scenario $\xi = (q, T, h)$ having been realized and a first stage setting of the variables x_0. It's easy to see that the expectation is taken with respect to ξ.

Subsequently, we consider the two stage stochastic optimization problem with recourse and an additional restriction: finitely many scenarios. This means that the future will be one of a finite set of possibilities (scenarios), and the parameters and probability of occurrence of each scenario are known in advance. Specifically, the two stage model has a restriction which is characterized by a finite set of m scenarios in the second stage. The constraint matrix, cost vector and requirement vector take on values T_k, q_k and h_k in the scenario k respectively, and the probability of the scenario k occurring is p_k. In the circumstances, the mathematical formulation of this model is given below, where x_k represents our choice if scenario k materializes as follows [15]:

$$\min \quad c'x_0 + \sum_{k=1}^{m} p_k(q_k)'x_k$$

$$\text{s.t.} \quad Ax_0 = b,$$
$$T_k(x_0, x_k) = h_k, \qquad\qquad \forall k = 1, 2, ..., m,$$
$$(x_0, x_k) \in P, \qquad\qquad \forall k = 1, 2, ..., m.$$

For the two stage stochastic optimization problem, many researchers made an important contribution. Schultz et al. [16] provided an excellent survey of the two stage stochastic integer programming. According to the recent work on scenario reduction by Heitsch and Römisch [10], the relevance of the finite scenario model becomes more pronounced. In order to get a more complete description of models of stochastic optimization and their uses, we advise the interested readers may consult any of the texts cited above or others.

2 Related Work

The Steiner tree problem has also attracted significant attention, and many relevant variants of the Steiner tree problem have been extensively studied. Goemans and Williamson [8,11] used the primal-dual scheme to derive a $(2 - \frac{1}{n-1})$-approximation for the rooted prize-collecting Steiner tree (PCSP), where $n = |V|$. Trying all possible choices for the root, they obtained a $(2 - \frac{1}{n-1})$-approximation for the unrooted PCST. The resulting algorithm, which we call GW, runs in time $O(n^3 logn)$. Johnson, Minkoff and Phillips [12] proposed a variant of the algorithm that runs the primal-dual scheme only once, resulting in a $O(n^2 logn)$ time bound. They claimed that their variant, which we refer to as JMP, and it achieves the same approximation ratio as algorithm GW. Cole et al. [6] proposed a faster implementation of the GW algorithm, which also runs the primal-dual scheme only once and produces a $(2 + \frac{1}{ploy(n)})$-approximation for the PCST. For the two stage stochastic Steiner tree problem, Gupta and Pál [9] gave a 4-approximation algorithm by boosted sampling technique; Kurz, Mutzel and Zey [14] showed this problem is in the class of fixed-parameter tractable problems (FPT), parameterized by the number of terminals.

The remainder of our paper is organized as follows. In Sect. 3, we introduce the stochastic prize-collecting Steiner tree problem. Furthermore, applying the primal-dual method, we propose a approximation algorithm for the problem and prove that the ratio is 3.

3 Stochastic Prize-Collecting Steiner Tree Problem

In this section, we will introduce the stochastic prize-collecting Steiner tree problem (SPCSTP) and present the relevant programs specifically. The SPCSTP can be described as follows. In the problem, there are two stages. We are given a potential edge set E in the first stage. In this stage, we can choose some edges for serving any client (terminal vertex) later. The purchasing cost of the edge e in the first stage is c_e^0. Until the second stage, all possible scenarios and the associated probabilities become known. In this paper, we consider the case of polynomial scenarios, that is to say, the number m of the scenarios is polynomial with respect to the input of the problem.

For a scenario $k \in \{1, 2, ..., m\}$ in the second stage, the probability of this scenario is p_k, the client (terminal vertex) set is denoted as D_k, the penalty cost for the unserved client set $T_k \subseteq D_k \setminus r$ is $h_k(T_k)$, which is a monotone linear function, and the purchasing cost of edge e in this scenario is c_e^k. The client (terminal vertex) in scenario k of the second stage can be served by an edge purchased in the first stage or the corresponding scenario; otherwise, the client (terminal vertex) is unserved. For ease of notation, we call (e, k) (for $e \in E, k = 0, 1, 2, ..., m$) an edge-scenario pair and denote the edge-scenario pair set as $\mathcal{E}$.

Similarly, each (S, k) (for $k = 0, 1, 2, ..., m, S \subseteq D_k$) represents a vertex subset-scenario pair, which is active in the k-th scenario. Note that the scenario is in the first stage when $k = 0$, and the scenario is in the second stage when $k = 1, 2..., m$. We denote the vertex subset-scenario pair set as $\mathcal{C}$. In scenario k, the set of the vertex subset-scenario pair (S, k) ($S \subseteq D_k$) is denoted as $\mathcal{C}_k$. Obviously, $\bigcup\limits_{k=1}^{m} \mathcal{C}_k = \mathcal{C}$.

We want to determine the set of edge-vertex subset pairs $\widehat{E}_0$ and $\widehat{E}_k$ to be purchased respectively in the first stage and in the k-th scenario of the second stage ($k = 1, 2, ..., m$), the set of client-scenario pairs $\widehat{T}_k$ ($k = 1, 2, ..., m$) that will incur penalties, and the expected edge purchasing cost $\sum\limits_{e \in \widehat{E}_0} c_e^0 + \sum\limits_{k=1}^{m} \sum\limits_{e \in \widehat{E}_k} c_e^k$, the expected penalty cost $\sum\limits_{k=1}^{m} p_k h_k(\widehat{T}_k)$, and the goal is to minimize the sum of expected edge purchasing cost and expected penalty cost.

If we set $p_0 = 1$ and $\sum\limits_{k=1}^{m} p_k = 1$, the SPCSTP can be formulated as the following linear integer program.

$$\min \sum_{(e,k) \in E} p_k c_e^k x_e^k + \sum_{k=1}^{m} \sum_{T_k \subseteq D_k \setminus r} p_k h_k(T_k) z_{T_k}$$

$$(IP) \quad \text{s.t.} \sum_{e \in \delta(S)} x_e^0 + \sum_{e \in \delta(S)} x_e^k + \sum_{T_k : S \subseteq T_k} z_{T_k} \geq 1, \forall (S, k) \in \mathcal{C}, r \notin S,$$

$$\sum_{T_k : T_k \subseteq D_k \setminus r} z_{T_k} \leq 1,$$

$$x_e^0, x_e^k, z_{T_k} \in \{0, 1\}, \forall e \in E, T_k \subseteq D_k \setminus r.$$

In the above formulation, all the variables are binary. Note that $h_k(T_k) = \sum_{i \in T_k} h_k(i)$. x_e^0 represents whether edge e is purchased in the first stage or not. If $x_e^0 = 1$, edge $e \in E$ is purchased in the first stage (i.e. edge-scenario pair $(e, 0)$ is purchased); otherwise, $x_e^0 = 0$. Similarly, x_e^k represents whether edge e is purchased in the k-th scenario of the second stage. If $x_e^k = 1$, edge $e \in E$ is purchased in the second stage for the k-th scenario (i.e. edge-scenario pair (e, k) is purchased); otherwise, $x_e^k = 0$. z_{T_k} indicates whether a set of clients $T_k \subseteq D_k \setminus r$ incurs penalties or not. The first constraint models that either there is an edge in $\delta(S)$, or S is a part of a vertex set that pays penalty (a vertex set T_k with $z_{T_k} = 1$) for each client-scenario pair (S, k).

A linear programming relaxation of the integer program can be created by replacing the integrality constraints with the constraints $x_e^0 \geq 0$, $x_e^k \geq 0$ and $z_{T_k} \geq 0$ and dropping the constraints $\sum_{T_k : T_k \subseteq D_k \setminus r} z_{T_k} \leq 1$ (in fact, including this constraint does not affect the optimal solution). So we obtain the LP relaxation and the corresponding dual linear program as follows.

$$\min \quad \sum_{(e,k) \in E} p_k c_e^k x_e^k + \sum_{k=1}^{m} \sum_{T_k \subseteq D_k \setminus r} p_k h_k(T_k) z_{T_k}$$

$$(LP) \quad \text{s.t.} \sum_{e \in \delta(S)} x_e^0 + \sum_{e \in \delta(S)} x_e^k + \sum_{T_k : S \subseteq T_k} \geq 1, \forall (S, k) \in \mathcal{C}, r \notin S,$$

$$x_e^0, x_e^k, z_{T_k} \geq 0, \forall e \in E, T_k \subseteq D_k \setminus r.$$

$$\max \quad \sum_{(S,k) \in \mathcal{C}: r \notin S} y_S^k$$

$$(DP) \quad \text{s.t.} \sum_{(S,k) \in \mathcal{C}: r \notin S} y_S^k \leq c_e^0, \qquad \forall e \in E,$$

$$\sum_{S \subseteq D_k \setminus r} y_S^k \leq p_k c_e^k, \qquad \forall e \in E, k = 1, 2, ..., m,$$

$$\sum_{S \subseteq T_k} y_S^k \leq p_k h_k(T_k), \qquad \forall T_k \subseteq D_k \setminus r, k = 1, 2, ..., m,$$

$$y_S^k \geq 0, \qquad \forall (S, k) \in \mathcal{C}.$$

In the above dual formulation, the variable y_S^k can be interpreted as the budget of client-scenario pair (S, k).

3.1   The Primal-Dual Algorithm

In this section, we propose a primal-dual algorithm for the SPCSTP. In fact, our algorithm is motivated by a procedure of dual ascent of the standard primal-dual algorithm for the deterministic prize-collecting Steiner tree problem [8].

For ease of notation, we define the following notations: $\widehat{T}_k$ $(k = 1, ..., m)$ represents the penalty client (terminal vertex) set in the k-th scenario.

$\overline{E}_0$ denotes the temporarily purchased edge in the first stage.

$\overline{E}_k$ denotes the temporarily purchased edge in the second stage with respect to k-th scenario.

Initially, all the dual variables are zero. All the clients (terminal vertices) are not punished. All the edges are not purchased.

In order to have a better understanding of the algorithm, we give some intuitively interpretation for Step 1 as follows. In Step 1.1, we define the property of components including active and inactive. Step 1.2 corresponds to three cases. Case 1 corresponds to purchasing a new edge in the first stage. Case 2 corresponds to purchasing a new edge in the second stage. Case 3 corresponds to the client subset which is punished.

It is worth noting that the corresponding penalty constraint must be tight for any $T_k \subseteq D_k \setminus r$ which is not connected by the resulting graph.

3.2 The Analysis of the Algorithm

From Algorithm 1, we obtain a feasible solution of the SPCSTP, which is denoted by SOL. Let $cost(SOL)$ denote the total cost of SOL.

$$cost(SOL) = \sum_{e \in \widehat{E}_0} c_e^0 + \sum_{k=1}^{m} \sum_{e \in \widehat{E}_k} p_k c_e^k + \sum_{k=1}^{m} p_k h_k(\widehat{T}_k).$$

We use OPT to denote the optimal value of the SPCSTP. In this section, we will prove the following two inequalities:

$$\sum_{k=1}^{m} p_k h_k(\widehat{T}_k) \leq OPT, \tag{1}$$

$$\sum_{e \in \widehat{E}_0} c_e^0 + \sum_{k=1}^{m} \sum_{e \in \widehat{E}_k} p_k c_e^k \leq 2OPT. \tag{2}$$

These two inequalities together imply the main result as follows.

Theorem 1. *Algorithm 1 is a 3-approximation algorithm for the SPCSTP.*

Let y_S^k denote the dual solution obtained from Algorithm 1. From our algorithm, the penalty constraint is tight for any T_k, where $k = 1, 2, ..., m$. It can be obtained by the following lemmas.

Lemma 1. *For each $\widehat{T}_k$, we have $\sum\limits_{S \subseteq \widehat{T}_k} y_S^k = p_k h_k(\widehat{T}_k)$, where $k = 1, 2, ..., m$.*

Proof. By the construction of $\widehat{E}_0$ and $\widehat{E}_k$, each vertex not spanned by $\widehat{E}_0$ and $\widehat{E}_k$ (i.e. the vertices in some $\widehat{T}_k$ for $k = 1, 2, ..., m$) lies in some component deactivated at some point in the algorithm. Moreover, if the vertex was in some

Algorithm 1. Primal-dual algorithm

1: Initialization

$y_S^k := 0$, $\widehat{T}_k := \emptyset$ $(k = 1, 2..., m)$, $\overline{E}_0 := \emptyset$, $\overline{E}_k := \emptyset$ $(k = 1, 2..., m)$, $\mathcal{C}_o :=$ $\{\{v, k\}|\forall(v, k) \in \mathcal{C}\}$, where $\mathcal{C}_o$ denotes all component sets.

2: Step 1 Constructing a dual feasible solution:

3:    Step 1.1. For each component $C \in \mathcal{C}_o$, let us define $\lambda(C) = 1$ if $\exists\, k \in \{1, 2, ...m\}$ s.t. $C \subseteq T_k \subseteq D_k \setminus r$ and $\sum_{S \subseteq T_k} y_S^k < P_k h_k(T_k)$; otherwise, $\lambda(C) = 0$. If $\lambda(C) = 1$, we let C active; if $\lambda(C) = 0$, we let C inactive.

4:    Step 1.2. Increase these y_S^k uniformly until one of the following three events happens:

   Case 1. There exists $C_p, C_q \in \mathcal{C}_o$ such that the first cost constraint of an edge $e = (i, j)$ becomes tight where $i \in C_p$ and $j \in C_q$.

   Case 2. There exist $C_{p'}, C_{q'} \in \mathcal{C}_o$ such that the second cost constraint of an edge $e = (i, j)$ becomes tight where $i \in C_{p'}$ and $j \in C_{q'}$.

   Case 3. There exists $k \in \{1, 2, ..., m\}$ such that the third penalty constraint of a vertex set T_k becomes tight.

5: **if** Case 1 happens **then**

6:    update $\overline{E}_0 = \overline{E}_0 \bigcup\{e\}$ and $\mathcal{C}_o = \mathcal{C}_o \bigcup\{C_p \bigcup C_q\} - \{C_p\} - \{C_q\}$ and if $r \in \{C_p \bigcup C_q\}$ then $\lambda(C_p \bigcup C_q) \leftarrow 0$ else $\lambda(C_p \bigcup C_q) \leftarrow 1$, go to Step 1.3.

7: **else**

8:    **if** Case 2 happens **then**

9:       update $\overline{E}_k = \overline{E}_k \bigcup\{e\}$ and $\mathcal{C}_o = \mathcal{C}_o \bigcup\{C_{p'} \bigcup C_{q'}\} - \{C_{p'}\} - \{C_{q'}\}$ and if $r \in \{C_{p'} \bigcup C_{q'}\}$ then $\lambda(C_{p'} \bigcup C_{q'}) \leftarrow 0$ else $\lambda(C_{p'} \bigcup C_{q'}) \leftarrow 1$, go to Step 1.3.

10:    **else**

11:       **if** Case 3 happens **then**

12:          there exists $k \in \{1, 2, ..., m\}$ such that the penalty constraint of a vertex set $T_k \subseteq D_k \setminus r$ becomes tight and go to Step 1.3.

13:       **end if**

14:    **end if**

15: **end if**

16: **if** multiple cases happen simultaneously **then**

17:    execute one of three cases arbitrarily.

18: **end if**

19:    Step 1.3. Update $\lambda(C)$ for all $C \in \mathcal{C}_o$ as in Step 1.1. If there exists $C \in \mathcal{C}_o$ such that $\lambda(C) = 1$, go to Step 1.2; otherwise, stop Step 1.

20: Step 2. (Constructing a primal integral feasible solution.)

21: **for** all $e \in \overline{E}_0 \bigcup \overline{E}_k$ **do**

22:    **if** removing e does not impact the connectivity of the resulting graph **then**

23:       remove e from the resulting graph and update $\overline{E}_0$ and $\overline{E}_k$ for each $k = 0, 1, 2, ..., m$.

24:    **end if**

25: **end for**

26: We choose edge set $\widehat{E}_0$ from $\overline{E}_0$ to purchase in the first stage. Then choose edge $\widehat{E}_k$ from $\overline{E}_k$ to purchase for each $k = 1, 2, ..., m$ in the second stage.

27: Let $\widehat{T}_k$ be the set of unserved clients in scenario k for each $k = 1, 2, ..., m$.

---

deactivated component C, then none of the vertices of C are spanned by $\widehat{E}_0$ and $\widehat{E}_k$. Using the observations, we can partition the vertices of $\widehat{T}_k$ into disjoint deactivated components $C_{k1}, C_{k2}, ..., C_{kl}$ for some k. These are the maximal vertex sets in $\widehat{T}_k$. Since each C_{kj} is a deactivated component, then $\sum_{S \in C_{kj}} y_S^k = p_k h_k(C_{kj})$, and thus $\sum_{k=1}^{m} p_k h_k(\widehat{T}_k) = \sum_{k=1}^{m} \sum_{j=1}^{l} p_k h_k(C_{kj}) = \sum_{k=1}^{m} \sum_{j=1}^{l} \sum_{S \subseteq C_{kj}} y_S^k \leq \sum_{k=1}^{m} \sum_{S \subseteq \widehat{T}_k} y_S^k \leq \sum_{k=1}^{m} \sum_{S \subseteq D_k \setminus r} y_S^k \leq OPT$, which concludes the proof of inequality (1).

Next, we will prove inequality (2).

Recall that the incurred cost constraint is tight for each $e \in \widehat{E}_0$. We have

$$c_e^0 = \sum_{(S,k)} y_S^k,$$

$$\sum_{e \in \widehat{E}_0} c_e^0 = \sum_{e \in \widehat{E}_0} \sum_{(S,k)} y_S^k = \sum_{(S,k)} |\widehat{E}_0 \cap \delta(S)| y_S^k.$$

Analogously, recall that the incurred cost constraint is tight for each $e \in \widehat{E}_k$. We have

$$p_k c_e^k = \sum_{S \subseteq D_k \setminus r} y_S^k,$$

$$\sum_{e \in \widehat{E}_k} p_k c_e^k = \sum_{e \in \widehat{E}_k} \sum_{S \subseteq D_k \setminus r} y_S^k = \sum_{S \subseteq D_k \setminus r} |\widehat{E}_k \cap \delta(S)| y_S^k.$$

Therefore, we have the above inequality (2) is equivalent to the following lemma.

Lemma 2. *The dual feasible solution y_S^k satisfies the following inequality*

$$\sum_{(S,k)} |\widehat{E}_0 \cap \delta(S)| y_S^k + \sum_{k=1}^{m} \sum_{S \subseteq D_k \setminus r} |\widehat{E}_k \cap \delta(S)| y_S^k \leq 2 \sum_{(S,k) \in \mathcal{C}} y_S^k. \qquad (3)$$

This lemma can be proved by induction on the main loop. At the beginning of Algorithm 1, $y_S^k = 0$ for any (S, k) (for $k = 0, 1, 2, ..., m$), inequality (3) of this lemma holds. In order to prove this lemma, it suffices to show that the incremental cost of the left side is bounded by the incremental cost of the right side in every iteration.

Let $\mathcal{C}_o$ denote all the components at the beginning of some iteration. We use $\mathcal{C}_o^{active}$ and $\mathcal{C}_o^{inactive}$ to denote the active components and inactive components in $\mathcal{C}_o$, respectively. In each iteration, we only increase the dual variables of all the active components. For every active component C belonging to $\mathcal{C}_o^{active}$, the algorithm will raise the corresponding dual variable y_S^k. Assume the incremental cost of any active set is ϵ. The left side of inequality (3) will increase by

$$\sum_{C \in \mathcal{C}_o^{active}} |\widehat{E}_0 \cap \delta(C)| \epsilon + \sum_{k=1}^m \sum_{C \in \mathcal{C}_o^{active}} |\widehat{E}_k \cap \delta(C)| \epsilon,$$

and the right side of inequality (3) will increase by

$$2|\mathcal{C}_o^{active}| \epsilon.$$

Below we prove that

$$\sum_{C \in \mathcal{C}_o^{active}} |\widehat{E}_0 \cap \delta(C)| + \sum_{k=1}^m \sum_{C \in \mathcal{C}_o^{active}} |\widehat{E}_k \cap \delta(C)| \le 2|\mathcal{C}_o^{active}|. \tag{4}$$

Now, we construct a new graph H as follows. Every component in $\mathcal{C}_o$ is a vertex in H. Every edge $e \in \widehat{E}_0 \cap \delta(C)$ or $e \in \widehat{E}_k \cap \delta(C)$ for some $C \in \mathcal{C}_o$ is an edge in H. We use $d(v)$ to denote the degree of vertex v in H. Let A denote the vertices in H which contract from $\mathcal{C}_o^{active}$ and let B denote the vertices in H which contract from $\mathcal{C}_o^{inactive}$ with nonzero degree. That is, we discard all isolated inactive vertices in H.

The left side of inequality (4) is $\sum_{v \in A} d(v) + \sum_{k=1}^m \sum_{v \in A} d_k(v)$, and the right side of inequality (4) is $2 \cdot |A|$. Rewriting inequality (4) in terms of the degree of vertices in H, we have

$$\sum_{v \in A} d(v) + \sum_{k=1}^m \sum_{v \in A} d_k(v) \le 2 \cdot |A|. \tag{5}$$

In order to prove inequality (5), we present the following lemmas.

Lemma 3. *The graph H is a forest.*

Proof. According to Algorithm 1, every edge is added between two components. Thus, the resulting graph is a forest. Recall that we obtain H by contracting components in the resulting graph. Hence, H is a forest.

Lemma 4. *There can be at most one inactive leaf in H, which must correspond to the component containing r.*

Proof. Suppose that v is an inactive leaf of H, adjacent to edge e, and suppose that C_v is the inactive component corresponding to v and C_v does not contain the root r, then C_v is deactivated by Algorithm 1. By the construction of the resulting graph, then e is not in the resulting graph, which is a contraction. So, the conclusion is right. That is, there can be at most one inactive leaf in H, which must correspond to the component containing r.

Lemma 5. *For every vertex $v \in B$ except one inactive leaf in H (if the leaf exists), we have $d(v) \geq 2$.*

Proof. It can be obtained by the above lemmas.

From Lemma 3, H has at most $|A| + |B| - 1$ edges. Thus, the degree of H is at most $2 \cdot (|A| + |B| - 1)$. Therefore, we have

$$\sum_{A \bigcup B} d(v) \leq 2(|A| + |B| - 1).$$

From Lemmas 4, we have

$$\sum_{v \in B} d(v) \geq 2|B| - 1.$$

Obviously, we have

$$\sum_{v \in A} d(v) + \sum_{k=1}^{m} \sum_{v \in A} d_k(v) = \sum_{v \in A \bigcup B} d(v) - \sum_{v \in B} d(v)$$
$$\leq 2 \cdot (|A| + |B| - 1) - (2 \cdot |B| - 1)$$
$$\leq 2 \cdot |A|.$$

Now we complete the proof.

4   Conclusions

Considering the stochastic prize-collecting Steiner tree problem, we present a primal-dual 3-approximation algorithm. Many researchers made substantial contribution in stochastic problems by considering the use of sampling, cost sharing function, and primal-dual, etc. In the future, we believe that there will be more substantial progress in approximation algorithms for the stochastic optimization problems, and it will be interesting to improve the approximation ratios of these problems.

Acknowledgements. The authors are supported by Natural Science Foundation of China (Nos.11871280, 11471003).

References

1. Bern, M., Plassmann, P.: The Steiner problem with edge lengths 1 and 2. Inf. Process. Lett. **32**, 171–176 (1989)
2. Byrka, J., Grandoni, F., Rothvoß, T., Sanità, L.: Steiner tree approximation via iterative randomized rounding. J. ACM **60**(1), 1–33 (2013)
3. Chopra, S., Rao, M.R.: The Steiner tree problem I: formulations, compositions and extension of facets. Math. Program. **64**, 20–229 (1994)

4. Chopra, S., Rao, M.R.: Properties and classes of facets: the Steiner tree problem II. Math. Program. **64**, 231–246 (1994)

5. Chopra, S., Tsai, C.Y.: Polyhedral approaches for the Steiner tree problem on graphs. In: Steiner Trees in Industry, pp. 175–202. Kluwer Academic Publishers (2001)

6. Cole, R., Hariharan, R., Lewenstein, M., Porat, E.: A faster implementation of the Goemans-Williamson clustering algorithm. In: Proceedings of the 12th annual ACM-SIAM Symposium on Discrete Algorithms, pp. 17–25 (2001)

7. Feofiloff, P., Fernandes, C.G., Ferreira, C.E., Pina, J.C.D.: Primal-dual approximation algorithms for the prize-collecting Steiner tree problem. Inf. Process. Lett. **103**(5), 195–202 (2007)

8. Goemans, M.X., Williamson, D.P.: A general approximation technique for constrained forest problems. SIAM J. Comput. **24**(2), 296–317 (1995)

9. Gupta, A., Pál, M., Ravi, R., Sinha A.: Boosted sampling: approximation algorithms for stochastic optimization problems. In: Proceedings of the 36th Annual ACM Symposium on Theory of Computing, pp. 417–426 (2004)

10. Heitsch, H., Römisch, W.: Scenario reduction algorithm in stochastic programming. Comput. Optim. Appl. **24**, 187–206 (2003)

11. Hochbaum, D.S.: Approximation Algorithms for NP-hard Problems. PWS Publishing Company, Boston (1997)

12. Johnson, D.S., Minkoff, M., Phillips, S.: The prize collecting Steiner tree problem: theory and practice. In: Proceedings of the 11th Annual ACM-SIAM Symposium on Discrete Algorithms, pp. 760–769 (2000)

13. Karp, R.M.: Reducibility among Combinatorial Problems. In: Miller, R.E., Thatcher, J.W., Bohlinger, J.D. (eds.) Complexity of Computer Computations, pp. 85–103. Springer, Boston (1972). https://doi.org/10.1007/978-1-4684-2001-2_9

14. Kurz, D., Mutzel, P., Zey, B.: Parameterized algorithms for stochastic Steiner tree problems. In: Kučera, A., Henzinger, T.A., Nešetřil, J., Vojnar, T., Antoš, D. (eds.) MEMICS 2012. LNCS, vol. 7721, pp. 143–154. Springer, Heidelberg (2013). https://doi.org/10.1007/978-3-642-36046-6_14

15. Ravi, R., Sinha, A.: Hedging uncertainty: approximation algorithms for stochastic optimization problems. **108**(1), 97–114 (2006)

16. Schultz, R., Stougie, L., van der Vlerk, M.H.: Two stage stochastic integer programming: a survey. Statistica Neerlandica **50**(3), 404–416 (1996)

A General Framework for Path Convexities

João Vinicius C. Thompson[1,2], Loana T. Nogueira[2], Fábio Protti[2],
Raquel S. F. Bravo[2], Mitre C. Dourado[3], and Uéverton S. Souza[2(✉)]

[1] Centro Federal de Educação Tecnológica Celso Suckow da Fonseca,
Rio de Janeiro, Brazil
`joao.thompson@cefet-rj.br`
[2] Universidade Federal Fluminense, Niterói, Brazil
`{loana,fabio,raquel,ueverton}@ic.uff.br`
[3] Universidade Federal do Rio de Janeiro, Rio de Janeiro, Brazil
`mitre@nce.ufrj.br`

Abstract. In this work we deal with the so-called *path convexities*, defined over special collections of paths. For example, the collection of the shortest paths in a graph is associated with the well-known *geodesic convexity*, while the collection of the induced paths is associated with the *monophonic convexity*; and there are many other examples. Besides reviewing the path convexities in the literature, we propose a general path convexity framework, of which most existing path convexities can be viewed as particular cases. Some benefits of the proposed framework are the systematization of the algorithmic study of related problems and the possibility of defining new convexities not yet investigated.

Keywords: Algorithmic complexity · Graph convexity ·
Path convexity

1 Introduction

A *finite convexity space* is a pair $(V, \mathcal{C})$ consisting of a finite set V and a family $\mathcal{C}$ of subsets of V such that $\emptyset \in \mathcal{C}$, $V \in \mathcal{C}$, and $\mathcal{C}$ is closed under intersection. Members of $\mathcal{C}$ are called *convex sets*.

Let $\mathscr{P}$ be a collection of paths of a graph G, and let $I_{\mathscr{P}} : 2^{V(G)} \to 2^{V(G)}$ be a function (called *interval function*) such that

$$I_{\mathscr{P}}(S) = S \cup \{z \notin S \mid \exists \ u, v \in S \text{ such that } z \text{ lies in an } uv\text{-path } P \in \mathscr{P}\}.$$

Distinct choices of $\mathscr{P}$ lead to interval functions of quite different behavior. Such functions, in turn, are naturally associated with special convexity spaces (the so-called *path convexities*). For instance, if $\mathscr{P}$ contains precisely all the shortest paths in a graph then the corresponding interval function is naturally associated with the well-known *geodesic convexity*; if $\mathscr{P}$ is the collection of induced

D.-Z. Du et al. (Eds.): AAIM 2019, LNCS 11640, pp. 272–283, 2019.
https://doi.org/10.1007/978-3-030-27195-4_25

paths then the corresponding interval function is associated with the *monophonic convexity*; and there are many other examples in the literature.

In this work we propose a general path convexity framework, of which most path convexities in the literature can be viewed as particular cases. Some benefits of the proposed framework are the systematization of the algorithmic study of related problems and the possibility of defining new path convexities not yet investigated.

Our contributions are concentrated mainly in Sect. 3, where we describe in detail our framework. The idea is to control the length of the paths in $\mathscr{P}$, as well as the types of chords allowed to exist in such paths. Such control can be done by means of four matrices that specify, for each pair (u, v) of vertices, the minimum/maximum length and minimum/maximum chord length in all uv-paths of $\mathscr{P}$. We prove hardness results for the more general approach, where the matrices are part of the input of the related computational problems. We also describe some polynomial cases by restricting the usage of such matrices, including linear-time methods for bounded treewidth graphs. In addition, we show how to define most existing path convexities in the literature within the proposed framework. Due to space constraints, some proofs are omitted.

2 Preliminaries

In this section we first provide all the necessary background. Next, we briefly review the main path convexities in the literature and list six fundamental computational problems in graph convexity that will be considered in this work. Finally, we prove two useful propositions.

All graphs are finite, simple, nonempty, and connected. Let G denote a graph with n vertices and m edges. The *length* of a path P in G, denoted by $|P|$, is its number of edges. A path P in G with endpoints u and v is an *uv-path*. An uv-path P in G is *shortest* if there is no uv-path P' in G such that $|P'| < |P|$. If an uv-path P is shortest then $|P|$ is the *distance* between u and v in G, and we write $|P| = dist_G(u, v)$. A *chord of length $l \geq 2$* in a path $P = (v_1, v_2, \ldots, v_{|P|})$ is an edge $v_i v_j \in E(G)$ such that $i, j \in \{1, \ldots, |P|\}$ and $|i - j| = l \geq 2$.

Let $\mathscr{P}$ be a collection of paths of a graph G, and let $I_{\mathscr{P}} : 2^{V(G)} \to 2^{V(G)}$ be the interval function associated with $\mathscr{P}$, i.e.,

$$I_{\mathscr{P}}(S) = S \cup \{z \notin S \mid \exists\, u, v \in S \text{ such that } z \text{ lies in an } uv\text{-path } P \in \mathscr{P}\}. \quad (1)$$

Define $\mathcal{C}_{\mathscr{P}}$ as the family of subsets of $V(G)$ such that $S \in \mathcal{C}_{\mathscr{P}}$ if and only if $I_{\mathscr{P}}(S) = S$. Then it is easy to see that $(V(G), \mathcal{C}_{\mathscr{P}})$ is a finite convexity space, whose convex sets are precisely the fixed points of $I_{\mathscr{P}}$.

Proposition 1. [48] $(V(G), \mathcal{C}_{\mathscr{P}})$ *is a finite convexity space.*

In order to ease the notation, we omit the subscript $\mathscr{P}$ whenever it is clear from the context.

2.1 Path Convexities in the Literature

By varying the choice of the collection $\mathscr{P}$, interval functions of different behavior can be defined using Eq. (1). The convexity spaces associated with such functions are called *path convexities*.

In Table 1 we list the main path convexities that appear in the literature. In the table, each convexity is defined by the collection of paths $\mathscr{P}$ considered.

Table 1. Some path convexities studied in the literature.

Convexity name	Collection of paths $\mathscr{P}$ considered
Geodesic [36,37,40]	Shortest paths
Monophonic [11,30,31]	Induced paths
g^3 [41]	Shortest paths of length at least three
m^3 [6,29]	Induced paths of length at least three
g_k [32]	Shortest paths of length at most k
P_3 [7,27,42]	Paths of length two
P_3^* [1]	Induced paths of length two
Triangle-path [10,13,14]	Paths allowing only chords of length two
Total [18]	Paths allowing only chords of length at least three
Detour [15–17]	Longest paths
All-path [12,34,46]	All paths

2.2 Computational Problems

In this work we focus on six computational problems that are usually studied in the field of convexity in graphs. The list, of course, is not complete and other important problems could also be considered.

We need some additional definitions. Let $S \subseteq V(G)$. If $I(S) = V(G)$ then S is an *interval set*. The *convex hull* $H(S)$ of S is the smallest convex set containing S. Write $I^0(S) = S$ and define $I^{i+1}(S) = I(I^i(S))$ for $i \geq 0$. Note that $I(S) = I^1(S)$ and there exists an index i for which $H(S) = I^i(S)$. If $H(S) = V(G)$ then S is a *hull set*. The *convexity number* $c(G)$ of G is the size of a maximum convex set $S \neq V(G)$. The *interval number* $i(G)$ of G is the size of a smallest interval set of G. The *hull number* $h(G)$ of G is the size of a smallest hull set of G. Now we are in position to state the six problems dealt with in this work:

CONVEX SET - CS
Input: A graph G and a set $S \subseteq V(G)$.
Question: Is S convex?

INTERVAL DETERMINATION - ID
Input: A graph G, a set $S \subseteq V(G)$, and a vertex $z \in V(G)$.
Question: Does z belong to $I(S)$?

CONVEX HULL DETERMINATION - CHD
Input: A graph G, a set $S \subseteq V(G)$, and a vertex $z \in V(G)$.
Question: Does z belong to $H(S)$?

CONVEXITY NUMBER - CN
Input: A graph G and a positive integer r.
Question: Is $c(G) \geq r$?

INTERVAL NUMBER - IN
Input: A graph G and a positive integer r.
Question: Is $i(G) \leq r$?

HULL NUMBER - HN
Input: A graph G and a positive integer r.
Question: Is $h(G) \leq r$?

The table below shows the complexity of the six problems listed in the preceding subsection for some convexity spaces. All the entries of the table correspond to results found in the literature, or to trivial results (indicated by '[t]').

Table 2. Problems vs convexities: complexity results.

	Geodesic		Monophonic		P_3		P_3^*		Triangle-path	
CS	P	[t]	P	[26]	P	[t]	P	[1]	P	[28]
ID	P	[t]	NPc	[26]	P	[t]	P	[1]	NPc	[28]
CHD	P	[t]	P	[26]	P	[t]	P	[1]	P	[28]
CN	NPc	[33]	NPc	[26]	NPc	[9]	NPc	[1]	P	[28]
IN	NPc	[3]	NPc	[26]	NPc	[8]	NPc	[1]	NPc	[28]
HN	NPc	[25]	P	[26]	NPc	[9]	NPc	[1]	P	[28]

2.3   Two Useful Facts

The next two propositions are useful. They say that if INTERVAL DETERMINATION or CONVEX SET can be solved in polynomial time for some convexity space then some other problems listed in Sect. 2.2 can also be solved in polynomial time, for the same convexity space.

Proposition 2. *Let $(V(G), \mathcal{C})$ be any convexity space. If* INTERVAL DETERMINATION *can be solved in polynomial time for $(V(G), \mathcal{C})$ then* CONVEX SET *and* CONVEX HULL DETERMINATION *can also be solved in polynomial time for $(V(G), \mathcal{C})$.*

Let $S \subseteq V(G)$. If S is not convex then an *augmenting set* of S is any set S' such that $S \subset S' \subseteq H(S)$ (where the symbol $\subset$ stands for proper inclusion).

Proposition 3. *Let $(V(G), \mathcal{C})$ be a convexity space. If there is a polynomial-time certification algorithm to solve* CONVEX SET *for $(V(G), \mathcal{C})$ that outputs an augmenting set when the problem has a negative answer then* CONVEX HULL DETERMINATION *can also be solved in polynomial time for $(V(G), \mathcal{C})$.*

Note that Propositions 2 and 3 can be used to fill some entries of Table 2. For example, since INTERVAL DETERMINATION is in P for the geodesic convexity, by Proposition 2 the problems CONVEX SET and CONVEX HULL DETERMINATION are also in P for such convexity. The same applies to the P_3- and P_3^*- convexities. On the other hand, Proposition 3 implies that CONVEX HULL DETERMINATION is in P for the monophonic convexity.

3   A General Framework for Path Convexities

In this section, we propose a general framework for the study of path convexities.

From now on, we assume that every n-vertex graph G has vertices labeled $1, 2, \ldots, n$. A *length matrix* is a symmetric $n \times n$ matrix M such that each entry $M(i, j)$, for $i, j \in V(G)$, is a natural number; in addition, all diagonal entries of M are zero.

Let A, B, C, D be four $n \times n$ length matrices. Suppose that $\mathscr{P}$ is the family of paths of G such that an ij-path P of G is a member of $\mathscr{P}$ if and only if:

(1) $|P| \geq A(i, j)$;
(2) $|P| \leq B(i, j)$;
(3) all the chords in P are of length at least $C(i, j)$;
(4) all the chords in P are of length at most $D(i, j)$.

Let $I_{\mathscr{P}} : 2^{V(G)} \to 2^{V(G)}$ be the interval function associated with $\mathscr{P}$, and let $\mathcal{C}_{\mathscr{P}}$ be the family of subsets of $V(G)$ such that $S \in \mathcal{C}_{\mathscr{P}}$ if and only if $I_{\mathscr{P}}(S) = S$. Since $\mathscr{P}$ is a particular collection of paths of G, by Proposition 1, we have that $(V(G), \mathcal{C}_{\mathscr{P}})$ is a finite convexity space, equipped with interval function $I_{\mathscr{P}}$. Let us say that such a convexity space defines a *matrix path convexity*.

Again, we omit the subscript $\mathscr{P}$ when it is clear from the context.

Say that an ij-path P *satisfies* matrices A, B, C, D if all the conditions (1) to (4) above are satisfied by P.

3.1   Putting the Matrices as Part of the Input

In the six problems listed in Sect. 2.2, the graph G is always part of the input; however, the rule that determines which collection of paths of G must be considered is *not* part of the input. More general versions of such problems are possible when the desired convexity space, expressed as a graph G together with a set of four length matrices, is part of the input. For example, consider the following version of CONVEX SET:

MATRIX CONVEX SET

Input: A graph G, four $n \times n$ length matrices A, B, C, D, and $S \subseteq V(G)$.

Question: Is S convex under the matrix path convexity ruled by A, B, C, D?

All the remaining problems listed in Sect. 2.2 can be restated analogously.

The next theorems say that such "matrix problems" are all hard. However, we shall see that restrictions on the matrices A, B, C, D lead to interesting cases. In this regard, some types of length matrices are of special interest. For a graph G, the *distance matrix* of G is the length matrix M_{dist} with entries $M_{dist}(i,j) = dist_G(i,j)$, for $i, j \in V(G)$. For a positive integer constant k, the $(n-k)$-*matrix* and the k-*matrix* are the length matrices M_{n-k} and M_k with off-diagonal entries all equal to, respectively, $n - k$ and k.

Theorem 1. MATRIX CONVEX SET *is co-NP-complete.*

Proof. A certificate for a negative answer to MATRIX CONVEX SET is a triple i, j, z (with $i, j \in S$ and $z \notin S$) and an ij-path P in G containing z such that P satisfies A to D. Such a certificate can be clearly checked in polynomial time. Therefore, MATRIX CONVEX SET is in coNP.

To prove that MATRIX CONVEX SET is co-NP-complete, we show a reduction from the following NP-complete problem [38]: given three distinct vertices i, j, z in a graph H, decide whether there is a chordless ij-path passing through z.

Let G be the graph obtained from H by replacing each edge (s, z) incident to z by an sz-path containing $n-1$ internal vertices of degree two, where $n = |V(H)|$. In other words, G is a subdivision of H obtained by subdividing each edge incident to z using $n - 1$ vertices. Set A and B as the length matrices with off-diagonal entries all equal to, respectively, $2n$ and $3n - 3$. Also, set $C = D = M_1$ (the k-matrix for $k = 1$). Finally, set $S = \{i, j\}$. Note that the collection of paths $\mathscr{P}$ defined by A, B, C, D contains the chordless paths with length at least $2n$ and at most $3n - 3$.

Suppose that there is a chordless ij-path P_H in H passing through z. Write $P_H = (s_0 = i, s_1, \ldots, s_{h-1}, s_h = z, s_{h+1}, \ldots, s_l = j)$. Then there is a chordless ij-path P_G in G obtained from P_H by subdividing edges (s_{h-1}, z) and (z, s_{h+1}) using $n - 1$ vertices of degree two for each edge. Note that $|P_G| = (l - 2) + 2n$. Since $2 \leq l \leq n - 1$, we have $2n \leq |P_G| \leq 3n - 3$. Therefore, P_G satisfies A, B, C, D, and its existence implies that S is not convex.

Conversely, suppose that S is not convex. Then there is a chordless ij-path P_G of length at least $2n$ passing through some vertex of G lying outside S. But, by the construction of G, all the ij-paths of length at least $2n$ must necessarily pass through z. Let P_{hz} be the subpath of P_G with length n that starts at a vertex h and ends at z. Similarly, let $P_{zh'}$ be the subpath of P_G with length n that starts at z and ends at a vertex h'. By replacing P_{hz} and $P_{zh'}$ by edges (h, z) and (z, h'), we obtain a chordless ij-path in H passing through z. This completes the proof. $\square$

Theorem 2. MATRIX INTERVAL DETERMINATION *is NP-complete.*

Theorem 3. MATRIX CONVEX HULL DETERMINATION *is NP-complete.*

Theorem 4. 1. MATRIX CONVEXITY NUMBER *is NP-hard.*
2. MATRIX INTERVAL NUMBER *is NP-complete.*
3. MATRIX HULL NUMBER *is NP-complete.*

3.2   Constant Matrices: The (a, b, c, d)-path Convexity

In this section, we study the case in which there are constants a, b, c, d such that
$A = M_a$, $B = M_b$, $C = M_c$, and $D = M_d$. In this scenario we can assume that
the matrices are *not* part of the input, because length restrictions are known in
advance. This gives rise to "constant matrix versions" of the problems studied
in the preceding subsection. For example, consider the following problems:

(a, b, c, d)-CONVEX SET
Input: A graph G, a set $S \subseteq V(G)$.
Question: Is S convex under the matrix path convexity ruled by A, B, C, D?
Equivalently: Is S convex under the path convexity defined by the collection
$\mathscr{P}(a, b, c, d)$ of paths of G whose length is at least a and at most b, and whose
chords have length at least c and at most d?

(a, b, c, d)-INTERVAL DETERMINATION
Input: A graph G, a set $S \subseteq V(G)$, and a vertex $z \in V(G)$.
Question: Does z belong to $I(S)$, where I is the interval function associated with
the collection $\mathscr{P}(a, b, c, d)$ of paths of G?

The remaining matrix problems can be restated analogously.
The path convexity for which the path/chord length restrictions are ruled by
four constants a, b, c, d as explained above is called (a, b, c, d)-*path convexity.*

Theorem 5. (a, b, c, d)-INTERVAL DETERMINATION *is in* P.

By Proposition 2, we have:

Corollary 1. (a, b, c, d)-CONVEX SET *and* (a, b, c, d)-CONVEX HULL DETER-
MINATION *are in* P.

As for the other three problems, (a, b, c, d)-CONVEXITY/INTERVAL/HULL NUM-
BER, we remark that the special cases

$$(a = 2, b = 2, c = 1, d = 2) \text{ and } (a = 2, b = 2, c = 1, d = 1)$$

correspond precisely to the P_3- and P_3^*- convexities, as indicated in Table 1. For
both convexities, all the three problems are NP-complete (see Table 2).

3.3 (a, b, c, d)-path Convexity and Bounded Treewidth Graphs

In this section, we investigate the complexity of the six (a, b, c, d)-path convexity problems in Sect. 3.2 when applied to bounded treewidth graphs. As we shall see, linear-time methods will be possible in this case.

Let G be a graph, T a tree, and $V = (V_t)_{t \in T}$ a family of vertex sets $V_t \subseteq V(G)$ indexed by the vertices t of T. The pair (T, V) is called a *tree-decomposition of* G if it satisfies the following three conditions [24]:

(T1) $V(G) = \bigcup_{t \in T} V_t$;
(T2) for every edge $e \in G$ there exists $t \in T$ such that both ends of e lie in V_t;
(T3) if V_{t_i} and V_{t_j} both contain a vertex v then $v \in V_{t_k}$ for all vertices t_k in the path between t_i and t_j.

The *width* of (T, V) is the number $\max\{|V_t| - 1 \mid t \in T\}$, and the *treewidth* $tw(G)$ of G is the minimum width of any tree-decomposition of G.

Graphs of treewidth at most k are called *partial k-trees*. Some graph classes with bounded treewidth include: forests (treewidth 1); pseudoforests, cacti, series-parallel graphs, and outerplanar graphs (treewidth at most 2); Halin graphs and Apollonian networks (treewidth at most 3) [4,5]. Control flow graphs arising in the compilation of structured programs also have bounded treewidth (at most 6) [47].

In 1990, Courcelle [19] stated that for any graph G with treewidth bounded by a constant k and for any graph property Π that can be formulated in CMSOL$_2$ (*Counting Monadic Second-Order Logic* where quantification over sets of vertices or edges and predicates testing the size of sets modulo constants are allowed), there is a linear-time algorithm that decides if G satisfies Π [19,21–23]. This result has been extended a number of times. In particular, Arnborg and Lagergren [2] study optimization problems over sets definable in Counting Monadic Second-Order Logic.

By meta-theorems based on CMSOL$_2$ [19,21,22], obtaining linear-time methods to solve the six problems of Sect. 3.2 on bounded treewidth graphs amounts to showing that the related properties are expressible in CMSOL$_2$.

Theorem 6. (a, b, c, d)-INTERVAL DETERMINATION *is solvable in linear time on bounded treewidth graphs.*

Proof. It is enough to show that the property "$z \in I(S)$" is CMSOL$_2$-expressible. Given G, S, and z, we construct $\varphi(G, S, z, a, b, c, d)$ such that $z \in I(S) \Leftrightarrow \varphi(G, S, z, a, b, c, d)$ as follows:

$$(\, z \in S \,) \vee$$
$$(\, \exists \, u, v, P \, (\, u, v \in S \, \wedge$$

P is an uv-path $\wedge$

$z \in P \, \wedge$

$Card(P) \geq a \, \wedge$

$Card(P) \leq b \, \wedge$ (2)

$\forall P' \, (\, (\, P' \subseteq P \, \wedge$

$Card(P') \geq 2 \, \wedge$

$\exists \, u', v'(P'$ is an $u'v'$-path $\wedge \; adj(u', v')))$

$\Rightarrow (Card(P') \geq c \; \wedge \; Card(P') \leq d) \,)$

$))$

In the above formula, paths are regarded as subsets of edges. Using this approach, the subformula "P is an uv-path" can be expressed in CMSOL$_2$ (see [22]). Note that a chord is expressed as an $u'v'$-subpath P' of P with length at least c and at most d such that u' is adjacent to v'. □

Corollary 2. (a, b, c, d)-CONVEX SET *can be solved in linear time on bounded treewidth graphs.*

Proof. The property "S is convex" is equivalent to "there is no z such that $z \notin S$ and $z \in I(S)$". By Theorem 6, "$z \in I(S)$" is CMSOL$_2$-expressible. Thus the result easily follows. □

Corollary 3. (a, b, c, d)-CONVEX HULL DETERMINATION *can be solved in linear time on bounded treewidth graphs.*

Proof. The property "$z \in H(S)$" is equivalent to "there exists S_1 such that: (a) S_1 is convex, (b) $S \subseteq S_1$, (c) $z \in S_1$, and (d) there is no S_2 such that S_2 is convex, $S \subseteq S_2$, and S_2 is properly contained in S_1". By Corollary 2, we can use CMSOL$_2$ to say that the sets S_1 and S_2 are convex. Thus the result follows. □

For the remaining three problems $((a, b, c, d)$-CONVEXITY, (a, b, c, d)-INTERVAL and (a, b, c, d)-HULL NUMBER), we consider their optimization versions (maximization in the case of CONVEXITY NUMBER, and minimization in the case of INTERVAL/HULL NUMBER). Note that the properties "S is a convex set distinct from $V(G)$", "S is an interval set", and "S is a hull set" can be expressed in CMSOL$_2$. Therefore the optimization versions ("find an optimal set satisfying the required property") are LinCMSOL$_2$ problems [2, 20].

Theorem 7. [2, 20] *Let k be a positive constant, and Π be a LinCMSOL$_2$ problem. Then Π can be solved in linear time on graphs of treewidth bounded by k (if the tree-decomposition is given with the input graph).*

Therefore:

Corollary 4. *The optimization versions of* (a, b, c, d)-CONVEXITY NUMBER, (a, b, c, d)-INTERVAL NUMBER, *and* (a, b, c, d)-HULL NUMBER *can be solved in linear time on bounded treewidth graphs.*

3.4 Particular Cases of the (a, b, c, d)-path Convexity

In this section we show that, by extending the meaning of the parameters a, b, c, d, most path convexities in the literature can be viewed as particular cases of the (a, b, c, d)-path convexity. In Table 3 below, the symbol 'σ' (resp.,'ℓ') means that the length of the shortest (resp., longest) path between each pair of distinct vertices must be considered. The symbol '∞' stands for no length restriction. For a constant k, the symbol '$k \mid \sigma$' means that, for each pair (i, j) of distinct vertices, the minimum value between k and the length of the shortest ij-path must be considered.

Table 3. Path convexities as particular cases of the (a, b, c, d)-path convexity. Note that putting $c = d = 1$ implies that all the paths of the considered collection $\mathscr{P}$ are chordless.

Convexity	a	b	c	d
Geodesic	σ	σ	1	1
Monophonic	2	∞	1	1
g^3	3	σ	1	1
g_k	σ	$k \mid \sigma$	1	1
m^3	3	∞	1	1
P_3	2	2	1	2
P_3^*	2	2	1	1
Triangle-path	2	∞	1	2
Total	2	∞	3	∞
Detour	ℓ	ℓ	1	∞
All-path	2	∞	1	∞

References

1. Araujo, R.T., Sampaio, R.M., Szwarcfiter, J.L.: The convexity of induced paths of order three. Discrete Math. **44**, 109–114 (2013)
2. Arnborg, S., Lagergren, J., Seese, D.: Easy problems for tree-decomposable graphs. J. Algorithms **12**(2), 308–340 (1991)
3. Atici, M.: Computational complexity of geodetic set. Int. J. Comput. Math. **79**, 587–591 (2002)
4. Bodlaender, H.L.: A partial k-arboretum of graphs with bounded treewidth. Theoret. Comput. Sci. **209**(1–2), 1–45 (1998)

5. Brandstädt, A., Le, V.B., Spinrad, J.P.: Graph Classes: A Survey, vol. 3. SIAM, Philadelphia (1999)
6. Cáceres, J., Oellermann, O.R., Puertas, M.L.: Minimal trees and monophonic convexity. Discuss. Math. Graph Theory **32**(4), 685–704 (2012)
7. Centeno, C.C., Dantas, S., Dourado, M.C., Rautenbach, D., Szwarcfiter, J.L.: Convex partitions of graphs induced by paths of order three. Discrete Math. **12**(5), 175–184 (2010)
8. Centeno, C.C., Dourado, M.C., Penso, L.D., Rautenbach, D., Szwarcfiter, J.L.: Irreversible interval of graphs. Theoret. Comput. Sci. **412**, 3693–3700 (2011)
9. Centeno, C.C., Dourado, M.C., Szwarcfiter, J.L.: On the convexity of paths of length two in undirected graphs. Electron. Notes Discrete Math. **32**, 11–18 (2009)
10. Changat, M., Mathew, J.: On triangle path convexity in graphs. Discrete Math. **206**, 91–95 (1999)
11. Changat, M., Mathew, J.: Induced path transit function, monotone and Peano axioms. Discrete Math. **286**(3), 185–194 (2004)
12. Changat, M., Klavzar, S., Mulder, H.M.: The all-paths transit function of a graph. Czechoslovak Mathematic J. **51**(2), 439–448 (2001)
13. Changat, M., Mulder, H.M., Sierksma, G.: Convexities related to path properties on graphs. Discrete Math. **290**(2–3), 117–131 (2005)
14. Changat, M., Narasimha-Shenoi, P.G., Mathews, J.: Triangle path transit functions, betweenness and pseudo-modular graphs. Discrete Math. **309**(6), 1575–1583 (2009)
15. Changat, M., Narasimha-Shenoi, P.G., Pelayo, I.: The longest path transit function of a graph and betweenness. Utilitas Mathematica **82**, 111–127 (2010)
16. Chartrand, G., Garry, L., Zhang, P.: The detour number of a graph. Utilitas Mathematica **64**, 97–113 (2003)
17. Chartrand, G., Escuadro, H., Zhang, P.: Detour distance in graphs. J. Combin. Math. Combin. Comput. **52**, 75–94 (2005)
18. Chepoi, V.: Peakless functions on graphs. Discrete Appl. Math. **73**(2), 175–189 (1997)
19. Courcelle, B.: The monadic second-order logic of graphs I. Recognizable sets of finite graphs. Inf. Comput. **25**(1), 12–75 (1990)
20. Courcelle, B.: The monadic second-order logic of graphs III: tree-decompositions, minor and complexity issues. Informatique Théorique et Appl. **26**, 257–286 (1992)
21. Courcelle, B., Mosbah, M.: Monadic second-order evaluations on tree-decomposable graphs. Theoret. Comput. Sci. **109**(1), 49–82 (1993)
22. Courcelle, B.: The expression of graph properties and graph transformations in monadic second-order logic. In: Handbook of Graph Grammars and Computing by Graph Transformations, vol. 1, pp. 313–400 (1997)
23. Courcelle, B., Engelfriet, J.: Graph Structure and Monadic Second-Order Logic. Cambridge University Press, Cambridge (2011)
24. Diestel, R.: Graph Theory, 3rd edn. Springer, Heidelberg (2005)
25. Dourado, M.C., Gimbel, J.G., Kratochvil, J., Protti, F., Szwarcfiter, J.L.: On the computation of the hull number of a graph. Discrete Math. **309**, 5668–5674 (2009)
26. Dourado, M.C., Protti, F., Szwarcfiter, J.L.: Complexity results related to monophonic convexity. Discrete Math. **158**, 1268–1274 (2010)
27. Dourado, M.C., Rautenbach, D., dos Santos, V.F., Schäfer, P.M., Szwarcfiter, J.L., Toman, A.: An upper bound on the P_3-radon number. Discrete Math. **312**(16), 2433–2437 (2012)
28. Dourado, M.C., Sampaio, R.M.: Complexity aspects of the triangle path convexity. Discrete Appl. Math. **206**, 39–47 (2016)

29. Dragan, F.F., Nicolai, F., Brandstädt, A.: Convexity and HHD-free graphs. SIAM J. Discrete Math. **12**, 119–135 (1999)
30. Duchet, P.: Convex sets in graphs, II. Minimal path convexity. J. Comb. Theory Ser. B **44**, 307–316 (1988)
31. Farber, M., Jamison, R.E.: Convexity in graphs and hypergraphs. SIAM J. Alg. Disc. Math. **7**(3), 433–444 (1986)
32. Farber, M., Jamison, R.E.: On local convexity in graphs. Discrete Math. **66**, 231–247 (1987)
33. Gimbel, J.G.: Some remarks on the convexity number of a graph. Graphs Comb. **19**, 357–361 (2003)
34. Gutin, G., Yeo, A.: On the number of connected convex subgraphs of a connected acyclic digraph. Discrete Appl. Math. **157**(7), 1660–1662 (2009)
35. Harary, F.: Convexity in graphs: achievement and avoidance games. Ann. Discrete Math. **20**, 323 (1984)
36. Harary, F., Nieminen, J.: Convexity in graphs. J. Differ. Geom. **16**, 185–190 (1981)
37. Harary, F., Loukakis, E., Tsouros, C.: The geodetic number of a graph. Math. Comput. Model. **17**(11), 89–95 (1993)
38. Haas, R., Hoffmann, M.: Chordless path through three vertices. Theoret. Comput. Sci. **351**, 360–371 (2006)
39. Kanté, M.M., Nourine, L.: Polynomial time algorithms for computing a minimum hull set in distance-hereditary and chordal graphs. In: van Emde Boas, P., Groen, F.C.A., Italiano, G.F., Nawrocki, J., Sack, H. (eds.) SOFSEM 2013. LNCS, vol. 7741, pp. 268–279. Springer, Heidelberg (2013). https://doi.org/10.1007/978-3-642-35843-2_24
40. Nebeský, L.: A characterization of the interval function of a connected graph. Czechoslovak Math. J. **44**(1), 173–178 (1994)
41. Nielsen, M.H., Oellermann, O.R.: Steiner trees and convex geometries. SIAM J. Discrete Math. **23**(2), 680–693 (2011)
42. Parker, D.B., Westhoff, R.F., Wolf, M.J.: Two-path convexity in clone-free regular multipartite tournaments. Australas. J. Combin. **36**, 177–196 (2006)
43. Pelayo, I.M.: Geodesic Convexity in Graphs. Springer, New York (2013). https://doi.org/10.1007/978-1-4614-8699-2
44. Parvathy, K.S.: Studies on convex structures with emphasis on convexity in graphs. Ph.D. thesis, Cochin University, Kochi (1995)
45. Peterin, I.: The pre-hull number and lexicographic product. Discrete Appl. Math. **312**, 2153–2157 (2012)
46. Sampathkumar, E.: Convex sets in a graph. Indian J. Pure Appl. Math. **15**(10), 1065–1071 (1984)
47. Thorup, M.: All structured programs have small tree width and good register allocation. Inf. Comput. **142**(2), 159–181 (1998)
48. van de Vel, M.L.J.: Theory of Convex Structures. North Holland, Amsterdam (1993)

An Approximation Algorithm for the Dynamic k-level Facility Location Problem

Limin Wang[1], Zhao Zhang[2], Dachuan Xu[3], and Xiaoyan Zhang[4(✉)]

[1] State Key Laboratory for Novel Software Technology, Nanjing University,
Nanjing 210023, Jiangsu, People's Republic of China
[2] College of Mathematics and Computer Science, Zhejiang Normal University,
Jinhua 321004, Zhejiang, People's Republic of China
[3] Beijing Institute for Scientific and Engineering Computing,
Beijing University of Technology, Beijing 100124, People's Republic of China
[4] School of Mathematical Science and Institute of Mathematics,
Nanjing Normal University, Nanjing 210023, Jiangsu, People's Republic of China
zhangxiaoyan@njnu.edu.cn

Abstract. In this paper, we consider the dynamic k-level facility location problem, which is a generalization of the uncapacitated k-level facility location problem when considering time factor. We present a combinatorial primal-dual approximation algorithm for the problem which finds a solution within 6 times the optimum. This approximation ratio under a dynamic setting coincides with that of the simple dual ascent 6-approximation algorithm for the (static) multilevel facility location problem (Bumb, 2001) with a weak triangle inequality property.

Keywords: Approximation algorithm · Primal-dual · Dynamic · Facility location

1 Introduction

Facility location has a wide range of important applications in many fields of modernization such as manufacturing, transportation, network, information, resource allocation, etc. It is a kind of problem with important research value. An important problem in facility location is to choose a set of facilities, such as plants, warehouses or wireless antenna towers, in order to minimize the total cost of opening facilities and meet the demands for some commodity [4].

The dynamic facility location problem (DFLP) [10] is a generalization of the uncapacitated facility location problem (UFLP). In each time period, the demand of a client is served by facilities which should be opened at this time

Supported by National Natural Science Foundation of China (Grant Nos. 61425024, 11531011, 11771013, 11871081, 11871280, 11471003), and National Thousand Young Talents Program, and Qing Lan Project.

period or earlier. For a potential facility, the opening cost varies in different time periods. For a client and a potential facility, the connection cost between them also varies in different time periods. The goal is to select a subset of facilities to open at each time period to minimize the total cost in all time periods such that the demand of each client is satisfied in each time period. Based on primal-dual scheme, a 3-approximation algorithm was proposed for DFLP in [13]. Then combining the techniques from [5,7], they further proposed a 1.86-approximation algorithm, which is so far the best approximate algorithm for DFLP.

In the k-level facility location problem (k-FLP), each client must be served by a path of k level sets of open facilities. The uncapacitated k-level facility location problem is a static problem; it concerns how many facilities to build and where to locate them. The first approximation algorithm for the multilevel FLP was developed in [1,9], they were based on rounding an LP solution to an integer one. The performance guarantees of these algorithms were 3.16 and 3, respectively. The first combinatorial algorithm for the multilevel FLP was developed in [8], and it finds a solution within $O(\log|D|)$ the optimum, where D is the set of demand points. Based on extending the primal-dual algorithm, Bumb [3] proposed a 6-approximation algorithm for the k-FLP. An improved combinatorial approximation algorithm for k-FLP was presented in [2], and the performance guarantee of their algorithm is 3.27.

In practical applications, the k-FLP problem is often applied to the supply chains. Due to the cost of facilities, service cost and the demand of each client will change with time, we consider the k-FLP problem in a dynamic settings. In this paper, we investigative the dynamic k-level facility location problem, which is a generalization of the uncapacitated k-level facility location problem when considering time factor. Being an extension of the uncapacitated facility location problem, which is known to be Max SNP-hard [5], this problem is Max SNP-hard as well. Motivated by the idea from [3,6,11,12], we present a simple dual ascent method for the k-DFLP that finds a solution within 6 times the optimum. This approximation ratio under a dynamic setting coincides with that of the simple dual ascent 6-approximation algorithm for the (static) multilevel facility location problem [3] with a weak triangle inequality property.

The paper is organized as follows. In Sect. 2, we present the formal definition of the k-DFLP problem and give its linear programming relaxation. In Sect. 3, we present our primal-dual algorithm for k-DFLP. The algorithm is analyzed in Sect. 4. Finally, we conclude the paper in Sect. 5.

2 k-DFLP

The k-DFLP problem also addresses the issue of when to build a facility. More precisely, in the k-DFLP problem, we are given a set of clients D, and the time periods numbered from 1 to T. Let $F = \bigcup_{l=1}^{k} F^l$ be the set of all facilities, where each F^l is the set of sites where facilities on level l ($1 \leq l \leq k$) are located and the sets $F^1, \ldots, F^l$ are pairwise disjoint. In the following, unless otherwise specified, we use i to represent the facility and j to represent the client. We refer

$p = (i_1, \ldots, i_k)(i_l \in F^l, l = 1, \ldots, k)$ to be a path of facilities. The set of all possible paths is denoted by P. At each time period t, a client $j \in D$ is specified by a demand d_j^t that can be served by facilities open at the beginning of the time period t, i.e., at time period t or earlier. A cost f_i^t is incurred when the facility $i \in F$ needs to be open at time period t, where f_i^t is ∞ if facility i is not available at time period t. A cost c_{ij}^t is incurred for supplying one unit demand of client j in time period t from facility i. The connection cost between facilities in two adjacent levels is $c_{i_l i_{l+1}}^t$ $(1 \leq l \leq k-1)$ in time period t. If client j is connected to a path p at time period t, we may say that client j is assigned to the path p or is served by the path p at time period t. The cost incurred by assigning client j to the path $p = (i_1, \ldots, i_k)$ is equal to $c_{jp}^t = c_{ji_1}^t + \sum_{l=1}^{k-1} c_{i_l i_{l+1}}^t$. The objective is to choose a subset of facilities F to open at each time period and assign each client j to a path p, such that all demands of the clients are satisfied and the total cost is minimized. Similar to the assumption in [13], we assume that the weak triangle inequality property holds, namely $c_{i_1 j_1}^t \leq c_{i_2 j_1}^t + c_{i_2 j_2}^{t'} + c_{i_1 j_2}^{t'}$ for any $i_1, i_2 \in F$, $j_1, j_2 \in D$ and time periods $1 \leq t, t' \leq T$. It is clear that when $T = 1$, k-DFLP becomes k-FLP.

More precisely, the k-DFLP can be formulated as an integer program as follows:

$$\text{minimize} \quad \sum_{t=1}^{T} \sum_{p \in P} \sum_{j \in D} d_j^t c_{jp}^t x_{jp}^t + \sum_{s=1}^{T} \sum_{i \in F} f_i^s y_i^s$$

$$\text{s.t.} \quad \sum_{p \in P} x_{jp}^t = 1 \quad \forall j \in D, 1 \leq t \leq T$$

$$\sum_{p : i \in p} x_{jp}^t \leq \sum_{s=1}^{t} y_i^s \quad \forall i \in F, j \in D, 1 \leq t \leq T$$

$$x_{jp}^t, y_i^s \in \{0,1\} \quad \forall i \in F, j \in D, p \in P, 1 \leq s, t \leq T$$

where $y_i^s = 1$ if facility i is open at time period s, $y_i^s = 0$ otherwise; $x_{jp}^t = 1$ if the demand of client j is supplied from path p at time period t. The first constraint indicates that the demand of each client j must be served by one path p at any period t. The second constraint indicates that if path p at time period t supplies the demand of a client, then $i \in p$ must be open at time period t.

Consider the linear programming relaxation of the above integer program:

$$\text{minimize} \quad \sum_{t=1}^{T}\sum_{p\in P}\sum_{j\in D} d_j^t c_{jp}^t x_{jp}^t + \sum_{s=1}^{T}\sum_{i\in F} f_i^s y_i^s$$

$$\text{s.t.} \quad \sum_{p\in P} x_{jp}^t = 1 \quad \forall j\in D, 1\le t\le T$$

$$\sum_{p:i\in p} x_{jp}^t \le \sum_{s=1}^{t} y_i^s \quad \forall i\in F, j\in D, 1\le t\le T$$

$$x_{jp}^t, y_i^s \ge 0 \quad \forall i\in F, j\in D, p\in P, 1\le s,t\le T$$

The dual of this linear programming relaxation is

$$\text{maximize} \quad \sum_{t=1}^{T}\sum_{j\in D} \alpha_j^t$$

$$\text{s.t.} \quad \alpha_j^t - d_j^t c_{jp}^t \le \sum_{i\in p} \beta_{ij}^t \quad \forall j\in D, p\in P, 1\le t\le T$$

$$\sum_{t=s}^{T}\sum_{j\in D} \beta_{ij}^t \le f_i^s \quad \forall i\in F, 1\le s\le T$$

$$\beta_{ij}^t \ge 0 \quad \forall i\in F, j\in D, 1\le t\le T$$

It is straightforward to show that the above dual is equivalent to the following linear program.

$$\text{maximize} \quad \sum_{t=1}^{T}\sum_{j\in D} d_j^t \alpha_j^t$$

$$\text{s.t.} \quad \alpha_j^t - c_{jp}^t \le \sum_{i\in p} \beta_{ij}^t \quad \forall j\in D, p\in P, 1\le t\le T$$

$$\sum_{t=s}^{T}\sum_{j\in D} d_j^t \beta_{ij}^t \le f_i^s \quad \forall i\in F, 1\le s\le T$$

$$\beta_{ij}^t \ge 0 \quad \forall i\in F, j\in D, 1\le t\le T$$

Intuitively, the dual variable α_j^t can be interpreted as a budget that client j is willing to spend to get one unit of its demand served in time period t, and the dual variable β_{ij}^t can be interpreted as the part of α_j^t that is contributed to pay for opening facility i in time period t.

3   Primal-Dual Algorithm

We present a dual ascent algorithm by integrating the techniques in [3,6,13]. The algorithm first constructs a feasible dual solution and then finds a feasible integer primal solution based on the dual solution.

We define the following three concepts before constructing a feasible dual solution.

(1) A facility $i_l \in F^l$ is temporarily open when $\sum_{t=s}^{T} \sum_{j \in D} d_j^t \beta_{ij}^t = f_i^s$. Denote by T_{i_l} the time when facility $i_l \in F^l$ becomes temporarily open.

(2) A client $j \in D$ at time period t reaches $i_l \in F^l$ if for some path $p = (i_1, i_2, \ldots, i_l)$ from i_1 to i_l, all facilities $i_1, i_2, \ldots, i_{l-1}$ are open and $\alpha_j^t = c_{jp}^t + \sum_{l'=1}^{l} \beta_{i_{l'}j}^t$

(3) If, in addition, also i_l is temporarily open, we say that j leaves i_l at time period t or, in case $l = k$, that j gets connected along p to $i_k \in F^k$ at time period t.

Phase 1. Construction of a dual feasible solution

We introduce a notion of time θ in the algorithm. Initially $\theta = 0$, all dual variables are set to be 0, all facilities i are closed, and all demand-period pairs (j, t) are said to be unfrozen. We start increasing dual variables α_j^t for all demand-period pairs (j, t) at the same rate as long as they are unfrozen, i.e., at time θ, $\alpha_j^t = \theta$ for all unfrozen demand-period pairs. When the demand-period pair $(j, t) \in D$ reaches some closed facility $i_l \in F^l$, the dual variable $\beta_{i_lj}^t$ will be increased at the same rate as α_j^t. When i_l is open, then freeze all the dual variables $\beta_{i_lj}^t$, $(j, t) \in D$. Keep increasing time θ until there is no unfrozen demand-period pairs.

For each temporarily open facility $i_l \in F^l$ ($l \geq 2$) in time period t, the predecessor of i_l will be the facility in the level $l - 1$ via which was for the first time reached by a demand-period pair, i.e.,

$$Pred(i_l, t) := argmin_{i \in F^{l-1}} \{T_i + c_{ii_l}^t\}$$

The predecessor of a temporarily open $i_1 \in F^1$ will be its closest client and we define the time $T_{Pred(i_1,t)} = 0$. As time increases, the following two cases may occur:

- Facility $i_k \in F^k$ is temporarily open. In this case, freeze those unfrozen demand-period pairs $(j, t) \in D$ with $\beta_{ji_k}^t > 0$ and connect them to facility i_k, which is called the connecting witness for (j, t). In addition, denote $p(i_k, t) = (i_1, \ldots, i_k; t)$ as the associated central path at time period t such that

$$i_l = Pred(i_{l+1}, t), \forall 1 \leq l \leq k - 1$$

and $(j, t)_{i_k}$ as the predecessor of i_1. We call the neighborhood of i_k the set of clients contributing to $p(i_k, t)$, i.e.,

$$N(i_k) = \{(j, t) \in D | \beta_{ji_l}^t \geq 0 \text{ for some } i_l \in p(i_k, t)\}$$

– If an unfrozen demand-period pair (j, t) reaches a temporarily open facility-period pair i_k, then freeze (j, t) and connect (j, t) to i_k, which is also called the connecting witness for (j, t).

When all demand-period pairs are frozen, the first phase terminates. If two events occur simultaneously, the algorithm executes them in an arbitrary order.

Phase 2. Construction of primal feasible solution

In this phase, we specify which facilities to open in each period, and how the demands should be served. Let $\tilde{F}^k \subseteq F^k$ be the set of tentatively open facility in the first phase. We say that facilities i_k and i'_k $(i_k, i'_k \in \tilde{F}^k)$ are dependent if there exists some demand-period pair $(j, t) \in D$ such that $(j, t) \in N(i_k) \cap N(i'_k)$. We pick a maximal independent set of facility-period pairs $\bar{F}^k \subseteq \tilde{F}^k$ such that for any $i_k \in \tilde{F}^k \backslash \bar{F}^k$, there exists $i'_k \in \bar{F}^k$ where $T_{i'_k} \leq T_{i_k}$, i_k and i'_k are dependent. Then we open the facility in $\bar{F}^k$, and also open the associated central path $p(i_k, t)$, $i_k \in \bar{F}^k$.

Assign demand-period pair $(j, t) \in D$ to an open facility $i_k \in \bar{F}^k$ at time period t by the following rule. If there is i_k such that $(j, t) \in N(i_k)$, then (j, t) is connected to i_k along the associated central path $p(i_k, t)$. Otherwise, (j, t) is connected to the closest open facility-period pair i_k along path $p(i_k, t)$.

4   Analysis of Algorithm

Firstly, we introduce a crucial lemmas for proving the approximation ratio of our algorithm.

Lemma 1. *Let $i_k \in \bar{F}^k$ be a temporarily open facility and its associated path $p(i_k, t) = (i_1, i_2, \ldots, i_k; t)$. For all $(j, t) \in D$ and $i_l \in p(i_k, t)$ with $\beta^t_{i_l j} > 0$, there exists a path p from F^1 to i_l such that*

$$c^t_{jp} + \sum_{l'=l}^{k-1} c^t_{i_{l'} i_{l'+1}} \leq T_{i_k}$$

Proof. Due to the definition of a predecessor, it implies that for each temporarily open facility $i \in \bar{F}^k$

$$c^t_{Pred(i,t)i} + T_{Pred(i,t)} \leq T_i$$

By adding the above inequalities for $i_{l+1}, \ldots, i_k$, one can obtain

$$\sum_{l'=l}^{k-1} c^t_{i_{l'} i_{l'+1}} + T_{i_l} \leq T_{i_k}$$

Since $\beta^t_{i_l j} > 0$, there exists a path p along which j reached i_l before T_{i_l}. Clearly, $c^t_{jp} \leq T_{i_l}$. Hence, we prove that the conclusion is correct.

Then, we consider the facility cost.

Lemma 2.

$$\sum_{t=1}^{T} \sum_{i_k \in \bar{F}^k} \sum_{i_l \in p(i_k)} f_{i_l}^t \leq \sum_{(j,t) \in D} d_j^t \alpha_j^t$$

Finally, we consider the connection cost of client $(j,t) \in D$.

Lemma 3. *If $(j,t) \in D$ is assigned to $p(i_k, t)$ in Phase 2 of Algorithm 1, then we have $c_{jp(i_k,t)}^t \leq 5\alpha_j^t$.*

Finally, we are ready to present the approximation ratio of Algorithm 1.

Theorem 1. *Algorithm 1 is a primal-dual 6-approximation combinatorial algorithm for the k-DFLP.*

Proof. Denote SOL as the solution of Algorithm 1, and let F_{SOL} be the facility cost of the solution, C_{SOL} be the connection cost of the solution. It follows from Lemmas 2 and 3 that the total cost of SOL is at most

$$cost(SOL) = F_{SOL} + C_{SOL}$$
$$\leq \sum_{(j,t) \in D} d_j^t \alpha_j^t + 5 \sum_{(j,t) \in D} d_j^t \alpha_j^t$$
$$\leq 6 \sum_{(j,t) \in D} d_j^t \alpha_j^t$$

5 Discussions

In this paper, we consider the dynamic k-level facility location problem, which is a generalization of the uncapacitated k-level facility location problem when considering time factor. We present a simple dual ascent method for the problem under a dynamic setting which finds a solution within 6 times the optimum. More challenging research directions are in solving dynamic facility location problems with capacity constraints and nonmetric service costs.

References

1. Aardal, K., Chudak, F.A., Shmoys, D.B.: A 3-approximation algorithm for the k-level uncapacitated facility location problem. Inf. Process. Lett. **72**(5–6), 161–167 (1999)
2. Ageev, A., Ye, Y., Zhang, J.: Improved combinatorial approximation algorithms for the k-level facility location problem. SIAM J. Discrete Math. **18**(1), 207–217 (2004)
3. Bumb, A., Kern, W.: A simple dual ascent algorithm for the multilevel facility location problem. In: Goemans, M., Jansen, K., Rolim, J.D.P., Trevisan, L. (eds.) APPROX/RANDOM -2001. LNCS, vol. 2129, pp. 55–63. Springer, Heidelberg (2001). https://doi.org/10.1007/3-540-44666-4_10

4. Cornuejols, G., Nemhauser, G.L., Wolsey, L.A.: The uncapacitated facility location problem. In: Mirchandani, P., Francis, R. (eds.) Discrete Location Theory, pp. 119–171. Wiley, New York (1990)

5. Guha, S., Khuller, S.: Greedy strikes back: improved facility location algorithms. J. Algorithms **31**(1), 228–248 (1999)

6. Jain, K., Vazirani, V.V.: Approximation algorithms for metric facility location and k-median problems using the primal-dual schema and Lagrangian relaxation. J. ACM (JACM) **48**(2), 274–296 (2001)

7. Mahdian, M., Ye, Y., Zhang, J.: Improved approximation algorithms for metric facility location problems. In: Jansen, K., Leonardi, S., Vazirani, V. (eds.) APPROX 2002. LNCS, vol. 2462, pp. 229–242. Springer, Heidelberg (2002). https://doi.org/10.1007/3-540-45753-4_20

8. Meyerson, A., Munagala, K., Plotkin, S.: Cost-distance: two metric network design. In: Proceedings 41st Annual Symposium on Foundations of Computer Science, pp. 624–630. IEEE (2000)

9. Shmoys, D.B., Aardal, K.I.: Approximation algorithms for facility location problems, vol. 1997. Utrecht University: Information and Computing Sciences (1997)

10. Van Roy, T.J., Erlenkotter, D.: A dual-based procedure for dynamic facility location. Manag. Sci. **28**(10), 1091–1105 (1982)

11. Wang, Z., Du, D., Xu, D.: A primal-dual approximation algorithm for the k-level stochastic facility location problem. In: Chen, B. (ed.) AAIM 2010. LNCS, pp. 253–260. Springer, Heidelberg (2010). https://doi.org/10.1007/978-3-642-14355-7_26

12. Xu, D., Du, D.: The k-level facility location game. Oper. Res. Lett. **34**(4), 421–426 (2006)

13. Ye, Y., Zhang, J.: An approximation algorithm for the dynamic facility location problem. In: Cheng, M.X., Li, Y., Du, D.Z. (eds.) Combinatorial Optimization in Communication Networks, vol. 18, pp. 623–637. Springer, Boston (2006). https://doi.org/10.1007/0-387-29026-5_22

Weighted Two-Dimensional Finite Automata

Qichao Wang[1(✉)], Yongming Li[1], and Wei Zhou[2]

[1] College of Computer Science, Shaanxi Normal University, Xi'an 710119, China
{wangqc,liyongm}@snnu.edu.cn
[2] HANA Platform Core, SAP SE, 69190 Walldorf, Germany
wei.zhou01@sap.com

Abstract. Two-dimensional finite automata (2D-FA) are a natural generalization of finite automata to two-dimension and used to recognize picture languages. In order to study quantitative aspects of computations of 2D-FA, we introduce weighted two-dimensional finite automata (W2D-FA), which can represent functions from some input alphabet into a semiring. In this work, we investigate some basic properties of these functions like upper bounds and closure properties. First, we prove that the value of such a function is bounded by $2^{O(n^2)}$. Then, we will see that this upper bound is actually sharp, and a deterministic W2D-FA of a restricted type already can compute a function that reaches this bound. Finally, we study the closure properties of the classes of functions that are computed by W2D-FA of various types under some rational operations, e.g., sum, Hadamard product, vertical (horizontal) multiplication, and scalar multiplication.

Keywords: Picture language · Weighted automaton · Semiring · Upper bound · Closure property

1 Introduction

Picture languages are two-dimensional formal languages over some alphabet, and their applications can be found in many areas such as pattern recognition and image processing [5,12]. In order to recognize a picture languages, various automata models and two-dimensional grammars have been introduced [10,13,15]. *Two-dimensional finite automata* (2D-FA for short) are a natural generalization of finite automata to two-dimension, which is one of the simplest two-dimensional automaton models, and some restricted types have been studied in [1,6,7,9]. Just like one-dimensional finite automata, a 2D-FA can only accept or reject the input, that is, such an automaton can be viewed as computing a Boolean function. However, we are often interested in quantitative aspects of computations of 2D-FA, e.g., the number of accepting computations, and the

This work is supported by the Fundamental Research Funds for the Central Universities under Grant GK201903094.

D.-Z. Du et al. (Eds.): AAIM 2019, LNCS 11640, pp. 292–303, 2019.
https://doi.org/10.1007/978-3-030-27195-4_27

minimal number of steps in an accepting computation. To do this, here we introduce *weighted two-dimensional finite automata* (W2D-FA for short). A W2D-FA $\mathcal{A} = (A, \omega)$ consists of a 2D-FA A with an input alphabet Σ and a weight function ω that assigns a quantitative value from a semiring S as a weight to each transition of A. For an input picture $p \in \Sigma^{**}$, by forming the product of the weights of all the transitions during a computation C on p, a weight $\omega(C)$ can be assigned to the computation C, and by summing the weights of all accepting computations, an element of S is associated with the input p. Therefore, the W2D-FA $\mathcal{A}$ can represent a function f_ω^A from pictures over Σ into a semiring. By using different semirings S, various quantitative aspects of a two-dimensional finite automaton can be expressed, e.g., the number of movement steps in a given direction, and even some special properties for picture languages can also be represented, e.g., the two-dimensional coordinate of a symbol.

This paper is structured as follows. In Sect. 2 we recall some basic notions concerning picture languages and two-dimensional finite automata, and here we also define the weighted two-dimensional finite automata and give some examples. In Sect. 3 we investigate the upper bounds of the functions that are computed by W2D-FA of various types, and in Sect. 4 we consider the closure properties of these function classes under some rational operations, such as sum, Hadamard product, vertical (horizontal) multiplication, and scalar multiplication. The paper closes with a short summary and some problems for future work.

2   Definitions and Examples

We assume that the reader is familiar with the standard notions and concepts of theoretical computer science, such as monoids, finite automata, and semirings. Throughout the paper we will use $\mathbb{N}$ to denote the set of all natural numbers, and $\mathbb{Z}$ to denote the set of all integers. Further, let S be a semiring. For $s \in S$ and $k \in \mathbb{N}$, we will use the notation s^k to denote the k-fold product $s \cdot s \cdot \ldots \cdot s$. In addition, $k \cdot s$ is used to denote the k-fold sum $s + s + \ldots + s$.

In this section we will give the formal definitions of picture languages and two-dimensional finite automata, and for them we mainly refer to *Handbook of Formal Languages* [4] as well as [3,7].

Definition 1. *A picture over some alphabet Σ is a two-dimensional rectangular array of elements of Σ. The set of all pictures over Σ is denoted by Σ^{**}. A picture language over Σ is a subset of Σ^{**}.*

For a picture $p \in \Sigma^{**}$, we use $row(p)$ to denote the number of rows of p and $col(p)$ to denote the number of columns of p. The size of a picture is denoted by $row(p) \times col(p)$. Let $\Sigma^{m \times n}$ denote the set of all pictures over Σ of size $m \times n$ $(m, n > 0)$. Further, the symbol with the coordinate (i, j) in p is denoted by $p(i, j)$ or $p_{i,j}$. Finally, we denote the number of a-symbols in picture p by $|p|_a$. Now we give some simple examples of picture languages.

Example 1. Let $\Sigma = \{0, 1\}$ be an alphabet. The set of pictures over Σ that contain at least one 1-symbol can be formally defined as $L_1 = \{p \in \Sigma^{**} \mid |p|_1 \geq 1\}$.

For example, the following picture belongs to L_1.

0	0	0	0	0	0
0	0	0	0	0	1
0	0	1	0	0	0
0	0	0	1	0	0
0	0	0	0	0	0
1	0	1	0	0	0

Example 2. Let $\Sigma = \{0,1\}$ be an alphabet, and let L_2 be a set of pictures over Σ that contain as many 0-symbols as 1-symbols. Then, L_2 can be formally defined as $L_2 = \{p \in \Sigma^{**} \mid |p|_0 = |p|_1\}$.

For example, the following picture belongs to L_2.

0	0	1	1	1	0
1	0	0	0	0	1
1	0	1	1	1	0
0	1	0	1	0	1
0	1	0	0	0	1
0	1	1	1	0	1

We continue with some concatenation operations for picture languages.

Definition 2. *For picture p of size $m \times n$ and picture q of size $m \times k$, the column concatenation of p and q is defined as*

$$p \, ① \, q = \begin{array}{ccc|ccc} p_{1,1} & \cdots & p_{1,n} & q_{1,1} & \cdots & q_{1,k} \\ \vdots & \ddots & \vdots & \vdots & \ddots & \vdots \\ p_{m,1} & \cdots & p_{m,n} & q_{m,1} & \cdots & q_{m,k} \end{array}$$

Further, for picture p of size $m \times n$ and picture q of size $k \times n$, the row concatenation of p and q is defined as

$$p \ominus q = \begin{array}{ccc} p_{1,1} & \cdots & p_{1,n} \\ \vdots & \ddots & \vdots \\ p_{m,1} & \cdots & p_{m,n} \\ \hline q_{1,1} & \cdots & q_{1,n} \\ \vdots & \ddots & \vdots \\ q_{k,1} & \cdots & q_{k,n} \end{array}$$

Finally, the empty picture λ is the neutral element for column and row concatenation.

Now we introduce some different types of 2D-FA. The general type of 2D-FA is *two-dimensional four-way finite automaton* (4NFA for short), which is extended from one-dimensional two-way finite automata [2].

Definition 3. *A* 4NFA *A* *is formally defined by a 7-tuple* $A = (\Sigma, Q, \Delta, q_0, q_a, q_r, \delta)$, *where* Σ *is an input alphabet,* Q *is a finite set of states,* $\Delta = \{R, L, U, D\}$ *is a set of directions of movement of a* 4NFA, $q_0 \in Q$ *is the initial state,* q_a *is the accepting state,* q_r *is the rejecting state,* $\delta \subseteq (Q \setminus \{q_a, q_r\} \times \Sigma) \times (Q \times \Delta)$ *is a transition relation.*

In order to recognize picture, starting from the position $(1, 1)$ in the initial state, the finite control of a 4NFA can move in four directions: *Right, Left, Up,* and *Down*. Note that the finite control is not allowed to be out of the input picture. In order to avoid this, we identify the boundary of a picture by using a mapping $\pi : \Sigma^{**} \to \Gamma^{**}$, where

$$\Gamma = \Sigma \cup \{[a]_X \mid a \in \Sigma, X \subseteq \{R, L, U, D\}\},$$

that is, the boundary symbols are labelled with some directions. For a picture $p \in \Sigma^{**}$ and a symbol $p(i, j)$ in p, if one or more of $i = 1$, $i = row(p)$, $j = 1$, and $j = col(p)$ hold, then the direction U, D, L and R will be added to the label set of the symbol $p(i, j)$, respectively. For a symbol of the form $[a]_X$, if the direction $d \in X$, then from this symbol a 4NFA is not allowed to move in the direction d. For a picture of size $m \times n$ $(m, n > 1)$, $\pi(p)$ is presented as follows.

$$\pi(p) = \begin{vmatrix} [p_{1,1}]_{\{U,L\}} & [p_{1,2}]_{\{U\}} & \cdots & [p_{1,n-1}]_{\{U\}} & [p_{1,n}]_{\{U,R\}} \\ [p_{2,1}]_{\{L\}} & p_{2,2} & \cdots & p_{2,n-1} & [p_{2,n}]_{\{R\}} \\ \vdots & \vdots & \ddots & \vdots & \vdots \\ [p_{m-1,1}]_{\{L\}} & p_{m-1,2} & \cdots & p_{m-1,n-1} & [p_{m-1,n}]_{\{R\}} \\ [p_{m,1}]_{\{D,L\}} & [p_{m,2}]_{\{D\}} & \cdots & [p_{m,n-1}]_{\{D\}} & [p_{m,n}]_{\{D,R\}} \end{vmatrix}$$

A computation of a 4NFA is finished in an accepting state or a rejecting state, and also finished if there is no applicable instruction. Note that a 4NFA does not need to read all the symbols in the input picture. If a four-way finite automaton can have at most one computation for each input picture, it is called deterministic (4DFA for short), that is, in this case the transition relation δ is a partial function. Unlike in the one-dimensional case, the class of languages accepted by 4DFA is a proper subset of the class of languages accepted by 4NFA [2]. There are some restricted types of 4NFA that have been introduced in [1,6,7,9]. A *two-dimensional three-way finite automaton* (3NFA for short) is not allowed to move up, and a *two-dimensional two-way finite automaton* (2NFA for short) is not allowed to move up and left, and the deterministic versions of them are denoted by 3DFA and 2DFA, respectively. By $\mathbb{L}(X)$ we denote the class of languages that are accepted by 2D-FA of type X. Now we present a simple example of 2D-FA.

Example 3. The picture language L_1 given in Example 1 can be accepted by a 3DFA A_1. For a picture p of size $m \times n$ $(m, n > 1)$, starting from the position $(1, 1)$ A_1 moves to the right end of the first row. On seeing the boundary symbol $[p_{1,n}]_{\{U,R\}}$, it moves down and then moves to the left end of the second row. In this way, A_1 can read all the symbols in p. If a b-symbol appears in the finite control, then A_1 enters the accepting state and halts.

Note that a 2NFA cannot read all the symbols in a picture of the size $m \times n$, if $m, n > 1$. Obviously, the language L_2 given in Example 2 cannot be accepted by a 4NFA, as it is not able to store the number of a- or b-symbols by using finitely many states. In order to study the quantitative aspects of a 2D-FA, here we introduce *weighted two-dimensional finite automata* (W2D-FA for short).

Definition 4. *Let $A = (Q, \Sigma, \Delta, q_0, q_a, q_r, \delta)$ be a 2D-FA, and let ω be a weight function that assigns an element of a semiring $S = (S, +, \cdot, 0, 1)$ as a weight to each transition of δ. The weight of a transition t is denoted by $\omega(t)$. A W2D-FA $\mathcal{A}$ is defined as a pair $\mathcal{A} = (A, \omega)$. The weight of a computation C of A on an input picture $p \in \Sigma^{**}$ is the product of the weights of all the transitions during the computation C, i.e., $\omega(C) = \omega(t_1) \cdot \omega(t_2) \cdot \cdots \cdot \omega(t_n)$ for $t_1, t_2, \cdots, t_n \in C$. By summing the weights of all the accepting computations, we can define a function*

$$f_\omega^A(p) = \left(\sum_{C \in A(p)} \omega(C) \right) \in S,$$

where $A(p)$ is the set of accepting computations of A on the input picture p. Let $\mathbb{F}(X, \Sigma, S)$ denote the set of all functions of the form f_ω^A that are computed by 2D-FA of type X.

We continue with some examples of W2D-FA.

Example 4. Let $\mathbb{N} = (\mathbb{N}, +, \cdot, 0, 1)$ be the semiring of natural numbers with addition and multiplication, and let $\mathcal{A}_1 = (A_2, \omega_1)$ be a W2D-FA over the semiring $\mathbb{N}$. Further, let $\omega_1(t) = 1$ for each transition t of A_2. In this way, the weight associated to each accepting computation is 1, and for each input picture $p \in \Sigma^{**}$, $f_{\omega_1}^{A_2}(p)$ represent the number of accepting computations of A_2 on p.

By using the direct product of semirings [16], we can describe some special quantitative properties for picture languages.

Example 5. Let $\overline{\mathbb{Z}} = (\mathbb{Z} \cup \{\infty\}, \min, +, \infty, 0)$ be the *tropical semiring*. For the 3DFA A_1 given in Example 3, we define the weight function ω_2 that assigns the weight $(0, 1)$, $(0, -1)$, and $(1, 0)$ to each move-right, move-left, and move-down step of A_1, respectively. Therefore, $\mathcal{A}_2 = (A_1, \omega_2)$ is a weighted 3DFA over $\overline{\mathbb{Z}} \times \overline{\mathbb{Z}}$, and for an input picture p, by using the function $f_{\omega_2}^{A_1}(p)$ we can determine the coordinate of the b-symbol of the first discovery in p.

3   Upper Bounds

In this section, we study the upper bounds of the functions that are computed by W2D-FA of various types. For this purpose we need to introduce some definitions that are actually given in [14].

Definition 5. *The semiring* $S = (S, +, \cdot, 0, 1)$ *is called* linearly ordered *with respect to an order* $\leq$, *if* $(S, +, 0)$ *is a linearly ordered monoid with respect to* $\leq$, *and if* $(s \cdot a) \leq (s \cdot b)$ *and* $(a \cdot s) \leq (b \cdot s)$ *for* $s \geq 0$ *and* $a \leq b$.

It is easily seen that $(\mathbb{N}, +, \cdot, 0, 1)$, $(\mathbb{Z}, +, \cdot, 0, 1)$, and $(\{0, 1\}, \vee, \wedge, 0, 1)$ are commutative semirings that are linearly ordered with respect to the natural order. If S is a linearly ordered semiring that is ordered with respect to a linear order $\leq$, and if $T \subseteq S$ is a finite non-empty subset, then the maximum and the minimum of T can be determined, that is, there are an element $a \in T$ such that $a \leq t$ for all $t \in T$, and an element $b \in T$ such that $b \geq t$ for all $t \in T$. In order to study the upper bounds of the functions of the form f_ω^A, we abstract these functions to a function from $\mathbb{N}$ into S.

Definition 6. *Let* S *be a linearly ordered semiring, let* A *be a* 2D-FA *of type* X *with input alphabet* Σ, *and let* ω *be a weight function that maps the transitions of* A *into* S. *We define the functions*

$$(1)\ \hat{f}_\omega^A(x, y) = \max\{f_\omega^A(p) \mid p \in \Sigma^{x \times y}\},$$
$$(2)\ \ \hat{f}_\omega^A(x) \ = \max\{f_\omega^A(p) \mid p \in \Sigma^{x \times x}\}.$$

Let $\hat{\mathbb{F}}(X, \Sigma, S)$ *and* $\hat{\mathbb{F}}_{sq}(X, \Sigma, S)$ *denote the set of all functions of the forms* $\hat{f}_\omega^A$: $\mathbb{N} \times \mathbb{N} \to S$ *and* $\hat{f}_\omega^A : \mathbb{N} \to S$, *respectively, where* A *is a* 2D-FA *of type* X *with input alphabet* Σ.

The function $\hat{f}_\omega^A$ represents the relation between the size of an input picture p and the value of the function $f_\omega^A(p)$. Note that a computation of a 2D-FA can be infinite, that is, the finite control may move between two positions without halting. As in the infinite case the input is not accepted, here we only consider finite computations.

Theorem 1. *Let* $S = (S, +, \cdot, 0, 1)$ *be a semiring that is ordered with respect to a linear order* $\leq$, *let* $A = (\Sigma, Q, \Delta, q_0, q_a, q_r, \delta)$ *be a* 2NFA, *and let* ω *be a weight function that assigns each transition of* A *an element from the subset* $T = \{ s \in S \mid s \geq 0 \}$ *of* S. *Then there exist some constants* $c_1, c_2 \in \mathbb{N}$ *and an element* $s \in T$ *such that*

$$(1)\ \hat{f}_\omega^A(x, y) \leq c_1^{x+y} \cdot s^{x+y},$$
$$(2)\ \ \hat{f}_\omega^A(x) \ \leq c_1^x \cdot s^{c_2 \cdot x}$$

hold for all $x, y \geq 1$.

Proof. First, we prove the inequality (1). As a 2NFA is only able to move right and down, for an input picture $p \in \Sigma^{x \times y}$ the automaton A can perform at most $y - 1$ steps of horizontal movement and at most $x - 1$ steps of vertical movement. It follows that a computation of A on $p \in \Sigma^{x \times y}$ consists of at most $x + y - 2$ steps. Further, let $s = \max\{ \omega(t) \mid t$ is a transition of $A \}$. Therefore, the upper bound of the weight that is associated with a computation of A is s^{x+y-2}.

Now, we turn to the maximal number of accepting computations of A on an input picture of size $x \times y$. Let $t : (p, a, q, m)$ be a transition of A, where $p \in Q \setminus \{q_a, q_r\}$, $a \in \Sigma$, $q \in Q$, and $m \in \Delta \setminus \{U, L\}$. It is easily seen that any configuration of A has at most $2|Q|$ many immediate successor configurations, and thus there are at most $2|Q|^{x+y-2}$ accepting computations.

From the arguments above, it follows that

$$\hat{f}_\omega^A(x, y) \le 2|Q|^{x+y-2} \cdot s^{x+y-2} \le 2|Q|^{x+y} \cdot s^{x+y}$$

for all $x, y \ge 1$. Hence, the inequality (1) holds with the constants $c_1 = 2|Q|$. Along the same line,

$$\hat{f}_\omega^A(x) \le 2|Q|^{2x-2} \cdot s^{2x-2} = (2|Q|^2)^{x-1} \cdot s^{2x-2} \le (2|Q|^2)^x \cdot s^{2x}$$

for all $x \ge 1$, and the inequality (2) can be obtained by taking $c_1 = 2|Q|^2$ and $c_2 = 2$. □

In the following we restrict our attention to the upper bound of the functions that are computed by weighted 4NFA. In a quite similar way, the following result can be derived.

Theorem 2. *Let $S = (S, +, \cdot, 0, 1)$ be a semiring that is ordered with respect to a linear order $\le$, let $A = (\Sigma, Q, \Delta, q_0, q_a, q_r, \delta)$ be a 4NFA, and let ω be a weight function that assigns each transition of A an element from the subset $T = \{ s \in S \mid s \ge 0 \}$ of S. Then there exist some constants $c_1, c_2 \in \mathbb{N}$ and an element $s \in T$ such that*

$$(1) \ \hat{f}_\omega^A(x, y) \le c_1^{xy} \cdot s^{c_2 \cdot xy},$$
$$(2) \ \hat{f}_\omega^A(x) \ \le c_1^{x^2} \cdot s^{c_2 \cdot x^2}$$

hold for all $x, y \ge 1$.

Proof. A 4NFA is able to move around in a picture, and thus it can scan a row or a column repeatedly. It is easily seen that during a finite computation the maximal number of times that a 4NFA scans a row or a column is bounded $k \cdot |Q|$ for some $k \in \mathbb{N}$. It follows that for an input picture $p \in \Sigma^{x \times y}$, A can perform at most $(y-1)k|Q|$ steps of horizontal movement on a row and at most $(x-1)k|Q|$ steps of vertical movement on a column. Therefore, a finite computation of A contains at most $(y-1)k|Q|x$ steps of horizontal movement and $(x-1)k|Q|y$ steps of vertical movement. If $s = \max\{ \omega(t) \mid t$ is a transition of $A \}$, then $\omega(C) \le s^{(y-1)k|Q|x+(x-1)k|Q|y}$ for a computation C of A. Further, it is clear that there are at most $4|Q|$ immediate successor configurations for any configuration of A, and thus for an input picture $p \in \Sigma^{x \times y}$ A has at most $4|Q|^{(y-1)k|Q|x+(x-1)k|Q|y}$ many accepting computations. From the arguments above, it follows that

$$\hat{f}_\omega^A(x,y) \leq 4|Q|^{(y-1)k|Q|x+(x-1)k|Q|y} \cdot s^{(y-1)k|Q|x+(x-1)k|Q|y}$$
$$= 4|Q|^{2k|Q|xy-k|Q|(x+y)} \cdot s^{2k|Q|xy-k|Q|(x+y)}$$
$$\leq 4|Q|^{2k|Q|xy} \cdot s^{2k|Q|xy}$$
$$= (4|Q|^{2k|Q|})^{xy} \cdot s^{2k|Q|xy}$$

for all $x, y \geq 1$. Hence, the inequality (1) holds by taking $c_1 = 4|Q|^{2k|Q|}$ and $c_2 = 2k|Q|$. Analogously, the inequality (2) can be also derived. $\square$

Actually, the upper bound given in the above theorem is sharp.

Theorem 3. *Let $S = (S, +, \cdot, 0, 1)$ be a linearly ordered semiring, let $s \in S$ such that $s \geq 0$, let Σ be a finite alphabet, and let $c_1, c_2 \in \mathbb{N}_+$. Then there exist a 3DFA A over alphabet Σ and a weight function ω for A such that*

$$(1) \quad \hat{f}_\omega^A(x,y) = c_1^{xy} \cdot s^{c_2 \cdot xy},$$
$$(2) \quad \hat{f}_\omega^A(x) = c_1^{x^2} \cdot s^{c_2 \cdot x^2}$$

hold for all $x, y > 1$.

Proof. Let A be a 3DFA that proceeds as follows. For an input picture $p \in \Sigma^{x \times y}$, A starts from the position $(1,1)$, and moves to the right end of the row 1. On seeing the boundary symbol in the position $(1,y)$ it moves down to the position $(2,y)$ and then moves to the left end of the row 2. In this way, A scans each symbol of p exactly once, and finally it enters the accepting state on the bottom-left or bottom-right corner. Obviously, such a computation consists of $xy - 1$ steps. Further, let ω be the weight function that is defined by taking

$$\omega(t) = \begin{cases} (c_1 \cdot s^{c_2})^2, & \text{if } t \text{ is a transition to the accepting state,} \\ c_1 \cdot s^{c_2}, & \text{otherwise.} \end{cases}$$

It follows that

$$f_\omega^A(p) = (c_1 \cdot s^{c_2})^{xy-2} \cdot (c_1 \cdot s^{c_2})^2 = c_1^{xy} \cdot s^{c_2 \cdot xy}$$

for all $p \in \Sigma^{**}$. Hence, the equality (1) holds, and in an analogous way the equality (2) can also be derived. $\square$

We have seen that a weighted 3DFA already can reach the upper bounds given in Theorem 2.

4 Closure Properties

The closure properties of the classes of languages accepted by 2D-FA have been studied in [7,8]. In this section we extend these results to the function classes $\mathbb{F}(X, \Sigma, S)$ for some rational operations that are introduced in [11]. We begin with the operation of *sum* $\oplus$ that is defined as

$$(f \oplus g)(p) = f(p) + g(p)$$

for all $p \in \Sigma^{++}$ and $f, g : \Sigma^{**} \to S$.

Theorem 4. *For all alphabets* Σ *and semirings* S, *the classes of functions* $\mathbb{F}(2\mathsf{NFA}, \Sigma, S)$, $\mathbb{F}(3\mathsf{NFA}, \Sigma, S)$ *and* $\mathbb{F}(4\mathsf{NFA}, \Sigma, S)$ *are closed under the operation of sum* $\oplus$.

Proof. Let A_1 and A_2 be 2NFAs with input alphabet Σ, and let ω_1 and ω_2 be weight functions that map the transitions of A_1 and of A_2 to the semiring $S = (S, 0, 1, +, \cdot)$. Now we construct a 2NFA A with input alphabet Σ and a weight function ω. For an input picture p, A non-deterministically simulates each computation of A_1 and A_2. Further, let ω be a weight function that is defined by taking

$$\omega(t) = \begin{cases} \omega_1(t), \text{ if } t \text{ is a transition of } A_1, \\ \omega_2(t), \text{ if } t \text{ is a transition of } A_2. \end{cases}$$

It follows that for all $p \in \Sigma^{**}$

$$\begin{aligned} f_\omega^A(p) &= \sum_{C \in A_1(p)} \omega_1(C) + \sum_{C \in A_2(p)} \omega_2(C) \\ &= f_{\omega_1}^{A_1}(p) + f_{\omega_2}^{A_2}(p), \end{aligned}$$

where $A_1(p)$ and $A_2(p)$ are the sets of computations of A_1 and A_2 on p, respectively. In the same way, the intended closure property can also be obtained for the types 3NFA and 4NFA. $\square$

We continue with the closure property under the operation of *Hadamard product* $\odot$ that is defined as

$$(f \odot g)(p) = f(p) \cdot g(p)$$

for all $p \in \Sigma^{++}$ and $f, g : \Sigma^{**} \to S$.

Theorem 5. *For all alphabets* Σ *and semirings* S, *the classes of functions* $\mathbb{F}(4\mathsf{DFA}, \Sigma, S)$ *and* $\mathbb{F}(4\mathsf{NFA}, \Sigma, S)$ *are closed under the operation of Hadamard product* $\odot$.

Proof. Let A_1 and A_2 be 4DFAs with input alphabet Σ, and let ω_1 and ω_2 be weight functions that map the transitions of A_1 and of A_2 to S. Now we construct a 4DFA A with input alphabet Σ and a weight function ω. For an input picture p A simulates the computations of A_1. Each transition of A during this phase is assigned the same weight as the corresponding transition of A_1. When A_1 accepts, A moves to the start position and these steps of movement have the weight 1. Then, A simulates the computation of A_2 starting from the position $(1, 1)$. When A_2 accepts, A halts in accepting state. During this simulation phase each transition is assigned the same weight as the corresponding transition of A_2. It is easily seen that A accepts if and only if both of A_1 and A_2 accept, and

$$\begin{aligned} f_\omega^A(p) &= \sum_{C \in A_1(p)} \omega_1(C) \cdot \sum_{C \in A_2(p)} \omega_2(C) \\ &= f_{\omega_1}^{A_1}(p) \cdot f_{\omega_2}^{A_2}(p), \end{aligned}$$

where $A_1(p)$ and $A_2(p)$ are the set of computations of A_1 and A_2 on p, respectively. Obviously, the intended closure property for 4NFA can be obtained in a similar way. $\square$

Horizontal multiplication and *vertical multiplication* are natural generalization of *Cauchy product* to two-dimension, and they are defined as

$$(f \oplus g)(p) = \sum_{p_1 \oplus p_2 = p} f(p_1) \cdot g(p_2)$$

and

$$(f \ominus g)(p) = \sum_{p_1 \ominus p_2 = p} f(p_1) \cdot g(p_2)$$

for all $p \in \Sigma^{++}$ and $f, g : \Sigma^{**} \to S$.

Theorem 6. *For all alphabets Σ and semirings S, the class $\mathbb{F}(3\mathsf{NFA}, \Sigma, S)$ is closed under the operation of vertical multiplication $\ominus$.*

Proof. Let A_1 and A_2 be 3NFAs with input alphabet Σ, and let ω_1 and ω_2 be weight functions that map the transitions of A_1 and of A_2 to the semiring $S = (S, 0, 1, +, \cdot)$. Now we construct a 3NFA A with input alphabet Σ and a weight function ω. For an input picture $p \in \Sigma^{x \times y}$, A simulates a computation of A_1 starting from the position $(1, 1)$, and ω assigns each transition during this phase the same weight as the corresponding transition of A_1. When A_1 accepts, A has to determine the factorization $p = p_1 \ominus p_2$. Note that during the picture recognition process a 2D-FA does not need to read all the symbols of the input picture. Therefore, there may be some unread rows in the picture p_1. Let (i, j) ($0 \leq i \leq x$ and $0 \leq j \leq y$) be the position where A_1 enters accepting state. In order to guess the possible factorizations $p = p_1 \ominus p_2$, A moves to the left end of the rows below the row i. During this guessing phase all the transitions of A have the weight 1. Then, A non-deterministically simulates the computations of A_2 starting from the position $(m, 1)$ for all $i + 1 \leq m \leq x$, and each transition during this simulation phase is assigned the same weight as the corresponding transition of A_2. It follows that for all $p \in \Sigma^{**}$

$$f_\omega^A(p) = \sum_{p_1 \ominus p_2 = p} f_{\omega_1}^{A_1}(p_1) \cdot f_{\omega_2}^{A_2}(p_2),$$

which completes this proof. □

It is easily seen that if a three-way finite automaton can move up, down and right, then $\mathbb{F}(3\mathsf{NFA}, \Sigma, S)$ is also closed under the operation of horizontal multiplication $\oplus$ By using the above simulation technique, we can also prove the closure property of the language class $\mathbb{L}(3\mathsf{NFA})$ under column concatenation. Note that the language classes $\mathbb{L}(4\mathsf{DFA})$ and $\mathbb{L}(4\mathsf{NFA})$ are not closed under the operations of row and column concatenation [4], as a two-dimensional four-way finite automaton is not able to determine the lowest position that it scanned. Obviously, the above simulation technique is also not applicable for 2DFAs and 2NFAs, since these automata cannot move left to the start point of p_2.

In the following we consider the closure properties under the operation of *scalar multiplication* $\cdot$ that is defined as

$$(s \cdot f)(p) = s \cdot f(p),$$

where $p \in \Sigma^{++}$, $f, g : \Sigma^{**} \to S$, and s is an element of a semiring S.

Theorem 7. *For all alphabets Σ, all commutative semirings S, and all $\mathsf{X} \in$ $\{2DFA, 2NFA, 3DFA, 3NFA, 4DFA, 4NFA\}$, the function class $\mathbb{F}(\mathsf{X}, \Sigma, S)$ is closed under the operation of scalar multiplication $\cdot$.*

Proof. Let A_1 be a 2D-FA with an input alphabet Σ, and let ω_1 be a weight functions that maps the transitions of A_1 to the semiring S. Let A be a 2D-FA that works exactly as A_1. Further, let ω be a weight function that is defined as

$$\omega(t) = \begin{cases} s \cdot \omega_1(t), & \text{if } t \text{ is a transition to the accepting state,} \\ \omega_1(t), & \text{otherwise.} \end{cases}$$

As during an accepting computation there can be at most one transition to the accepting state, and the semiring S is commutative, it follows that for all $p \in \Sigma^{**}$

$$f_\omega^A(p) = s \cdot \sum_{C \in A_1(p)} \omega_1(C) = s \cdot f_{\omega_1}^{A_1}(p),$$

where $A_1(p)$ is the set of accepting computations of A_1 on p. □

We close this section with a summary of above closure properties that is given in Table 1.

Table 1. Summary of closure properties of classes of functions computed by above two-dimensional finite automata for all input alphabets Σ and semirings S.

	$\oplus$	$\odot$	$\ominus$	$\cdot$
$\mathbb{F}(2DFA, \Sigma, S)$	?	?	?	√
$\mathbb{F}(2NFA, \Sigma, S)$	√	?	?	√
$\mathbb{F}(3DFA, \Sigma, S)$	?	?	?	√
$\mathbb{F}(3NFA, \Sigma, S)$	√	?	√	√
$\mathbb{F}(4DFA, \Sigma, S)$	?	√	?	√
$\mathbb{F}(4NFA, \Sigma, S)$	√	√	?	√

5   Conclusion

We have introduced the notion of weighted two-dimensional finite automata in order to study quantitative properties of a given picture p and quantitative aspects of a computation of a two-dimensional finite automata on the picture p. Such an automaton is defined as a pair (A, ω), where A is a 2D-FA, and ω is a weight function that assigns an element of a semiring to each transition of A. Therefore, a W2D-FA can represent a function $f_\omega^A : \Sigma^{**} \to S$. We proved that the value of a function that is computed by a weighted 4NFA is bounded by $2^{O(n^2)}$. Further, we have seen that the above upper bound is actually sharp, and a weighted 3DFA already can compute a function that reaches this upper bound. Finally, we investigated the closure properties of these functions under some operations including sum, Hadamard product, vertical multiplication, and scalar multiplication, and summarized them in Table 1. However, some problems remain open. For example, the closure properties under Hadamard product and vertical multiplication are still unknown for many types of W2D-FA.

Actually, if S is a semiring of formal languages over some alphabet Γ, then a W2D-FA transforms a picture $p \in \Sigma^{**}$ into the string languages over Γ. Further, if S is a semiring of picture languages over some alphabet Γ, then the function f_ω^A is essentially a transformation from Σ^{**} into Γ^{**}, that is, a W2D-FA realizes a *picture transducer*. To do this, further study is called for.

References

1. Anselmo, M., Giammarresi, D., Madonia, M.: New operations and regular expressions for two-dimensional languages over one-letter alphabet. Theor. Comput. Sci. **340**(1), 408–431 (2005)
2. Blum, M., Hewitt, C.: Automata on a two-dimensional tape. In: IEEE Symposium on Switching and Automata Theory, pp. 155–160 (1967)
3. Giammarresi, D., Restivo, A.: Two-dimensional finite state recognizability. Fundam. Inform. **25**(3), 399–422 (1996)
4. Giammarresi, D., Restivo, A.: Two-dimensional languages. In: Rozenberg, G., Salomaa, A. (eds.) Handbook of Formal Languages, pp. 215–267. Springer, Heidelberg (1997). https://doi.org/10.1007/978-3-642-59126-6_4
5. Han, Y.-S., Průša, D.: Template-based pattern matching in two-dimensional arrays. In: Brimkov, V.E., Barneva, R.P. (eds.) IWCIA 2017. LNCS, vol. 10256, pp. 79–92. Springer, Cham (2017). https://doi.org/10.1007/978-3-319-59108-7_7
6. Hirakawa, H., Inoue, K., Ito, A.: Three-way two-dimensional deterministic finite automata with rotated inputs. IEICE Trans. Inf. Syst. **88-D**(1), 31–38 (2005)
7. Inoue, K., Takanami, I.: A survey of two-dimensional automata theory. Inf. Sci. **55**(1–3), 99–121 (1991)
8. Inoue, K., Takanami, I., Nakamura, A.: A note on two-dimensional finite automata. Inf. Process. Lett. **7**(1), 49–52 (1978)
9. Inoue, K., Takanami, I., Vollmar, R.: Three-way two-dimensional finite automata with rotated inputs. Inf. Sci. **38**(3), 271–282 (1986)
10. Krtek, L., Mráz, F.: Two-dimensional limited context restarting automata. Fundam. Inform. **148**(3–4), 309–340 (2016)
11. Mäurer, I.: Recognizable and rational picture series. In: Conference on Algebraic Informatics, pp. 141–155. Aristotle University of Thessaloniki Press (2005)
12. Messerschmidt, H., Stommel, M.: Church-rosser picture languages and their applications in picture recognition. J. Automata Lang. Comb. **16**(2–4), 165–194 (2011)
13. Otto, F., Mráz, F.: Deterministic ordered restarting automata for picture languages. Acta Inf. **52**(7–8), 593–623 (2015)
14. Otto, F., Wang, Q.: Weighted restarting automata. Soft Comput. **22**(4), 1067–1083 (2018)
15. Průša, D.: Non-recursive trade-offs between two-dimensional automata and grammars. Theor. Comput. Sci. **610**, 121–132 (2016)
16. Wang, Q.: On the expressive power of weighted restarting automata. In: Freund, R., Mráz, F., Průša, D. (eds.) Ninth Workshop on Non-Classical Models of Automata and Applications. books@ocg.at, vol. 329, pp. 227–241. Österreichische Computer Gesellschaft (2017)

Improved Parameterized Algorithms for Mixed Domination

Mingyu Xiao$^{(\boxtimes)}$ and Zimo Sheng

University of Electronic Science and Technology of China, Chengdu, China
myxiao@gmail.com, 1491858607@qq.com

Abstract. A mixed domination of a graph $G = (V, E)$ is a mixed set D of vertices and edges such that for every edge or vertex, if it is not in D, then it is adjacent or incident to at least one vertex or edge in D. The MIXED DOMINATION problem is to check whether there is a mixed domination of size at most k in a graph. Mixed domination is a mixture concept of vertex domination and edge domination, and the mixed domination problem has been studied from the view of approximation algorithms, parameterized algorithms, and so on. In this paper, we give a branch-and-search algorithm with running time bound of $O^*(4.172^k)$, which improves the previous bound of $O^*(7.465^k)$. For kernelization, it is known that the problem parameterized by k in general graphs is unlikely to have a polynomial kernel. We show the problem in planar graphs allows linear kernels by giving a kernel of $11k$ vertices.

Keywords: Parameterized algorithms · Mixed Domination · Graph algorithms · Kernelization · Branch-and-search

1 Introduction

Domination is an important concept in graph theory. There are some well-known **NP**-hard problems that are related to domination, such as VERTEX DOMINATION and EDGE DOMINATION. These problems have many applications in the real world and been studied extensively in exact algorithms and parameterized algorithms. As for parameterized algorithms with the parameter being the solution size k, VERTEX DOMINATION is $W[2]$-hard and not likely to have a polynomial kernel [4], while EDGE DOMINATION is fixed parameter tractable [2] and allows polynomial kernels [12].

In this paper we study a mixed variant of VERTEX DOMINATION and EDGE DOMINATION, called MIXED DOMINATION, that is to dominate all vertices and edges in a graph by using a minimum number of vertices and edges. In a graph, a vertex *dominates* itself, all of its neighbors, and all of edges incident to it. An edge *dominates* itself, its two endpoints and all edges sharing one endpoint with it. A subset of vertices and edges in a graph is called a *mixed domination* if it dominates all vertices and edges in the graph. In other words, for a graph $G = (V, E)$, a mixed domination D is a union of a vertex subset V_D and an edge

© Springer Nature Switzerland AG 2019
D.-Z. Du et al. (Eds.): AAIM 2019, LNCS 11640, pp. 304–315, 2019.
https://doi.org/10.1007/978-3-030-27195-4_28

subset E_D such that any vertex not appearing in D is adjacent to a vertex in V_D and any edge not in E_D is incident to at least one vertex or one edge in D. The optimization version of MIXED DOMINATION is to find a mixed domination of minimum size in a given graph. In this paper, we consider the parameterized version of MIXED DOMINATION that is to check whether a graph has a mixed domination of size at most k in a given graph, where the parameter is k. So we formally define our problem as follows.

MIXED DOMINATION (MD)   **Parameter:** k
Input: An undirected graph $G = (V, E)$, and a positive integer k.
Question: Does there exist a mixed domination of size at most k in G?

Based on some specific application scenarios, MIXED DOMINATION was first introduced with the name TOTAL COVERING in 1977 by Alavi et al. [1]. Another direct application of MIXED DOMINATION in system control was introduced by Zhao et al. [13]. MIXED DOMINATION is a rich problem in algorithms and computational complexity. It is known that MIXED DOMINATION is **NP**-complete on even split graphs [13], bipartite graphs, chordal graphs and planar bipartite graphs of maximum degree 4 [6,9]. It is easy to see that it can be solved in polynomial time in trees. Lan and Chang [8] gave a linear time algorithm for MD on cacti. On the aspect of approximation algorithms, Hatami [5] showed that there is a 2-approximation algorithm for MD in general graphs. Approximation upper and lower bounds for a weighted version of MD were studied in [11]. As for parameterized algorithms, the parameter of the treewidth $tw(G)$ has been studied. The problem can be solved in $O^*(6^{tw(G)})$ time [7,10]. For the parameter being the solution size k, Jain et al. [7] proved that the problem does not admit a polynomial kernel unless coNP $\subseteq$ **NP**/poly and gave an algorithm with running time bound $O^*(7.456^k)$. This is the best known parameterized algorithm for the problem in general graphs.

In this paper, we further study MIXED DOMINATION with the parameter being k. We will improve the running time bound to $O^*(4.172^k)$ and show that the problem in planar graphs allows linear kernels by giving a kernel of $11k$ vertices. Our parameterized algorithm is a branch-and-search algorithm based on deep analysis of the graph structures. The main idea is to enumerate the possible candidates of the vertex set V_D in the mixed domination D and the set of vertices $V(E_D)$ appearing in the edge set E_D of D. If we find the correct candidates, we construct an edge set E'_D from $V(E_D)$ by finding a maximum matching in $V(E_D)$ and for each unmatched vertex adding an arbitrary edge incident on it. The set $V_D \cup E'_D$ will be a satisfying mixed domination. For the linear kernel in planar graphs, we will use the Euler formula for planar graphs to bound the number of vertices in the graph.

The following parts of the paper are organized as follows. Section 2 introduces some basic notions and basic properties. Sections 3 to 6 give the parameterized algorithm for MIXED DOMINATION in general graphs. Section 7 designs the linear kernel for MIXED DOMINATION in planar graphs. Finally, Sect. 8 makes some concluding remarks. Some proofs are omitted due to the space limitation, which can be found in the full version of this paper.

2   Preliminaries

Let $G = (V, E)$ be an undirected graph with $|V| = n$ vertices and $|E| = m$ edges. For a vertex $v \in V$, we let $N_G(v)$ denote the set of neighbors of v, $d_G(v) = |N_G(v)|$ and $N_G[v] = N_G(v) \cup \{v\}$. When the graph G is clear from the context, we may omit the subscript and simply use $N(v)$, $d(v)$ and $N[v]$. For a graph or an edge set G', we use $V(G')$ to denote the set of vertices appearing in G' and use $E(G')$ to denote the set of edges appearing in G'. For a vertex subset $V' \subseteq V$, we use $G[V']$ to denote the subgraph of G induced by V'. In a graph, *contracting* an edge vu into a new vertex v^* is to first introduce a new vertex v^* that is adjacent to all vertices in $N_G(v) \cup N_G(u)$ and then delete vertices v and u (and all edges incident on them) from the graph. A connected component of a graph is called an *edge component* if this connected component contains exactly one edge.

For a mixed set D of vertices and edges, we may always use V_D to denote the set of *vertex elements* in D and V_E to denote the set of *edge elements* in D. Recall that a mixed set $D = V_D \cup E_D$ of vertices and edges is called a *mixed domination* if any vertex not in $V(D)$ is adjacent to at least one vertex in V_D and any edge not in E_D is incident to at least one vertex in $V(D)$.

In our branch-and-search algorithm, we will guess some vertices as part of the solution. So in most steps, we are going to solve a constrained problem that is to find a mixed domination containing a given set of vertices. The problem is formally defined as follows.

CONSTRAINED MIXED DOMINATION(CMD)   **parameter** k
Input: An undirected graph $G = (V, E)$, two disjoint vertex subsets $V_v, V_e \in V$, and a positive integer k.
Question: Does there exist a mixed domination $D = V_D \cup E_D$ of size $|V_D| + |E_D| \leq k$ such that $V_v \subseteq V_D$ and $V_e \subseteq V(E_D)$?

Since MIXED DOMINATION is a special case of CONSTRAINED MIXED DOMINATION with $V_v = V_e = \emptyset$, for parameterized algorithms we will consider CONSTRAINED MIXED DOMINATION directly. For an instance of CONSTRAINED MIXED DOMINATION (G, V_v, V_e, k), we will always use V_r to denote $V \setminus (V_v \cup V_e)$, G_r to denote $G[V_r]$, $N_r(v)$ to denote $N_{G[V_r]}(v)$ for a vertex $v \in V_r$. We also let $d_r(v) = |N_r(v)|$. For a vertex $v \in V_r$, we say v is *dominated* if v is adjacent to some vertices in V_v and *undominated* otherwise.

By the definition of mixed domination, it is easy to see the following property.

Proposition 1. *A mixed set $D = V_D \cup E_D$ of vertices and edges is a mixed domination of the graph $G = (V, E)$ if and only if the remaining vertex set $V \setminus V(D)$ is an independent set and each vertex in $V \setminus V(D)$ is adjacent to at least one vertex in V_D.*

3   Branch-and-Search Paradigms

As mentioned above, our algorithm is a branch-and-search algorithm, which may branch on the current instance into several smaller instances to search for a solution. For branch-and-search algorithms, one important step is to evaluate the running time bound. In order to do this, we usually need to analyze the size of the "search tree" generated by the algorithm.

Assume that we use a parameter p as the measure to evaluate the instances. Then we use $T(p)$ to denote the maximum number of leaves in the search tree generated by the algorithm for any instance with the parameter being at most p. When the parameter becomes zero or less than zero, the instance can be solved directly. The parameter can be chosen as the number of vertices (or edges) of the graph, the size of the solution and so on.

In a branching operation, the algorithm branches on the current instance with parameter p into l branches. If in the i-th branch the parameter of the subinstance decreases by at least a_i, i.e., the i-th subinstance has the parameter $p_i = p - a_i$, then we obtain a recurrence

$$T(p) \leq T(p - a_1) + T(p - a_2) + \cdots + T(p - a_l).$$

The largest root of the function $f(x) = 1 - \sum_{i=1}^{l} x^{-a_i}$ is called the *branching factor* of the recurrence. The above recurrence can also be represented by the *branching vector* $[a_1, a_2, \ldots, a_l]$. Let γ be the maximum branching factor among all branching factors in the algorithm. The size of the search tree that represents the branching process of the algorithm applied to an instance with parameter p is given by $O(\gamma^p)$. More details about the analysis and how to solve recurrences can be found in the monograph [3].

For two branching vectors $[a_1, a_2, \ldots, a_l]$ and $[b_1, b_2, \ldots, b_l]$, if $a_i \geq b_i$ holds for all ($i = 1, 2, \ldots, l$), then the branching factor of the first one is not greater than this of the second one. This property will be used in the analysis.

3.1   Applied to Constrained Mixed Domination

For an instance of CONSTRAINED MIXED DOMINATION (G, V_v, V_e, k), our idea is to iteratively branch on vertices in $V_r = V \setminus (V_v \cup V_e)$ by including them to V_v or including them to V_e or keeping them in V_r. To analyze the running time bound, we need to set a measure. In this paper, we will let the measure be

$$p := k - |V_v| - 0.5|V_e|.$$

Note that it always holds that $p \leq k$, and when the parameter $p < 0$, the instance has no constrained mixed domination of size at most k. During the algorithm, the measure p will never increase and in each subbranch of a branching operation, the measure p will decrease. Thus, we can get a bounded search tree for the algorithm.

Next, we give a simple branch-and-search algorithm. Our branching rule is designed based on the following observation. For each edge vu in the induced graph $G[V_r]$ and any mixed set $D = V_D \cup E_D$ of the graph that dominates the edge vu, one of the following four cases must hold: $v \in V_D$, $u \in V_D$, $v \in V(E_D)$ and $u \in V(E_D)$. So for each edge vu in $G[V_r]$, we branch into four branches by including v to V_v, including u to V_v, including v to V_e, or including u to V_e. Note that the parameter p decreases by 1, 1, 0.5 and 0.5 in the four subbranches, respectively. This branching operation leads to a recurrence with a branching vector

$$[1, 1, 0.5, 0.5],$$

the corresponding branching factor is 7.465. The algorithm can iteratively apply the above branching operation until $p \leq 0$ or the induced graph $G[V_r]$ has no edge. For the latter case, the instance can be solved in polynomial time (we will show this in Theorem 2 later). The size of the search tree representing the branching process of the algorithm is $O(7.465^p)$, where $p \leq k$. So this algorithm runs in $O^*(7.465^k)$ time, which has the same running time as that of the previous algorithm in [7].

The above algorithm is simple. We refine the algorithm by using more careful analysis and branching rules.

4   A Polynomial-Time Solvable Case

In the above simple algorithm, we need to use a property that the instance can be solved in polynomial time when the induced graph $G[V_r]$ has no edge, where $V_r = V \setminus (V_v \cup V_e)$. In this section, we will prove a stronger result that the instance is polynomially solvable when then induced graph $G[V_r]$ has a degree at most 1. To prove this result, we first give some structural properties of this problem.

Lemma 1 ($\star$). *Let $(G = (V, E), V_v, V_e, k)$ be a* CONSTRAINED MIXED DOMI-NATION *instance and v be an undominated degree-0 vertex in $G[V_r]$.*

(i) *If v is of degree 0 in G, then instance(G, V_v, V_e, k) is a yes-instance if and only if $I' = (G, V_v \cup \{v\}, V_e, k)$ is a yes-instance;*

(ii) *If v is of degree at least 1 in G, than instance $(G = (V, E), V_v, V_e, k)$ is a yes-instance if and only if $I' = (G, V_v, V_e \cup \{v\}, k)$ is a yes-instance.*

The proofs of lemmas marked with ($\star$) can be found in the full version.

Recall that an edge component is a connected component of one edge. The following lemma shows that we can deal with edge components in $G[V_r]$ by moving (part of) them to V_e.

Lemma 2 ($\star$). *Let $I = (G = (V, E), V_v, V_e, k)$ be a* CONSTRAINED MIXED DOMINATION *instance and vu be an edge component in $G[V_r]$.*

(i) *If both of v and u are undominated, then $I = (G, V_v, V_e, k)$ is a yes-instance if and only if $I' = (G, V_v, V_e \cup \{v, u\}, k)$ is a yes-instance;*

(ii) *If exactly one of v and u, say v is undominated, then $I = (G, V_v, V_e, k)$ is a yes-instance if and only if $I' = (G, V_v, V_e \cup \{v\}, k)$ is a yes-instance;*

(iii) *If both of v and u are dominated, then $I = (G, V_v, V_e, k)$ is a yes-instance if and only if $I' = (G^*, V_v, V_e \cup \{v^*\}, k)$ is a yes-instance, where G^* is obtained from G by contracting edge vu into a new vertex v^*.*

Lemma 3 (⋆). *For a CONSTRAINED MIXED DOMINATION instance $(G = (V, E), V_v, V_e, k)$, if any vertex in V_r is dominated and V_r is an independent set in $G[V_r]$, then it can be solved in polynomial time.*

Based on the above lemmas, we get the following result.

Theorem 2. *Given a CONSTRAINED MIXED DOMINATION instance $(G = (V, E), V_v, V_e, k)$ with $V_r = V \setminus (V_v \cup V_e)$. If the induced graph $G[V_r]$ has degree at most 1, then the problem can be solved in polynomial time.*

5 Some Branching Rules

Our algorithm is a branch-and-search algorithm. The branching operations will play an important role in the algorithm and the running time analysis. Before introducing our algorithm, we first introduce some general branching rules, which are based on the structures of the graph and will be used in several steps of the algorithm.

The simplest branching rule in our algorithm is that

Branching Rule (B1): *Branch on a vertex $v \in V_r$ to generate $2^{d_r(v)} + 2$ sub instances by either (i) including v to V_v or (ii) including v to V_e or (iii) including V' to V_v and including $N_r(v) \setminus V'$ to V_e for all subsets $V' \subseteq N_r(v)$.*

This branching rule will not lose an optimal solution. We can observe this based on the following observation. For any domination set $D = V_D \cup E_D$, if $v \notin V_D \cup V(E_D)$, then any neighbor of v must be in one of V_D and $V(E_D)$.

The branching factor of (B1) is related to the degree of the vertex v. The branching factor of (B1) is 7.465 when $d_r(v) = 1$, while the branching factor of (B1) is 5.367 when $d_r(v) = 2$. The branching factor of (B1) will decrease when the value of $d_r(v)$ increases. By Theorem 2, we only need to deal with vertices with a degree at least 2 in $G[V_r]$ with branching operation (B1) and solve the remaining instance in polynomial time. Thus, we can solve the problem in $O^*(5.367^k)$ time. This is already an improvement over the previous result of $O^*(7.465^k)$ [7]. We are not satisfied with this result. In order to get further improvements, we will use more branching rules.

The next two branching rules are designed to deal with special cases of (B1), where we can ignore one subbranch to increase the running time bound. We have the following lemma.

Lemma 4. *Given a* CONSTRAINED MIXED DOMINATION *instance* $(G = (V, E), V_v, V_e, k)$ *and an undominated vertex* $v \in V_r$. *There is a minimum mixed domination* $D = V_D \cup E_D$ *such that if* $v \notin V_D \cup V(E_D)$ *then*

$$N_r(v) \subseteq V_D \cup V(E_D) \quad and \quad N_r(v) \cap V_D \neq \emptyset.$$

Note that if we remove the condition "$N_r(v) \cap V_D \neq \emptyset$" from the above lemma, then the lemma clearly holds by the above observation for (B1). We can add this new condition in this lemma because v is an undominated vertex and v must be dominated in any mixed domination. Based on Lemma 4, we get the following branching rule.

Branching Rule (B2): *Branch on an undominated vertex* $v \in V_r$ *to generate* $2^{d_r(v)} + 1$ *sub instances by either (i) including* v *to* V_v *or (ii) including* v *to* V_e *or (iii) including* V' *to* V_v *and including* $N_r(v) \setminus V'$ *to* V_e *for all nonempty subsets* $V' \subseteq N_r(v)$.

We consider undominated degree-1 vertices.

Lemma 5 (⋆). *Given a* CONSTRAINED MIXED DOMINATION *instance* $(G = (V, E), V_v, V_e, k)$ *and an undominated vertex* $v \in V_r$ *with* $d_r(v) = 1$. *There is a minimum mixed domination* $D = V_D \cup E_D$ *such that either*

$$v \in V(E_D) \quad or \quad u \in V_D,$$

where u *is the unique neighbor of* v.

Based on Lemma 5, we have a branching rule below

Branching Rule (B3): *Branch on an undominated degree-1 vertex* $v \in V_r$ *to generate 2 sub instances by either (i) including* v *to* V_e *or (ii) including* u *to* V_v, *where* u *is the unique neighbor of* v.

Lemma 6 (⋆). *Given a* CONSTRAINED MIXED DOMINATION *instance* $(G = (V, E), V_v, V_e, k)$. *Assume that* $v \in V_r$ *is a vertex with at least* $d_r(v) - 1$ *dominated neighbors in* $N_r(v)$. *There is a minimum mixed domination* $D = V_D \cup E_D$ *such that either* $v \in V(E_D)$ *or* $N_r(v) \subseteq V_D \cup V(E_D)$.

Instead of using Lemma 6 directly to get a branching rule, our algorithm will use the following branching rule obtained by combining Lemma 6 and Lemma 4.

Branching Rule (B4): *Let* $v \in V_r$ *be an undominated vertex with at least* $d_r(v) - 1$ *dominated neighbors in* $N_r(v)$. *Branch on* $v \in V_r$ *to generate* $2^{d_r(v)}$ *sub instances by either (i) including* v *to* V_e *or (ii) including* V' *to* V_v *and including* $N_r(v) \setminus V'$ *to* V_e *for all nonempty subsets* $V' \subseteq N_r(v)$.

The fifth branching rule is to deal with the graph $G[V_r]$ where all vertices are dominated vertices.

Lemma 7 (⋆). *Given a* CONSTRAINED MIXED DOMINATION *instance* $(G = (V, E), V_v, V_e, k)$. *If the graph* $G[V_r]$ *has no undominated vertex, then there is a minimum mixed domination* $D = V_D \cup E_D$ *such that* $V_D \cap V_r = \emptyset$.

Branching Rule (B5): *If $G[V_r]$ has no undominated vertex, we branch on an vertex $v \in V_r$ to generate two sub instances by either (i) including v to V_e or (ii) including $N_r(v)$ to V_e.*

6   A Refined Parameterized Algorithm

Now we describe the whole parameterized algorithm, which is denoted by $mds(G, V_v, V_e, k)$. The algorithm contains serval reduction rules and branching rules. The main idea of our algorithm is to deal the vertices in V_r by moving them to either V_v or V_e. The algorithm will first deal with vertices of degree at least 5, then deal with undominated vertices of degree 1, 2, 3 and 4, and last deal with the graph having no undominated vertices. Below, when we introduce a step of the algorithm, we assume that all previous steps can not be applied to the current instance anymore.

Step 3. *If $p < 0$, return "no".*

Step 4. *If $G[V_r]$ has maximum degree at most 1, then apply the polynomial-time algorithm in Theorem 2 to find a minimum solution D, and return "no" if $|D| > k$ and return "yes" otherwise.*

Step 5 (Vertices of Degree ≥ 5). *If there is a vertex v with $d_r(v) \geq 5$ in $G[V_r]$, we branch on v with Branching Rule (B1) by returning one of the follows with the maximum value*

$$mds(G, V_v \cup \{v\}, V_e, k),\ mds(G, V_v, V_e \cup \{v\}, k)\ and$$

$$mds(G, V_v \cup V', V_e \cup (N_r(v) \setminus V'), k)\ for\ all\ V' \subseteq N_r(v).$$

We will get a recurrence relation

$$T(p) \leq T(p-1) + T(p-0.5) + \sum_{i=0}^{d} \binom{d}{i} T(p - \frac{i}{2} - (d-i)),$$

where $d = d_r(v) \geq 5$. The branching factor of the above recurrence relation will decrease when the value of d increases. For the worst case that $d = 5$, the branching factor is 3.962.

Next, we assume that the maximum degree of $G[V_r]$ is at most 4.

Step 6 (Undominated Vertices of Degree 3 or 4). *If there is an undominated vertex v with $d_r(v) = \{3, 4\}$ in $G[V_r]$, we branch on v with Branching Rule (B2) by returning one of the follows with the maximum value*

$$mds(G, V_v \cup \{v\}, V_e, k),\ mds(G, V_v, V_e \cup \{v\}, k)\ and$$

$$mds(G, V_v \cup V', V_e \cup (N_r(v) \setminus V'), k)\ for\ all\ \emptyset \neq V' \subseteq N_r(v).$$

We analyze the branching factor according to the degree of V. When $d_r(v) = 3$, the recurrence relation is

$$T(p) \leq T(p-1) + T(p-0.5) + 3T(p-2) + 3T(p-2.5) + T(p-3),$$

and the branching factor is 4.172.

When $d_r(v) = 4$, the recurrence relation is

$$T(p) \leq T(p-1) + T(p-0.5) + 4T(p-2.5) + 6T(p-3) + 4T(p-3.5) + T(p-4),$$

and the branching factor is 4.013.

Step 7 (Undominated Vertices of Degree 1). *If there is an undominated vertex v with $d_r(v) = 1$ in $G[V_r]$, where u is the unique neighbor of v, we branch on v with Branching Rule (B3) by returning one of the follows with the maximum value*

$$mds(G, V_v, V_e \cup \{v\}, k) \quad and \quad mds(G, V_v \cup \{u\}, V_e, k).$$

We get a recurrence relation

$$T(p) \leq T(p-0.5) + T(p-1),$$

the branching factor of which is 2.619.

After this step, in the graph $G[V_r]$ all undominated vertices must be degree-2 vertices.

Step 8 (Undominated Vertices of Degree 2 With At Most One Undominated Neighbor). *If there is an undominated vertex $v \in V_r$ with $d_r(v) = 2$ and at most one undominated neighbor, we branch on v with Branching Rule (B4) by returning one of the follows with the maximum value*

$$mds(G, V_v, V_e \cup \{v\}, k) \quad and$$
$$mds(G, V_v \cup V', V_e \cup (N_r(v) \setminus V'), k) \text{ for all } \emptyset \neq V' \subseteq N_r(v).$$

We get

$$T(p) \leq T(p-0.5) + 2T(p-1.5) + T(p-2),$$

the branching factor of which is 3.220.

Step 9 (Undominated Vertices of Degree 2 With Two Undominated Neighbors). *If there is an undominated vertex $v \in V_r$ with $d_r(v) = 2$, where the two neighbors u_1 and u_2 are also undominated, we branch on u_1 with Branching Rule (B2), and in the sub branch where u_1 is included to V_e, we get an undominated degree-1 vertex v and further branch on v with Branching Rule (B3). We will return one of the follows with the maximum value*

$$mds(G, V_v \cup \{u_1\}, V_e, k), \; mds(G, V_v, V_e \cup \{u_1, v\}, k), \; mds(G, V_v \cup \{u_2\}, V_e \cup \{u_1\}, k),$$

$$and \; mds(G, V_v \cup V', V_e \cup (N_r(u_1) \setminus V'), k) \text{ for all } \emptyset \neq V' \subseteq N_r(u_1).$$

Note that u_1 is an undominated vertex and now it must hold that $d_r(u_1) = 2$. Therefore, we have the following recurrence relation

$$T(p) \le T(p - 0.5 - 0.5) + T(p - 0.5 - 1) + T(p - 1) + 2T(p - 1.5) + T(p - 2),$$

the branching factor of which is 3.802.

After Step 9, the induced graph $G[V_r]$ has no undominated vertex.

Step 10 (Graph $G[V_r]$ Containing Only Dominated Vertices). *Pick a vertex v of maximum degree in $G[V_r]$ and branch on v with Branching Rule (B5). We will return one of the follows with the maximum value*

$$mds(G, V_v, V_e \cup \{v\}, k) \ and \ mds(G, V_v, V_e \cup \{N_r(v)\}, k).$$

This step will lead to recurrence relation

$$T(p) \le T(p - 0.5) + T(p - 0.5d_r(v)),$$

where $d_r(v) \ge 2$, because Step 2 cannot be applied now. For worst case $d_r(v) = 2$, the branching factor of the recurrence is 2.619.

Among all the branching factors in the algorithm, the largest one is 4.172. So the algorithm runs in $O^*(4.172^p)$ time, where $p \le k$.

Theorem 11. MIXED DOMINATION *and* CONSTRAINED MIXED DOMINATION *can be solved in* $O^*(4.172^k)$ *time.*

7   Kernels in Planar Graphs

In this section, we consider kernelization of our problems. In terms of kernelization, MIXED DOMINATION and CONSTRAINED MIXED DOMINATION may be different. We need to reduce an instance of a problem to an instance of the same problem. For an instance of MIXED DOMINATION, we can not claim a kernel by reducing it to an instance of CONSTRAINED MIXED DOMINATION. It is already known that MIXED DOMINATION with the solution size k being the parameter does not allow polynomial kernels unless coNP $\subseteq$ NP/poly [7]. We will show that when the graph is restricted to planar graphs, MIXED DOMINATION allows a kernel of at most $11k$ vertices.

For planar graphs, we will use the following important property to bound the size of the graph, which can be derived from the famous Euler's formula for planar graphs.

Proposition 12. *For a single planar graph $G = (V, E)$, it always holds that*

$$|E| \le 3|V| - 6 \tag{1}$$

and, for a single bipartite planar graph $G' = (V', E')$, it always holds that

$$|E'| \le 2|V'| - 4. \tag{2}$$

7.1 Planar Mixed Domination

The kernelization algorithm for MIXED DOMINATION in planar graphs contains three reduction rules to reduce degree-1 and degree-2 vertices and one reduction rule to reduce the whole size.

Reduction Rule 1: (Degree-1 Vertices) *If there are more than two degree-1 vertices sharing a common neighbor, delete one of the degree-1 vertices from the graph.*

Reduction Rule 2: (Degree-2 Vertices Between Two Vertices) *If there are more than three degree-2 vertices sharing two common neighbors, delete one of the degree-2 vertices from the graph.*

Reduction Rule 3: (Degree-2 Vertices Between A Vertex and An Edge) *If there are a vertex v and an edge uw such that v and u have at least one common degree-2 neighbor, v and w have at least two common degree-2 neighbors, delete one degree-2 vertex adjacent to v and w from the graph.*

The proofs of the correctness of the three reduction rules can be found in the full version.

Reduction Rule 4: (Size Bound) *If the graph has more than $11k - 16$ vertices, return '⊥' to indicate the instance is a 'no'-instance; else return the current graph.*

Note that we only apply Reduction Rule 4 when none of the first three reduction rules can be applied anymore. The correctness of Reduction Rule 4 is based on the following lemma.

Lemma 8 ($\star$). *For a planar graph G having a mixed domination of size at most k, if none of the conditions in the first three reduction rules holds, then G has at most $11k - 16$ vertices.*

Our algorithm is to iteratively apply the first three reduction rules unless none of them can be applied anymore and then apply the last reduction rule. Since each application of the first three reduction rules will delete one vertex from the graph, we know that the algorithm will run at most n times of these reduction operations, each of them can be executed in linear time. So the algorithm will stop in $O(n^2)$ time. Reduction Rule 4 guarantees a bound of $11k - 16$ on the number of vertices.

Theorem 13. MIXED DOMINATION *in planar graphs allows a linear kernel of $11k - 16$ vertices.*

8 Conclusion

In this paper, we study MIXED DOMINATION and CONSTRAINED MIXED DOMINATION from the viewpoint of parameterized algorithms with the parameter k being the solution size. By developing deep structural properties, we show that MIXED DOMINATION and CONSTRAINED MIXED DOMINATION can be solved in $O^*(4.172^k)$ time, improving the previous result of $O^*(7.465^k)$. Furthermore, we show that MIXED DOMINATION in planar graphs allows a linear kernel of at most $11k - 16$ vertices.

Acknowledgements. This work was supported by the National Natural Science Foundation of China, under grants 61772115 and 61370071.

References

1. Alavi, Y., Behzad, M., Lesniak-Foster, L.M., Nordhaus, E.: Total matchings and total coverings of graphs. J. Graph Theory **1**(2), 135–140 (1977)
2. Fernau, H.: EDGE DOMINATING SET: efficient enumeration-based exact algorithms. In: Bodlaender, H.L., Langston, M.A. (eds.) IWPEC 2006. LNCS, vol. 4169, pp. 142–153. Springer, Heidelberg (2006). https://doi.org/10.1007/11847250_13
3. Fomin, F.V., Kratsch, D.: Exact Exponential Algorithms. Springer, Heidelberg (2010). https://doi.org/10.1007/978-3-642-16533-7
4. Garey, M.R., Johnson, D.S.: Computers and Intractability: A Guide to the Theory of NP-Completeness. W. H. Freeman, New York (1979)
5. Hatami, P.: An approximation algorithm for the total covering problem. Discussiones Mathematicae Graph Theory **27**(3), 553–558 (2007)
6. Hedetniemi, S.M., Hedetniemi, S.T., Laskar, R., McRae, A., Majumdar, A.: Domination, independence and irredundance in total graphs: a brief survey. In: Graph Theory, Combinatorics and Applications: Proceedings of the 7th Quadrennial International Conference on the Theory and Applications of Graphs, vol. 2, pp. 671–683. Wiley, New York (1995)
7. Jain, P., Jayakrishnan, M., Panolan, F., Sahu, A.: MIXED DOMINATING SET: a parameterized perspective. In: Bodlaender, H.L., Woeginger, G.J. (eds.) WG 2017. LNCS, vol. 10520, pp. 330–343. Springer, Cham (2017). https://doi.org/10.1007/978-3-319-68705-6_25
8. Lan, J.K., Chang, G.J.: On the mixed domination problem in graphs. Theoret. Comput. Sci. **476**, 84–93 (2013)
9. Manlove, D.F.: On the algorithmic complexity of twelve covering and independence parameters of graphs. Discrete Appl. Math. **91**(1–3), 155–175 (1999)
10. Rajaati, M., Hooshmandasl, M.R., Dinneen, M.J., Shakiba, A.: On fixed-parameter tractability of the mixed domination problem for graphs with bounded tree-width. Discrete Math. Theoret. Comput. Sci. **20**(2), 1–25 (2018)
11. Xiao, M.: Upper and lower bounds on approximating weighted mixed domination. In: COCOON 2019 (2019, to appear)
12. Xiao, M., Kloks, T., Poon, S.H.: New parameterized algorithms for the edge dominating set problem. Theoret. Comput. Sci. **511**, 147–158 (2013)
13. Zhao, Y., Kang, L., Sohn, M.Y.: The algorithmic complexity of mixed domination in graphs. Theoret. Comput. Sci. **412**(22), 2387–2392 (2011)

New Results on the Zero-Visibility Cops and Robber Game

Yuan Xue$^{(\boxtimes)}$, Boting Yang, and Sandra Zilles

Department of Computer Science, University of Regina, Regina, Canada
{xue228,boting,zilles}@cs.uregina.ca

Abstract. We study the zero-visibility cops and robber game, a variant of the cops and robber game in which the robber is invisible. We give a method for proving lower bounds on the zero-visibility cop number. Using this method, we investigate graph joins, lexicographic products of graphs, complete multipartite graphs and split graphs. Lower bounds and upper bounds, along with rigorous proofs, are given for the zero-visibility cop number of each of these types of graphs.

1 Introduction

Pursuit evasion games on graphs typically model situations in which a set of searchers is trying to capture a fugitive moving in a given graph. An example of such a game is the cops and robber game [12,14], in which a robber and cops are located in vertices of the given graph and can move along the edges of the graph. The zero-visibility cops and robber game was introduced by Tošić [16]. As a variant of the classic cops and robber game, the only difference in the game setting is that the cops have "zero" information about the robber throughout the game, i.e., the robber is invisible to the cops. As a consequence of limiting the information presented to the cops, more cops are needed to ensure the capture of the robber. Already for trees, the zero-visibility cop number can exceed the cop number in the classic cops and robber game, and in general the difference between these two parameters can be arbitrarily large.

The cops and robber game has been widely studied. Nowakowski and Winkler [12] and Quilliot [14] introduced the game and they gave a characterization of those graphs for which a single cop is sufficient for capturing the robber. Aigner and Fromme [1] introduced the cop number, which is the smallest number of cops required for capturing the robber on a graph. More results on the cops and robber game can be found in [2–5,8,10,11,13].

However, there are not many results on the zero-visibility cops and robber game. Tošić [16] characterized the graphs for which one cop is sufficient. Jeliazkova [9] gave several constructions of the graphs on which two cops are sufficient in the game. Dereniowski et al. [7] proved that the zero-visibility cop number of a graph is bounded above by its pathwidth. They also defined a monotonic version of the zero-visibility cops and robber model, and proved an upper bound and a lower bound on the monotonic zero-visibility cop number. Tang

© Springer Nature Switzerland AG 2019
D.-Z. Du et al. (Eds.): AAIM 2019, LNCS 11640, pp. 316–328, 2019.
https://doi.org/10.1007/978-3-030-27195-4_29

[15] gave a quadratic time algorithm for computing the zero-visibility cop number of a tree, which was later improved by Dereniowski et al. [6] by presenting a linear-time algorithm. Dereniowski et al. also considered the computational complexity of the zero-visibility cops and robber game, and proved that the problem of determining the zero-visibility cop number of a graph is NP-complete.

In this paper, we concentrate on the zero-visibility cops and robber game. By establishing specific conditions on strategies, we introduce a method for finding lower bounds on the zero-visibility cop number. Using this method, we consider the zero-visibility cops and robber game on the graph join of two arbitrary graphs. Then we investigate a few extensions and special cases of graph joins including lexicographic products of graphs, complete multipartite graphs and split graphs. Lower bounds and upper bounds, along with rigorous proofs, are given for the zero-visibility cop number of each of the types of graphs mentioned above.

2   Preliminaries

Let $G = (V, E)$ denote a graph with vertex set V and edge set E. We also use $V(G)$ and $E(G)$ to refer to V and E, respectively. For a subset $V' \subseteq V$, we use $G[V']$ to denote the subgraph induced by V', which consists of all vertices of V' and all the edges of G between vertices in V'. We use $G - V'$ to denote the induced subgraph $G[V \setminus V']$. Let $pw(G)$ denote the pathwidth of a graph G.

A *matching* in G is a set of edges of G that share no common vertices. A *perfect matching* in G is a matching that includes all vertices of G. A *maximum matching* in G is a matching in G which contains the largest possible number of edges among all the matchings in G. We use $\mathcal{M}(G)$ to denote a maximum matching in G. The number of vertices of G that are not matched by edges in $\mathcal{M}(G)$ is equal to $|V(G) \setminus V(\mathcal{M}(G))|$. Thus, $|V(G)| = 2|\mathcal{M}(G)| + |V(G) \setminus V(\mathcal{M}(G))|$. Let $\mu(G) = |\mathcal{M}(G)| + |V(G) \setminus V(\mathcal{M}(G))| = |V(G)| - |\mathcal{M}(G)|$.

We now give a formal definition of the zero-visibility cops and robber game. There are two players involved in the game on graph G: the *cop player* and the *robber player*. The cop player controls a set of cop pieces while the robber player controls a single robber piece. The game is played in a sequence of rounds. At each round i, where $i \geq 0$, the cop player plays first, followed by the robber player. At round 0, both players place their pieces on some vertices of G. More than one piece may occupy the same vertex. All the cop pieces are visible to the robber player, while the robber piece is invisible to the cop player. In the following rounds, each cop piece either moves from the vertex currently occupied to one of its neighbors or stays still, then the robber piece does the same. The *cops win* if a cop piece and the robber piece occupy the same vertex after a finite number of rounds. The *robber wins* if such situation never happens. A *cop strategy* is a sequence of actions for the cop player; a *cop-win strategy* is a cop strategy that leads the cop player to win irrespective of the robber player's sequence of actions. The *cop number* of G, denoted by $c_0(G)$, is the minimum number of cops required in a cop-win strategy for G. A cop-win strategy for G

is *optimal* if it uses $c_0(G)$ cops to capture the robber on G. We say a cop *visits* a vertex $v \in V$ at round i, if the cop occupies v at the beginning or at the end of round i. We call a vertex *contaminated* if it may contain the robber, and we call a vertex *cleared* if the cop player can be certain that it does not contain the robber. We refer to any step that turns an unoccupied cleared vertex into a contaminated vertex as recontamination. For simplicity, we use *the cops* to denote the cop pieces and *the robber* to denote the robber piece.

Note that, for space constraints, several proofs of formal results in this paper are omitted.

3   Lower Bounds

Let u and v be two vertices of G. We use $d_G(u, v)$ to denote the distance between u and v, which is the number of edges in a shortest path connecting u and v on G. Let H be a subgraph of G. We say H is an *isometric subgraph* of G if $d_H(u, v) = d_G(u, v)$, for any two vertices $u, v \in V(H)$. Lemma 1 appears in [15].

Lemma 1. *If H is an isometric subgraph of G, then $c_0(H) \leq c_0(G)$.*

Corollary 1. *If G contains a clique with m vertices, then $c_0(G) \geq \lceil \frac{m}{2} \rceil$.*

Lemma 2. *In the zero-visibility cops and robber game on G, for any cop strategy, the number of cops that can visit all vertices of G within a single round of the game is at least $\mu(G)$.*

Proof. Consider a round of the game on G in which all vertices in $V(G)$ are visited by cops. Let m_1 be the number of cops that visit two vertices of G and let m_2 be the number of cops that visit exactly one vertex of G. It is easy to see that $|V(G)| = 2m_1 + m_2$. Since $\mathcal{M}(G)$ is a maximum matching in G, hence, $m_1 \leq |\mathcal{M}(G)|$. Note that $\mu(G) = |V(G)| - |\mathcal{M}(G)|$. Hence, we have $\mu(G) \leq |V(G)| - m_1 = (2m_1 + m_2) - m_1 = m_1 + m_2$.   □

In the following, we establish specific conditions on strategies, and show that these conditions must be met in every round of a strategy when insufficient cops are used for capturing the robber. This method is later used for finding lower bounds on the cop number of several types of graphs.

Definition 1. *For a strategy S of a graph G, let $\mathcal{P}_{G,S}(i)$ be a propositional function such that at the end of round i, every cleared vertex is occupied by at least one cop and the number of cleared vertices is less than $|V(G)|$. When G and S are clear from the context, we drop the subscript and simply use $\mathcal{P}(i)$.*

The following proposition is straightforward.

Proposition 1. *For a cop-win strategy for G, there must exist a round satisfying that:*

1. *all vertices of G are occupied by cops after the cop's turn, or*
2. *at least one unoccupied vertex of G is cleared after the robber's turn.*

3.1 Graph Joins

Let G and H be two graphs. The *graph join*, denoted as $G + H$, is the graph whose vertex set is $V(G) \cup V(H)$, and two vertices u and v are adjacent in $G + H$ if and only if $uv \in E(G)$, or $uv \in E(H)$, or $u \in V(G)$ and $v \in V(H)$.

Theorem 1. $c_0(G + H) \geq \min\{\mu(G), \mu(H)\}$.

Proof. Suppose that $\mu(G) \leq \mu(H)$. We will use Definition 1 to show that $\mu(G) - 1$ cops are insufficient for clearing $G + H$. Assume that $c_0(G + H) < \mu(G)$. Consider a strategy for $G + H$ that uses at most $\mu(G) - 1$ cops. We will show that at the end of each round, only occupied vertices are cleared.

At round 0, since cops are placed on vertices in $V(G + H)$, we know cleared vertices are those occupied ones. Hence, $\mathcal{P}(0)$ holds. Suppose that $\mathcal{P}(i)$ holds for some $i \geq 0$. From Lemma 2, we know that $\mu(G)$ cops are insufficient for visiting all the vertices in $V(G)$, not to speak of all vertices in $V(H)$. It is very easy to see that both G and H contain contaminated vertices when the cop's turn is finished at round $i + 1$. Hence, all unoccupied cleared vertices will get recontaminated during the robber's turn at round $i + 1$. We know $\mathcal{P}(i + 1)$ holds. Thus, from Proposition 1, we have $c_0(G + H) \geq \mu(G)$. $\qquad\square$

The lower bound in Theorem 1 is tight. For example, if G and H are both independent vertex sets, then $G + H$ is a complete bipartite graph. It is easy to see that $c_0(G + H) = \min\{|V(G)|, |V(H)|\}$. Since $\mu(G) = |V(G)|$ and $\mu(H) = |V(H)|$, we have $c_0(G + H) = \min\{\mu(G), \mu(H)\}$.

Let K_n denote a complete graph of n vertices, and let $\overline{K_n}$ denote an independent vertex set of n vertices.

Corollary 2. *If* $|V(G)| \leq n$, *then* $c_0(G + \overline{K_n}) \geq \mu(G)$.

Theorem 2. *Let* G *and* H *be two graphs such that* $2 \leq |V(G)| \leq |V(H)|$. *If* G *has a perfect matching, then* $c_0(G + H) \geq \mu(G) + 1$.

Proof. Since G has a perfect matching, we have $\mu(G) = \frac{|V(G)|}{2}$. We will use Definition 1 to show that $\mu(G)$ cops are insufficient. Suppose that $\mu(G)$ cops can clear $G + H$. At the end of round 0, it is easy to see that the cleared vertices are those occupied ones. Hence, $\mathcal{P}(0)$ holds. Suppose that $\mathcal{P}(i)$ holds for some $i \geq 0$. Obviously, there are at most $2\mu(G) = |V(G)|$ cleared vertices after the cops' turn in round $i + 1$. Consider the beginning moment of round $i + 1$.

Case 1. All cleared vertices are contained either in $V(G)$ or in $V(H)$. Without loss of generality, assume that all cops stay on G at the beginning moment of round $i + 1$.

Case 1.1. If all cops are still on G at the end of round $i + 1$, then all vertices in $V(H)$ must be contaminated throughout round $i + 1$. Hence, all unoccupied vertices in $V(G)$ must be contaminated at the end of round $i + 1$. Thus, $\mathcal{P}(i + 1)$ holds.

Case 1.2. If there is any cop sliding from G to H in round $i + 1$, then both G and H must contain a vertex that remains contaminated throughout round

$i + 1$. Hence, all unoccupied vertices in $V(G)$ and $V(H)$ must be contaminated at the end of round $i + 1$. Thus, $\mathcal{P}(i + 1)$ holds.

Case 2. Both G and H contain cleared vertices. It is easy to see that both G and H contain a vertex that remains contaminated throughout round $i + 1$. Hence, all cleared vertices at the end of round $i + 1$ are those occupied ones. We know $\mathcal{P}(i + 1)$ also holds.

Therefore, all cleared vertices in $V(G + H)$ at the end of each round are those occupied ones. From Proposition 1, we know that $\mu(G)$ cops are insufficient for clearing $G + H$. This completes the proof.   □

Cone graph is the graph join of a cycle C_m and an independent vertex set $\overline{K_n}$, where $m \geq 3$ and $n \geq 1$. Let $V(C_m) = \{u_1, \ldots, u_m\}$ and $V(\overline{K_n}) = \{v_1, \ldots, v_n\}$.

Theorem 3. *If $n \leq 2$, then $c_0(C_m + \overline{K_n}) \geq \min\{\lceil \frac{m+1}{2} \rceil, n + 1\}$. If $n \geq 3$, then $c_0(C_m + \overline{K_n}) \geq \min\{\lceil \frac{m+1}{2} \rceil, n\}$.*

Proof. If $n = 1$ or $n = 2$ and $m = 3$, then $\min\{\lceil \frac{m+1}{2} \rceil, n+1\} = 2$. Since $C_m + \overline{K_n}$ contains a cycle of length 3, it follows from Lemma 1 that $c_0(C_m + \overline{K_n}) \geq 2$.

If $n = 2$ and $m \geq 4$, then $\min\{\lceil \frac{m+1}{2} \rceil, n + 1\} = 3$. We will use Definition 1 to show that 2 cops are insufficient for clearing $C_m + \overline{K_n}$. Assume that $c_0(C_m + \overline{K_2}) \leq 2$. Consider a strategy for $C_m + \overline{K_2}$ that uses at most 2 cops. Note that $c_0(C_m + \overline{K_2}) \leq 2 < \frac{m+2}{2}$. Obviously, $\mathcal{P}(0)$ holds. Suppose that $\mathcal{P}(i)$ holds for some $i \geq 0$. Note that there are at most 2 cleared vertices at the beginning of round $i+1$. After the cops' turn at round $i+1$, we know there are at most 4 cleared vertices. It is easy to see that all those cleared vertices must have a contaminated neighbor. Thus, after the robber's turn at round $i + 1$, all unoccupied cleared vertices become recontaminated. Hence, $\mathcal{P}(i+1)$ also holds. From Proposition 1, we know that 2 cops are insufficient for clearing $C_m + \overline{K_2}$. Therefore, $c_0(C_m + \overline{K_2}) \geq 3$.

If $n \geq 3$, it follows from Theorem 1 that $c_0(C_m + \overline{K_n}) \geq \min\{\lceil \frac{m+1}{2} \rceil, n\}$.   □

Theorem 4. *For the graph join of C_m and C_n, where $m \leq n$, if m is odd, then $c_0(C_m + C_n) \geq \frac{m+1}{2} + 1$; if m is even, then $c_0(C_m + C_n) \geq \frac{m}{2} + 2$.*

Proof. We first consider the case when m is odd. We will use Definition 1 to show that $\frac{m+1}{2}$ cops are insufficient for clearing $C_m + C_n$. Assume that $c_0(C_m + C_n) \leq \frac{m+1}{2}$. Consider a strategy for $C_m + C_n$ that uses at most $\frac{m+1}{2}$ cops. Note that $c_0(C_m + C_n) \leq \frac{m+1}{2} < \frac{m+n}{2}$. So $\mathcal{P}(0)$ holds. Suppose that $\mathcal{P}(i)$ holds for some $i \geq 0$. Consider the moment when the cops' turn is finished at round $i + 1$. It is easy to see that all cleared vertices have a contaminated neighbor. After the robber's turn at round $i + 1$, all unoccupied cleared vertices will become recontaminated. Thus, $\mathcal{P}(i+1)$ also holds. Hence, $c_0(C_m + C_n) \geq \frac{m+1}{2} + 1$ when m is odd.

Similarly, we can prove that $c_0(C_m + C_n) \geq \frac{m}{2} + 2$ when m is even. This completes the proof.   □

3.2 Lexicographic Products of Graphs

Let G and H be two connected graphs. The *lexicographic product* of G and H, denoted as $G \cdot H$, is the graph whose vertex set is the cartesian product $V(G) \times V(H)$, and two vertices (a, a') and (b, b') are adjacent in $G \cdot H$ if and only if either $ab \in E(G)$ or $a = b$ and $a'b' \in E(H)$. Note that the lexicographic products are not commutative.

Lemma 3. *Let P_m be a path with m vertices. For $m \geq 4$, $c_0(P_m \cdot G) \geq |V(G)|$.*

Proof. For convenience, we use $G^1, G^2, \ldots, G^m$ to denote the m copies of G in $P_m \cdot G$. Assume that $c_0(P_m \cdot G) < |V(G)|$. Consider a strategy for $P_m \cdot G$ that uses at most $|V(G)| - 1$ cops. Let i be the index of the round satisfying the following two conditions: (i) when the cops' turn is finished at round i, there are at most two copies of G that are free of the robber; (ii) when the cops' turn is finished at round $i + 1$, there are at least three copies of G that are free of the robber.

Consider the moment when the cops' turn is done at round i. If there is only one copy of G that is free of the robber, say G^k, then we know each vertex in $P_m \cdot G$ must be adjacent to at least one contaminated vertex. Hence, at the end of round i, all unoccupied vertices are contaminated. Note that $c_0(G) \leq |V(G)| - 1$. We know there are at most $2|V(G)| - 2$ cleared vertices after the cop's turn at round $i + 1$. Since $2|V(G)| - 2 < 3|V(G)|$, there do not exist three copies of G that are free of the robber after the cop's turn at round $i + 1$. This contradicts condition (ii). If there are exactly two copies of G that are free of the robber, then there are two cases:

Case 1. The two copies of G are not consecutive. Obviously, each vertex in $V(P_m \cdot G)$ must be adjacent to at least one contaminated vertex. Hence, only occupied vertices are cleared at the end of round i. Thus, the total number of cleared vertices when the cops' turn is finished at round $i + 1$ is at most $2|V(G)| - 2$. This contradicts condition (ii).

Case 2. The two copies of G are consecutive. Let G^k and G^{k+1}, where $k \geq 0$, denote the two consecutive copies of G. There are two subcases.

Case 2.1. $k = 1$ or $k = m - 1$. We consider the case when $k = 1$ in the next, and the case when $k = m - 1$ can be proved in a similar way. Note that G^j contains at least one vertex that is contaminated after the cops' turn at round i, for every j in $\{3, \ldots, m\}$. All unoccupied vertices in $V(P_m \cdot G) \setminus V(G^1)$ are contaminated at the end of round i. Since there are at most $2|V(G)| - 1$ cleared vertices at the end of round i, we know the total number of cleared vertices in $P_m \cdot G$ is at most $3|V(G)| - 2 < 3|V(G)|$ when the cop's turn is finished at round $i + 1$. This contradicts condition (ii).

Case 2.2. $2 \leq k \leq m - 2$. It is easy to see that each vertex in $V(P_m \cdot G)$ is adjacent to at least one contaminated vertex. Hence, all unoccupied vertices are contaminated at the end of round i. Further, the total number of cleared vertices when the cop's turn is finished at round $i + 1$ is at most $2|V(G)| - 2$ and this contradicts condition (ii).

Hence, round i does not exist and this contradicts that $|V(G)| - 1$ cops are sufficient for clearing $P_m \cdot G$. This completes the proof.   □

Consider $P_2 \cdot G$. If G is a cycle of n vertices where n is even, then we can clear $P_2 \cdot G$ with at most $\frac{n}{2} + 2$ cops. Let G^1 and G^2 denote the two copies of G in $P_2 \cdot G$. The following briefly describes a strategy that clears $P_2 \cdot G$ utilizing $\frac{n}{2} + 2$ cops: (1) let $\frac{n}{2}$ cops vibrate on all vertices of G^1 such that every vertex of G^1 is visited by a cop in each round; (2) let two additional cops clear all vertices of G^2. Consider $P_3 \cdot G$. If G is a cycle of n vertices where n is even, we can also clear $P_3 \cdot G$ with at most $\frac{n}{2} + 2$ cops. Let G^1, G^2 and G^3 denote the three copies of G in $P_3 \cdot G$. Similar to the above, let $\frac{n}{2}$ cops vibrate on all vertices of G^2, and let two additional cops clear all vertices of G^1 and G^2 respectively. Hence, Lemma 3 is not always held when $m = 2$ or 3.

Corollary 3. *Let W_m be a graph obtained from $P_m \cdot G$ ($m \geq 4$) by replacing each copy of G by an arbitrary graph with $|V(G)|$ vertices. Then $c_0(W_m) \geq |V(G)|$.*

Lemma 4. *Let P_m be a path with m vertices. For $m \geq 4$, $c_0(P_2 \cdot P_m) \geq \lceil \frac{m}{2} \rceil + 1$.*

Proof. We will use Definition 1 to show that $\lceil \frac{m}{2} \rceil$ cannot clear $P_2 \cdot P_m$. Without loss of generality, suppose m is odd. The other case when m is even can be proved in a similar way. Assume that $c_0(P_2 \cdot P_m) \leq \lceil \frac{m}{2} \rceil$. Consider a strategy that clears $P_2 \cdot P_m$ using at most $\lceil \frac{m}{2} \rceil$ cops. Note that $c_0(P_2 \cdot P_m) \leq \frac{m+1}{2} < 2m$. Obviously, $\mathcal{P}(0)$ holds. Suppose that $\mathcal{P}(i)$ holds for some $i \geq 0$. Consider the moment when the cops' turn is finished at round $i + 1$. Since $c_0(P_2 \cdot P_m) \leq \frac{m+1}{2}$, there are at most $m + 1$ cleared vertices in $V(P_2 \cdot P_m)$. Further, if there exists a copy of P_m that is free of the robber, then the other copy of P_m contains at most one cleared vertex. Hence, every cleared vertex must be adjacent to a contaminated vertex. Thus, all unoccupied cleared vertices become recontaminated immediately after the robber's turn at round $i + 1$. Hence, $\mathcal{P}(i + 1)$ holds. From Proposition 1, we know that $\lceil \frac{m}{2} \rceil$ cops are insufficient for clearing $P_2 \cdot P_m$. This completes the proof.   □

3.3   Complete Multipartite Graphs

Let $K_{n_1,\dots,n_k} = (V_1, \dots, V_k, E)$ denote a complete k-partite graph, where $n_1 \leq \dots \leq n_k$. Clearly, complete multipartite graphs can be defined recursively using graph join operations. For example, $K_{n_1,\dots,n_k}$ can also be defined as $K_{n_1,\dots,n_{k-1}} + \overline{K_{n_k}}$. Since the problem of determining the cop number of complete bipartite graphs has been solved, we only consider complete k-partite graphs with $k \geq 3$ in this paper. Let $H_j = K_{n_1,\dots,n_k} - V_j$, where $1 \leq j \leq k$. Let $\mu_{\min} = \min\{\mu(H_j) | 1 \leq j \leq k\}$. The next corollary is from Lemma 2.

Corollary 4. *For $1 \leq i \leq k$, the minimum number of cops that can visit all vertices of H_i in a round is $\mu(H_i)$.*

From Corollary 4, we know that all vertices of $H_x - V(\mathcal{M}(H_x))$ must be contained in some vertex set V_i of $K_{n_1,\dots,n_k}$, where $1 \leq i \neq x \leq k$.

Lemma 5. $\frac{1}{2}\sum_{i=1}^{k-1} n_i \leq \mu_{\min} \leq \frac{1}{2}\sum_{i=1}^{k} n_i$, and both upper bound and lower bound are tight.

Proof. We first consider the lower bound. Note that $n_i \leq n_{i+1}$ for all $1 \leq i \leq k-1$. We have $\sum_{i=1}^{k-1} n_i \leq \sum_{i=1}^{k} n_i - n_x = |V(H_x)|$, where $1 \leq x \leq k$. Let j be the index such that $\mu_{\min} = \mu(H_j)$. Hence, it follows from Corollary 4 that $2\mu_{\min} \geq |V(H_j)| \geq \sum_{i=1}^{k-1} |V_i|$. Further, the equality holds if there is a perfect matching in H_k.

We now consider the upper bound. Since $n_k \geq n_i$ for all $1 \leq i \leq k$, we know $n_k \geq |V(H_k) \setminus V(\mathcal{M}(H_k))|$. Hence, $2\mu(H_k) = |V(H_k)| + |V(H_k) \setminus V(\mathcal{M}(H_k))| \leq |V(H_k)| + n_k = \sum_{i=1}^{k} n_i$. Therefore, we have $\mu_{\min} \leq \mu(H_k) \leq \frac{1}{2}\sum_{i=1}^{k} n_i$. Further, if $|V_k| = 1$ and k is even, then $\mu_{\min} = \frac{1}{2}\sum_{i=1}^{k} n_i$. $\quad\square$

Lemma 6. $c_0(K_{n_1,\dots,n_k}) \geq \mu_{\min}$.

Proof. We will use Definition 1 to show that $\mu_{\min} - 1$ cops cannot clear $K_{n_1,\dots,n_k}$. Assume that $c_0(K_{n_1,\dots,n_k}) \leq \mu_{\min} - 1$. Consider a strategy for $K_{n_1,\dots,n_k}$ that uses at most $\mu_{\min} - 1$ cops. From Lemma 5, we have $c_0(K_{n_1,\dots,n_k}) < \frac{1}{2}\sum_{i=1}^{k} n_i$. It is easy to see that $\mathcal{P}(0)$ holds. Suppose that $\mathcal{P}(i)$ holds for some $i \geq 0$. Consider round $i + 1$. Since $c_0(K_{n_1,\dots,n_k}) < \mu_{\min}$, it follows from Lemma 2 that $\mu_{\min} - 1$ cops are insufficient for visiting all vertices of any $k - 1$ vertex sets of $K_{n_1,\dots,n_k}$ within one round. Hence, there must exist two vertex sets V_p and V_q, where $1 \leq p \neq q \leq k$, such that both of which contain vertices that remain contaminated throughout round $i + 1$. In the robber's turn at round $i + 1$, all unoccupied cleared vertices thus become recontaminated. Hence, $\mathcal{P}(i+1)$ holds. From Proposition 1, we know that $\mu_{\min} - 1$ cops are insufficient for clearing $K_{n_1,\dots,n_k}$. Hence, $c_0(K_{n_1,\dots,n_k}) \geq \mu_{\min}$. $\quad\square$

Lemma 7. If there is a perfect matching in H_k, then $c_0(K_{n_1,\dots,n_k}) \geq \frac{1}{2}\sum_{i=1}^{k-1} n_i + 1$.

Proof. Since there is a perfect matching in H_k, then $|V(H_k)|$ is even and $\mu(H_k) = \frac{1}{2}\sum_{i=1}^{k-1} n_i$. From Lemma 5, we know $\mu(H_k) = \frac{1}{2}\sum_{i=1}^{k-1} n_i = \mu_{\min}$. From Lemma 6, we have $c_0(K_{n_1,\dots,n_k}) \geq \mu_{\min} = \frac{1}{2}\sum_{i=1}^{k-1} n_i$.

We will use Definition 1 to show that $\frac{1}{2}\sum_{i=1}^{k-1} n_i$ cops cannot clear $K_{n_1,\dots,n_k}$. Assume that $c_0(K_{n_1,\dots,n_k}) \leq \frac{1}{2}\sum_{i=1}^{k-1} n_i$. Consider a strategy for $K_{n_1,\dots,n_k}$ that uses at most $\frac{1}{2}\sum_{i=1}^{k-1} n_i$ cops. Note that $c_0(K_{n_1,\dots,n_k}) \leq \frac{1}{2}\sum_{i=1}^{k-1} n_i < \frac{1}{2}\sum_{i=1}^{k} n_i$. Obviously, $\mathcal{P}(0)$ holds. Suppose that $\mathcal{P}(i)$ holds for some $i \geq 0$. Consider round $i + 1$. Note that there are at least n_k contaminated vertices that remain unoccupied throughout round $i + 1$. Hence, there exist at least n_k vertices that remain contaminated throughout round $i+1$. These contaminated vertices must be contained in at least one vertex set of $K_{n_1,\dots,n_k}$. Note that $n_k \geq n_i$ for $1 \leq i \leq k$. If these contaminated vertices are contained in one vertex set, then the vertex set must contain n_k vertices.

Case 1. There exists a vertex set V_p, where $n_p = n_k$, which contains all the vertices that remain contaminated throughout round $i + 1$. Then all cops are on

$V(K_{n_1,\ldots,n_k}) \setminus V_p$ throughout round $i+1$. It is easy to see that after the robber's turn at round $i+1$, all unoccupied cleared vertices get recontaminated. Hence, $\mathcal{P}(i+1)$ holds.

Case 2. There exist two or more vertex sets containing vertices that remain contaminated throughout round $i+1$. Then we can also show that all unoccupied cleared vertices get recontaminated in the robber's turn at round $i+1$. Hence, $\mathcal{P}(i+1)$ holds.

Note that $\mathcal{P}(i+1)$ holds for both cases. From Proposition 1, we know that $\frac{1}{2}\sum_{i=1}^{k-1} n_i$ cops are insufficient. Thus, $c_0(K_{n_1,\ldots,n_k}) \geq \frac{1}{2}\sum_{i=1}^{k-1} n_i + 1$. □

Lemma 8. $|\mathcal{M}(H_i)| \leq |\mathcal{M}(H_k)| + n_k - n_i$, where $1 \leq i \leq k-1$.

Lemma 9. If there is no perfect matching in H_k, then $c_0(K_{n_1,\ldots,n_k}) \geq \mu(H_k)$.

Proof. We first prove that $\mu(H_k) = \mu_{\min}$. Consider a maximum matching in H_k. Since there is no perfect matching in H_k, we have $\mu(H_k) > |\mathcal{M}(H_k)|$. Note that $\mu(H_k) = |\mathcal{M}(H_k)| + \sum_{i=1}^{k-1} n_i - 2|\mathcal{M}(H_k)|$ and $\mu(H_i) = |\mathcal{M}(H_i)| + \sum_{j=1}^{k} n_j - n_i - 2|\mathcal{M}(H_i)|$. Then, $\mu(H_k) - \mu(H_i) = |\mathcal{M}(H_i)| - n_k + n_i - |\mathcal{M}(H_k)|$. It follows from Lemma 8 that $\mu(H_k) - \mu(H_i) \leq 0$ for all $1 \leq i \leq k$. Hence, $\mu(H_k) = \min\{\mu(H_i)|1 \leq i \leq k\}$. It follows from Lemma 6 that $c_0(K_{n_1,\ldots,n_k}) \geq \mu_{\min} = \mu(H_k)$. □

By combining Lemmas 1 and 9, we obtain the next Theorem.

Theorem 5. Let G be a graph such that $K_{n_1,\ldots,n_k}$ is an induced subgraph of G, where $k \geq 3$. We have $c_0(G) \geq \mu(H_k)$.

3.4 Split Graphs

A graph is a *split graph* if its vertex set can be partitioned into a clique and an independent set. Let $S_{m,n} = (C,I)$ denote a split graph, where C is a clique of m vertices and I is an independent vertex set of n vertices.

Theorem 6. Let $S_{m,n} = (C,I)$ be a split graph with $|V(C)| = m$ and $|V(I)| = n$. If m is odd or $V(C)$ has a pair of vertices that share no common neighbor in $V(I)$, then $c_0(S_{m,n}) \geq \lceil\frac{m}{2}\rceil$. If m is even and every pair of vertices in $V(C)$ have a common neighbor in $V(I)$, then $c_0(S_{m,n}) \geq \frac{m}{2} + 1$.

Proof. (i) Since $S_{m,n}$ has a clique of size m, it follows from Corollary 1 that $c_0(S_{m,n}) \geq \lceil\frac{m}{2}\rceil$.

(ii) Assume that $c_0(S_{m,n}) \leq \frac{m}{2}$. Consider a strategy for $S_{m,n}$ that uses at most $\frac{m}{2}$ cops. Let $\mathcal{P}'(i)$ be the proposition that there exists a contaminated vertex in $V(C)$ whose unoccupied neighbors are all contaminated at the end of round i. We will use induction to show that the proposition holds for one of any two consecutive rounds.

Base case: $\mathcal{P}'(0)$ obviously holds.

Induction step: Suppose that $\mathcal{P}'(i)$ holds for some $i \geq 0$. Consider the beginning moment of round $i+1$. There are two cases.

Case 1. Some cops stay on vertices in V_2. Note that $c_0(S_{m,n}) \leq \frac{m}{2}$. Clearly, there must be a vertex $u \in V(C)$ that remains contaminated throughout round $i+1$. After the robber's turn at round $i+1$, it is easy to see that all unoccupied neighbors of u become contaminated. Therefore, $\mathcal{P}'(i+1)$ holds.

Case 2. All cops stay on vertices in V_1. Let $u \in V(C)$ be a contaminated vertex whose unoccupied neighbors are all contaminated. If some cop slides to $V(I)$ at round $i+1$, similar to Case 1, we can show that $\mathcal{P}'(i+1)$ holds. Assume that all cops are on vertices in $V(C)$ throughout round $i+1$. Then all neighbors of u in $V(I)$ remain contaminated throughout round $i+1$. Note that every pair of vertices in $V(C)$ have a common neighbor in $V(I)$. Hence, all unoccupied cleared vertices in $V(C)$ get recontaminated at the end of round $i+1$. If u is unoccupied after the cops' turn at round $i+1$, then $\mathcal{P}'(i+1)$ holds; if u is occupied by a cop after the cops' turn at round $i+1$, then we consider the moment when the cops' turn is finished at round $i+2$.

Case 2.1. All the cops are still on vertices in $V(C)$. If u is unoccupied after the cops' turn at round $i+2$, then u gets recontaminated in the following robber's turn and $\mathcal{P}'(i+2)$ holds. If u is occupied by a cop, then there must exist a vertex in $V(C)$ which remains contaminated throughout round $i+2$. Similar to Case 1, we can show that $\mathcal{P}'(i+2)$ holds.

Case 2.2. Some cops slide to $V(I)$ at round $i+2$. Similar to Case 1, we can show that $\mathcal{P}'(i+2)$ holds.

Above all, we have shown that the proposition holds for one of any two consecutive rounds. This contradicts the assumption that $\frac{m}{2}$ cops are sufficient for clearing $S_{m,n}$. Therefore, $c_0(S_{m,n}) \geq \frac{m}{2} + 1$. □

4 Matching Upper Bounds

Using the lower bounds presented in Sect. 3, we now investigate the cop number of graph joins, lexicographic products of graphs, complete multipartite graphs and split graphs in the following subsections.

4.1 Graph Joins

Let G and H be two graphs. Recall that $G + H$ is the graph join of G and H. Let $G_1, \ldots, G_m$ be the maximal connected components of G and let $H_1, \ldots, H_n$ be the maximal connected components of H. Let $\gamma_1 = \max\{\max\{\mathrm{pw}(H_i)|1 \leq i \leq n\} + 1 - (\mu(G) - |\mathcal{M}(G)|), 0\}$ and let $\gamma_2 = \max\{\max\{\mathrm{pw}(G_i)|1 \leq i \leq m\} + 1 - (\mu(H) - |\mathcal{M}(H)|), 0\}$.

Lemma 10. $c_0(G + H) \leq \min\{\mu(G) + \gamma_1, \mu(H) + \gamma_2\}$.

In combination of Theorem 1 and Lemma 10, we give the next theorem.

Theorem 7. $\min\{\mu(G), \mu(H)\} \leq c_0(G + H) \leq \min\{\mu(G) + \gamma_1, \mu(H) + \gamma_2\}$.

Theorem 8. If $n \leq 2$, then $c_0(C_m + \overline{K_n}) = \min\{\lceil \frac{m+1}{2} \rceil, n + 1\}$; if $n \geq 3$, then $c_0(C_m + \overline{K_n}) = \min\{\lceil \frac{m+1}{2} \rceil, n\}$.

Proof. Let $C_m = \{u_1, \ldots, u_m\}$ and $\overline{K_n} = \{v_1, \ldots, v_n\}$. We first show that $c_0(C_m + \overline{K_n}) \leq \lceil \frac{m+1}{2} \rceil$. Consider the case when m is odd. To clear $C_m + \overline{K_n}$, place cop λ_i on u_{2i-1}, where $1 \leq i \leq \frac{m+1}{2}$. Hence, we use $\frac{m+1}{2}$ cops in total. Let cop λ_i, where $1 \leq i \leq \frac{m-1}{2}$, vibrate on u_{2i-1} and u_{2i} throughout the strategy until all vertices are cleared. Let cop $\lambda_{\frac{m+1}{2}}$ vibrate on u_m and each vertex of $\overline{K_n}$ to clear all the vertices of $\overline{K_n}$. When m is even, we can use $\frac{m}{2} + 1$ cops to clear $C_m + \overline{K_n}$ in a similar way. Therefore, $c_0(C_m + \overline{K_n}) \leq \lceil \frac{m+1}{2} \rceil$.

We then show that $c_0(C_m + \overline{K_n}) \leq n + 1$ if $n \leq 2$. If $n = 1$, then we place two cops on u_1 initially. Let one cop vibrate on u_1 and v_1 until all vertices are cleared. Let the second cop slide around C_m to clear all its vertices. If $n = 2$, then we place cop λ_i on u_i, where $1 \leq i \leq 3$. Let cop λ_i, where $1 \leq i \leq 2$, vibrate on u_i and v_i until all vertices are cleared. Let cop λ_3 slide around C_m to clear all its vertices.

In the last, we show that $c_0(C_m + \overline{K_n}) \leq n$ if $n \geq 3$. Place cop λ_i on v_i, where $1 \leq i \leq n$. Let cop λ_1 vibrate on v_1 and u_1 until all vertices are cleared. Slide λ_{n-1} to u_2 and slide λ_n to u_3. If $m \leq 3$, then all vertices are cleared. If $m \geq 4$, then (1) slide λ_{n-1} back to v_{n-1} and λ_n back to v_n, and (2) slide λ_{n-1} to u_3 and slide λ_n to u_4. All vertices of C_m can be cleared in a similar way.

In combination of the above and Theorem 3, we have: (1) if $n \leq 2$, then $c_0(C_m + \overline{K_n}) = \min\{\lceil \frac{m+1}{2} \rceil, n + 1\}$, and (2) if $n \geq 3$, then $c_0(C_m + \overline{K_n}) = \min\{\lceil \frac{m+1}{2} \rceil, n\}$.

$\square$

Theorem 9. *For $3 \leq m \leq n$, if m is odd, then $c_0(C_m + C_n) = \frac{m+1}{2} + 1$; if m is even, then $c_0(C_m + C_n) = \frac{m}{2} + 2$.*

Theorem 10. *If $n = 1$, then $c_0(K_m + P_n) = \lceil \frac{m+1}{2} \rceil$; if $n \geq 2$, then $c_0(K_m + P_n) = \lceil \frac{m}{2} \rceil + 1$.*

4.2 Lexicographic Products of Graphs

Theorem 11. *For $m \geq 4$, $c_0(P_m \cdot G) = |V(G)|$.*

Lemma 11. *For $m \geq 4$, $c_0(P_2 \cdot P_m) = \lceil \frac{m}{2} \rceil + 1$.*

Proof. From Lemma 4, we have $c_0(P_2 \cdot P_m) \geq \lceil \frac{m}{2} \rceil + 1$. So we only need to give a strategy that clears $P_2 \cdot P_m$ with $\lceil \frac{m}{2} \rceil + 1$ cops. Let P_m^1 and P_m^2 be the two copies of P_m in $P_2 \cdot P_m$. To clear $P_2 \cdot P_m$, we can use $\lceil \frac{m}{2} \rceil$ cops and let them vibrate on vertices in $V(P_m^1)$ to ensure that every vertex in $V(P_m^1)$ remains cleared in at least one of any two consecutive rounds. Then we can use one additional cop to clear all vertices in $V(P_m^2)$.

$\square$

4.3 Complete Multipartite Graphs

Theorem 12. *If there is a perfect matching in $K_{n_1, \ldots, n_k} - V_k$, then $c_0(K_{n_1, \ldots, n_k}) = \frac{1}{2} \sum_{i=1}^{k-1} |V_i| + 1$.*

Proof. From Lemma 7, we have $c_0(K_{n_1,\ldots,n_k}) \geq \frac{1}{2}\sum_{i=1}^{k-1}|V_i| + 1$. To complete the proof, we describe a cop-win strategy for $K_{n_1,\ldots,n_k}$ that uses $\frac{1}{2}\sum_{i=1}^{k-1}|V_i|+1$ cops.

1. For each edge in $\mathcal{M}(H_k)$, place a cop on one endpoint of the edge, and let it vibrate between two endpoints of the edge until all vertices in $V(K_{n_1,\ldots,n_k})$ are cleared.
2. Select a vertex $v \in V_1$ and place a cop λ on it. Then, slide λ between v and V_k to clear all vertices in V_k.

$\square$

Theorem 13. *If there is no perfect matching in $K_{n_1,\ldots,n_k} - V_k$, then $c_0(K_{n_1,\ldots,n_k}) = \mu(H_k)$.*

4.4 Split Graphs

Lemma 12. *For a split graph $S_{m,n} = (C, I)$, if m is odd, or there is a pair of vertices in $V(C)$ which share no common neighbor in $V(I)$, then $c_0(S_{m,n}) \leq \lceil \frac{m}{2} \rceil$.*

Lemma 13. *For a split graph $S_{m,n}$, if m is even and every pair of vertices in V_1 have a common neighbor in V_2, then $c_0(S_{m,n}) \leq \frac{m}{2} + 1$.*

Proof. To clear $S_{m,n}$, we place cop λ_i on u_i, where $1 \leq i \leq \frac{m}{2}$, and let the cop vibrate on u_i and $u_{i+\frac{m}{2}}$ until the end of the strategy; then we place another cop λ' on u_1, and use it to clear all u_1's neighbors in V_2; in the next, we slide λ' to u_i, for $i = 2, \ldots, m$, to clear all its contaminated neighbors in V_2. Obviously, we clear $S_{m,n}$ using a total of $\frac{m}{2} + 1$ cops. $\square$

By combining Theorem 6, Lemmas 12 and 13, we give the cop number of split graph in the next theorem.

Theorem 14. *Let $S_{m,n} = (C, I)$ be a split graph with $|V(C)| = m$ and $|V(I)| = n$. (i) $c_0(S_{m,n}) = \lceil \frac{m}{2} \rceil$, if m is odd or $V(C)$ has a pair of vertices that share no common neighbor in $V(I)$; (ii) $c_0(S_{m,n}) = \frac{m}{2} + 1$, if m is even and every pair of vertices in $V(C)$ have a common neighbor in $V(I)$.*

5   Conclusions

In this paper, we considered the zero-visibility cops and robber game on graph joins, as well as a few other types of graphs and graph products. We gave a method for finding lower bounds on the zero-visibility cop number by establishing conditions on search strategies. The challenge in establishing lower bounds on various graph search parameters typically lie in finding non-trivial lower bounds that apply to more than just a few special graphs. The method that we have used in this paper can be seen as a general approach to find lower bounds on the zero-visibility cop number of a large class of graphs. We believe that this technique can inspire more lower bound results for other classes of graphs, and potentially also for other graph search parameters.

References

1. Aigner, M., Fromme, M.: A game of cops and robbers. Discrete Appl. Math. **8**(1), 1–12 (1984)
2. Bonato, A.: Conjectures on cops and robbers. In: Gera, R., Hedetniemi, S., Larson, C. (eds.) Graph Theory. PBM, pp. 31–42. Springer, Cham (2016). https://doi.org/10.1007/978-3-319-31940-7_3
3. Bonato, A., Gordinowicz, P., Hahn, G.: Cops and robbers ordinals of cop-win trees. Discrete Math. **340**(5), 951–956 (2017)
4. Bonato, A., MacGillivray, G.: Characterizations and algorithms for generalized cops and robbers games. Contrib. Discrete Math. **12**(1), 110–122 (2017)
5. Bonato, A., Nowakowski, R.: The Game of Cops and Robbers on Graphs. American Mathematical Society, Providence (2011)
6. Dereniowski, D., Dyer, D., Tifenbach, R.M., Yang, B.: The complexity of zero-visibility cops and robber. Theoret. Comput. Sci. **607**, 135–148 (2015)
7. Dereniowski, D., Dyer, D., Tifenbach, R.M., Yang, B.: Zero-visibility cops and robber and the pathwidth of a graph. J. Comb. Optim. **29**(3), 541–564 (2015)
8. Fitzpatrick, S.L., Larkin, J.P.: The game of cops and robber on circulant graphs. Discrete Appl. Math. **225**, 64–73 (2017)
9. Jeliazkova, D.: Aspects of the cops and robber game played with incomplete information. Master's thesis, Acadia University (2006)
10. Kinnersley, W.B.: Cops and robbers is exptime-complete. J. Comb. Theory Ser. B **111**, 201–220 (2015)
11. Mamino, M.: On the computational complexity of a game of cops and robbers. Theoret. Comput. Sci. **477**, 48–56 (2013)
12. Nowakowski, R., Winkler, P.: Vertex-to-vertex pursuit in a graph. Discrete Math. **43**(2–3), 235–239 (1983)
13. Offner, D., Ojakian, K.: Variations of cops and robber on the hypercube. Australas. J. Comb. **59**(2), 229–250 (2014)
14. Quilliot, A.: Jeux et pointes fixes sur les graphes. Ph.D. thesis, Université de Paris VI (1978)
15. Tang, A.: Cops and robber with bounded visibility. Master's thesis, Dalhousie University (2004)
16. Tošić, R.: Vertex-to-vertex search in a graph. Graph Theory (Dubrovnik 1985), pp. 233–237 (1985)

A Two-Stage Constrained Submodular Maximization

Ruiqi Yang[1], Shuyang Gu[2], Chuangen Gao[3], Weili Wu[2], Hua Wang[3],
and Dachuan Xu[1(✉)]

[1] Department of Operations Research and Scientific Computing,
Beijing University of Technology, Beijing 100124, People's Republic of China
yangruiqi@emails.bjut.edu.cn, xudc@bjut.edu.cn
[2] Department of Computer Science, University of Texas at Dallas,
Richardson, TX 75080, USA
{Shuyang.Gu,weiliwu}@utdallas.edu
[3] School of Computer Science and Technology, Shandong University,
Jinan 250101, People's Republic of China
gaochuangen@gmail.com, wanghua@sdu.edu.cn

Abstract. We consider a two-stage submodular maximization under p-matroid (or p-extendible) constraints. In the model, we are given a collection of submodular functions and some p-matroid (or extendible) system constraints for each of these functions, one need to choose a representative set with a cardinality constraint and simultaneously select a series of subsets that are restricted to the representative set for all functions, the aim is to maximize the average of the summarization of these function values. We extend the two-stage submodular maximization under single matroid to handle p-matroid (or p-extendible) constraints, and derive constant approximation ratio algorithms for the two problems, respectively. In the end, we empirically demonstrate the efficiency of our method on some datasets.

Keywords: Submodular maximization · Approximation algorithms · Independence system constraints

1 Introduction

The submodular maximization has many applications, such as document summarization [5,11], recommender systems [13,14,17], and other applications [9,19], etc. Formally, it can be modeled as $\max_{S \subseteq \Omega: S \in \mathcal{I}} f(S)$, where f is a submodular function defined on ground set Ω and $\mathcal{I}$ is some specific constraint. In the text, we will give a brief summary of the submodular maximization.

For the submodular maximization under a cardinality constraint, [15] provided a $(1 - 1/e)$-approximation. Under more general p-matroid constraint, [7] got a deterministic $1/(p+1)$-approximation algorithm. In particular, if $p = 1$, it reduced to a $1/2$-approximation algorithm for monotone submodular maximization under a single matroid constraint (SMMC). [10] improved the ratio from

© Springer Nature Switzerland AG 2019
D.-Z. Du et al. (Eds.): AAIM 2019, LNCS 11640, pp. 329–340, 2019.
https://doi.org/10.1007/978-3-030-27195-4_30

$1/(p + 1)$ to $1/(p + \epsilon)$ for monotone submodular maximization under p-matroid constraints (SMMC-p). Combining continuous greedy process and pipage rounding technique, [4] obtained a random $(1 - 1/e)$-approximation algorithm for the monotone SMMC. As the hardness of approximation ratio of the above models is $(1 - 1/e + \epsilon)$, it is still a long history of closing the gap of approximation ratio for the monotone SMMC by a deterministic algorithm. The breakthrough result was presented by [3], who gave the first deterministic 0.5008-approximation algorithm for the monotone SMMC-p. [12] introduced a more general p-extendible system constraints, which captures a class of constraints, such as p-matroid constraints, b-matching, maximum profit scheduling and maximum asymmetric traveling salesman problem. He presented a $1/p$-approximation algorithm based on greedy for a monotone submodular maximization under p-extendible constraint. For a non-monotone submodular maximization under p-extendible system constraint, [6] presented a $p/(p + 1)^2$-approximation algorithm.

Motivated by the tasks of multi-objective summarization, [2] introduced the *two-stage submodular maximization problem*. In the model, we are given a ground set Ω of size n, integers ℓ, k, and multiple submodular functions $f_1, ..., f_m$, the goal is to choose a subset S of $|S| \leq \ell$ such that the sum of $\max_{T_i \subseteq S, |T_i| \leq k} f_i(T_i)$ is maximum. Obviously, this problem reduces to classical cardinality submodular maximization problem if $m = 1$, or if $\ell = k$. Combining the techniques of continuous greedy and dependent rounding, they firstly presented an approximation arbitrary close to $1 - 1/e$ as $k \to \infty$. Secondly, under the case of that each f_i, $i \in [m]$ is a coverage function, they obtained a $1/2(1 - 1/e)$-approximation by a local search, while the query complexity is bounded by $O(km\ell n^2 \log n)$. [17] considered a the two-stage submodular maximization with general matroid constraint, that is, $\sum_{i=1}^m \max_{T_i \in \mathcal{I}(S)} f_i(T_i)$, where $(S, \mathcal{I}(S))$ is a matroid. They derived a $1/2(1 - 1/e^2)$-approximation algorithm, and its query complexity is at most $O(rm\ell n)$, where r is the matroid rank. [14] first studied this problem under streaming and distributed settings, respectively. In the streaming setting, they derived a one pass, $1/7$-approximation algorithm, while its memory complexity is bounded by $O(\ell \log(\ell)/\epsilon)$ and the query complexity is at most $O(kmn \log(\ell)/\epsilon)$. In the distributed setting, they got two $1/4(1 - 1/e^2)$ and 0.107-approximation algorithms, respectively. The query complexities of the above two distributed algorithms are bounded by $O(kmn\ell/M + Mkm\ell^2)$ and $O(kmn \log \ell/M + Mkm\ell^2 \log \ell)$, respectively, where M is the number of the machines. We first consider the two-stage submodular maximization under more general constraints. Specifically, we aim to maximize $\sum_{i=1}^m \max_{T_i \in \mathcal{I}^i(S)} f_i(T_i)$, where $(S, \mathcal{I}^i(S))$ is a p-matriod (or p-extendible) for any i. For the two-stage submodular maximization under p-matroid system constraint, we propose a $1/(p+1)(1-1/e^2)$-approximation algorithm, while its query complexity is bounded by $O(\ell mnr^p)$, where r is the maximum independence set size. Under more general p-extendible constraints, we yield a $1/(r+1)(1-1/e^2)$-approximation algorithm with the same query complexity. Finally, we demonstrate the efficiency of our algorithm on some datasets.

The rest of our paper is organized as follows. We present some necessary preliminaries in Sect. 2. In Sect. 3, we introduce the two-stage submodular maximization under p-matroid system constraints and provide a $1/(p+1)(1 - 1/e^2)$-approximation algorithm. In addition, we present a $1/(r+1)(1 - 1/e^2)$-approximation algorithm for two-stage submodular maximization under p-extendible system constraints in Sect. 4. In Sect. 5, we show the results of some numerical experiments of our algorithm. Finally, we give a conclusion in Sect. 6.

2   Preliminaries

In our setting, we are given an element ground set Ω of size n, and a collection $\mathcal{F} = \{f_1, ..., f_m\}$ of non-negative monotone submodular functions that are defined on the ground set Ω. For any $i \in [m] = \{1, ..., m\}$, $f_i : 2^\Omega \to R_+$ is a *submoduar function*, i.e.,

$$f_i(A) + f_i(B) \geq f_i(A \cup B) + f_i(A \cap B), \forall A, B \subseteq \Omega.$$

For any $i \in [m]$, there exists a constraint $\mathcal{I}^i$, the objective is to find a representative set $S \subseteq \Omega$ with $|S| \leq \ell (\ll n)$, such that the average of the summarization of the optimum of $f_i, i \in [m]$ restricted to S is maximum. Let $G_m(S) = \frac{1}{m} \sum_{i=1}^m \max_{T_i \in \mathcal{I}^i(S)} f_i(T_i)$. Then our model can be defined as

$$\max_{S \subseteq \Omega, |S| \leq \ell} G_m(S) = \max_{S \subseteq \Omega, |S| \leq \ell} \frac{1}{m} \sum_{i=1}^m \max_{T_i \in \mathcal{I}^i(S)} f_i(T_i), \tag{1}$$

where $\mathcal{I}^i(S)$ denotes the constraint $\mathcal{I}^i$ restricted to S for any $i \in [m]$.

In order to have a better understand of our model and constraints, we restate some necessary notations and definitions as follows. Given a finite element set Ω, and a collection $\mathcal{I}$ of subsets of Ω. A two-tuples $M = (\Omega, \mathcal{I})$ is defined as an *independent system*, if it has for any subset $T \in \mathcal{I}$, then any subset $S \subseteq T$ such that $S \in \mathcal{I}$. Each subset of $\mathcal{I}$ is named as *independence set*. The independent system $M = (\Omega, \mathcal{I})$ is a *matroid* if it also satisfies that if for any independence sets $S, T \in \mathcal{I}$ with $|S| > |T|$, then there exists an element $e \in S \setminus T$ such that $T \cup \{e\} \in \mathcal{I}$. A maximal independent subset $A \in \mathcal{I}$ is called a *base* of the independent system $M = (\Omega, \mathcal{I})$. Given an integer p, Let $M_j = (\Omega, \mathcal{I}_j)$ be a matroid according to $j \in [p]$, then we call the intersection of these p matroids $(\Omega, \cap_{j=1}^p \mathcal{I}_j)$ as p-matroid. Given any subsets $S, T \in \mathcal{I}$, we say T is an *extension* of S if $S \subseteq T$. We restate the definition of p-extendible system as follows.

Definition 1 [6,12]. *An independence system $M = (\Omega, \mathcal{I})$ is p-extendible system if for every independent set $S \in \mathcal{I}$, an extension T of S and an element $e \notin S$ obeying $S \cup \{e\} \in \mathcal{I}$ there must exist a subset $Y \subseteq T \setminus S$ with $|Y| \leq p$ such that $T \setminus Y \cup \{e\} \in \mathcal{I}$. Specially, if the independent set S is maximal i.e., $T = S$, then we can reduce the definition by setting $Y = \emptyset$.*

In our p-matroid constraints model, for $i \in [m]$ and $S \subseteq \Omega$, any subset $T_i \subseteq S$ is feasible if $T_i \in \mathcal{I}^i = \cap_{j=1}^p \mathcal{I}_j^i$, where $M^i = (S, \cap_{j=1}^p \mathcal{I}_j^i(S))$ is a p-matroid system restricted to S. Similarly, for the p-extendible system constraint, $M^i = (S, \mathcal{I}^i)$ is defined as p-extendible system. We say subset T_i is feasible if $T_i \in \mathcal{I}^i$. We also assume there are value and independence oracles, i.e., for any $i \in [m]$ and subset A, we can obtain the value of $f_i(A)$ and know if $A \in \mathcal{I}^i$ or not.

3 P-Matroid System Constraints

In this section, we extend the ReplacementGreedy algorithm introduced by [17] (for comparison, we say their algorithm as One-to-One ReplacementGreedy) for a single matroid constraint to address the two-stage submodular maximization under p-matroid system constraints.

Algorithm 1. One-to-Many ReplacementGreedy

1: $S \leftarrow \emptyset$, $T_i \leftarrow \emptyset$ for all $i \in [m]$
2: **for** $t \in [\ell]$ **do**
3:     $x^* \leftarrow \arg\max_{x \in \Omega} \frac{1}{m} \sum_{i=1}^m \nabla_i'(x, T_i)$
4:     $S \leftarrow S \cup x^*$
5:     **for** all $i \in [m]$ **do**
6:         **if** $\nabla_i'(x, T_i) > 0$ **then**
7:             $T_i \leftarrow T_i \cup \{x^*\} \setminus \text{Rep}_i'(x^*, T_i)$
8:         **end if**
9:     **end for**
10:     $t \leftarrow t + 1$
11: **end for**
12: Return S and $\{T_i\}_{i \in [m]}$

3.1   Algorithm

In order to have a better understand of our One-to-Many ReplacementGreedy, we investigate the One-to-One ReplacementGreedy in the first. The *replacement gain* is denoted as $\nabla_i(x, A)$, who characterizes how much they can increase the value of $f_i(A)$ by either adding x to A or replacing x with one element of A while preserving the independence of A. We restate the related notations as follows. Set $\Delta_i(x, A) = f_i(A \cup \{x\}) - f_i(A)$ as the *marginal gain* of adding x to A for any $i \in [m]$. We restate the *replace gain* of deleting an element $y \in A$ and replacing it with x as $\nabla_i(x, y, A) = f_i(A \cup \{x\} \setminus \{y\}) - f_i(A)$. As maintaining the independence of solution in each iteration is a very important point in One-to-One ReplacementGreedy algorithm, let $\mathcal{I}(x, A)$ be the set of feasible candidate $y \in A$, i.e., $\mathcal{I}(x, A) = \{y \in A : A \cup \{x\} \setminus \{y\} \in \mathcal{I}\}$. Finally, we formally redefine the replacement gain as

$$\nabla_i(x, A) = \begin{cases} \Delta_i(x, A), & \text{if } A \cup \{x\} \in \mathcal{I} \\ \max\{0, \max_{y \in \mathcal{I}(x,A)} \nabla_i(x, y, A)\}, \text{o.w.} \end{cases}$$

To specific say the element with the maximum replacement gain due to x, they define

$$\text{Rep}_i(x, A) = \begin{cases} \emptyset, & \text{if } A \cup \{x\} \in \mathcal{I} \\ \arg\max_{y \in \mathcal{I}(x,A)} \nabla_i(x, y, A), \text{o.w.} \end{cases}$$

In our p-matroid system constraint setting, we define $\nabla_i(x, Y, A) = f_i(A \cup \{x\} \setminus Y) - f_i(A)$ as the new *replacement gain*. To keep the independence in each loop, we set $\mathcal{I}'^{,i}(x, A) = \{Y \subseteq A : |Y| \leq p, A \cup \{x\} \setminus Y \in \mathcal{I}^i (= \cap_{j=1}^p \mathcal{I}_j^i)\}$ as the new collection of candidate subsets. Let

$$\nabla_i'(x, A) = \begin{cases} \Delta_i(x, A), & \text{if } A \cup \{x\} \in \mathcal{I}^i \\ \max\{0, \max_{Y \subseteq \mathcal{I}'^{,i}(x,A)} \nabla_i(x, Y, A)\}, \text{o.w.} \end{cases}$$

Similarly, we set

$$\text{Rep}_i'(x, A) = \begin{cases} \emptyset, & \text{if } A \cup \{x\} \in \mathcal{I} \\ \arg\max_{Y \subseteq \mathcal{I}'^{,i}(x,A)} \nabla_i(x, Y, A), \text{o.w.} \end{cases}$$

In the One-to-Many ReplacementGreedy, we greedily choose an element x^* with the maximum average new replacement gain in each iteration until the size of S increases to ℓ. In each iteration, for any current substitute solution set T_i, $i \in [m]$, if the new replacement gain $\nabla_i'(x, A) > 0$, we will update T_i by removing $\text{Rep}_i'(x^*, T_i)$. The main pseudo codes are presented by Algorithm 1.

3.2   Theoretical Analysis

In this section, we will analyze the performance ratio of One-to-Many ReplacementGreedy. For clarity, we adopt the notations provided by [2,14,16]. Let

$$S^{m,\ell} = \arg\max_{S \subseteq \Omega, |S| \leq \ell} \left\{ \frac{1}{m} \sum_{i=1}^m \max_{T_i \subseteq S, T_i \in \mathcal{I}^i(S)} f_i(T_i) \right\}$$

be any optimal solution, and set $S_i^{m,\ell} = \arg\max_{T \in \mathcal{I}^i(S^{m,\ell})} f_i(T)$ to be the independent subset of $S^{m,\ell}$ with maximum f_i value for $i \in [m]$. Let T_i^t be the solution at the end of iteration t. Then our main result can be summarized as following theorem.

Theorem 1. *For any fixed $p \geq 1$, the One-to-Many ReplacementGreedy is a $1/(p+1)(1-1/e^2)$-approximation algorithm for the two-stage submodular maximization with a p-matroid system constraint $M^i = (\Omega, \cap_{j=1}^p \mathcal{I}_j^i)$ for each $i \in [m]$.*

Proof. Let X_t be the total value $\frac{1}{m}\sum_{i=1}^{m} f_i(T_i^t)$ in the end of t iteration, then the increment of value during t iteration can be lower bounded by as follows.

$$X_{t+1} - X_t \geq \frac{G_m(S^{m,\ell})}{\ell} - \frac{(p+1)X_t}{\ell}. \tag{2}$$

From inequality (2), we could complete the main proof by induction. We assume

$$X_t \geq \frac{1}{p+1}\left(1 - (1 - \frac{1}{\ell})^{2t}\right)G_m(S^{m,\ell}).$$

In basis step, if $t = 1$, we have $X_1 \geq \frac{1}{\ell}G_m(S^{m,\ell}) \geq \frac{2}{(p+1)\ell}G_m(S^{m,\ell})$, where the first inequality follows from the line 3 in Algorithm 1. In the reduction step, as

$$X_{t+1} \geq \frac{1}{\ell}G_m(S^{m,\ell}) + \left(1 - \frac{p+1}{\ell}\right)X_t$$

$$\geq \frac{1}{\ell}G_m(S^{m,\ell}) + \left(1 - \frac{p+1}{\ell}\right) \cdot \left[\frac{1}{p+1}\left(1 - (1 - \frac{1}{\ell})^{2t}\right)G_m(S^{m,\ell})\right]$$

$$= \frac{1}{p+1} \cdot \left[1 - (1 - \frac{1}{\ell})^{2t}(1 - \frac{p+1}{\ell})\right] \cdot G_m(S^{m,\ell})$$

$$\geq \frac{1}{p+1}\left(1 - (1 - \frac{1}{\ell})^{2t+2}\right) \cdot G_m(S^{m,\ell}).$$

By the above reduction process, we complete the assumption. At the end of ℓ iteration, we have

$$\frac{1}{m}\sum_{i=1}^{m} f_i(T_i^\ell) = X_\ell \geq \frac{1}{p+1}(1 - (1 - \frac{1}{\ell})^{2\ell}) \cdot G_m(S^{m,\ell})$$

$$\geq \frac{1}{p+1}(1 - \frac{1}{e^2}) \cdot G_m(S^{m,\ell}),$$

where the second inequality is derived by the inequality of $e^{-x} \geq 1 - x$.

In order to prove the inequality (2), we provide two technical lemmas as follows.

Lemma 1. *For any $i \in [m]$, given any two independent sets $A, B \in \mathcal{I}^i = \cap_{j=1}^{p}\mathcal{I}_j$, there exists a mapping of elements in $B \setminus A$ to $[A \setminus B]^{\leq p}$ (namely, a collection of subsets included into $A \setminus B$ of size at most p), such that each element $u \in A \setminus B$ appears in at most p subsets.*

Proof. Refer to the full version of this paper.

Lemma 2. *For any $i \in [m], t \in [\ell]$, let $\pi_t^i : S_i^{m,\ell} \setminus T_i^t \to [T_i^t \setminus S_i^{m,\ell}]^{\leq p}$ be the mapping derived by Lemma 1, then we have*

$$\sum_{x \in S_i^{m,\ell} \setminus T_i^t} \Delta_i(\pi_t^i(x), T_i^t \setminus \pi_t^i(x)) \leq p \cdot f_i(T_i^t).$$

Proof. Refer to the full version of this paper.

To prove the increment of value during t iteration we have the following

$$\sum_{i=1}^{m}\nabla_i'(x^*,T_i^t) \geq \frac{1}{\ell}\sum_{x\in S^{m,\ell}}\sum_{i=1}^{m}\nabla_i'(x,T_i^t) \tag{3}$$

$$\geq \frac{1}{\ell}\sum_{i=1}^{m}\sum_{x\in S_i^{m,\ell}\setminus T_i^t}\nabla_i'(x,T_i^t) \tag{4}$$

$$\geq \frac{1}{\ell}\sum_{i=1}^{m}\sum_{x\in S_i^{m,\ell}\setminus T_i^t}f_i(T_i^t\cup\{x\}\setminus\pi_t^i(x))-f_i(T_i^t) \tag{5}$$

$$= \frac{1}{\ell}\sum_{i=1}^{m}\sum_{x\in S_i^{m,\ell}\setminus T_i^t}\Delta_i(x,T_i^t)-\Delta_i(\pi_t^i(x),T_i^t\cup\{x\}\setminus\pi_t^i(x)) $$

$$\geq \frac{1}{\ell}\sum_{i=1}^{m}\sum_{x\in S_i^{m,\ell}\setminus T_i^t}\Delta_i(x,T_i^t)-\Delta_i(\pi_t^i(x),T_i^t\setminus\pi_t^i(x)) \tag{6}$$

$$\geq \frac{1}{\ell}\sum_{i=1}^{m}f_i(S_i^{m,\ell})-(p+1)f_i(T_i^t). \tag{7}$$

The inequality (3) is obtained by the selection of x^* from the ground set Ω in each iteration. The inequality (4) follows from the fact of on-negativity of $\nabla_i'(x,T_i)$ for all $i\in[m]$. The inequality (5) is derived by

$$\frac{1}{\ell}\sum_{i=1}^{m}\sum_{x\in S_i^{m,\ell}}\nabla_i'(x,T_i^t) = \frac{1}{\ell}\sum_{i=1}^{m}\sum_{x\in S_i^{m,\ell}}f_i(T_i^t\cup\{x\}\setminus\text{Rep}_i'(x,T_i^t))-f_i(T_i^t)$$

$$\geq \frac{1}{\ell}\sum_{i=1}^{m}\sum_{x\in S_i^{m,\ell}}f_i(T_i^t\cup\{x\}\setminus\pi_t^i(x))-f_i(T_i^t),$$

where the inequality holds because the $\text{Rep}_i'(x,T_i)$ is the maximum subset and the $\pi_t^i(x)$ is an feasible subset of $\mathcal{I}^i(x,T_i^t)$. The inequality (6) is implied by the submodularity. By the additivity of submodularity of any $f_i, i\in[m]$, we have

$$\sum_{x\in S_i^{m,\ell}\setminus T_i^t}\Delta_i(x,T_i^t) \geq f_i(S_i^{m,\ell})-f_i(T_i^t). \tag{8}$$

Combining Lemma 2 and inequality (8), we obtain the inequality (7). The inequality (2) can be directly obtained by the above process.

The main result can be described as the following theorem.

Theorem 2. *For any fixed $p \geq 1$, the query complexity of the One-to-Many ReplacementGreedy algorithm is upper bounded by $O(\ell mnr^p)$.*

Proof. It concludes that the main time computation is the greedy chosen of line 3 in Algorithm 1. Given any iteration $t \in [\ell]$ and $i \in [m]$, it needs to check at most $O(n)$ elements to find the element x^* while it also needs at most $O(r^p)$ function evaluations by enumerating all candidate subsets, where r is the maximum size of feasible subsets belong to p-matroid. Then the total query complexity (i.e., the number of function evaluations) of Algorithm 1 is bounded by $O(\ell mnr^p)$.

4   *P*-Extendible System Constraints

In this section, we extend our algorithm for p-matroid system constraints to dealing with p-extendible system constraints. As discussed in the work of [6, 12], the p-extendible system constraint is a generalization of p-matroid system constraint. In our model, we choose a set S of size at most ℓ, while we also select a set $T_i \subseteq S$ such that $T_i \in \mathcal{I}^i$ for each $i \in [m]$, where $M^i = (S, \mathcal{I}^i)$ is a p-extendible system. The aim is to maximize the average of summarization of their function values.

Algorithm 2. Generalized One-to-Many ReplacementGreedy

1: $S, T_i, E(T_i) \leftarrow \emptyset$ for all $i \in [m]$
2: **for** $t \in [\ell]$ **do**
3:       $x^* \leftarrow \arg\max\limits_{x \in \Omega} \frac{1}{m} \sum\limits_{i=1}^{m} \tilde{\nabla}_i(x, T_i)$
4:       $S \leftarrow S \cup x^*$
5:       **for all** $i \in [m]$ **do**
6:            **if** $\tilde{\nabla}_i(x, T_i) > 0$ **then**
7:                 $T_i \leftarrow E(T_i) \cup \{x^*\} \setminus \tilde{\text{Rep}}_i(x^*, T_i)$
8:            **end if**
9:       **end for**
10:      compute an extension $E(T_i)$ of T_i
11:      $t \leftarrow t + 1$
12: **end for**
13: Return S and $\{E(T_i)\}_{i \in [m]}$

4.1   Algorithm

Following from the definition of p-extendible system, we have if $A \subseteq B \in \mathcal{I}$ and $A \cup \{x\} \in \mathcal{I}$, then there exists a subset $Y \subseteq B \setminus A$ with $|Y| \leq p$ such that $B \setminus A \cup \{x\} \in \mathcal{I}$. Given a p-extendible system $M^i = (\Omega, \mathcal{I}^i)$ for each $i \in [m]$. The goal is to select a set S of size at most ℓ, such that the average of the summary of the optimum of f_i restricted to S according to a p-extendible system $M^i = (\Omega, \mathcal{I}^i)$ for all $i \in [m]$ is maximum.

In our setting, to keep the independence of $T_i, i \in [m]$ in each iteration under p-extendible system constraint, we modify the One-to-Many ReplacementGreedy to a Generalized One-to-Many ReplacementGreedy, the main pseudo codes are presented in Algorithm 2.

For each $i \in [m]$, we define $\tilde{\nabla}_i(x, A)$ as the new replacement gain, in specific, $\tilde{\nabla}_i(x, Y, A) = f_i(E(A) \cup \{x\} \setminus Y) - f_i(E(A))$, where $E(A)$ is an extension of A. Let $\tilde{\mathcal{I}}^i(x, A) = \{Y \subseteq E(A) \setminus A : |Y| \le p, A \cup \{x\} \in \mathcal{I}^i, E(A) \cup \{x\} \setminus Y \in \mathcal{I}^i\}$ be the candidate set. Let

$$\tilde{\nabla}_i(x, A) = \begin{cases} \Delta_i(x, A), & \text{if } A \cup \{x\} \in \mathcal{I}^i \\ \max\{0, \max_{Y \in \tilde{\mathcal{I}}^i(x,A)} \tilde{\nabla}_i(x, Y, A)\}, & \text{o.w.} \end{cases}$$

Simultaneously, let

$$\tilde{\text{Rep}}_i(x, A) = \begin{cases} \emptyset, & \text{if } A \cup \{x\} \in \mathcal{I} \\ \arg\max_{Y \in \tilde{\mathcal{I}}^i(x,A)} \tilde{\nabla}_i(x, A), & \text{o.w.} \end{cases}$$

4.2 Theoretical Analysis

In this section, we present the analyses of the Generalized One-to-Many ReplacementGreedy. The setting under p-extendible system constraints differ from the p-matroid constraints, that is, there is not such similar mapping presented by Lemma 2. We notice that there is interesting property provided by the following lemma.

Lemma 3. *For any $i \in [m]$, let $\{Y_j^i\}_{j=1}^q$ be a collection of $E(T_i) \setminus T_i$ such that each element of $E(T_i) \setminus T_i$ appears in at most r of these subsets, where r is the size of maximal independence set in p-extendible system. Then we have*

$$\sum_{j=1}^q (f(E(T_i)) - f_i(E(T_i) \setminus Y_j^i)) \le r \cdot (f(E(T_i)) - f(T_i)).$$

Proof. Refer to the full version of this paper.

Theorem 3. *For any fixed $p \ge 1$, the Generalized One-to-Many Replacement-Greedy is a $1/(r+1)(1-1/e^2)$-approximation algorithm, while the query complexity is upper bounded by $O(\ell m n r^p)$, for the two-stage submodular maximization with p-extendible system constraints $M^i = (\Omega, \mathcal{I}^i)$ for each $i \in [m]$, where r is the size of the maximum independence set in $\mathcal{I}^i$.*

Proof. Refer to the full version of this paper.

5 Experiments

In this section, we run Algorithm 2, generalized one-to-many ReplaceGreedy (say, G-REPLACEGREEDY), on the application exemplar-based clustering with two dataset and consider the following benchmarks:

- Random selection (i.e., Random): the output is randomly k elements chosen for each function f_i, $i \in [m]$.
- Greedy-Sum (i.e., Greedy-SUM): the output is greedily k elements selected for each function f_i, $i \in [m]$, and return the union as S.

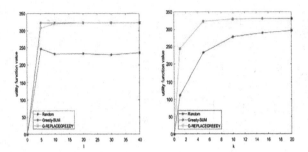

Fig. 1. Performances of Algorithm 2 comparing with Random and Geedy-Sum on Census.

We consider the application of exemplar-based clustering on Census 1990 [1], which has $24,581$ elements with 68 attributes. Let the first 10 attributes as our classified genres, such as, age, ancestry, citizen etc. In the experiment, we first choose a dataset Ω of 500 different people by reservoir sampling [18]. We denote Ω_i as the set of people containing genre $i \in [m]$. All people are expressed by their features vector, the distance of any two vectors is calculated by Euclidean distance and $d(v, S) = \min_{u \in S} d(u, v)$ denotes the distance of element v to set S. The goal is to choose a subset $S \subseteq \Omega$ of size at most ℓ, such that each genre $i \in [m]$ has a good expression of size limit k. For each genre $i \in [m]$, the utility function $f_i(S)$ is defined by Exemplar Based Clustering [1,17]. We restate as follows

$$f_i(S) = L_i(\{e_0\}) - L_i(S \cup \{e_0\}),$$

where e_0 is an auxiliary vector (w.l.o.g., $e_0 = \mathbf{0}$), $S_i = S \cap \Omega_i$ is the set of people with genre i, and $L_i(S) = \frac{1}{|\Omega_i|} \sum_{x \in \Omega_i} d(x, S_i)$. As the submodularity and non-negativity of utility function have been discussed by [1,8], we omit the proof here. The left of Fig. 1 shows the performance of Algorithm 2 on census application, when k is fixed as 5 and the right figure shows that the performance of Algorithm 2 with fixing $\ell = 20$. We observe that if k is fixed, then the function value goes to some assured values with the increasing of ℓ and our Algorithm 2 performs similarly to Greedy-Sum.

We also consider a classification application that feature vectors are generated from a random distribution. Specifically, we generate a 500×10 feature matrix which each component is randomly chosen from the range $[0, 3]$. Figure 2 shows the performance of Algorithm 2 on the classification, when k and ℓ are fixed, respectively. We observe that our Algorithm 2 still matches the Greedy-

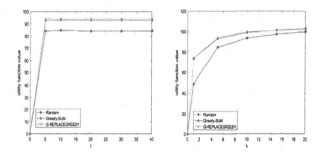

Fig. 2. Performances of Algorithm 2 comparing with Random and Geedy-Sum on Classification.

Sum algorithm, and performs better than Random algorithm. As the generation process of data, we also observe that the three algorithms perform similarly as ℓ increases.

6   Conclusion

We consider the two-stage submodular maximization under p-matroid and p-extendible system constraints for each sub-functions, respectively. Specifically, for the front model, we derive a $1/(p+1)(1-1/e^2)$-approximation algorithm, which needs $O(\ell mnr^p)$ function evaluations. For the second setting, we obtain a $1/(r+1)(1-1/e^2)$-approximation algorithm with the same query complexity. In the end, we show the performance of our generalized One-to-Many algorithm on some datasets.

Acknowledgements. The first and sixth authors are supported by Natural Science Foundation of China (Nos. 11531014, 11871081). The second and fourth authors are supported by Natural Science Foundation (No. 1747818).

References

1. Badanidiyuru, A., Mirzasoleiman, B., Karbasi, A., Krause, A.: Streaming submodular maximization: massive data summarization on the fly. In: Proceedings of SIGKDD, pp. 671–680 (2014)
2. Balkanski, E., Mirzasoleiman, B., Krause, A., Singer, Y.: Learning sparse combinatorial representations via two-stage submodular maximization. In: Proceedings of ICML, pp. 2207–2216 (2016)
3. Buchbinder, N., Feldman, M., Garg, M.: Deterministic $(1/2+\varepsilon)$-approximation for submodular maximization over a matroid. In: Proceedings of SODA, pp. 241–254 (2019)
4. Calinescu, G., Chekuri, C., Pál, M., Vondrák, J.: Maximizing a monotone submodular function subject to a matroid constraint. SIAM J. Comput. **40**(6), 1740–1766 (2011)

5. Dasgupta, A., Kumar, R., Ravi, S.: Summarization through submodularity and dispersion. In: Proceedings of ACL, pp. 1014–1022 (2013)
6. Feldman, M., Harshaw, C., Karbasi, A.: Greed is good: near-optimal submodular maximization via greedy optimization. arXiv: 1704.01652 (2017)
7. Fisher, M.L., Nemhauser, G.L., Wolsey, L.A.: An analysis of approximations for maximizing submodular set functions-II. In: Balinski, M.L., Hoffman, A.J. (eds.) Polyhedral Combinatorics, pp. 73–87. Springer, Heidelberg (1978). https://doi.org/10.1007/BFb0121195
8. Gomes, R., Krause, A.: Budgeted nonparametric learning from data streams. In: Proceedings of ICML, pp. 391–398 (2010)
9. Kempe, D., Kleinberg, J., Tardos, É.: Maximizing the spread of influence through a social network. In: Proceedings of SIGKDD, pp. 137–146 (2003)
10. Lee, J., Sviridenko, M., Vondrák, J.: Submodular maximization over multiple matroids via generalized exchange properties. Math. Oper. Res. **35**(4), 795–806 (2010)
11. Lin, H., Bilmes, J.: A class of submodular functions for document summarization. In: Proceedings of ACL, pp. 510–520 (2011)
12. Mestre, J.: Greedy in approximation algorithms. In: Proceedings of ESA, pp. 528–539 (2006)
13. Mitrovic, S., Bogunovic, I., Norouzi-Fard, A., Tarnawski, J.M., Cevher, V.: Streaming robust submodular maximization: a partitioned thresholding approach. In: Proceedings of NIPS, pp. 4557–4566 (2017)
14. Mitrovic, M., Kazemi, E., Zadimoghaddam, M., Karbasi, A.: Data summarization at scale: a two-stage submodular approach. arXiv preprint arXiv:1806.02815 (2018)
15. Nemhauser, G.L., Wolsey, L.A., Fisher, M.L.: An analysis of approximations for maximizing submodular setfunctions-I. Math. Program. **14**(1), 265–294 (1978)
16. Sarpatwar, K.K., Schieber, B., Shachnai, H.: Constrained submodular maximization via greedy local search. Oper. Res. Lett. **47**(1), 1–6 (2019)
17. Stan, S., Zadimoghaddam, M., Krause, A., Karbasi, A.: Probabilistic submodular maximization in sub-linear time. In: Proceedings of ICML, pp. 3241–3250 (2017)
18. Vitter, J.S.: Random sampling with a reservoir. ACM Trans. Math. Softw. **11**(1), 37–57 (1985)
19. Wu, W.L., Zhang, Z., Du, D.Z.: Set function optimization. J. Oper. Res. Soc. China (2018). https://doi.org/10.1007/s40305018-0233-3

Local Search Approximation Algorithms for the Spherical k-Means Problem

Dongmei Zhang[1], Yukun Cheng[2(✉)], Min Li[3], Yishui Wang[4], and Dachuan Xu[5]

[1] School of Computer Science and Technology, Shandong Jianzhu University, Jinan 250101, People's Republic of China
zhangdongmei@sdjzu.edu.cn
[2] Suzhou Key Laboratory for Big Data and Information Service, School of Business, Suzhou University of Science and Technology, Suzhou 215009, People's Republic of China
ykcheng@amss.ac.cn
[3] School of Mathematics and Statistics, Shandong Normal University, Jinan 250014, People's Republic of China
liminEmily@sdnu.edu.cn
[4] Shenzhen Institutes of Advanced Technology, Chinese Academy of Sciences, 1068 Xueyuan Avenue, Shenzhen University Town, Shenzhen 518055, People's Republic of China
ys.wang1@siat.ac.cn
[5] Department of Operations Research and Scientific Computing, Beijing University of Technology, Beijing 100124, People's Republic of China
xudc@bjut.edu.cn

Abstract. In this paper, we study the spherical k-means problem (SKMP) which is one of the most well-studied clustering problems. In the SKMP, we are given an n-client set $\mathcal{D}$ in d-dimensional unit sphere $\mathbb{S}^d$, and an integer $k \le n$. The goal is to open a center subset $F \subset \mathbb{S}^d$ with $|F| \le k$ that minimizes the sum of cosine dissimilarity measure for each client in $\mathcal{D}$ to the nearest open center. We give a $(2(4 + \sqrt{7}) + \varepsilon)$-approximation algorithm for this problem using local search scheme.

Keywords: Spherical k-means · Local search · Approximation algorithm

1 Introduction

In modern data analysis, it is a very important task to cluster text documents. The spherical k-means clustering is one of the most representative data mining tools to obtain useful information. Dhillon and Modha [12] present the primitive spherical k-means clustering with cosine similarities based on k-means clustering by [21,22]. Honik et al. [16] introduce the standard spherical k-means clustering with the so-called cosine dissimilarities.

In the spherical k-means problem (SKMP), we are given an n-point set $\mathcal{D}$ in d-dimensional unit sphere $\mathbb{S}^d$, and an integer $k \le n$. We select at most k points

© Springer Nature Switzerland AG 2019
D.-Z. Du et al. (Eds.): AAIM 2019, LNCS 11640, pp. 341–351, 2019.
https://doi.org/10.1007/978-3-030-27195-4_31

in $\mathbb{S}^d$ to be cluster centers and then assign each input point $j \in \mathcal{D}$ to the nearest selected center. If point j is assigned to a center i, it incurs a cost of the so-called cosine dissimilarity measure between i and j. The goal is to select the k centers so as to minimize the sum of the assignment costs.

Endo and Miyamoto [13] adapt the famous k-means++ algorithm of [3] with $O(\log k)$-approximation for the SKMP with a slight different cosine dissimilarity measure. Li et al. [19] prove that the above algorithm is $O(\log k)$-approximation for the SKMP itself and constant approximation for the SKMP with separable sets.

Since the SKMP is closely related to the k-means and k-median problems, we briefly review the corresponding literatures for the k-means and k-median problems. Aloise et al. [2] and Dasgupta [11] have shown that the k-means problem is NP-hard. Moreover, Mahajan et al. [23] find that even the planar k-means problem is also NP-hard. The first constant 108-approximation algorithm is given by Jain and Vazirani [17]. By using the approximate centroid set of Matoušek [25], Kanungo et al. [18] obtain a local search $(9+\varepsilon)$-approximation algorithm. Recently, Ahmadian et al. [1] present the currently best $(6.357 + \varepsilon)$-approximation algorithm using primal-dual technique. Moreover, the bi-criteria approximation algorithm for k-means is also given by Makarychev et al. [24]. The literatures for the k-median problem are summarized as follows. Charikar et al. [7] present the first constant $6\frac{2}{3}$-approximation algorithm based on LP rounding technique. Jain and Vazirani [17] obtain a 6-approximation using the primal-dual schema and Lagrangian relaxation. Charikar and Guha [6] improve the above ratio to 4. Arya et al. [4] give a local search $(3+\varepsilon)$-approximation algorithm. Li and Svensson [20] offer a bipoint rounding $(1 + \sqrt{3} + \varepsilon)$-approximation algorithm. Byrka et al. [5] further improve the above ratio to $2.675 + \varepsilon$ which is the currently best ratio. The 6-approximation of Jain and Vazirani [17] also holds for the k-facility location problem which is a generalization of the k-median problem. Zhang [26] improves the above ratio 6 to $(2+\sqrt{3}+\varepsilon)$ using local search technique. Local search is widely used in the approximation algorithm area and understanding the properties of local optima is an important topic itself. For more results on k-means and k-median problems based on local search schema, we refer to [8–10,14,15] and references therein.

It is easy to verify that any γ-approximation algorithm for k-means problem can be adapted to a 2γ-approximation algorithm for the SKMP. Therefore, the currently best ratio for the SKMP is $12.714 + \varepsilon$ based on the $(6.357 + \varepsilon)$-approximation of [1] for k-means problem using the involved primal-dual and Lagrangian relaxation techniques. If we focus on the local search technique, the local search $(9 + \varepsilon)$-approximation of [18] for k-means problem can be adapted to a $(18 + \varepsilon)$-approximation for the SKMP. In this paper, we give a local search $(8(2+\sqrt{3})+\varepsilon)$-approximation algorithm with single-swap operation along with a local search $(2(4+\sqrt{7})+\varepsilon)$-approximation algorithm with multi-swap operation for the SKMP by exploring the spherical structure. The ratio $(2(4 + \sqrt{7}) + \varepsilon)$ improves the simple $(18 + \varepsilon)$-approximation adaption. This direct local search algorithm and its analysis can help us to understand deeply the properties of local optima for the SKMP.

The organization of this paper is as follows. We introduce some basic notations in Sect. 2. In Sect. 3, we offer a local search $(8(2 + \sqrt{3}) + \varepsilon)$-approximation algorithm for the SKMP by using single-swap operation. We further improve the above approximation ratio to $2(4 + \sqrt{7}) + \varepsilon$ by using multi-swap operation in Sect. 4. In Sect. 5, we show the result of some numerical experiments for the local search algorithm. Some discussions are given in Sect. 6.

All the proofs are deferred to the journal version.

2 Preliminaries

In this paper, we always consider the points with the unit length in $\mathbb{R}^d$, which can be denoted by

$$\mathbb{S}^d = \{s \in \mathbb{R}^d | \ \|s\| = 1\}.$$

Given any two points $a, b \in \mathbb{S}^d$, the cosine dissimilarity measure between them is denoted by

$$\Delta(a, b) := 1 - \cos(a, b) = \frac{1}{2} \|a - b\|^2.$$

Given a set $U \subseteq \mathbb{S}^d$ and a point $c \in \mathbb{S}^d$, we define the total sum of cosine dissimilarity measure of U with respect to c and the spherical centroid of U as follows,

$$\Delta(c, U) := \sum_{j \in U} \Delta(c, j) = \sum_{j \in U} (1 - c^T j),$$

$$\mathrm{sc}(U) := \frac{\sum_{j \in U} j}{\left\| \sum_{j \in U} j \right\|}.$$

With the above notations, Endo and Miyamoto [13] give an important property of spherical centroidal solution in the following lemma.

Lemma 1 ([13]). *For any subset $U \subseteq \mathbb{S}^d$ and a point $c \in \mathbb{S}^d$, we have*

$$\Delta(c, U) = \Delta(\mathrm{sc}(U), U) + \left\| \sum_{j \in U} j \right\| \Delta(\mathrm{sc}(U), c). \tag{1}$$

Since each solution of the SKMP is determined by the corresponding center subset, we use this center subset to denote the solution. Let S be a feasible solution and O be a global optimal solution of the SKMP. Without loss of generality, suppose that $|S| = |O|$. For each client $j \in \mathcal{D}$, we introduce the following notations,

$$\Delta_j(S) := \min_{s \in S} \Delta(j, s), \ s_j := \arg\min_{s \in S} \Delta(j, s),$$

$$\Delta_j(O) := \min_{o \in O} \Delta(j, o), \ o_j := \arg\min_{o \in O} \Delta(j, o).$$

For each center $o \in O$, if

$$s_o := \arg\min_{s \in S} \Delta(o, s),$$

we say that s_o captures o. For any s in S, if there exists a center point $o \in O$ captured by s, we also call s as bad center. Otherwise, we call s as good center. All the bad centers constitute a subset $\text{Bad}(S)$. Denote $m := |\text{Bad}(S)|$. All elements of $\text{Bad}(S)$ are listed as

$$\text{Bad}(S) = \{s_1, ..., s_m\}.$$

For each $i \in \{1, ..., m\}$, let

$$S_i := \{s_i\},$$

$$O_i := \{o \in O | o \text{ is captured by } s_i\},$$

and

$$m_i = |O_i|.$$

For each i, arbitrarily add $m_i - 1$ good centers $\{s_i^2, ..., s_i^{m_i}\}$ in $S \backslash \cup_{i=1,...,m} S_i$ to S_i. Thus, we partition S and O into two sets of groups $S_1, S_2, ..., S_m$ and $O_1, O_2, ..., O_m$ with $|S_i| = |O_i|$ for each $i \in \{1, 2, ..., m\}$. Then, We construct (s, o) pairs as follows (cf. Fig. 1).

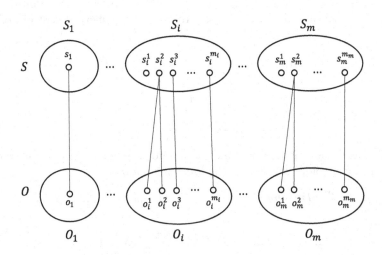

Fig. 1. Partition S and O.

Procedure 1.

(1) For each i with $m_i = 1$, the pair (s_i, o_i) is defined with $S_i = \{s_i\}$ and $O_i = \{o_i\}$.

(2) For each i with $m_i \geq 2$, denote $S_i := \{s_i, s_i^2, ..., s_i^{m_i}\}$ and $O_i := \{o_i^1, o_i^2, ..., o_i^{m_i}\}$. The pairs $(s_i^2, o_i^1), (s_i^2, o_i^2), (s_i^3, o_i^3), ..., (s_i^{m_i}, o_i^{m_i})$ are constructed.

Furthermore, for each $s \in S$ (or $o \in O$), we give the following notations, which can imply different partitions of the clients set $\mathcal{D}$.

$$\mathcal{D}_S(s) := \{j \in \mathcal{D} | s_j = s\} \, (\text{or} \, \mathcal{D}_O(o) := \{j \in \mathcal{D} | o_j = o\}).$$

The total cost of $\Delta_j(S)$ (or $\Delta_j(O)$) over $\mathcal{D}$ is denoted as

$$\text{cost}(S) := \sum_{j \in \mathcal{D}} \Delta_j(S) = \sum_{s \in S} \Delta(s, \mathcal{D}_S(s))$$

$$(\text{or} \, \text{cost}(O) := \sum_{j \in \mathcal{D}} \Delta_j(O) = \sum_{o \in O} \Delta(o, \mathcal{D}_O(o))).$$

It follows from the optimality of O and the definition of spherical centroid that $o = \text{sc}(\mathcal{D}_O(o))$ for each $o \in O$.

3 A Local Search $((8(2 + \sqrt{3}) + \varepsilon)$-Approximation Algorithm

There are exponential potential spherical centroid points which correspond to all the subsets of $\mathcal{D}$. To guarantee the polynomial running time for each swap operation, we impose that all candidate centers are chosen from $\mathcal{D}$. For any feasible solution S, the single-swap operation $\text{swap}(a, b)$ with $a \in S$ and $b \in \mathcal{D} \backslash S$ is defined to delete a from S and add b to S. The neighborhood of S associated with the $\text{swap}(a, b)$ is defined as

$$\text{Ngh}_1(S) := \{S \backslash \{a\} \cup \{b\} | a \in S, b \in \mathcal{D} \backslash S\}.$$

Now we are ready to present our single-swap local search algorithm in Algorithm 1.

Algorithm 1. The single-swap local search algorithm for spherical k-means problem

---

Step 0. (Initialization) Arbitrarily choose a feasible solution S from $\mathcal{D}$.
Step 1. (Local search) Compute

$$S_{\min} := \arg \min_{S' \in \text{Ngh}_1(S)} \text{cost}(S').$$

Step 2. (Stop criterion) If $\text{cost}(S_{\min}) \geq \text{cost}(S)$, output S. Otherwise, set $S := S_{\min}$ and go to Step 1.

To proceed the analysis, we need the following technical lemma.

Lemma 2. *Let S and O be a local optimal solution and a global optimal solution to the SKMP, respectively. We have,*

$$\sum_{j \in \mathcal{D}} \sqrt{\Delta_j(S)\Delta_j(O)} \leq \sqrt{\sum_{j \in \mathcal{D}} \Delta_j(S)} \cdot \sqrt{\sum_{j \in \mathcal{D}} \Delta_j(O)}, \tag{2}$$

and

$$\sum_{j \in \mathcal{D}} \Delta(s_{o_j}, j) \leq \sum_{j \in \mathcal{D}} \Delta_j(S) + 2 \sum_{j \in \mathcal{D}} \Delta_j(O) + 2 \sum_{j \in \mathcal{D}} \sqrt{\Delta_j(S)\Delta_j(O)}. \tag{3}$$

We give some high level illustration for our analysis. The analysis for local search type algorithm is usually proceeded as follows. Since the algorithm outputs a local optimal solution S, we bound the cost of S by using the local optimality. Each solution in the neighbour of S corresponds to a swap operation which can not improve the currently solution S. In fact, there is an inequality implied in each swap operation mentioned above. We carefully choose these swap operations to bound the cost of S by means of the cost of O. Usually, we need to consider the swap(s, o) for each $s \in S$ and $o \in O$. The difficulty is that o may be not in $\mathcal{D}$ and we can not perform the swap(s, o) in the local search algorithm. To overcome this difficulty, we carefully choose a point $\hat{o} \in \mathcal{D}$ to approximate o. We consider the swap$(\hat{o}, s)$ instead of swap(o, s).

We introduce a center $\hat{o} \in \mathcal{D}$ associated with each $o \in O$. For each $o \in O$, let us define

$$\hat{o} := \arg \min_{j \in \mathcal{D}_O(o)} \Delta(o, j). \tag{4}$$

It follows from the triangle inequality for $\sqrt{\Delta(\cdot, \cdot)}$ and (4), one can easily get the following results.

$$\begin{aligned}
\Delta(\hat{o}, \mathcal{D}_O(o)) &= \sum_{j \in \mathcal{D}_O(o)} \Delta(\hat{o}, j) \\
&\leq \sum_{j \in \mathcal{D}_O(o)} \left(\sqrt{\Delta(o, j)} + \sqrt{\Delta(o, \hat{o})} \right)^2 \\
&\leq 2 \sum_{j \in \mathcal{D}_O(o)} (\Delta(o, j) + \Delta(o, \hat{o})) \\
&\leq 4 \sum_{j \in \mathcal{D}_O(o)} \Delta(o, j) \\
&= 4\Delta(o, \mathcal{D}_O(o)). \tag{5}
\end{aligned}$$

Then, we consider the constructed pair (s, o) and obtain the following result. One can see the proof in the supplementary material.

Lemma 3. *Assume S and O be a feasible and a global optimal solution of the SKMP, then for each pair (s, o) constructed in Procedure 1, the following result is satisfied.*

$$\sum_{j \in \mathcal{D}_S(s)} (\Delta(s_{o_j}, j) - \Delta_j(S)) + \sum_{j \in \mathcal{D}_O(o)} (4\Delta_j(O) - \Delta_j(S)) \geq 0. \tag{6}$$

Thus, we can estimate the cost of S in the following theorem.

Theorem 1. *Algorithm 1 produces a local optimal solution S satisfying*

$$\text{cost}(S) \leq 8 \left(2 + \sqrt{3}\right) \text{cost}(O).$$

By applying the standard technique of [4,6], one can similarly get a local search algorithm in polynomial time, which should sacrifice any given $\varepsilon' > 0$ in the approximation ratio.

4 Improved Local Search $(2(4 + \sqrt{7}) + \varepsilon)$-Approximation Algorithm

For any feasible solution S, we define the so-called multi-swap operation swap(A, B) as follows. In the operation, we are given two subsets $A \subseteq S$ and $B \subseteq \mathcal{D} \setminus S$ with $|A| = |B| \leq p$, where p is a fixed integer. All centers in A are deleted from S. Meanwhile, all centers in B are added to S.

We define the neighborhood of S with respect to the above multi-swap operation as follows,

$$\text{Ngh}_p(S) := \{S \setminus A \cup B | A \subseteq S, B \subseteq \mathcal{D} \setminus S, |A| = |B| \leq p\}.$$

For any given $\varepsilon > 0$, we present our multi-swap local search algorithm in Algorithm 2.

Algorithm 2. The multi-swap local search algorithm for spherical k-means problem

---

Step 0. (Initialization) Set

$$p := \left\lceil \frac{18}{\varepsilon} \right\rceil.$$

Arbitrarily choose a feasible solution S from $\mathcal{D}$.
Step 1. (Local search) Compute

$$S_{\min} := \arg \min_{S' \in \text{Ngh}_p(S)} \text{cost}(S').$$

Step 2. (Stop criterion) If $\text{cost}(S_{\min}) \geq \text{cost}(S)$, output S. Otherwise, set $S := S_{\min}$ and go to Step 1.

Theorem 2. *For any given $\varepsilon > 0$, Algorithm 2 produces a local optimal solution S satisfying*

$$\text{cost}(S) \leq \left(2 \left(4 + \sqrt{7}\right) + \varepsilon\right) \text{cost}(O).$$

5   Numerical Tests

We test our local search algorithm on some real document-term datasets from the CLUTO website (http://glaros.dtc.umn.edu/gkhome/fetch/sw/cluto/datasets.tar.gz). Note that our algorithm outputs a local optimal solution of the discrete spherical k-means problem (that is, the set of centers is contained in $\mathcal{D}$). The discrete spherical k-means problem can be formulated as the following 0-1 integer programming,

$$\min \sum_{i,j \in \mathcal{D}} \Delta(i,j) x_{ij}$$

$$\text{s.t.} \sum_{i \in \mathcal{F}} x_{ij} = 1, \quad \forall j \in \mathcal{D},$$

$$x_{ij} \leq y_i, \quad \forall i,j \in \mathcal{D},$$

$$\sum_{i \in \mathcal{D}} y_i = k,$$

$$x_{ij}, y_i \in \{0,1\} \quad \forall i,j \in \mathcal{D},$$

where the variable y_i indicates whether the point i is selected as a center, and x_{ij} indicates whether the point j is belong to the cluster with the center i. The corresponding LP relaxation is

$$\min \sum_{i,j \in \mathcal{D}} \Delta(i,j) x_{ij}$$

$$\text{s.t.} \sum_{i \in \mathcal{F}} x_{ij} = 1, \quad \forall j \in \mathcal{D},$$

$$x_{ij} \leq y_i, \quad \forall i,j \in \mathcal{D},$$

$$\sum_{i \in \mathcal{D}} y_i = k,$$

$$x_{ij}, y_i \geq 0 \quad \forall i,j \in \mathcal{D}.$$

We first do some numerical experiments to compare the local optimal value produce by the single-swap local search algorithm and the optimal value of the LP relaxation.

Since the computational skill is limited, we select randomly 50 and 100 points in the original data and do the experiment on these points. The initial solution of the local search algorithm is set as k points drawn uniformly from $\mathcal{D}$. Because of the randomness of the algorithm, we test 10 instances for each size and dataset, and show the average ratio of the local optimal value and global optimal value in Table 1. The result shows that the local optimal solution is very near to the global optimal solution of the discrete spherical k-means problem, implying that single-swap is better than multi-swap for these datasets since single-swap can produce a solution which is close to multi-swap in less time.

Table 1. Compare to the LP relaxation

Data name	50 points		100 points	
	k	ratio	k	ratio
cacmcisi	2	1.0000	2	1.0000
classic	4	1.0000	4	1.0000
cranmed	2	1.0000	2	1.0000
fbis	15	1.0015	15	1.0016
hitech	6	1.0000	6	1.0000
k1a	16	1.0012	19	1.0002
k1b	6	1.0005	6	1.0003
la1	6	1.0000	6	1.0000
la2	6	1.0000	6	1.0000
la12	6	1.0000	6	1.0004
mm	2	1.0000	2	1.0000
new3	25	1.0018	34	1.0016
ohscal	10	1.0005	10	1.0002
re0	11	1.0004	12	1.0006
re1	15	1.0004	22	1.0026
reviews	5	1.0000	5	1.0000
tr11	8	1.0000	9	1.0007
tr12	7	1.0000	8	1.0000
tr23	5	1.0012	6	1.0000
tr31	6	1.0000	6	1.0013
tr41	10	1.0000	10	1.0000
tr45	10	1.0006	9	1.0000
wap	13	1.0000	17	1.0002

6   Discussions

In this paper, we study the SKMP and present $(8(2+\sqrt{3})+\varepsilon)$- and $(2(4+\sqrt{7})+\varepsilon)$-approximation algorithms using single-swap and multi-swap respectively. Since there is a primal-dual 6.357-approximation algorithm of Ahmadian et al. [1], which improves the previous local search $(9+\varepsilon)$-approximation algorithm given by Kanungo et al. [18] for the classic k-means problem, it is natural to ask whether our local search $(2(4+\sqrt{7})+\varepsilon)$-approximation can be further improved by using primal-dual technique.

Acknowledgements. The first author is supported by Natural Science Foundation of China (No. 11871081). The second author is supported by Natural Science Foundation of China (No. 11871366). The third author is the Higher Educational Science and Technology Program of Shandong Province (No. J17KA171). The fourth author is supported by Natural Science Foundation of China (No. 61433012), Shenzhen Research Grant (KQJSCX2018033017 0311901, JCYJ20180305180840138 and GFW2017073114031767), and Hong Kong GRF 17210017. The fifth author is supported by Natural Science Foundation of China (No. 11531014).

References

1. Ahmadian, S., Norouzi-Fard, A., Svensson, O., Ward, J.: Better guarantees for k-means and Euclidean k-median by primal-dual algorithms. In: Proceedings of FOCS, pp. 61–72 (2017)
2. Aloise, D., Deshpande, A., Hansen, P., Popat, P.: NP-hardness of Euclidean sum-of-squares clustering. Mach. Learn. **75**, 245–249 (2009)
3. Arthur, D., Vassilvitskii, S.: k-means++: the advantages of careful seeding. In: Proceedings of SODA, pp. 1027–1035 (2007)
4. Arya, V., Garg, N., Khandekar, R., Meyerson, A., Munagala, K., Pandit, V.: Local search heuristics for k-median and facility location problems. SIAM J. Comput. **33**, 544–562 (2004)
5. Byrka, J., Pensyl, T., Rybicki, B., Srinivasan, A., Trinh, K.: An improved approximation for k-median, and positive correlation in budgeted optimization. ACM Trans. Algorithms **13**(2) (2017). Article No. 23
6. Charikar, M., Guha, S.: Improved combinatorial algorithms for the facility location and k-median problems. In: Proceedings of FOCS, pp. 378–388 (1999)
7. Charikar, M., Guha, S., Tardos, É., Shmoys, D.B.: A constant-factor approximation algorithm for the k-median problem. J. Comput. Syst. Sci. **65**, 129–149 (2002)
8. Cohen-Addad, V., Mathieu, C.: Effectiveness of local search for geometric optimization. In: Proceedings of SoCG, pp. 329–343 (2015)
9. Cohen-Addad, V., Klein, P.N., Mathieu, C.: Local search yields approximation schemes for k-means and k-median in Euclidean and minor-free metrics. In: Proceedings of FOCS, pp. 353–364 (2016)
10. Cohen-Addad, V., Schwiegelshohn, C.: On the local structure of stable clustering instances. In: Proceedings of FOCS, pp. 49–60 (2017)
11. Dasgupta, S.: The hardness of k-means clustering. Technical report CS2007-0890, University of California, San Diego (2007)
12. Dhillon, I.S., Modha, D.S.: Concept decompositions for large sparse text data using clustering. Mach. Learn. **42**, 143–175 (2001)
13. Endo, Y., Miyamoto, S.: Spherical k-means++ clustering. In: Proceedings of MDAI, pp. 103–114 (2015)
14. Friggstad, Z., Rezapour, M., Salavatipour, M.R.: Local search yields a PTAS for k-means in doubling metrics. In: Proceedings of FOCS, pp. 365–374 (2016)
15. Friggstad, Z., Zhang, Y.: Tight analysis of a multiple-swap heuristic for budgeted red-blue median. In: Proceedings of ICALP, pp. 75:1–75:13 (2016)
16. Hornik, K., Feinerer, I., Kober, M., Buchta, C.: Spherical k-means clustering. J. Stat. Softw. **50**(10), 1–22 (2012)
17. Jain, K., Vazirani, V.V.: Approximation algorithms for metric facility location and k-median problems using the primal-dual schema and Lagrangian relaxation. J. ACM **48**, 274–296 (2001)

18. Kanungo, T., Mount, D.M., Netanyahu, N.S., Piatko, C.D., Silverman, R., Wu, A.Y.: A local search approximation algorithm for k-means clustering. Comput. Geom. Theory Appl. **28**, 89–112 (2004)
19. Li, M., Xu, D., Zhang, D., Zou, J.: The seeding algorithms for spherical k-means clustering. J. Glob. Optim. 1–14 (2019). https://doi.org/10.1007/s10898-019-00779-w
20. Li, S., Svensson, O.: Approximating k-median via pseudo-approximation. SIAM J. Comput. **45**, 530–547 (2016)
21. Lloyd, S.: Least squares quantization in PCM. Technical report, Bell Laboratories (1957)
22. Lloyd, S.: Least squares quantization in PCM. IEEE Trans. Inf. Theory **28**, 129–137 (1982)
23. Mahajan, M., Nimbhorkar, P., Varadarajan K.: The planar k-means problem is NP-hard. In: Proceedings of WALCOM, pp. 274–285 (2009)
24. Makarychev, K., Makarychev, Y., Sviridenko, M., Ward, J.: A bi-criteria approximation algorithm for k-means. In: Proceedings of APPROX/RONDOM, pp. 14:1–14:20 (2016). Article No. 14
25. Matoušek, J.: On approximate geometric k-clustering. Discrete Comput. Geom. **24**, 61–84 (2000)
26. Zhang, P.: A new approximation algorithm for the k-facility location problem. Theor. Comput. Sci. **384**, 126–135 (2007)

Author Index

Printed in the United States
By Bookmasters